Bringing Bayesian Models to Life

Chapman & Hall/CRC
Applied Environmental Statistics

Series Editors

Douglas Nychka, *Colorado School of Mines*
Alexandra Schmidt, *Universidade Federal do Rio de Janero*
Richard L. Smith, *University of North Carolina*
Lance A. Waller, *Emory University*

Recently Published Titles

For more information about this series, please visit:
https://www.crcpress.com/Chapman--HallCRC-Applied-Environmental-Statistics/book-series/ CRCAPPENVSTA

Bringing Bayesian Models to Life

Mevin B. Hooten
U.S. Geological Survey
Colorado Cooperative Fish and Wildlife Research Unit
Colorado State University
Department of Fish, Wildlife, and Conservation Biology
Department of Statistics

Trevor J. Hefley
Kansas State University
Department of Statistics

CRC Press
Taylor & Francis Group
Boca Raton London New York

CRC Press is an imprint of the
Taylor & Francis Group, an **informa** business

A CHAPMAN & HALL BOOK

CRC Press
Taylor & Francis Group
6000 Broken Sound Parkway NW, Suite 300
Boca Raton, FL 33487-2742

International Standard Book Number-13: 978-0-367-19848-0 (Hardback)

Library of Congress Cataloging-in-Publication Data

Names: Hooten, Mevin B., 1976- author. | Hefley, Trevor, author.
Title: Bringing Bayesian models to life / authors: Mevin B. Hooten, Trevor
Hefley.
Description: Boca Raton, FL : CRC Press, Taylor & Francis Group, 2019. |
Includes bibliographical references.
Identifiers: LCCN 2019003460 | ISBN 9780367198480 (hardback : alk. paper)
Subjects: LCSH: Bayesian statistical decision theory. | Environmental
sciences--Mathematical models.
Classification: LCC QA279.5 .H66 2019 | DDC 519.5/42--dc23
LC record available at https://lccn.loc.gov/2019003460

Visit the Taylor & Francis Web site at
http://www.taylorandfrancis.com

and the CRC Press Web site at
http://www.crcpress.com

Contents

SECTION I   Background

SECTION II   Basic Models and Concepts

SECTION III   Intermediate Models and Concepts

SECTION IV Advanced Models and Concepts

SECTION V   Expert Models and Concepts

Preface

Bayesian statistical methods have become essential in ecological and environmental science. Ecological and environmental inference is now made based on Bayesian posterior distributions and credible intervals almost as regularly as p-values and confidence intervals. The rise of Bayesian methods in ecology and environmental science is due, in large part, to the associated coherent framework for constructing statistical models that mimic processes that generate data observed in the field. In contrast to the strict recipe-based approach to calculating summary statistics and performing hypothesis tests that appears in many entry level textbooks and statistics courses, the Bayesian approach naturally reconciles with the scientific method and scientific thinking—an approach that many find liberating. The ability to build custom statistical models that are suited to particular ecological questions rather than the other way around appeals to scientists. While the mathematics associated with implementing Bayesian models require a bit more study than more traditional statistical methods, we are fortunate that several excellent Bayesian references now exist at all technical levels.

From a computing perspective, we are also fortunate that several software packages exist that translate Bayesian model statements into algorithms to fit the models to data. The fact that this can even be done automatically is amazing and speaks to the intuitive nature of both Bayesian models and the algorithms used to fit them to data. While many users will go on to apply these automatic software packages to fit Bayesian models, some seek to understand what happens behind the scenes. This book complements other resources on both Bayesian modeling concepts and computing by providing an in-depth perspective on the nuts and bolts of Bayesian algorithms. It provides critical details about numerical and stochastic computing techniques to fit a suite of Bayesian models "from scratch." In fact, all computer code and data used in this book are available at the authors websites (we're easy to find!).

Readers will be able to take this set of basic recipes and fundamental principles we describe and extend them to fit their own custom models to ecological and environmental data. These techniques are helpful for reducing the time to obtain inference, especially for models that will be used repetitively. The material in this book also empowers the user to fit models that cannot be implemented in existing automatic Bayesian software.

With great power, comes great responsibility. As you embark on this Bayesian programming journey, remember to follow the rules. Properly implemented Bayesian algorithms follow a strict set of rules that ensure they will provide the correct inference. When modifying or constructing these algorithms from scratch, please remember to adhere to the formal set of rules we describe in what follows. Happy computing!

Acknowledgments

The authors acknowledge the following funding sources: NSF DMS 1614392, NSF EF 1241856, and NOAA AKC188000, NSF DEB 1754491, USGS G18AC00317, and USGS G16AC00413. In alphabetical order, the authors are grateful to Res Altwegg, Ali Arab, Brian Avila, Larissa Bailey, Anne Ballmann, Jarrett Barber, Hannah Birgé, Alice Boyle, Randy Brehm, Kristin Broms, Brian Brost, Franny Buderman, Jim Clark, Paul Conn, Evan Cooch, Noel Cressie, David Clancy, Bob Dorazio, Gabriele Engler, Andy Finley, Bailey Fosdick, Marti Garlick, Alan Gelfand, Brian Gerber, Ephraim Hanks, Tom Hobbs, Jennifer Hoeting, Gina Hooten, Liying Jin, Devin Johnson, Bill Kendall, Bill Link, Lucy Lu, Kumar Mainali, Brett McClintock, Narmadha Mohankumar, Juan Morales, Barry Noon, Kiona Ogle, John Organ, Chris Paciorek, Aaron Pearse, Jim Powell, Andy Royle, Viviana Ruiz-Gutierrez, Robin Russell, Henry Scharf, John Thompson, John Tipton, Jay Ver Hoef, Nelson Walker, Lance Waller, Dan Walsh, Rekha Warrier, Kevin Whalen, Gary White, Chris Wikle, Perry Williams, Ander Wilson, Ken Wilson, Dana Winkelman, Trevor Witt, Haoyu Zhang, Congxing Zhu, and Jun Zhu for various engaging discussions about Bayesian models and software, assistance, collaboration, and support during this project. Any use of trade, firm, or product names is for descriptive purposes only and does not imply endorsement by the U.S. Government. Although the software herein has been used by the U.S. Geological Survey (USGS), no warranty, expressed or implied, is made by the USGS or the U.S. Government as to the accuracy and functioning of the program and related program material nor shall the fact of distribution constitute any such warranty, and no responsibility is assumed by the USGS in connection therewith. Cover illustration by Daisy Chung (daisychung.com).

Authors

Mevin B. Hooten is a professor in the Departments of Fish, Wildlife, & Conservation Biology and Statistics at Colorado State University Fort Collins. He is also assistant unit leader of the Colorado Cooperative Fish and Wildlife Research Unit (U.S. Geological Survey) and a fellow of the American Statistical Association. He earned a PhD in Statistics at the University of Missouri and focuses on the development of statistical methodology for spatial and spatio-temporal ecological and environmental processes.

Trevor J. Hefley is an assistant professor in the Department of Statistics at Kansas State University. He earned a PhD in Statistics and Natural Resource Science at the University of Nebraska-Lincoln and focuses on developing and applying spatio-temporal statistical methods to inform conservation and management of fish and wildlife populations.

Section I

Background

1 Bayesian Models

1.1 INTRODUCTION AND STATISTICAL NOTATION

Bayesian statistical models provide a useful way to obtain inference and predictions for unobserved quantities in nature based on a solid foundation of mathematical rules pertaining to probability. Furthermore, Bayesian methods naturally allow us to account for and express uncertainty in the systems we seek to model. Fundamentally, a Bayesian statistical model is a formal mathematical description of stochastic data-generating mechanisms. We can use the resulting model to simulate new data and use existing data to learn about the data-generating mechanisms.

Bayesian models are inherently parametric (including so-called nonparametric Bayesian models) because they involve components that are specified as known probability distributions, the parameters of which are often of primary interest. This method is in contrast to truly nonparametric statistical methods that rely on a less restrictive set of assumptions, but that also limit their application to simpler settings.

Bayesian statistical methods allow us to break up a complex joint problem into a set of simpler conditional elements. We verify this statement directly through the definition of conditional probability and Bayes rule. In fact, working with conditional distributions is a large part of common Bayesian modeling activities.

To construct Bayesian model fitting algorithms from scratch, it is critical that we understand and are able to work with probability distributions analytically. We provide a basic review of probability and Bayesian modeling in this introductory chapter, but it is essential to acquaint ourselves with other, more detailed, references. For example, the textbook by Hobbs and Hooten (2015) provides a comprehensive intuition and set of Bayesian model building concepts for ecologists. It is accessible and affordable, yet concise enough to allow the reader to work through it, perhaps concurrently while using other more advanced books (e.g., Gelman et al. 2013) as ongoing desk references.

This book contains a substantial amount of mathematical expressions and computer syntax. We use a consistent set of mathematical notation throughout so that each new concept and set of models builds on the previous set sequentially. We show a summary of our notation, which is somewhat standard in statistical ecology literature, in Table 1.1. With few exceptions, we use a standard Bayesian notation where lowercase (e.g., y, ϕ) and uppercase (i.e., N, J) unbolded letters represent scalar variables, lowercase bolded letters (e.g., $\mathbf{y}$, $\boldsymbol{\psi}$) represent vectors, and uppercase bolded letters (e.g., $\mathbf{M}$, $\boldsymbol{\Sigma}$) represent matrices. We use script unbolded letters to denote sets (e.g., $\mathscr{T}$).

Table 1.1

Statistical Notation

Notation	Definition
i	Observation index for $i = 1, \ldots, n$ total observations.
t	Time point at which the data or process occurs (in the units of interest).
$\mathscr{T}$	The set of times at which the process exists; typically compact interval in continuous time such that $t \in \mathscr{T}$.
t_i	Time associated with observation i.
T	Either the largest time in observations or process, or upper temporal endpoint in study, depending on context.
y_i	Univariate response variable (i.e., observation) in a statistical model.
$\mathbf{y}$	Vector of observations $\mathbf{y} \equiv (y_1, \ldots, y_i, \ldots, y_n)'$. Sometimes $\mathbf{z}$ is used when observations are binary.
$\mathbf{X}$	An $n \times p$ "design" matrix of covariates, which will often be decomposed into rows $\mathbf{x}_i$ for row i, depending on the context in which it is used.
$\boldsymbol{\beta}$	Vector of regression coefficients (i.e., $\boldsymbol{\beta} \equiv (\beta_1, \beta_2, \ldots, \beta_p)'$), where p is the number of columns in $\mathbf{X}$.
$\boldsymbol{\beta}'$	The "prime" symbol ($'$) denotes a vector or matrix transpose (e.g., converts a row vector to a column vector).
σ^2	Variance component associated with a data or process model.
$\boldsymbol{\Sigma}$	Covariance matrix for either a parameter vector such as $\boldsymbol{\beta}$ (if subscripted) or for the data or process models.
$f(\cdot), [\cdot]$	Probability density or mass function. $p()$, $P()$, and $\pi()$ are used in other literature. The $[\cdot]$ has become a Bayesian convention for probability distributions.
$E(y)$	Expectation of random variable y; an integral if y is continuous and sum if y is discrete.
$\propto$	Proportional symbol. Often used to say that one probability distribution is proportional to another (i.e., only differs by a scalar multiplier).
$\equiv$	Definition symbol. Often used as a way to rename a variable or function to simplify a mathematical expression (e.g., $\mathbf{x} \equiv (1, x_1, x_2)'$).

1.2   PROBABILITY CONCEPTS

Bayesian models rely on probability distributions and the concept of random variables. Stochastic Bayesian algorithms also rely on properties of random variables. Thus, as a brief introduction in preparation for Bayesian model building and implementation, we review a few critical rules of probability. First, most Bayesian models deal with random variables that are either continuous or discrete valued (i.e., their "support" refers to the values they can assume). For example, suppose the random variable a can take on any real number as a value (i.e., $a \in \Re$). The random variable a arises from a probability distribution that we can characterize in several ways. The most useful way to characterize probability distributions in Bayesian models is with probability density functions (PDFs); or probability mass functions (PMFs) for discrete random variables. Probability textbooks will often use $a \sim f(a)$ to denote that the random variable a arises from PDF $f(a)$. In this book, when we specify parametric statistical models in the next section, we use $a \sim [a]$ to denote the same thing (where $[a]$ denotes the PDF of a). We rarely need to distinguish between random variables and their associated fixed realizations, so we do not often use notation like $P(A = a)$ or $f_A(a)$ as is common in many probability textbooks.

The main properties of PDFs are that they are positive and integrate to one over the support of the random variable (i.e., $[a] > 0$ and $\int_{\Re} [a]da = 1$). We can express moments of the probability distribution as integrals when the random variables have continuous support. For example, we write the expected value of a as $E(a) = \int_{\Re} a[a]da$ and the variance of a as $\text{Var}(a) = \int_{\Re} (a - E(a))^2[a]da$. We return to moments and approaches for computing them in the next section on computing strategies.

When concerned with multiple random variables simultaneously, we rely on probability rules involving joint and conditional distributions—key elements in working with Bayesian models. Now suppose we have another random variable of interest b associated with the same setting in which we are considering a. The variables a and b may share information and thus it may be valuable to consider how they covary in terms of their stochastic properties. Therefore, we can characterize the joint probability distribution for a and b by the joint PDF $[a, b]$.

Bayesian models rely heavily on joint and conditional distributions. When we know something about one of the variables—say we know the value b as an outcome of our study—then the conditional probability distribution provides insight on the remaining variable given our knowledge of the other. In our example, we write the conditional PDF of a, given knowledge of b, as $[a|b]$. A key rule of probability relates conditional distributions with joint distributions and marginal distributions (i.e., the distribution involving one variable only). That is, we can express the conditional probability distribution a given b as

$$[a|b] = \frac{[a, b]}{[b]}, \tag{1.1}$$

where, in the context of two continuous random variables, we can write the marginal distribution of b as $[b] = \int [a, b]da$. We can interpret this integral expression for $[b]$ as removing (or "integrating out") a from the joint distribution of a and b to arrive at the marginal distribution. The conditional probability equation (1.1) represents the foundation for Bayesian models. To make that more clear, note that we can write the joint distribution as a product of the conditional and the marginal with respect to either a or b. This method is called the multiplication rule of probability and is expressed as

$$[a, b] = [a|b][b] = [b|a][a]. \tag{1.2}$$

Thus, we can substitute this product (1.2) into the right-hand-side of the equation for conditional probability (1.1) to get the Bayes rule

$$[a|b] = \frac{[b|a][a]}{\int [b|a][a]da}. \tag{1.3}$$

The power of the Bayes rule is that it allows us to express the conditional distribution of a given b in terms of the conditional distribution of b given a and the marginal distribution of a (or vise versa). Thus, if we know one of the conditional distributions and a marginal distribution, then we can find the other conditional distribution. Using this result as a statistical model, we can learn about an unknown variable a in nature given observable information b. In Bayesian models, a represents a model parameter and b represents data. We expand on this idea in the next section and throughout the book.

In a situation with three or more random variables of interest, we can extend the basic ideas in (1.1)–(1.3). For example, for a, b, and c, we can express the joint distribution as $[a,b,c] = [a,b|c][c]$ or $[a,b,c] = [a|b,c][b|c][c]$ (among many other permutations) and the full-conditional distribution of any one particular variable, say b, we denote often as $[b|\cdot] \propto [a|b,c][b|c]$. The full-conditional notation is short for $[b|\cdot] = [b|a,c]$ and it is only proportional to $[a|b,c][b|c]$ because we need to divide this product by $[a,c] = \int [a,b,c]db$ so that it will integrate to one as required. The full-conditional distribution is the analytical (i.e., pencil and paper) workhorse essential for creating many algorithms to fit Bayesian models. Thus, we need to know how to analytically derive (i.e., find with pencil and paper) full-conditional distributions for models involving multiple parameters. Implementing Bayesian models from scratch by constructing algorithms usually involves-analytical and numerical work. This method is a fundamental difference between specifying and fitting models using automatic software like BUGS, JAGS, or STAN. There *will* be math; and it will usually pay off! It will pay off because often we can make custom algorithms run faster than those used in automatic software and, in some cases, we may not even be able to specify the model of interest using automatic software.

1.3 MODELING CONCEPTS

With a review of probability under our belt, we can translate those concepts to help us formulate statistical models that involve parameters that we wish to learn about. Bayesians often refer to probability models for the data as "data models" so that is where we begin. We write a generic data model statement as $y_i \sim [y_i|\boldsymbol{\theta}]$, where y_i are the observations for $i = 1,\ldots,n$, $\boldsymbol{\theta}$ is a vector of p data model parameters ($\boldsymbol{\theta} \equiv (\theta_1,\ldots,\theta_p)'$), and the bracket notation "$[\cdot]$" represents a probability distribution (or PDF/PMF as needed) like before. Bayesians often refer to the data model as the "likelihood," which takes the same form as the likelihood used in the maximum likelihood estimation (MLE), which is the joint distribution of the data conditioned on the parameters. When the observations are conditionally independent, we often write the likelihood as $[\mathbf{y}|\boldsymbol{\theta}] = \prod_{i=1}^{n}[y_i|\boldsymbol{\theta}]$, where we can multiply individual distributions for each observation to obtain the joint distribution because of conditional independence. That is, conditioned on the parameters $\boldsymbol{\theta}$, the observations are independent of each other. As a reminder, to fit the model using MLE, we condition the likelihood on the observed data and usually maximize it numerically to find the optimal parameter values $\hat{\boldsymbol{\theta}}$.

The Bayesian approach involves the specification of a probability model for the parameters, $\boldsymbol{\theta} \sim [\boldsymbol{\theta}]$, that depend on fixed hyperparameters that we assume are known. The prior probability distribution should usually contain information about the parameters that is known before we collect the data. Rather than maximizing the likelihood, the Bayesian approach finds the conditional distribution of the parameters given the data (i.e., the posterior distribution)

$$[\boldsymbol{\theta}|\mathbf{y}] = \frac{[\mathbf{y}|\boldsymbol{\theta}][\boldsymbol{\theta}]}{\int [\mathbf{y}|\boldsymbol{\theta}][\boldsymbol{\theta}]d\boldsymbol{\theta}}, \tag{1.4}$$

where $\mathbf{y}$ is a vector notation for all the observations and the denominator in (1.4) is a p-dimensional integral that equates to a scalar constant (also called a "normalizing" constant) after we have observed the data. For complicated models, we cannot obtain the multidimensional integral in the denominator of (1.4) analytically (i.e., exactly by pencil and paper) and we must calculate it numerically or avoid it by using a stochastic simulation procedure. For example, Markov chain Monte Carlo (MCMC; Gelfand and Smith 1990) is a stochastic simulation procedure that allows us to obtain samples from the posterior distribution while avoiding the calculation of the normalizing constant in the denominator of (1.4). MCMC algorithms have many advantages (e.g., easy to develop), but also limitations (e.g., can be time consuming to run), as we show in Chapter 4.

Hierarchical models are comprised of a sequence of nested probability distributions for the data, the process, and the parameters (Berliner 1996). For example, a basic Bayesian hierarchical model is

$$y_{i,j} \sim [y_{i,j}|z_i, \boldsymbol{\theta}], \tag{1.5}$$

$$z_i \sim [z_i|\boldsymbol{\beta}], \tag{1.6}$$

$$\boldsymbol{\theta} \sim [\boldsymbol{\theta}], \tag{1.7}$$

$$\boldsymbol{\beta} \sim [\boldsymbol{\beta}], \tag{1.8}$$

where z_i is an underlying process for observation i and $y_{i,j}$ are repeated measurements for each observation ($j = 1, \ldots, J$). Notice that the process model parameters $\boldsymbol{\beta}$ also require a prior distribution if the model is Bayesian. The posterior for this model is a generalized version of (1.4) such that

$$[\mathbf{z}, \boldsymbol{\theta}, \boldsymbol{\beta}|\mathbf{y}] = \frac{[\mathbf{y}|\mathbf{z}, \boldsymbol{\theta}][\mathbf{z}|\boldsymbol{\beta}][\boldsymbol{\theta}][\boldsymbol{\beta}]}{\int \int \int [\mathbf{y}|\mathbf{z}, \boldsymbol{\theta}][\mathbf{z}|\boldsymbol{\beta}][\boldsymbol{\theta}][\boldsymbol{\beta}]d\mathbf{z}d\boldsymbol{\theta}d\boldsymbol{\beta}}. \tag{1.9}$$

Throughout the remainder of this book, we use both hierarchical and nonhierarchical Bayesian model specifications to obtain statistical inference using ecological data of various types. Ecologists and environmental scientists have welcomed Bayesian hierarchical modeling in part because it meshes with how they think about the data-generating mechanisms and many complicated hierarchical models are easier to implement from a Bayesian perspective.

1.4 ADDITIONAL CONCEPTS AND READING

Generating data (or simulating data) is an important concept in statistical modeling. In fact, generative statistical models (Talts et al. 2018) provide an explicit way to generate (i.e., simulate) data. We can use simulated data to test algorithms and check the assumptions of models, among other things. Consider a Bayesian statistical model based on a Bernoulli distribution specified as $y_i \sim \text{Bern}(p)$ for $i = 1, \ldots, n$, with $p \sim \text{Unif}(0,1)$, for example. If we first draw p from a uniform distribution on $(0,1)$, then draw n independent realizations for y_i for $i = 1, \ldots, n$, then we have generated a data set based on a fully specified statistical model. We use the statistical software R (R Core Team 2018) for all computer examples throughout this book (all R code and data are available at the authors websites). For example, for $n = 5$, we can generate a data set for our simple Bernoulli model in R as

BOX 1.1 SIMULATE BERNOULLI DATA

```
1 > n=5
2 > set.seed(101)
3 > p=runif(1)
4 > p
5 [1]  0.3721984
6 > y=rbinom(n,1,p)
7 > y
8 [1]  0 1 1 0 0
```

This type of simulation procedure is useful in many ways.[1] First, by fitting our statistical model to simulated data that we generated using our model specification, we can assess how well the model is able to recover the parameters in the model (Cook et al. 2006). Second, generating data from a particular distribution is a useful concept in Bayesian prediction and we return to that topic in Chapter 12.

Hobbs and Hooten (2015) provide an accessible description of hierarchical and nonhierarchical Bayesian models and model building strategies as well as an overview of basic probability and fundamental approaches for fitting models. Their focus is on conceptually understanding how Bayesian models work and how to capture ecological mechanisms with models. Gelman et al. (2013) provide more detail

---

[1] We display R code in two different ways throughout this book: (1) As it appears here where the ">" symbols indicate that the code has been processed by R, usually to emphasize a particular aspect of the output and (2) as it begins to appear in Chapter 3 where the R code simply exists as a script or function not yet processed. In neither case do we expect readers to copy and paste the code directly from the book because the full data and code are accessible online. The R code that is shown explicitly here is to provide readers with a visual understanding of how we construct functions and algorithms using precise R syntax. Moreover, the code available online contains additional comments and non-essential options to functions that may enhance the appearance of the graphics. We withheld optional syntax to simplify the appearance of the R code here only.

on a variety of aspects related to Bayesian modeling and implementation with a more general targeted audience.

For details on models pertaining to wildlife ecology, Royle and Dorazio (2008) and Link and Barker (2010) have become canonical references. Parent and Rivot (2012) provide a good introduction to Bayesian modeling focused more on fish ecology and management and Clark (2007) provides a thorough volume for more general environmental scientists and ecologists.

2 Numerical Integration

2.1 BAYESIAN INTEGRALS

Bayesian methods for continuous random variables involve integrals. For example, to calculate the expectation of the continuous random variable a with prior PDF $[a]$, we need to evaluate the integral

$$E(a) = \int a[a]da, \qquad (2.1)$$

where the limits of integration are defined by the support of a. In the case where a is a uniform random variable with support $(0,1)$ (i.e., $a \sim \text{Unif}(0,1)$), the expectation in (2.1) becomes

$$E(a) = \int_0^1 a[a]da, \qquad (2.2)$$

$$= \int_0^1 a1da, \qquad (2.3)$$

$$= \left(\frac{a^2}{2}\right)\Big|_0^1, \qquad (2.4)$$

$$= \frac{1}{2} - \frac{0}{2}, \qquad (2.5)$$

$$= \frac{1}{2}. \qquad (2.6)$$

The result in (2.6) matches our understanding of the mean of a uniform distribution. But, how could we approximate this integral numerically if we cannot evaluate it analytically? We show how to approximate this integral numerically in the next section.

As another example, consider the denominator in the Bayes rule from (1.3)

$$\int_0^1 [b|a][a]da, \qquad (2.7)$$

where $a \sim \text{Unif}(0,1)$ is specified as before and the conditional distribution of b given a is

$$[b|a] = \text{Bern}(a). \qquad (2.8)$$

where we define $[b|a]$ as the Bernoulli probability mass function (PMF) with parameter a.

We often refer to the integral in (2.7) as a normalizing constant in Bayesian posterior distributions; it is also the main crux of Bayesian inference because we must either calculate it or sidestep it to find the posterior distribution.

Based on our specifications, the integral in (2.7) becomes

$$\int_0^1 [b|a][a]da = \int_0^1 a^b(1-a)^{1-b}1da,\tag{2.9}$$

$$= \frac{\Gamma(b+1)\Gamma(2-b)}{\Gamma(3)},\tag{2.10}$$

$$= \frac{b!(1-b)!}{2!},\tag{2.11}$$

$$= \frac{1}{2},\tag{2.12}$$

which implies that the marginal PDF of b evaluates to $1/2$ regardless of the value for b. The relationship between the desired integral and the beta function is a key feature of our derivation; without that, the problem would be much more difficult. As before, how could we arrive at this result if the integral were analytically intractable? In fact, most of the integrals we encounter in Bayesian models are intractable, so we need a way to evaluate them numerically or sidestep them altogether. In the two sections that follow, we show how to numerically approximate integrals using numerical quadrature or stochastic sampling. In the section after that, we introduce a method that allows us to avoid calculating some integrals (e.g., marginal distributions) completely.

2.2   NUMERICAL QUADRATURE

The concept of numerical quadrature is to break up the area under a curve into several narrow rectangles like in a Riemann sum. Using numerical quadrature in one dimension, we form narrow rectangles under the curve by first specifying a regular set of points along the x-axis and, for each one, multiply the width of the rectangle by the height where it meets the curve resulting in the area of the rectangles. Then we sum up the areas to approximate the integral. If we need more accuracy, we increase the number of points along the x-axis (so that their width is narrower).

More formally, suppose we wish to approximate the integral $\int_{a_l}^{a_u} f(a)da$ using numerical quadrature. We first select a set of m equally spaced points between a_l and a_u, a_j for $j = 1,\ldots,m$. Then, we calculate the function $f(a_j)$ at each point for $j = 1,\ldots,m$. If the points a_j are equally spaced, then they are all separated by distance Δa. Thus, to approximate the integral we compute

$$\int_{a_l}^{a_u} f(a)da \approx \sum_{j=1}^m \Delta a f(a_j).\tag{2.13}$$

Returning to the integral in (2.1) from the previous section, suppose we wish to approximate the first moment of a using numerical quadrature when $[a] \equiv \text{Unif}(0,1)$. After selecting m equally spaced points a_j, for $j = 1,\ldots,m$ such that $0 \le a_j \le 1$, we calculate

$$E(a) = \int_0^1 a[a]da \approx \sum_{j=1}^m (\Delta a)a_j[a_j], \qquad (2.14)$$

$$\approx \Delta a \sum_{j=1}^m a_j, \qquad (2.15)$$

because $[a_j] = 1$ for all $j = 1,\ldots,m$. In R, we calculated the required sum in (2.15), for three values of $m = 100$, $m = 1000$, and $m = 10000$:

BOX 2.1   QUADRATURE FOR UNIFORM

```
 1 > m=100
 2 > a=seq(0,1,,m)
 3 > Delta.a=a[2]-a[1]
 4 > Delta.a*sum(a*dunif(a,0,1))
 5 [1] 0.5050505
 6 >
 7 > m=1000
 8 > a=seq(0,1,,m)
 9 > Delta.a=a[2]-a[1]
10 > Delta.a*sum(a*dunif(a,0,1))
11 [1] 0.5005005
12 >
13 > m=10000
14 > a=seq(0,1,,m)
15 > Delta.a=a[2]-a[1]
16 > Delta.a*sum(a*dunif(a,0,1))
17 [1] 0.50005
```

As we can see, the numerical approximation more closely approximates the correct solution (i.e., $1/2$) as the number of points m increases. In this example, the approximate value is always larger than the truth because the rectangles associated with our selected values for a_l and a_u include small regions outside the support $(0,1)$. As the number of points m increases, the rectangles become more narrow and the error associated with these non-support regions reduces.

   We also can approximate our second example from the previous section, involving an integral (2.12) representing the marginal distribution $[b]$, using numerical quadrature as

$$\int_0^1 [b|a][a]da = \int_0^1 a^b(1-a)^{1-b}da, \qquad (2.16)$$

$$\approx \Delta a \sum_{j=1}^m [b|a_j], \qquad (2.17)$$

where $[b|a_j]$ is the Bernoulli PMF evaluated at a_j. In R, we calculated the required sum in (2.17) for $b = 0$ (recall that the integral is the same for both values of b) and the same three values for m we used in the previous example:

BOX 2.2   QUADRATURE FOR BERNOULLI

```
1  > m=100
2  > a=seq(0,1,,m)
3  > b=0
4  > Delta.a=a[2]-a[1]
5  > Delta.a*sum(dbinom(b,1,a))
6  [1] 0.5050505
7  >
8  > m=1000
9  > a=seq(0,1,,m)
10 > b=0
11 > Delta.a=a[2]-a[1]
12 > Delta.a*sum(dbinom(b,1,a))
13 [1] 0.5005005
14 >
15 > m=10000
16 > a=seq(0,1,,m)
17 > b=0
18 > Delta.a=a[2]-a[1]
19 > Delta.a*sum(dbinom(b,1,a))
20 [1] 0.50005
```

Thus, again, using numerical quadrature we are able to approximate the integral up to a desired level of accuracy with the integral in (2.17) computed as 0.50005 based on $m = 10000$.

   Numerical quadrature is definitely useful in Bayesian model implementations and, while we can implement it for multivariate situations involving multiple integrals, it becomes more challenging to program. While some automatic Bayesian software (e.g., integrated nested Laplace approximation; INLA; Rue et al. 2009) heavily rely on numerical quadrature concepts, most Bayesian model fitting algorithms rely on stochastic sampling to approximate integrals. We explore stochastic approaches to numerically approximate integrals in the following chapter.

2.3   ADDITIONAL CONCEPTS AND READING

Numerical integration is a massive topic in mathematics in its own right. The approach we presented to do numerical integration is just the tip of the iceberg. There are entire software packages devoted to providing algorithms for efficiently approximating integrals using deterministic methods (e.g., Mathematica, Maple), but many methods involve modifications of the basic quadrature concept that we presented in this chapter. Givens and Hoeting (2013) describe a suite of numerical approximation approaches for integral equations.

Despite the rise in stochastic sampling-based methods for fitting Bayesian models to data, numerical techniques are still relevant; for example, the INLA software uses numerical integration to approximate marginal posterior distributions (Rue et al. 2009). However, as Liu (2004) pointed out, while numerical methods perform well in low dimensions, they require impossibly high resolution quadrature sets as the dimension of the probability distribution increases. Betancourt (2017) provided a good overview of the challenges associated with numerical integration for Bayesian modeling and some of the history associated with alternative approaches.

3 Monte Carlo

3.1 SAMPLING

As an alternative to the deterministic quadrature approach to approximating integrals, we also can use stochastic methods. In addition to providing a means to perform integration, as we discuss in the following section, stochastic methods provide a way to characterize entire probability distributions. Stochastic methods rely on random number generation. For example, we often refer to generating independent random realizations from a known distribution as Monte Carlo (MC) sampling. By sampling numerous realizations from a known probability distribution, we can use those samples to help us approximate several important things including the probability distribution itself, moments of the distribution, and similar characteristics pertaining to functions of the random variable (i.e., derived quantities).

When using stochastic methods to obtain realizations from a known probability distribution, we often use a superscript notation to clarify that the realizations are associated with a computational implementation and not part of the model or data themselves. For example, for a uniform random variable a, we would use $a^{(k)} \sim [a] \equiv$ Unif$(0, 1)$ for $k = 1, \ldots, K$ to denote the K realizations that we independently sample from a uniform distribution with support $(0, 1)$. We refer to the set of realizations $\{a^{(1)}, \ldots, a^{(k)}, \ldots, a^{(K)}\}$ as a MC sample.

We can approximate the PDF that generated our sample with a histogram. For example, we used the following R code to generate the histograms in Figure 3.1 based on $K = 100$, $K = 1000$, and $K = 10000$ MC samples with the true uniform distribution overlaid (Box 3.1):

BOX 3.1   UNIFORM MONTE CARLO

```
1  set.seed(105)
2  layout(matrix(1:3,1,3))
3  hist(runif(100),xlab="a",ylab="[a]",prob=TRUE,
4      main="a",ylim=c(0,2))
5  abline(h=1,lty=2)
6  mtext("K=100",line=.25)
7  hist(runif(1000),xlab="a",ylab="[a]",prob=TRUE,
8      main="b",ylim=c(0,2))
9  abline(h=1,lty=2)
10 mtext("K=1000",line=.25)
```

(*Continued*)

BOX 3.1 (Continued)   UNIFORM MONTE CARLO

```
11 hist(runif(10000),xlab="a",ylab="[a]",prob=TRUE,
12     main="c",ylim=c(0,2))
13 abline(h=1,lty=2)
14 mtext("K=10000",line=.25)
```

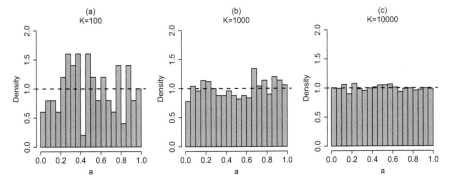

Figure 3.1 Histograms based on MC samples of size (a) $K = 100$, (b) $K = 1000$, and (c) $K = 10000$, from a uniform distribution with support $(0,1)$. Dashed line represents the true uniform PDF.

As we can see in Figure 3.1, the histogram approximates the true distribution better when K is larger. Such graphics are commonplace in Bayesian statistics because we often seek to characterize posterior and prior distributions visually based on samples. However, it is less well-known that the MC histogram approximation to the true distribution is made possible because of numerical (stochastic) integration. We describe how this works in the next section.

3.2   MONTE CARLO INTEGRATION

To approximate integrals using a MC sample, we first draw the sample, say $a^{(k)} \sim [a]$ for $k = 1, \ldots, K$ (assuming we are interested in an integral with respect to the first moment of a function $f(a)$ of the random variable a), and then calculate a sample mean as

$$\int f(a)[a]da \approx \frac{\sum_{k=1}^{K} f(a^{(k)})}{K}. \tag{3.1}$$

Thus, we can use (3.1) to compute prior and posterior means for a function of a if we have samples from those distributions. Moreover, the MC integration will become more accurate as the MC sample size increases. For example, suppose the prior for a is a uniform distribution on $(0,1)$. Then, letting $f(a) = a$, we can approximate the prior mean (i.e., expectation) for $K = 100$, $K = 1000$, and $K = 10000$, using R as (Box 3.2)

BOX 3.2   UNIFORM MONTE CARLO INTEGRATION

```
1 > set.seed(1)
2 > a=runif(100)
3 > mean(a)
4 [1] 0.5178471
5 > a=runif(1000)
6 > mean(a)
7 [1] 0.4973043
8 > a=runif(10000)
9 > mean(a)
10 [1] 0.5004932
```

Notice that the accuracy of the integral approximation improves with increasing MC sample size and approaches the true prior mean we derived in the previous chapter (i.e., $E(a) = 1/2$).

MC integration is a powerful tool that we can use to calculate moments of functions of random variables (or derived quantities). Without MC integration, we would require techniques like the delta method (Dorfman 1938; Powell 2007; Ver Hoef 2012) to obtain inference for derived quantities of random variables, but such methods become increasingly complicated as the complexity of the function f increases. For illustration, consider the quantity a^2 where the random variable a arises from a uniform distribution as before. What is the mean of a^2? Hint: it is not $(1/2)^2$, simply the mean squared. In fact, $E(a^2) = 1/3$ because

$$E(a^2) = \int_0^1 a^2 [a] da, \tag{3.2}$$

$$= \int_0^1 a^2 1 da, \tag{3.3}$$

$$= \left(\frac{a^3}{3} \right) \Big|_0^1, \tag{3.4}$$

$$= \frac{1}{3} - \frac{0}{3}, \tag{3.5}$$

$$= \frac{1}{3}. \tag{3.6}$$

To verify that the expected value of a^2 is $1/3$ using MC integration and $K = 10000$, we compute its approximation using R as (Box 3.3)

BOX 3.3   UNIFORM SQUARED MONTE CARLO EXPECTATION

```
1 > set.seed(1)
2 > a=runif(10000)
3 > mean(a^2)
4 [1] 0.3348666
```

---

which is accurate to the second decimal place. We can extend this method for characterizing derived quantities to multiple variables. We often refer to a derived quantity that is the weighted sum of coefficients in a linear regression model as a contrast. We can obtain contrasts easily, and inference pertaining to them, using MC methods.

We can compute higher order moments in the same way. For example, continuing with the same example, we can approximate the prior variance using MC integration as

$$\int (a - E(a))^2 [a] da \approx \frac{\sum_{k=1}^{K} (a^{(k)} - E(a))^2}{K}, \tag{3.7}$$

where we approximate $E(a)$ as described in (3.1). Considering a MC sample size of $K = 10000$, the R code to compute the prior variance using MC integration is (Box 3.4)

BOX 3.4   UNIFORM MONTE CARLO VARIANCE

```
1 > set.seed(1)
2 > a=runif(10000)
3 > var(a)
4 [1] 0.08470703
```

---

which results in a value near the true variance of $1/12 \approx 0.083$ (which can be derived analytically like we derived the mean in the previous chapter). A critical aspect of MC integration is that we require large MC samples to approximate higher order moments, especially in cases with more complicated functions of the random variable. However, as $K \to \infty$, the MC approximation becomes exact.

So far, we have shown that we can use MC integration techniques to compute moments of probability distributions. Can we use MC methods to compute more complicated integrals that are relevant for Bayesian statistics? For example, how could we use MC integration to calculate the marginal distribution $[b]$ from the previous chapter, given knowledge about the form of $[b|a]$ and $[a]$?

As an example, consider the marginal distribution of b that we approximated in the previous chapter where $[b|a]$ is Bernoulli with probability a, and a has a marginal uniform distribution with support $(0, 1)$. In this case, we can show that, for a particular value of b, we can write the required integral to calculate $[b]$ as

$$[b] = \int_0^1 [b|a][a]da, \tag{3.8}$$

$$= \int_0^1 a^b(1-a)^{1-b}[a]da, \tag{3.9}$$

$$\approx \frac{\sum_{k=1}^K (a^{(k)})^b(1-a^{(k)})^{1-b}}{K}, \tag{3.10}$$

where $a^{(k)} \sim \text{Unif}(0,1)$ for $k = 1, \ldots, K$ comprises a MC sample. Using R, the code to compute this MC approximation is (Box 3.5)

BOX 3.5   MONTE CARLO MARGINAL FOR b

```
> set.seed(1)
> a=runif(10000)
> b=0
> mean(dbinom(b,1,a))
[1] 0.499832
> b=1
> mean(dbinom(b,1,a))
[1] 0.500168
```

Note that, for both values $b = 0$ and $b = 1$, we approximate the integral near the true value $[b] = 1/2$ that we derived analytically. The MC integral approximation in (3.10) works because, for a given value of b, the Bernoulli PMF $[b|a]$ is a derived quantity of a. Marginal distributions like that in (3.8) are incredibly important in the Bayesian context because we can use them to calculate posterior model probabilities, which are fundamental elements of Bayesian model averaging (Hoeting et al. 1999; Hooten and Hobbs 2015). In fact, some statisticians refer to marginal distributions of the data as the model "evidence" because their values provide a quantitative way to compare models. As previously mentioned, the marginal distribution in the denominator of Bayes rule is usually the crux in fitting Bayesian models to data. Thus, despite the importance of the marginal distribution for inference, most contemporary methods for fitting Bayesian models to data sidestep computing the marginal distribution altogether. We explain how this works in the next chapter.

Finally, to explain the concept of a histogram as an approximation of a PDF for a continuous random variable a, we note that a histogram partitions the support of a into a disjoint set of m intervals $\mathscr{A}_j$ for $j = 1, \ldots, m$ where m represents the number of bins. For a given interval $\mathscr{A}_j$, the area under the curve is $\int_{\mathscr{A}_j} [a]da$, which we can rewrite as the convolution of an indicator function and $[a]$ over the entire support of a: $\int 1_{\{a \in \mathscr{A}_j\}}[a]da$. Thus, we can use the MC average $\sum_{k=1}^K 1_{\{a^{(k)} \in \mathscr{A}_j\}}/K$ to approximate the integral because the indicator function is a derived quantity of a and serves as the function f in (3.1). The MC approximation provides the areas (i.e., area$_j$) of the histogram bins and we can calculate the histogram bin heights (i.e., $h_j = \text{area}_j/w$) because we know the bin widths (w) corresponding to $\mathscr{A}_j$.

3.3 ADDITIONAL CONCEPTS AND READING

Stochastic methods for approximating integrals is a massive area of research and they have been extended in a number of directions to handle more complicated integrals. Aside from the various introductions in more general books on Bayesian statistics (e.g., Gelman et al. 2013), more focused texts include Liu (2004), Gamerman and Lopes (2006), and Givens and Hoeting (2013) (among many others).

We focused on how to use MC methods to calculate posterior moments in this chapter; however, we did not discuss why we would choose to use a posterior mean for inference. If we are interested in a summary of the posterior distribution that provides us with an understanding of the location (i.e., point estimate) of the parameter of interest, three obvious choices are the posterior mean, median, or mode. The question of how and why to select among them depending on our inferential goals relates to decision theory. Formal decision theory helps us choose the optimal point estimate given that we can quantify the loss associated with a poor estimate. Williams and Hooten (2016) provide an accessible introduction to statistical decision theory and some intuition based on loss functions for certain types of estimates (and more general decisions) based on Bayesian models.

On the topic of approximating marginal distributions directly, Gelfand and Dey (1994) suggested that the straightforward MC approximation as described in this chapter can be unstable when the prior $[a]$ is diffuse relative to the conditional distribution $[b|a]$. Thus, they suggested a generalization of a different approach presented by Newton and Raftery (1994) where the marginal distribution of $[b]$ is approximated as

$$[b] \approx \left(\frac{\sum_{k=1}^{K} \frac{g(a^{(k)})}{[b|a^{(k)}][a^{(k)}]}}{K} \right)^{-1}, \tag{3.11}$$

where $a^{(k)} \sim \text{Unif}(0,1)$ for $k = 1,\dots,K$ as before in our continued example. The function $g(\cdot)$ in (3.11) is arbitrary except that it must be positive and integrate to one. While this approach is more stable, it requires careful selection of the function $g(\cdot)$ and Carlin and Louis (2009) note that it may not perform well in cases where a is high dimensional. Thus, despite ongoing research to improve the approximation of the denominator in Bayes rule, characterizing Bayesian posterior distributions using numerical integration methods is challenging (Chib 1995; Park and Haran 2018).

Two other stochastic sampling approaches are frequently mentioned in the Bayesian context: rejection sampling and slice sampling. Both approaches are useful for sampling from probability distributions that are otherwise not trivial to sample from directly. In rejection sampling, we first obtain a MC realization $a^{(*)}$ from a temporary probability distribution that is easy to sample from, has the same support as the target distribution, and whose PDF (or PMF) can be made greater than the target PDF by multiplication of a known constant. Then we decide whether to accept or reject $a^{(*)}$ based on an acceptance probability that is the ratio of the target PDF and the product of the known constant and the temporary PDF, both evaluated at $a^{(*)}$. The set of accepted realizations forms a sample from the target distribution and we obtain the desired sample size by sampling from the temporary distribution as much as necessary until we reach the desired accepted sample size.

Slice sampling, by contrast, relies on a sample of auxiliary variables to obtain a sample from the target distribution of interest (Higdon 1998; Neal 1999). A basic slice sampling algorithm proceeds as follows:

1. Set $k = 0$
2. Specify an initial value $a^{(k)}$
3. Sample $u^{(k)} \sim \text{Unif}(0, [a^{(k)}])$
4. Increment $k = k + 1$
5. Sample $a^{(k)} \sim \text{Unif}(\mathscr{A}^{(k)})$, where $\mathscr{A}^{(k)}$ is the subset of support for a where $[a] \geq u^{(k-1)}$
6. Repeat steps 3–5 for $k = 1, \ldots, K$

This slice sampling algorithm provides a sample from $[a]$ and we can combine it with other types of stochastic sampling. However, there are also situations where slice sampling is not easy to implement (i.e., such as when the target PDF $[a]$ is not invertible). For further details about rejection sampling and slice sampling see Givens and Hoeting (2013).

4 Markov Chain Monte Carlo

4.1 METROPOLIS-HASTINGS

MCMC is a stochastic sampling method like MC, but is Markov because it results in dependent samples rather than independent samples (as in MC methods). MCMC methods allow us to obtain a sample from a posterior distribution without knowing its normalizing constant (i.e., marginal data distribution). We obtain a MCMC sample (i.e., a correlated sample of random numbers resulting from a MCMC algorithm) by constructing a time series in which an accept-reject algorithm governs its dynamics. Typical accept-reject algorithms involve a procedure whereby we propose and evaluate a tentative realization of the random variable of interest based on a criterion to decide whether we should retain or reject it.

We refer to a general form of MCMC as Metropolis-Hastings (M-H). We begin our introduction to MCMC with M-H algorithms, first from a single parameter perspective, then extending it to multiple parameter cases. We then show how to arrive at special cases of M-H, resulting in the well-known Gibbs sampler, among others.

Assuming the following Bayesian model involving data y and parameter θ

$$y \sim [y|\theta], \tag{4.1}$$

$$\theta \sim [\theta], \tag{4.2}$$

we seek to characterize the posterior distribution $[\theta|y] \propto [y|\theta][\theta]$ by obtaining a sample that coincides with it. We use the sample to summarize moments or other aspects of the posterior distribution as described in the previous chapter.

To construct a MCMC algorithm based on M-H updates[1] for this simple Bayesian model, we use the following procedure:

1. Assign an initial value for $\theta^{(0)}$
2. Set $k = 1$
3. Sample a proposed value for the parameter using $\theta^{(*)} \sim [\theta^{(*)}|\theta^{(k-1)}]$
4. Calculate the M-H ratio: $mh = \frac{[y|\theta^{(*)}][\theta^{(*)}][\theta^{(k-1)}|\theta^{(*)}]}{[y|\theta^{(k-1)}][\theta^{(k-1)}][\theta^{(*)}|\theta^{(k-1)}]}$
5. Let $\theta^{(k)} = \theta^{(*)}$ with probability $\min(mh, 1)$, otherwise retain the previous value $\theta^{(k)} = \theta^{(k-1)}$
6. Set $k = k + 1$, go to step 3; repeat until enough MCMC samples have been obtained to approximate the quantities of interest well

---

[1] We often use the term "update" to refer to the action of obtaining a realization, or random number for the parameter, in a MCMC algorithm.

Although this procedure is surprisingly simple, it depends on several critical elements that we describe herein. The initial value described in step 1 of the algorithm is user specified and we should choose it to be close to regions of high density in the posterior distribution, but technically it does not affect the final inference with infinite computing time. However, in practice, with finite computing time, a poorly chosen initial value (i.e., one that is in low density regions of the posterior distribution) can slow convergence of the algorithm. The concept of convergence is critical in MCMC because, unlike with MC samples, the Markov chain (i.e., sequence of random numbers) resulting from a MCMC algorithm may require many iterations to arrive in high density regions of the posterior distribution; a period referred to as "burn-in." After the burn-in period, all following realizations of $\theta^{(k)}$ will also arise from that distribution.

The literature on stochastic processes contains a much more rigorous description of Markov chains that involves terms like stationarity, ergodicity, and limiting distributions. For details about the theory that underpins MCMC, M-H, and how we can use these ideas to characterize posterior distributions of interest, see Gelfand and Smith (1990), Casella and George (1992), Chib and Greenberg (1995), and Dobrow (2016). For our purposes herein, we treat MCMC output containing realizations of parameters as if they arise directly from the posterior distribution of interest, with serial autocorrelation, because that is how we will use them summarizing posterior quantities for inference.

An important key to M-H algorithms is the proposal distribution $[\theta^{(*)}|\theta^{(k-1)}]$, which we write in a way that implies it may depend on previous values of the parameter in the chain $\theta^{(k-1)}$. In fact, we may tune the proposal distribution such that it results in an efficient algorithm, but it need not depend on previous values for the parameter. Perhaps the simplest proposal distribution is a known probability distribution that has the same support as the parameter. In our example that follows, we use a uniform distribution for the proposal, but we describe several alternatives in the next section.

The M-H ratio described in step 4 of the M-H algorithm is comprised of a product of three components in the numerator and three in the denominator. The first two components in the numerator of the M-H ratio correspond to the data model (or likelihood) $[y|\theta]$ and prior $[\theta]$, both evaluated at the proposed value $\theta^{(*)}$. The third component is the proposal, but written such that the proposed value and the previous value for the parameter are swapped. In the denominator of the M-H ratio (step 4 of the M-H algorithm), we evaluate the first two terms at the previous parameter value in the chain $\theta^{(k-1)}$ and the ordering of the proposed value and previous value for the parameter is as specified when generating the proposal. The M-H ratio acts as a criterion for accepting or rejecting the proposed value based on how well it suits the likelihood and prior. However, because we drew the proposal from an arbitrary distribution, we must correct for any bias that we induced in our samples due to the choice of proposal distribution. Bias in our proposed values can occur because the proposal distribution may be more likely to generate random numbers in one region of the support than another.

In step 5 of the M-H algorithm, we use the M-H ratio to decide if we should keep the proposed value as the current value in the chain or retain the previous value. Heuristically, we interpret the M-H algorithm as follows: If the proposed value for the parameter $\theta^{(*)}$ results in a larger numerator in the M-H ratio (i.e., suits the likelihood and prior well, after correcting for the proposal), then we keep it as the next value in the chain. On the other hand, if the proposed value results in a similar, but not larger, numerator as the denominator of the M-H ratio, then we keep the proposal with probability proportional to the M-H ratio. If the proposed value is far from high density regions of the posterior distribution, then we retain the previous parameter value with large probability. The fact that we will not always reject a poorly proposed value for the parameter in the M-H algorithm allows the chain to explore the posterior distribution without getting stuck at the "best" value (i.e., the highest posterior density, or mode). This stochastic process for simulating values for parameters contrasts with deterministic optimization algorithms (e.g., Newton-Raphson, steepest descent, etc.) that are used to find maximum likelihood estimates, for example.

To clarify how the M-H algorithm works in practice, we show how to fit the single-parameter Bayesian model in (4.1)–(4.2) using MCMC. In this case, we specify a Bernoulli data model for $[y|\theta]$ and a uniform prior for $[\theta]$. To construct a MCMC algorithm with M-H updates for the parameter θ, we select a uniform proposal. The uniform proposal may not result in an efficient algorithm if the posterior is very precise relative to the prior (also uniform), but it provides a simple first example and does not require tuning (a topic we discuss in the next section). Based on our chosen model and proposal distribution, we can calculate the M-H ratio as

$$mh = \frac{(\theta^{(*)})^y(1-\theta^{(*)})^{1-y}}{(\theta^{(k-1)})^y(1-\theta^{(k-1)})^{1-y}} \qquad (4.3)$$

because the prior and proposal are proportional to one and thus cancel out of the M-H ratio. Notice that the M-H ratio in (4.3) is just a ratio of the likelihoods, which are Bernoulli PMFs in this case. The M-H ratio will not always be a likelihood ratio, but if the prior and proposal cancel out, it simplifies the calculation of the M-H ratio.

The R code for an appropriate MCMC algorithm based on $K = 1000$ M-H updates with uniform proposal and assuming $y = 0$ is

---

BOX 4.1   METROPOLIS-HASTINGS FOR θ

```
1 y=0
2 K=1000
3 theta=0.5 # initial value
4 theta.save=rep(0,K)
5 theta.save[1]=theta
```

(*Continued*)

BOX 4.1 (Continued)   METROPOLIS-HASTINGS FOR θ

```
6  set.seed(1)
7  for(k in 2:K){
8    theta.star=runif(1)
9    mh=dbinom(y,1,theta.star)/dbinom(y,1,theta)
10   if(rbinom(1,1,min(mh,1))==1){
11     theta=theta.star
12   }
13   theta.save[k]=theta
14 }
15
16 layout(matrix(c(1,1,2),1,3))
17 plot(theta.save,type="l")
18 hist(theta.save,prob=TRUE)
19 segments(0,2,1,0,lty=2)
```

This R code produces the graphic in Figure 4.1. In the R code on line 9, we calculate the M-H ratio and, on line 11, we accept the proposed parameter value with probability $\min(mh, 1)$. Figure 4.1a shows the resulting MCMC sample for $K = 1000$ as a "trace plot," the name given to the sequence of MCMC realizations of the parameter. Figure 4.1b shows the resulting approximation of the posterior PDF for θ in comparison to the true posterior indicated as a dashed line. We showed in the previous chapters that the normalizing constant (i.e., the marginal distribution of y), when $y = 0$ for our model is $[y] = 1/2$. Thus, substituting the likelihood $[y|\theta]$, the prior $[\theta]$, and the marginal distribution $[y]$ into the Bayes rule (1.3) yields the true posterior PDF: $[\theta|y = 0] = 2(1 - \theta)$.

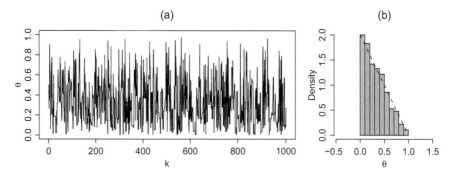

Figure 4.1   Trace plot (a) and histogram (b) based on a MCMC sample of size $K = 1000$ resulting from M-H updates to fit the Bayesian model in (4.1)–(4.2). The histogram (b) shows the true posterior PDF for θ as a dashed line.

With the MCMC sample in hand, we can use MC integration to approximate posterior quantities of interest. For example, the following R code computes the posterior mean $E(\theta|y)$ and variance $\text{Var}(\theta|y)$ based on data $y = 0$ and our model fit using $K = 1000$ MCMC iterations

---

BOX 4.2   MONTE CARLO MEAN AND VARIANCE FOR $\theta|y$

```
1 > mean(theta.save)
2 [1] 0.3201355
3 > var(theta.save)
4 [1] 0.05373195
```

Both the approximated posterior mean and variance using MCMC are fairly close to the true posterior mean $(1/3)$ and variance $(1/18)$ in this case, but we can improve the accuracy of these approximations by increasing the number of MCMC iterations (K). This method brings up an important set of questions regarding diagnostics for MCMC algorithms which we cover in the next section.

4.2   METROPOLIS-HASTINGS IN PRACTICE

We showed in the previous section that constructing a MCMC algorithm based on M-H updates is trivial and, at least in our simple example, only required a few lines of R code. However, there are several important details regarding the algorithm and resulting Markov chains that we need to discuss before moving on to more complicated models and algorithms. We begin with the algorithm.

The first important thing to notice in the MCMC algorithm in the previous section is on line 9 where we calculate the M-H ratio. Coding the M-H ratio as a direct quotient of the likelihoods matches the mathematical form we showed in (4.3), but it can be unstable numerically for very large or small values of the likelihood. Thus, it is much better practice to always compute the log of the M-H ratio numerator and denominator separately and subtract them. For example, a better alternative to line 9 in our MCMC algorithm is

---

BOX 4.3   METROPOLIS-HASTINGS RATIO USING NATURAL LOGARITHM

```
1 mh=exp(dbinom(y,1,theta.star,log=TRUE)-dbinom
2   (y,1,theta,log=TRUE))
```

---

because the difference of the logs will be much less sensitive to numerical underflow or overflow. Using the log=TRUE option will calculate the natural logarithm of the density and mass functions in R directly and yield a much more stable calculation of the M-H ratio.

Another feature of the MCMC algorithm is how we decide to accept or reject the proposed value for $\theta^{(*)}$. On line 10 of the MCMC algorithm, we computed a binary yes/no decision based on a Bernoulli sample (i.e., a coin flip) with probability $\min(mh, 1)$ as required. If the Bernoulli sample equals one, then we accept the proposed value and otherwise retain the previous value in the chain. Another common way to compute the accept-reject decision is using a uniform random variable. For example, we could use the following R code as an alternative to line 10 of the MCMC algorithm from the previous section:

BOX 4.4   ALTERNATIVE METROPOLIS-HASTINGS ACCEPTANCE CONDITION

```
1 if(mh > runif(1)){
```

---

which results in a slightly cleaner implementation of the algorithm. In the code, `mh > runif(1)` provides the correct condition to accept with probability $\min(mh, 1)$ because of the following result. Suppose a random variable u arises from a uniform distribution with support $(0, 1)$. The cumulative distribution function for u is $F(u) = \int_0^u 1 dv = u$. Thus, the probability that the M-H ratio (a number bounded below by zero) is greater than u can be written as $P(u < mh)$ in our notation. This probability is equal to the uniform CDF evaluated at the M-H ratio, and $F(mh) = \min(mh, 1)$ because $F(mh) = mh$ for $0 \leq mh \leq 1$ and $F(mh) = 1$ for $1 < mh$. Thus, the probability of this event occurring (i.e., $P(u < mh) = \min(mh, 1)$) is the same as the probability of a Bernoulli success. Therefore, both approaches result in the correct M-H accept-reject decision. In many of the algorithms we present throughout the later chapters of this text, we use the uniform random variable approach to decide whether we should accept or reject the proposal in M-H updates.

The second main M-H theme that arises in practice concerns aspects of the resulting Markov chains. The trace plot shown in Figure 4.1a contains several important features that we need to discuss. The five characteristics to consider in trace plots displaying MCMC samples are burn-in, convergence, mixing, thinning, and length. Many of these characteristics are related and we describe each herein.

Burn-in refers to the "warm up" period at the beginning of the Markov chain before it reaches its stationary distribution. The burn-in period is important to recognize and prune off of the MCMC sample because it does not represent the posterior distribution. With infinite computing time, a properly constructed MCMC algorithm is guaranteed to yield a Markov chain that has reached its stationary distribution; however, with only finite computing time, the burn-in period may depend on the starting value for the chain. If the user-chosen starting value is too far from the high density region of the stationary distribution, then it can sometimes require longer to reach the stationary distribution (especially in complicated multi-parameter models). In Figure 4.1a, we chose a reasonable starting value that is near most of the posterior mass, so there is no effective burn-in period and, hence, we do not need to truncate the MCMC sample to discard the burn-in period. We usually identify an appropriate burn-in period by eye and it is typically easy to spot depending on the other characteristics, particularly convergence and mixing.

Convergence of a Markov chain occurs when the chain reaches the stationary distribution. During the burn-in period, especially when the starting value is in a low density region of the posterior, the Markov chain has not yet converged. In this sense, stochastic sampling methods like MCMC are fundamentally different than optimization methods because the Markov chain converges to a distribution whereas an optimization trajectory converges to a single value (to find an MLE, for example). In Figure 4.1a, the trace plot indicates nearly instant convergence.

A well-mixed Markov chain explores the posterior distribution efficiently. Better mixing typically corresponds to less autocorrelation in the chain. Coincidentally, well-mixed chains tend to have quick burn-in periods and resemble MC samples. We often can adjust mixing by tuning the proposal distribution such that it results in less correlated proposals. The trace plot in Figure 4.1a indicates good mixing because the chain is not smooth over large intervals of MCMC iterations.

We correlate Markov chains by definition. Therefore, MCMC samples contain values of the parameter that are autocorrelated. The autocorrelation is the main disadvantage that comes as a result of having such a powerful method for obtaining samples from the posterior when the analytical form of the posterior is unknown. Autocorrelation in MCMC samples can affect the precision associated with MC integration and therefore the inference that is made based on these methods for fitting statistical models to data. Thus, general guidance suggests that we assess the autocorrelation (e.g., using autocorrelation functions (ACFs) common in time series analyses) and then thin the Markov chains at a constant interval to obtain an approximately uncorrelated sample to use for inference. The following R code

---

BOX 4.5   AUTOCORRELATION FUNCTION FOR θ

```
1 acf(theta.save)
```

---

computes the ACF for the MCMC sample shown in Figure 4.1a. The resulting ACF plot is shown in Figure 4.2 and indicates correlation structure out to lag 3. Based on the estimated correlation structure in the MCMC sample for $\theta|y$, we can thin the sample such that we retain every 4th iteration (i.e., using the R command: `theta.save[seq(1,K,4)]`) and recalculate our posterior histogram, mean, and variance.

Thinning reduces the length of the MCMC sample and thus reduces the power to approximate the posterior quantities of interest. Thus, the length of the MCMC sample is critical for obtaining accurate posterior summaries and inference. In fact, Link and Eaton (2012) suggested that it may be better to use the entire MCMC sample (rather than thinning) in some cases. Methods exist to compute the "effective" sample

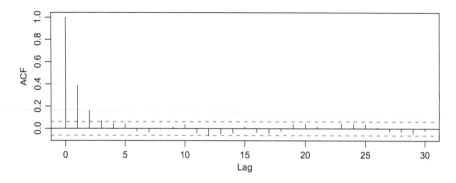

Figure 4.2 ACF plot for the MCMC sample for $\theta|y$ in Figure 4.1a.

size of MCMC samples that take into account autocorrelation, but a rule of thumb is to use more MCMC iterations (i.e., increase the length of the chain) in cases where MCMC samples are strongly autocorrelated. Another technique is to compute the posterior quantities of interest (e.g., the posterior mean) iteratively for every k only including values for all iterations less than k. That way the quantities of interest can be monitored for stability as the number of MCMC iterations increases. When the posterior quantity has converged to a fixed value (within some chosen tolerance), then typically the MCMC sample size is large enough. For example, suppose we are interested in monitoring the posterior mean and variance of θ in our ongoing example. The following R code uses the MCMC sample to compute the convergence trajectories in Figure 4.3:

BOX 4.6   CONVERGENCE TRAJECTORIES

```
1  theta.mean=rep(0,K)
2  theta.var=rep(0,K)
3  for(k in 1:K){
4     theta.mean[k]=mean(theta.save[1:k])
5     theta.var[k]=var(theta.save[1:k])
6  }
7
8  matplot(cbind(theta.mean,theta.var),type="l",
9     lty=1,col=1:2)
10 abline(h=c(1/3,1/18),col=1:2,lty=2)
```

Notice that, for both quantities in Figure 4.3, the MCMC-based values are unstable at fewer than 200 MCMC iterations, but then stabilize and converge to the exact posterior values as the MCMC sample length increases. Thus, the length of the MCMC chain needs to be at least 500 iterations long to approximate these quantities well.

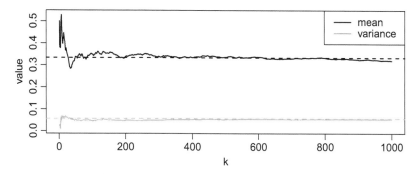

Figure 4.3 Convergence trajectories based on the previous MCMC sample for $\theta|y$ in Figure 4.1a. The running posterior mean is shown in black and the running posterior variance is shown in gray. The exact posterior mean and variance are shown in the dashed lines for comparison.

4.3 PROPOSAL DISTRIBUTIONS

Another important aspect to consider in the MCMC algorithm when M-H updates are used is the proposal. The proposal we used for $\theta^{(*)}$ (i.e., uniform) in our previous example was fixed and thus required no tuning. On the other hand, we were able to tune it to improve convergence and mixing. In our original description of the M-H procedure, we used a much more general notation for the proposal distribution so that it could be a function of previous values in the chain (i.e., $\theta^{(*)} \sim [\theta^{(*)}|\theta^{(k-1)}]$), or at least we could tune it so that it results in a more efficient MCMC algorithm. In this section, we present a variety of options for proposal distributions, including one that results in the so-called Gibbs sampler.

A simple way to generalize the proposal distribution for $\theta^{(*)}$ in our example is to use a beta distribution instead of a uniform. The beta distribution relies on two parameters that affect its shape and scale. We can adjust those tuning parameters so that they affect the Markov chain convergence and mixing. With infinite computing time, the choice of proposal distribution will not affect our posterior inference. Therefore, we can use whatever distribution we like for the proposal as long as its support matches that of the model parameter and the form of the proposal is consistent throughout the chain.[2]

---

[2] We can relax this assumption that the proposal must be the same for the entire chain if we follow certain rules about how it can change. This leads to adaptive MCMC methods and is a more advanced topic that we describe in the last chapter.

More generally, consider three different types of proposal distributions, each with their own advantages and disadvantages:

1. Prior: In this case, we specify the proposal distribution as the prior, $\theta^{(*)} \sim [\theta]$. This approach has the advantage that it is easy to specify and it simplifies the M-H ratio because the proposal and prior cancel out in the numerator and denominator:

$$mh = \frac{[\mathbf{y}|\theta^{(*)}][\theta^{(*)}][\theta^{(k-1)}|\theta^{(*)}]}{[\mathbf{y}|\theta^{(k-1)}][\theta^{(k-1)}][\theta^{(*)}|\theta^{(k-1)}]}, \tag{4.4}$$

$$= \frac{[\mathbf{y}|\theta^{(*)}][\theta^{(*)}][\theta^{(k-1)}]}{[\mathbf{y}|\theta^{(k-1)}][\theta^{(k-1)}][\theta^{(*)}]}, \tag{4.5}$$

$$= \frac{[\mathbf{y}|\theta^{(*)}]}{[\mathbf{y}|\theta^{(k-1)}]}. \tag{4.6}$$

Thus, the M-H ratio becomes a likelihood ratio and is easier to program. However, the prior may not result in an efficient MCMC algorithm and will have a low acceptance rate if it is very diffuse with respect to the posterior. We cannot tune the prior as a proposal, so it may not be suitable in all scenarios.

2. Symmetric: A symmetric proposal has the property that $[\theta^{(k-1)}|\theta^{(*)}] = [\theta^{(*)}|\theta^{(k-1)}]$. For example, if $\theta \in \Re$, then we can specify a symmetric proposal such that $\theta^{(*)} \sim N(\theta^{(k-1)}, \sigma^2_{\text{tune}})$, where σ^2_{tune} is a tuning parameter that is user specified. This proposal is historically referred to as a random walk or Metropolis proposal because of its original use and is symmetric because the kernel of the normal distribution has the property $\exp(-(\theta^{(*)} - \theta^{(k-1)})^2 / \sigma^2_{\text{tune}}) = \exp(-(\theta^{(k-1)} - \theta^{(*)})^2 / \sigma^2_{\text{tune}})$. The symmetry property is convenient because it simplifies the M-H ratio:

$$mh = \frac{[\mathbf{y}|\theta^{(*)}][\theta^{(*)}][\theta^{(k-1)}|\theta^{(*)}]}{[\mathbf{y}|\theta^{(k-1)}][\theta^{(k-1)}][\theta^{(*)}|\theta^{(k-1)}]}, \tag{4.7}$$

$$= \frac{[\mathbf{y}|\theta^{(*)}][\theta^{(*)}]}{[\mathbf{y}|\theta^{(k-1)}][\theta^{(k-1)}]}. \tag{4.8}$$

However, unlike the prior as a proposal, we can tune the symmetric proposal to yield an efficient MCMC algorithm and a better mixed sample by adjusting σ^2_{tune}. The disadvantage is that the support of the symmetric normal proposal may not match that of the prior and thus may not be directly useful in all situations. We return to this concept and how to modify M-H updates to use symmetric proposals, even when the support does not match, in Chapter 6.

3. Posterior: The best proposal distribution to use would perfectly match the posterior distribution we seek to obtain samples from. However, if we knew

the posterior distribution, then we would not need MCMC at all. As a thought experiment, suppose we did use the posterior as a proposal. In this case, the M-H ratio becomes

$$mh = \frac{[\mathbf{y}|\theta^{(*)}][\theta^{(*)}][\theta^{(k-1)}|\theta^{(*)}]}{[\mathbf{y}|\theta^{(k-1)}][\theta^{(k-1)}][\theta^{(*)}|\theta^{(k-1)}]}, \tag{4.9}$$

$$= \frac{[\mathbf{y}|\theta^{(*)}][\theta^{(*)}][\theta^{(k-1)}|\mathbf{y}]}{[\mathbf{y}|\theta^{(k-1)}][\theta^{(k-1)}][\theta^{(*)}|\mathbf{y}]}, \tag{4.10}$$

$$= \frac{[\mathbf{y}|\theta^{(*)}][\theta^{(*)}][\mathbf{y}|\theta^{(k-1)}][\theta^{(k-1)}]/c}{[\mathbf{y}|\theta^{(k-1)}][\theta^{(k-1)}][\mathbf{y}|\theta^{(*)}][\theta^{(*)}]/c}, \tag{4.11}$$

$$= \frac{1}{1}, \tag{4.12}$$

$$= 1, \tag{4.13}$$

where c represents the normalizing constant in the Bayes rule. Thus, if we use the posterior as the proposal distribution, then everything cancels in the numerator and denominator of the M-H ratio and we always accept the proposal as the next value in the Markov chain. This is called a Gibbs update and, when used in a multiparameter MCMC algorithm, we often refer to it as Gibbs sampler.

4.4   GIBBS SAMPLING

Gibbs sampling is useful for multiparameter Bayesian models. Up to this point, we have only considered single parameter statistical models. Extending the MCMC concept to two parameters, say θ_1 and θ_2, and data set $\mathbf{y}$, we have the posterior distribution

$$[\theta_1, \theta_2|\mathbf{y}] = \frac{[\mathbf{y}|\theta_1, \theta_2][\theta_1, \theta_2]}{[\mathbf{y}]}, \tag{4.14}$$

$$\propto [\mathbf{y}|\theta_1, \theta_2][\theta_1, \theta_2], \tag{4.15}$$

$$\propto [\mathbf{y}|\theta_1, \theta_2][\theta_1][\theta_2], \tag{4.16}$$

the last step of which is only valid if we assume that θ_1 and θ_2 are independent *a priori*; a common, but not necessary, assumption in Bayesian models. Assuming that the normalizing constant is not analytically tractable, we may want to use MCMC to fit the model to data. The standard approach to construct a MCMC algorithm for this setting involves updating the two parameters sequentially. That is, if we pretend to know one parameter (and the data) while sampling the other parameter, the M-H procedure reduces to the one described in the previous section. However, because we assume one parameter is known, we need to sample from the so-called full-conditional distribution for the other parameter. The full-conditional distribution is the distribution for the parameter of interest, assuming we know everything else in the model and write it as $[\theta_1|\cdot] \equiv [\theta_1|\mathbf{y}, \theta_2]$ in our two-parameter model if our focus is on θ_1 first. Similarly, the full-conditional distribution for θ_2 is $[\theta_2|\cdot] \equiv [\theta_2|\mathbf{y}, \theta_1]$. The full-conditional distribution acts like a temporary posterior distribution for the parameter of interest and is often easier to find analytically than the multivariate posterior distribution involving all parameters in the model.

The MCMC algorithm allows us to break up a hard problem (i.e., the posterior) into a sequence of smaller and more tractable problems (i.e., the full-conditional distributions). If we can derive analytically the full-conditional distributions given the model specification, then we can construct a fully Gibbs MCMC algorithm. In the fully Gibbs MCMC algorithm, we sequentially sample from each full-conditional distribution assuming that we know the rest of the parameters as their most recent values in the Markov chain. That way all the parameters have the chance to influence each other, but in a stochastic Markov updating scheme. For our general two parameter model, the algorithm involves the following steps:

1. Assign initial value for $\theta_2^{(0)}$
2. Set $k = 1$
3. Sample θ_1 from its full-conditional distribution that depends on the latest value for θ_2: $\theta_1^{(k)} \sim [\theta_1|\cdot] \equiv [\theta_1|\mathbf{y}, \theta_2^{(k-1)}]$
4. Sample θ_2 from its full-conditional distribution that depends on the latest value for θ_1: $\theta_2^{(k)} \sim [\theta_2|\cdot] \equiv [\theta_2|\mathbf{y}, \theta_1^{(k)}]$
5. Set $k = k + 1$, go to step 3; repeat until the MCMC sample is large enough to approximate the posterior quantities well

If we can derive analytically the full-conditional distributions, then there is no need for M-H updates, we can simply draw the next values in the chain directly from their full-conditional distributions. This is equivalent to using the temporary posterior (i.e., the full-conditional) as the proposal. If, for one or more of the parameters, we cannot find the full-conditional distribution, but we know what it is proportional to, then we can use M-H updates to sample those parameters (and use Gibbs updates for the cases where we can derive the full-conditional distributions).

The technique to derive a full-conditional distribution is to omit all of the distributions in the right-hand side of the posterior expression (i.e., the joint distribution) that do not contain the parameter of interest and then find the normalizing constant as if we are deriving a posterior distribution. It is not an actual posterior distribution however, because we omitted terms from the joint distribution that do not contain the parameter of interest.

Based on the generic two parameter model we introduced in this section, we first write the posterior and what it is proportional to as

$$[\theta_1, \theta_2|\mathbf{y}] \propto [\mathbf{y}|\theta_1, \theta_2][\theta_1][\theta_2], \tag{4.17}$$

assuming *a priori* independence for θ_1 and θ_2. Then we find the full-conditional distribution for θ_1 by omitting the distributions that do not involve θ_1 from the right-hand side of (4.17) to yield

$$[\theta_1|\cdot] \propto [\mathbf{y}|\theta_1,\theta_2][\theta_1]. \tag{4.18}$$

We can either find the normalizing constant that ensures (4.18) integrates to one and then use a Gibbs update in step 3 of our algorithm, or we use M-H to update θ_1 as we did in the previous section.

We find the full-conditional distribution for θ_2 in the same way. We omit the prior for θ_1 from the joint distribution resulting in

$$[\theta_2|\cdot] \propto [\mathbf{y}|\theta_1,\theta_2][\theta_2]. \tag{4.19}$$

As with the full-conditional distribution for θ_1, if we can analytically derive $[\theta_2|\cdot]$, then we can use a Gibbs update for it in our MCMC algorithm.

Extending the MCMC sampling approach from our two parameter Bayesian model to the case where we have J parameters, we have a similar set of steps as before:

1. Assign initial values for $\theta_2^{(0)}, \theta_3^{(0)}, \dots, \theta_J^{(0)}$
2. Set $k = 1$
3. Sample θ_1 from its full-conditional distribution that depends on the latest value for the other parameters: $\theta_1^{(k)} \sim [\theta_1|\cdot] \equiv [\theta_1|\mathbf{y},\theta_2^{(k-1)},\dots,\theta_J^{(k-1)}]$
4. Sample θ_2 from its full-conditional distribution that depends on the latest value for other parameters: $\theta_2^{(k)} \sim [\theta_2|\cdot] \equiv [\theta_2|\mathbf{y},\theta_1^{(k)},\theta_3^{(k-1)},\dots,\theta_J^{(k-1)}]$
5. Sample parameters $\theta_3, \dots, \theta_{J-1}$ from their full-conditional distributions
6. Sample θ_J from its full-conditional distribution that depends on the latest value for other parameters: $\theta_J^{(k)} \sim [\theta_J|\cdot] \equiv [\theta_J|\mathbf{y},\theta_1^{(k)},\theta_2^{(k)},\dots,\theta_{J-1}^{(k)}]$
7. Set $k = k+1$, go to step 3; repeat until the MCMC sample is large enough to approximate the posterior quantities well

As before, if we cannot derive analytically any of the full-conditional distributions, then we perform the updates for those using M-H.

The MCMC algorithm we described is similar to what is used to fit the vast majority of Bayesian models, although there are additional ways to update parameters besides Gibbs and M-H. We discuss MCMC algorithms more specifically in the descriptions of specific models in following chapters.

4.5 ADDITIONAL CONCEPTS AND READING

While a primary reference for MCMC in the statistics literature remains Gelfand and Smith (1990), the original developments date back to decades earlier (Metropolis 1987; e.g., Metropolis et al. 1953; Hastings 1970). In particular, Geman and Geman (1984) is commonly cited as an earlier inspiring reference for modern Bayesian computing using MCMC. Regardless of the origins, MCMC methods have no doubt transformed the way we fit Bayesian models. Much of the literature associated with MCMC methods is technical (e.g., Tierney 1994), but several gentle overviews also exist. For example, Green et al. (2015) provided an excellent overview of Bayesian computation in general, focusing on stochastic sampling methods such as MCMC. Robert and Casella (2011) summarize the history and influences of MCMC, whereas Chib and Greenberg (1995) and Casella and George (1992) provided accessible introductions to M-H and Gibbs samplers themselves.

Several more recent stochastic sampling methods have been developed based on MCMC to improve computational efficiency. These new developments often involve smarter ways to specify proposal distributions and explore the parameter space. For example, adaptive MCMC methods involve tuning proposal distributions to improve efficiency in the resulting Markov chains. These adaptive methods are often used to optimize acceptance probabilities or higher-order moments of proposal distributions to improve the mixing and convergence of Markov chains (Roberts and Rosenthal 2009). Givens and Hoeting (2013) provide a good introduction to adaptive MCMC methods.

Other approaches to improve the efficiency of MCMC methods involve simulating proposals based on dynamics that correspond to the target distribution itself. Early developments in this area include approaches like Metropolis-adjusted Langevin algorithms (Roberts and Tweedie 1996) and have evolved into more modern implementations of Hamiltonian Monte Carlo (Neal 2011). We describe Hamiltonian Monte Carlo in more detail in Chapter 29.

Finally, MCMC methods have been adapted to provide inference for models with intractable likelihoods (Blum 2010). In such cases, a distance metric is used in place of the likelihood in the MCMC algorithm resulting in an approach known as approximate Bayesian computation (ABC). Beaumont (2010) provided a good overview of ABC approaches for ecologists and evolutionary biologists.

5 Importance Sampling

M-H updates are useful for sampling parameters when we cannot derive the posterior distribution (or full-conditional) analytically, but they are not the only option for this situation. Importance sampling is another approach that we can use when the posterior distribution is intractable (Geweke 1989).

The classical use of importance sampling in the Bayesian settings results in a way to find characteristics of posterior distributions (e.g., moments) by using samples from another distribution and weighting them based on their "importance" for contributing to the quantity of interest. For example, in the simple one-parameter Bayesian model we used before, where $y \sim [y|\theta]$ and $\theta \sim [\theta]$, suppose we are interested in the posterior mean $E(\theta|y)$. Importance sampling exploits the fact that we can write the posterior mean as

$$E(\theta|y) = \int \theta[\theta|y]d\theta, \tag{5.1}$$

$$= \int \theta \frac{[\theta|y]}{[\theta]_*}[\theta]_* d\theta, \tag{5.2}$$

$$= E_* \left(\theta \frac{[\theta|y]}{[\theta]_*} \right), \tag{5.3}$$

where $[\theta]_*$ represents a new probability distribution (the importance distribution) with respect to θ, and E_* represents an expectation with respect to $[\theta]_*$.

Thus, we can use MC integration to approximate the expectation of interest in (5.3) based on a sample from the importance distribution. The first step is to sample $\theta^{(k)} \sim [\theta]_*$. The second step is to compute

$$E(\theta|y) \approx \frac{\sum_{k=1}^{K} \theta^{(k)} \frac{[\theta^{(k)}|y]}{[\theta^{(k)}]_*}}{K}. \tag{5.4}$$

This approach works when we know the analytical form of the posterior $[\theta|y]$, but in the more typical situation where we only know the joint distribution $[y|\theta][\theta]$ and not the normalizing constant $[y]$, we need to use a slightly different approach.

Self-normalized importance sampling involves calculating importance weights as $w^{(k)} = [y|\theta^{(k)}][\theta^{(k)}]/[\theta^{(k)}]_*$ for $k = 1,\ldots,K$ and then computing

$$E(\theta|y) \approx \frac{\sum_{k=1}^{K} \theta^{(k)} w^{(k)}}{\sum_{k=1}^{K} w^{(k)}}. \tag{5.5}$$

In this expression, the normalizing constants cancel in the numerator and denominator, and we only have to know the form of the likelihood, prior, and importance distribution.

In principle, we can use any distribution with support that matches the parameter of interest for the importance distribution, but it may be convenient to use the prior $[\theta]_* \equiv [\theta]$. If we use the prior for the importance distribution, the weight simplifies to include the likelihood only and we can write the MC integral approximation for the posterior mean as

$$E(\theta|y) \approx \frac{\sum_{k=1}^{K} \theta^{(k)} [y|\theta^{(k)}]}{\sum_{k=1}^{K} [y|\theta^{(k)}]}. \tag{5.6}$$

For example, in the case where $y \sim \text{Bern}(\theta)$ and $\theta \sim \text{Unif}(0,1)$, and if we use the prior as the importance distribution and assume $y = 0$, then we can use importance sampling to approximate the posterior mean in R as

---

BOX 5.1   IMPORTANCE SAMPLING MEAN FOR θ

```
1 > set.seed(1)
2 > theta=runif(1000)
3 > w=dbinom(0,1,theta)
4 > theta.mean=sum(theta*w)/sum(w)
5 > theta.mean
6 [1] 0.3336263
```

Recall from previous chapters, that the posterior mean for θ under this model when $y = 0$ is $E(\theta|y) = 1/3$, thus for $K = 1000$ importance samples, the self-normalized importance sampling approach is quite accurate.

The prior may not always serve as the best importance distribution if it is too different from the posterior because the MC approximation will not be efficient if few samples are close to the mean. Importance sampling can also suffer from degeneracy problems when one importance sample results in a large weight $w^{(k)}$ and the rest results in small weights. This difference will often lead to an erroneous approximation of the posterior characteristic of interest. Methods have been developed to adjust the importance weights to reduce the chance of degenerate importance samples.

Another useful approach is importance sampling resampling. In this approach, we sample the parameter from the importance distribution as before and compute the importance weights according to our choice of importance distribution. However, instead of averaging the samples for θ, we resample θ from the current importance sample, but with weights equal to $w^{(k)}$.

We can use importance sampling resampling for θ in place of M-H or Gibbs updates in a MCMC algorithm. As an alternative to M-H when the full-conditional is not analytically tractable, we can use importance sampling resampling as long as there are no degeneracy problems. In this case, we only need to resample a single θ value on each MCMC iteration.

However, if we resample θ using the importance weights with replacement, we can summarize the posterior distribution (or full-conditional distribution) if we have a large enough sample from the importance distribution to begin with. In our ongoing example, we could use the importance weights to resample with replacement from the original importance sample to yield posterior histograms (for $K = 1000$, $K = 10000$, and $K = 100000$) in Figure 5.1 using the following R code:

BOX 5.2   SAMPLING IMPORTANCE RESAMPLING FOR θ

```
set.seed(1)
theta1=runif(1000)
theta2=runif(10000)
theta3=runif(100000)

w1=dbinom(0,1,theta1)
w2=dbinom(0,1,theta2)
w3=dbinom(0,1,theta3)

theta1.resamp=sample(theta1,1000,replace=TRUE,
    prob=w1)
theta2.resamp=sample(theta2,10000,replace=TRUE,
    prob=w2)
theta3.resamp=sample(theta3,100000,replace=TRUE,
    prob=w3)

layout(matrix(1:3,1,3))
hist(theta1.resamp,col=8,prob=TRUE,main="a")
segments(0,2,1,0,lty=2)
hist(theta2.resamp,col=8,prob=TRUE,main="b")
segments(0,2,1,0,lty=2)
hist(theta3.resamp,col=8,prob=TRUE,main="c")
segments(0,2,1,0,lty=2)
```

Notice that the importance sampling histograms approach the true posterior PDF as the importance sample size (K) increases.

Overall, importance sampling is an underutilized approach in Bayesian model fitting. Aside from its utility in models with intractable posterior or full-conditional distributions, another main advantage of importance sampling is that it is based on independent samples (no Markov chain necessary). Thus, we can parallelize the computation involved in generating the importance sample, yielding much faster algorithms than methods like MCMC.

The choice of importance distribution presents an obvious challenge, especially in more complicated situations. Importance distributions that resemble the posterior perform very well, and those that do not resemble the posterior may be problematic.

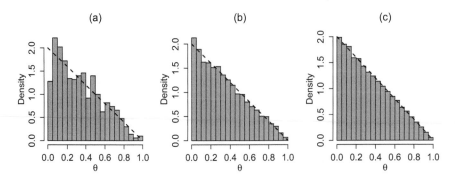

Figure 5.1 Posterior histograms resulting from: (a) $K = 1000$, (b) $K = 10000$, and (c) $K = 100000$ resampled importance samples. True posterior PDF shown as dashed line.

Thus, to some extent, the importance distribution plays the same role as the proposal distribution in MCMC.

5.1 ADDITIONAL CONCEPTS AND READING

Sequential Monte Carlo (SMC) is a procedure related to importance sampling (Doucet et al. 2001; Cappé 2007). Another name for SMC is "particle filtering" (Gordon et al. 1993), which is named thus because it extends the importance sampling idea to trajectories of connected variables.

The general concept of importance sampling we discussed in this chapter has many benefits, including easy parallelization and use in MC integration. However, conventional importance sampling requires us to recompute the importance weights based on all the data, even if we fit the model of interest to a subset of the data previously. In cases where we acquire new data regularly, it is often of interest to assimilate them as they become available without having to reanalyze the entire data set. Often we refer to this type of sequential inference as recursive Bayesian filtering (e.g., Hooten et al. 2019) and we can achieve it using SMC methods.

SMC is one of many approaches that allow us to learn about latent temporally dynamic state variables in hierarchical statistical models. Knape and de Valpine (2012) illustrated how to use SMC methods to fit Bayesian population models. In fact, they used combined SMC and MCMC approaches to improve computational efficiency in a method referred to as particle MCMC (Andrieu et al. 2010). We present Bayesian models for time series data in Chapter 16, and commonly we use SMC methods for temporally structured time series models, but they also apply to other types of models for dependent data (Chopin 2002).

Section II

Basic Models and Concepts

6 Bernoulli-Beta

Binary data are among the most common type of data collected in ecological studies. Examples of ecological questions that lead to binary data include: was the species detected, was the individual detected, was the individual female or male, was the disease present, was the landtype forest, and was the pixel aquatic or terrestrial? Binary data models serve as the foundation for occupancy and capture-recapture models (Chapters 23 and 24), two of the most commonly used statistical models in wildlife ecology.

As we described in the opening chapters of this book, we can model binary data with Bernoulli distributions. The natural parameter of a Bernoulli distribution is a probability (θ) associated with a "success" in a Bernoulli trial (i.e., usually $y = 1$). Thus, inference resulting from Bayesian statistical models for binary data involves posterior distributions for the parameter θ.

Generalizing what we presented in the first chapters on this topic, we now consider a set of binary data y_i for $i = 1, \ldots, n$ that are conditionally Bernoulli and depend on probability θ. To allow for more *a priori* information about θ in the model, we use a beta prior distribution for θ resulting in the full model statement

$$y_i \sim \text{Bern}(\theta), \ i = 1, \ldots, n, \tag{6.1}$$

$$\theta \sim \text{Beta}(\alpha, \beta), \tag{6.2}$$

where α and β are hyperparameters (i.e., terminal parameters in the model that are assumed fixed and known) and the form of the prior PDF is

$$[\theta] \equiv \frac{\Gamma(\alpha+\beta)}{\Gamma(\alpha)\Gamma(\beta)} \theta^{\alpha-1}(1-\theta)^{\beta-1}, \tag{6.3}$$

for $\alpha, \beta > 0$ and $\Gamma(\cdot)$ refers to the gamma function.

The Bernoulli-Beta model is almost always the first model described in Bayesian textbooks and thus we do not dwell on the fact that we can derive the posterior analytically here. In short, because the data model and parameter model (i.e., prior) have kernels in which we can add the exponents when multiplied, we can see the product of the likelihood and prior as a beta distribution for the posterior

$$[\theta|\mathbf{y}] \propto \left(\prod_{i=1}^{n} [y_i|\theta] \right) [\theta], \tag{6.4}$$

$$\propto \left(\prod_{i=1}^{n} \theta^{y_i}(1-\theta)^{1-y_i} \right) \theta^{\alpha-1}(1-\theta)^{\beta-1}, \tag{6.5}$$

$$\propto \theta^{\sum_{i=1}^{n} y_i} (1-\theta)^{\sum_{i=1}^{n} (1-y_i)} \theta^{\alpha-1} (1-\theta)^{\beta-1}, \qquad (6.6)$$

$$\propto \theta^{(\sum_{i=1}^{n} y_i)+\alpha-1} (1-\theta)^{(\sum_{i=1}^{n} (1-y_i))+\beta-1}, \qquad (6.7)$$

where the final term in (6.7) must integrate to $(\Gamma((\sum_{i=1}^{n} y_i) + \alpha) \Gamma((\sum_{i=1}^{n} (1-y_i)) + \beta)) / \Gamma((\sum_{i=1}^{n} y_i) + \alpha + (\sum_{i=1}^{n} (1-y_i)) + \beta)$ so that the posterior integrates to one. Thus, the posterior is a beta distribution: $[\theta | \mathbf{y}] = \text{Beta}((\sum_{i=1}^{n} y_i) + \alpha, (\sum_{i=1}^{n} (1-y_i)) + \beta)$.

When the hyperparameters in the prior are $\alpha = 1$ and $\beta = 1$ (i.e., a uniform prior) and $n = 1$, then the posterior matches what we described in the previous chapters. Regardless of the hyperparameters or data, we do not need to use MCMC to fit this model because we have an analytical form for the posterior. We can view the posterior PDF for $\theta | \mathbf{y}$ directly and characterize it by its moments and credible interval.

For example, consider the plant present/absence data analyzed by Hooten et al. (2003) and Hefley et al. (2017a) that contain binary measurements of *Desmodium glutinosum* at a set of 216 sites. The species was observed as present at 69 sites and absent at the other 147 sites. Thus, $\sum_{i=1}^{216} y_i = 69$ and $\sum_{i=1}^{216} (1-y_i) = 147$ in these data. Our resulting posterior distribution $[\theta | \mathbf{y}]$ for the probability of presence is $\text{Beta}(69 + \alpha, 147 + \beta)$ and is depicted in Figure 6.1 for $\alpha = 2$ and $\beta = 5$ (assuming *a priori* that the plant has mostly less than a 60% chance of occurring on any given plot). The following R code results in the posterior inference shown in Figure 6.1.

BOX 6.1   POSTERIOR FOR θ

```
1 y.sum=69
2 oneminusy.sum=147
3 alpha=2
4 beta=5
5 curve(dbeta(x,y.sum+alpha,oneminusy.sum+beta),
6     from=0,to=1)
7 curve(dbeta(x,alpha,beta),from=0,to=1,lty=2,
8     add=TRUE)
```

We can calculate the posterior mean, variance, and 95% equal tail credible interval for $\theta | \mathbf{y}$ in R as

---

BOX 6.2   POSTERIOR MEAN AND VARIANCE FOR θ

```
1 > alpha.new=y.sum+alpha
2 > beta.new=oneminusy.sum+beta
```

(Continued)

BOX 6.2 (Continued)   POSTERIOR MEAN AND VARIANCE FOR θ

```
3 > (alpha.new)/(alpha.new+beta.new)
4 [1] 0.3183857
5 > (alpha.new*beta.new)/((alpha.new+beta.new)^2
6   *(alpha.new+beta.new+1))
7 [1] 0.0009688224
8 > qbeta(c(0.025,0.975),alpha.new,beta.new)
9 [1] 0.2589954 0.3808615
```

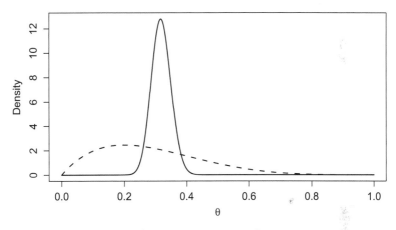

Figure 6.1   Posterior PDF for $\theta|y$ based on $\alpha = 2$ and $\beta = 5$. For comparison, the figure shows the prior PDF for θ as a dashed line.

because of the known properties of the beta probability distribution. For comparison, the MLE for θ in this example is $\sum_{i=1}^{n} y_i/n = 0.3194$, which is slightly larger than the posterior mean because the posterior is shrunk toward the prior mean of $2/7 = 0.286$.

6.1   MCMC WITH SYMMETRIC PROPOSAL DISTRIBUTIONS

The Bernoulli-Beta model provides a good example to demonstrate different approaches for MCMC because we can compare our numerical results to the analytical posterior directly. We illustrated already how to construct a MCMC algorithm for a slightly simpler model in Chapter 4, but suppose we seek a more efficient MCMC algorithm to fit the Bernoulli-Beta model.

One approach would be to use M-H updates based on a proposal distribution that we can tune to be more efficient. Recall that we used a uniform proposal distribution in our previous MCMC example. A generalization of the uniform is the beta distribution (e.g., $\theta^{(*)} \sim \text{Beta}(\alpha_{\text{tune}}, \beta_{\text{tune}})$). We can tune the beta distribution by adjusting its parameters (α_{tune} and β_{tune}) based on trial runs of the algorithm and monitoring the acceptance rate. The closer this proposal distribution is to the posterior, the more efficient the algorithm. In this case, because the form of the proposal actually matches that of the posterior, we could eventually tune the proposal to match the posterior which would result in a Gibbs update. However, it may be challenging to explore the parameter space for optimal proposals in more complicated models than the Bernoulli-Beta model.

As an alternative approach, we could use a conditional proposal that depends on the previous parameter value in the chain ($\theta^{(k-1)}$). For example, a commonly used conditional proposal is $\theta^{(*)} \sim \text{N}(\theta^{(k-1)}, \sigma^2_{\text{tune}})$, where σ^2_{tune} serves as the tuning parameter. An advantage of this proposal distribution is that it is symmetric (as discussed in Chapter 4), but the problem with it is that it does not have the same support as the parameter in this case because although $\theta \in (0,1)$, a normal random variable can assume any real number. Thus, a remedy for the mismatch in support is to first sample the unconstrained proposal for $\theta^{(*)}$ and, if the proposed value is outside the support for θ, then skip the update on that step of the chain (thereby retaining the previous value for $\theta^{(k)} = \theta^{(k-1)}$; Gelfand et al. 1992). Intuitively, this rejection procedure works because a proposed value outside the support for θ will result in $[\theta^{(*)}] = 0$ which, in turn, yields $mh = 0$ and a rejection of the proposed value. Using this M-H procedure and our ongoing binary data example pertaining to presence or absence of *Desmodium glutinosum*, we can write the MCMC algorithm as follows in R:

BOX 6.3   MCMC FOR θ

```
y=c(rep(1,y.sum),rep(0,oneminusy.sum))
K=1000
theta=0.5   # initial value
theta.save=rep(0,K)
theta.save[1]=theta
s2.tune=.1

set.seed(1)
for(k in 2:K){
    theta.star=rnorm(1,theta,sqrt(s2.tune))
    if((theta.star > 0) & (theta.star < 1)){
        mh.1=sum(dbinom(y,1,theta.star,log=TRUE))
            +dbeta(theta.star,alpha,beta,log=TRUE)
        mh.2=sum(dbinom(y,1,theta,log=TRUE))
            +dbeta(theta,alpha,beta,log=TRUE)
        mh=exp(mh.1-mh.2)
```

(Continued)

BOX 6.3 (Continued)   MCMC FOR θ

```
17      if (mh>runif (1) ) {
18          theta=theta.star
19      }
20    }
21    theta.save [k]=theta
22  }
23
24  layout (matrix (c (1,1,2) ,1,3) )
25  plot (theta.save ,type ="l" ,lwd=1 ,ylim=c (0,1) ,
26        main="a")
27  hist (theta.save ,prob=TRUE ,col=8 ,xlim=c (0,1) ,
28        main="b")
29  curve (dbeta (x ,y.sum+alpha ,oneminusy.sum+beta) ,
30          from=0 ,to=1 ,n=1000 ,lwd=1.5 ,add=TRUE)
31  curve (dbeta (x ,alpha ,beta) ,from=0 ,to=1 ,n=1000 ,
32          lty=2 ,lwd=1.5 ,add=TRUE)
```

which produces Figure 6.2 showing the trace plot (based on a MCMC sample size of $K = 1000$) and resulting posterior histogram, as compared with the analytical posterior and prior. Several features are worth noting in Figure 6.2. First, the flat spots in the Markov chain depicted in Figure 6.2a are due to a low acceptance probability in the algorithm. In fact, out of the $K = 1000$ proposals, our algorithm was only able to accept 112. In this algorithm, we used a tuning parameter of $\sigma_{\text{tune}}^2 = 0.1$. This value

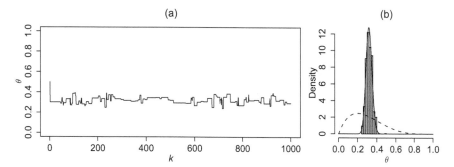

Figure 6.2   Trace plot (a) and histogram (b) based on a MCMC sample of size $K = 1000$ resulting from M-H updates using normal proposal with rejection sampling. In the histogram (b) the true posterior PDF for θ is a solid line and the prior PDF is a dashed line.

is perfectly acceptable, but we recommend obtaining a larger MCMC sample before using it for actual inference.

If we want to accept more proposed values of θ, we can reduce the tuning parameter. For example, Figure 6.3 shows the resulting trace plots for decreasing values of $\sigma_{\text{tune}}^2 = 0.1$, $\sigma_{\text{tune}}^2 = 0.01$, and $\sigma_{\text{tune}}^2 = 0.0001$. Smaller values of σ_{tune}^2 result in a higher acceptance rate, but also induce slower mixing and a longer burn-in period for the Markov chain. In the Markov chains shown in Figure 6.3, we accepted 118, 338, and 868 proposals, respectively. A variety of acceptance rates ranging from 0.2 to 0.4 have been proposed as optimal for various classes of MCMC algorithms (e.g., Bedard 2008). In our case, the Markov chain shown in Figure 6.3b had an acceptance rate closest to that range among the three shown (at $\approx 34\%$).

The normal proposal that centered on the previously accepted value $\theta^{(k-1)}$ with rejection sampling for proposed values outside $(0, 1)$ is an alternative to using a truncated normal proposal distribution where $\theta^{(*)} \sim \text{TN}(\theta^{(k-1)}, \sigma_{\text{tune}}^2)_0^1$. Although the symmetric proposal with rejection sampling implies that we may not evaluate every proposed value (due to rejection of some that fall outside the support for θ), it allows us to avoid formally accounting for the proposal in the M-H ratio explicitly (which is something we would have to do if we explicitly used the truncated normal distribution as a proposal).

6.2   MCMC WITH PROPOSAL DISTRIBUTIONS BASED ON TRANSFORMATIONS

The symmetric proposal with rejection sampling discussed in the previous section is definitely useful, but it is wastes proposed values that fall outside the support for θ (i.e., proposed values for which $mh = 0$). We discussed the fact that we could have used a truncated normal proposal instead because it obeys the support of θ in the proposed values. In fact, we could use any probability distribution for continuous random variables as long as it can be sensibly truncated on the support of θ.

Alternatively, we could specify a proposal for a transformation of θ that results in proposed values that, when transformed back to the space of θ, have the correct support. An example is the proposal $\text{logit}(\theta^{(*)}) \sim \text{N}(\text{logit}(\theta^{(k-1)}), \sigma_{\text{tune}}^2)$. Recall that the logit function is defined as $\text{logit}(\theta) = \log(\theta/(1-\theta))$. Thus, this type of proposal is like a random walk proposal in the space pertaining to the logit transformation of θ. Because the logit function has support on the real numbers, it is a good match for the normal proposal distribution and it constrains the inverse logit function of $\text{logit}(\theta)$ to be in $(0, 1)$. This is a perfectly valid way to propose values for $\text{logit}(\theta^{(*)})$ however, it is not symmetric with respect to θ and, thus, will not cancel in the M-H ratio.

The key to remember when selecting proposal distributions is that we need to represent them as distributions for θ in the M-H ratio, not as a transformation of θ. To convert the distribution for $\text{logit}(\theta)$ to a proposal distribution for θ, we rely on the

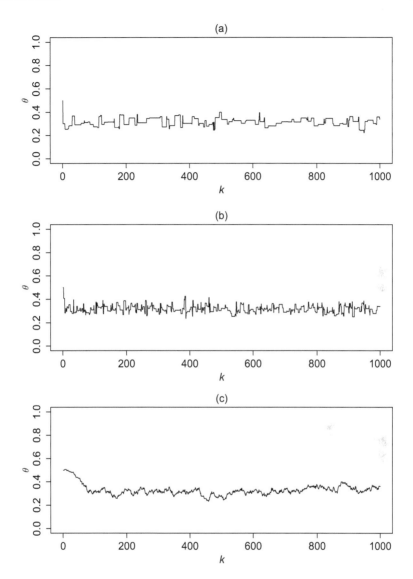

Figure 6.3   Trace plots for θ with (a) $\sigma^2_{\text{tune}} = 0.1$, (b) $\sigma^2_{\text{tune}} = 0.01$, and (c) $\sigma^2_{\text{tune}} = 0.0001$.

change of variables technique from calculus, where $g(\theta)$ is a generic transformation of θ, and we obtain the correct proposal distribution for θ by

$$[\theta] = [g(\theta)] \left| \frac{d}{d\theta} g(\theta) \right| . \tag{6.8}$$

If we know the distribution for the transformation of θ and can differentiate the transformation with respect to θ, then we can find the appropriate form for the proposal distribution on θ, even though we proposed a transformation of it. Based on our desired transformation $\text{logit}(\theta)$, the require derivative is

$$\frac{d}{d\theta}\text{logit}(\theta) = \frac{d}{d\theta}\log\left(\frac{\theta}{1-\theta}\right), \tag{6.9}$$

$$= \frac{1}{\theta} - \frac{d}{d\theta}\log(1-\theta), \tag{6.10}$$

$$= \frac{1}{\theta} + \frac{1}{1-\theta}, \tag{6.11}$$

$$= \frac{1}{\theta(1-\theta)} . \tag{6.12}$$

Substituting (6.12) and the proposal distribution for the transformation into (6.8) results in the correct induced proposal distribution for θ:

$$[\theta^{(*)}|\theta^{(k-1)}] = \frac{[\text{logit}(\theta^{(*)})|\text{logit}(\theta^{(k-1)})]}{\theta^{(*)}(1-\theta^{(*)})}. \tag{6.13}$$

Thus, if we want to propose values for θ in our MCMC algorithm by first sampling $\text{logit}(\theta)$ and then back-transforming to get θ, we need to use the following M-H ratio

$$mh = \frac{\left(\prod_{i=1}^{n}[y_i|\theta^{(*)}]\right)[\theta^{(*)}][\text{logit}(\theta^{(k-1)})|\text{logit}(\theta^{(*)})]/(\theta^{(k-1)}(1-\theta^{(k-1)}))}{\left(\prod_{i=1}^{n}[y_i|\theta^{(k-1)}]\right)[\theta^{(k-1)}][\text{logit}(\theta^{(*)})|\text{logit}(\theta^{(k-1)})]/(\theta^{(*)}(1-\theta^{(*)}))} . \tag{6.14}$$

The following R code contains the MCMC algorithm to fit our Bernoulli-Beta model based on a proposal for $\text{logit}(\theta)$:

BOX 6.4   MCMC FOR θ USING LOGIT NORMAL PROPOSAL

```
1  y=c(rep(1,y.sum),rep(0,oneminusy.sum))
2  K=1000
3  theta=0.5  # initial value
4  theta.save=rep(0,K)
5  theta.save[1]=theta
6  s2.tune=.1
7
8  logit<-function(p){log(p/(1-p))}
9  logit.inv<-function(x){exp(x)/(1+exp(x))}
10
11 set.seed(1)
12 theta.acc=1
13 for(k in 2:K){
```

(Continued)

BOX 6.4 (Continued)   MCMC FOR θ USING LOGIT NORMAL PROPOSAL

```
14  theta.star=logit.inv(rnorm(1,logit(theta),
15      sqrt(s2.tune)))
16  mh.1=sum(dbinom(y,1,theta.star,log=TRUE))
17      +dbeta(theta.star,alpha,beta,log=TRUE)
18      -log(theta)-log(1-theta)
19  mh.2=sum(dbinom(y,1,theta,log=TRUE))
20      +dbeta(theta,alpha,beta,log=TRUE)
21      -log(theta.star)-log(1-theta.star)
22  mh=exp(mh.1-mh.2)
23  if(mh>runif(1)){
24    theta=theta.star
25    theta.acc=theta.acc+1
26  }
27  theta.save[k]=theta
28 }
29
30 layout(matrix(c(1,1,2),1,3))
31 plot(theta.save,type="l",ylim=c(0,1),main="a")
32 hist(theta.save,prob=TRUE,col=8,xlim=c(0,1),
33      main="b")
34 curve(dbeta(x,y.sum+alpha,oneminusy.sum+beta),
35      from=0,to=1,add=TRUE)
36 curve(dbeta(x,alpha,beta),from=0,to=1,lty=2,
37      add=TRUE)
```

Note that we had to create the logit and logit^{-1} functions because they do not exist in base R. Also notice that a portion of the proposal (i.e., $[\text{logit}(\theta^{(*)})|\text{logit}(\theta^{(k-1)})]$) cancels in the numerator and denominator in this case because it is symmetric. This algorithm results in the trace plot and posterior histogram shown in Figure 6.4. The trace plot and posterior histogram in Figure 6.4 resemble those we obtained using other approaches. Based on a tuning parameter (in logit space) of $\sigma^2_{\text{tune}} = 0.1$, the algorithm yields an acceptance rate of 46%. If we had canceled the symmetric proposal for $\text{logit}(\theta)$ in the M-H ratio without using the change of variables technique to find the correct proposal distribution for θ directly, we would have ended up with a MCMC sample from the incorrect posterior distribution. Thus, if we prefer to use a proposal based on a transformation of the original parameter, then we need to remember to correct for that in the M-H ratio when constructing our MCMC algorithm.

The approach we presented in this section to obtain the correct M-H update procedure using a proposal of a transformation of θ may be useful in some situations, especially when we must use a particular prior. However, if we are flexible

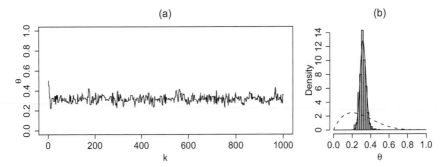

Figure 6.4 Trace plot (a) and histogram (b) based on a MCMC sample of size $K = 1000$ resulting from M-H updates using proposal based on a logit transformation of θ. In the histogram (b), the true posterior PDF for θ is a solid line and the prior PDF is a dashed line.

on the specific form we use as a prior and what we consider as the original parameter, then we could choose a prior to match the transformation of the proposal. For example, if we wanted to use the proposal $\text{logit}(\theta^{(*)}) \sim N(\text{logit}(\theta^{(k-1)}), \sigma^2_{\text{tune}})$, then we could avoid the change of variables altogether by specifying the prior directly as $[\text{logit}(\theta)] \equiv N(\text{logit}(\theta)|\mu_\theta, \sigma^2_\theta)$. This change of prior results in a different model than the Bernoulli-Beta model we have focused on in this chapter, but we can choose the new prior to match the beta prior closely and the new M-H ratio becomes

$$mh = \frac{\prod_{i=1}^{n}[y_i|\theta^{(*)}][\text{logit}(\theta^{(*)})]}{\prod_{i=1}^{n}[y_i|\theta^{(k-1)}][\text{logit}(\theta^{(k-1)})]}, \tag{6.15}$$

because the proposal distribution cancels in the numerator and denominator due to symmetry.

In our continued example, using a normal prior with mean of -1 and variance 0.81 (which closely matches the previous beta prior), the R code for the MCMC algorithm to fit the model is as follows:

BOX 6.5   MCMC FOR θ WITH NORMAL PRIOR

```
1  y=c(rep(1,y.sum),rep(0,oneminusy.sum))
2  K=1000
3  theta=0.5   # initial value
4  theta.save=rep(0,K)
5  theta.save[1]=theta
6  s2.tune=.1
```

(Continued)

BOX 6.5 (Continued)   MCMC FOR θ WITH NORMAL PRIOR

```
7  mu.theta=-1
8  s2.theta=.9^2
9
10 logit<-function(p){log(p/(1-p))}
11 logit.inv<-function(x){exp(x)/(1+exp(x))}
12
13 set.seed(1)
14 theta.acc=1
15 for(k in 2:K){
16   theta.star=logit.inv(rnorm(1,logit(theta),
17       sqrt(s2.tune)))
18   mh.1=sum(dbinom(y,1,theta.star,log=TRUE))
19       +dnorm(logit(theta.star),mu.theta,
20         sqrt(s2.theta),log=TRUE)
21   mh.2=sum(dbinom(y,1,theta,log=TRUE))
22       +dnorm(logit(theta),mu.theta,
23         sqrt(s2.theta),log=TRUE)
24   mh=exp(mh.1-mh.2)
25   if(mh>runif(1)){
26     theta=theta.star
27     theta.acc=theta.acc+1
28   }
29   theta.save[k]=theta
30 }
31
32 layout(matrix(c(1,1,2),1,3))
33 plot(theta.save,type="l",lwd=1,ylim=c(0,1),
34     main="a")
35 hist(theta.save,prob=TRUE,col=8,xlim=c(0,1),
36     main="b")
37 curve(dbeta(x,y.sum+alpha,oneminusy.sum+beta),
38     from=0,to=1,add=TRUE)
39 curve(dnorm(logit(x),mu.theta,sqrt(s2.theta))/x/
40     (1-x),from=0,to=1,lty=2,add=TRUE)
```

The resulting trace plot and posterior histogram are shown in Figure 6.5. Thus, all of these approaches to building models and constructing MCMC algorithms yield similar results and provide a suite of options for obtaining a sample that we can use to characterize the posterior distribution of interest. As we work through more examples, it will be helpful to recall these various approaches to sample parameters in Bayesian models.

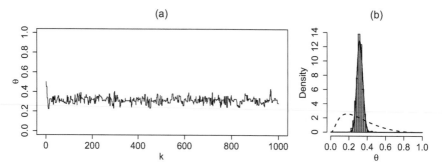

Figure 6.5 Trace plot (a) and histogram (b) based on a MCMC sample of size $K = 1000$ resulting from M-H updates using proposal and prior based on a logit transformation of θ. The histogram (b) shows the posterior PDF based on the previously used beta prior for θ as a solid line (for comparison) and the logit normal prior PDF as a dashed line.

6.3   ADDITIONAL CONCEPTS AND READING

Binary observations are a very common type of data collected in ecological studies. Thus, in several chapters that follow, we generalize the basic concepts presented in this chapter to accommodate various forms of heterogeneity, measurement error, and dependence in binary data. For example, in Chapter 20, we allow the Bernoulli probability θ to vary by observation (i.e., θ_i) and be associated with known predictor variables. We often refer to this basic model extension as logistic regression and it falls in the broader class of generalized linear models.

In Chapter 21, we demonstrate some equivalencies among certain types of binary data models and count data models. We also discuss models that rely on some replication among binary observations as well as heterogeneity at different scales, either in time or space. Conveniently, if $y_j \sim \text{Bern}(\theta)$ are independent and identically distributed for a set of repeated measurements, then $y = \sum_{j=1}^{J} y_j \sim \text{Binom}(J, \theta)$. Thus, we can use a binomial model for this type of data. Under the binomial data model there is a single measurement y that is the number of ones in a set of J binary replicates. We apply this type of model to estimate population abundance in Chapters 21 and 24.

Ecological data are often subject to missingness, at least some of which is caused by the failure to detect an organism when it was actually present in a region of time or space. We sometimes can account for this type of measurement error using a data model that accommodates extra zeros. We often refer to these models as zero-inflated models and are predominantly used in wildlife ecology. Chapters 22 through 24 focus mainly on zero-inflated models for binary data. Depending on the type of data and goals of the study, we may refer to these zero-inflated models as occupancy or abundance models. We show how to extend simple measurement error models for binary data in Chapter 25.

It is straightforward to account for dependence and heterogeneity among binary observations using a generalized linear model framework when there are known predictor variables. However, many forms of environmental and ecological data also contain latent dependence that we cannot characterize by observable predictor variables. However, this latent dependence may be structured in time and/or space. Thus, we describe ways to account for dependence in binary data using spatially explicit occupancy and abundance models in Chapters 26 and 27.

Binary data contain only minimal information compared to continuous data and counts. This lack of information can present challenges for acquiring sufficient inference to learn about ecological and environmental quantities of interest. However, because binary data are often easier to collect, they can sometimes be more readily available over larger regions of space or time. In Chapter 25, we show how to combine the strengths of binary data sources with other data, such as counts, to improve desired inference.

Finally, in some of the chapters that follow, we represent Bernoulli data models such as $y_i \sim \text{Bern}(\theta)$ using a mixture or cases notation

$$y_i = \begin{cases} 1, & \text{with probability } \theta \\ 0, & \text{with probability } 1 - \theta \end{cases}. \tag{6.16}$$

We describe stochastic mixture models in Chapter 10, and will not return to this type of notation until later chapters. However, from an implementation perspective, it is sometimes helpful to specify binary data models using the mixture notation in (6.16).

7 Normal-Normal

In the previous chapters, we focused on Bayesian models with Bernoulli likelihoods and beta priors. Most books on Bayesian statistics begin with Bernoulli-Beta models because they provide an intuitive and accessible set of examples to learn how the Bayes rule works and understand the fundamentals of Bayesian computing. However, as we work toward other types of Bayesian models and situations with multiple parameters, it will be helpful to learn a few tricks for dealing with additional distributions.

Thus, suppose we seek to learn about the mean of a certain population μ. Estimating μ is a primary topic in an entry-level statistics course. Suppose also that we collect continuously valued data y_i, for $i = 1, \ldots, n$ observations, from the population of interest. Then, if the data are approximately normally distributed with mean μ and independent with known variance σ^2 (for now), a Bayesian model to help us learn about μ is

$$y_i \sim \mathrm{N}(\mu, \sigma^2), \text{for } i = 1, \ldots, n, \tag{7.1}$$

$$\mu \sim \mathrm{N}(\mu_0, \sigma_0^2), \tag{7.2}$$

where the normal prior for μ is a natural choice because it has the correct support and it is flexible enough such that we can specify it to accommodate a wide range of prior knowledge.[1]

The posterior distribution associated with this Normal-Normal model is

$$[\mu \,|\, \mathbf{y}] \propto \prod_{i=1}^{n} [y_i | \mu][\mu], \tag{7.3}$$

where the product arises in the likelihood because of the independence assumption.[2] Thus, to find the posterior distribution, we need to derive the normalizing constant that makes the posterior integrate to one. The technique here is to write out the functional form of the joint distribution (i.e., the product of the likelihood and prior), simplify it by omitting multiplicative terms that do not involve μ, combine like terms that do involve μ, and then recognize what remains as the kernel of a distribution we already know.

---

[1] In this case, because the data model implies that the observations can assume any real number, the corresponding support for the mean parameter μ can also be any real number.

[2] The σ^2 parameter does not appear in the posterior or joint distribution because we assume it is known for now.

For the Normal-Normal model, we can work out the posterior distribution as

$$[\mu|\mathbf{y}] \propto \prod_{i=1}^{n}[y_i|\mu][\mu], \tag{7.4}$$

$$\propto \left(\prod_{i=1}^{n}\frac{1}{\sqrt{2\pi\sigma^2}}\exp\left(-\frac{1}{2\sigma^2}(y_i-\mu)^2\right)\right)\frac{1}{\sqrt{2\pi\sigma_0^2}}\exp\left(-\frac{1}{2\sigma_0^2}(\mu-\mu_0)^2\right), \tag{7.5}$$

$$\propto \exp\left(-\frac{1}{2\sigma^2}\sum_{i=1}^{n}(y_i-\mu)^2\right)\exp\left(-\frac{1}{2\sigma_0^2}(\mu-\mu_0)^2\right), \tag{7.6}$$

$$\propto \exp\left(-\frac{1}{2}\left(-2\left(\frac{\sum_{i=1}^{n}y_i}{\sigma^2}+\frac{\mu_0}{\sigma_0^2}\right)\mu+\left(\frac{n}{\sigma^2}+\frac{1}{\sigma_0^2}\right)\mu^2\right)\right), \tag{7.7}$$

$$\propto \exp\left(-\frac{1}{2}\left(-2b\mu+a\mu^2\right)\right), \tag{7.8}$$

where $a=\frac{n}{\sigma^2}+\frac{1}{\sigma_0^2}$ and $b=\frac{\sum_{i=1}^{n}y_i}{\sigma^2}+\frac{\mu_0}{\sigma_0^2}$. Then, by "completing the square," the posterior is $[\mu|\mathbf{y}] = N(a^{-1}b, a^{-1})$. Being able to recognize this form is incredibly useful for Bayesian models involving normal distributions. Because the form of the prior matches the form of the posterior, we refer to it as conjugate.

Completing the square is a phrase used in the Bayesian context to denote finding the normalizing constant that allows us to write the kernel for the posterior as $\exp\left(-\frac{1}{2a^{-1}}(\mu-a^{-1}b)^2\right)$. Recall that, we can decompose a square as $(c-d)^2 = c^2-2cd+d^2$, and in the exponent of our posterior we have

$$-2\left(\frac{\sum_{i=1}^{n}y_i}{\sigma^2}+\frac{\mu_0}{\sigma_0^2}\right)\mu+\left(\frac{n}{\sigma^2}+\frac{1}{\sigma_0^2}\right)\mu^2. \tag{7.9}$$

To complete the square, we let $c=\left(\frac{n}{\sigma^2}+\frac{1}{\sigma_0^2}\right)^{\frac{1}{2}}\mu$, which implies that $\mu=\left(\frac{n}{\sigma^2}+\frac{1}{\sigma_0^2}\right)^{-\frac{1}{2}}a$. Then we substitute this equation for μ into (7.9) to yield

$$-2\left(\frac{\sum_{i=1}^{n}y_i}{\sigma^2}+\frac{\mu_0}{\sigma_0^2}\right)\left(\frac{n}{\sigma^2}+\frac{1}{\sigma_0^2}\right)^{-\frac{1}{2}}a+a^2. \tag{7.10}$$

Now we set $d=\left(\frac{\sum_{i=1}^{n}y_i}{\sigma^2}+\frac{\mu_0}{\sigma_0^2}\right)\left(\frac{n}{\sigma^2}+\frac{1}{\sigma_0^2}\right)^{-\frac{1}{2}}$ and, using the original relationship $(c-d)^2 = c^2-2cd+d^2$, the square is

$$\left(\mu \left(\frac{n}{\sigma^2} + \frac{1}{\sigma_0^2} \right)^{\frac{1}{2}} - \left(\frac{\sum_{i=1}^n y_i}{\sigma^2} + \frac{\mu_0}{\sigma_0^2} \right) \left(\frac{n}{\sigma^2} + \frac{1}{\sigma_0^2} \right)^{-\frac{1}{2}} \right)^2, \qquad (7.11)$$

which equals $\mathrm{Var}(\mu|\mathbf{y})^{-1}(\mu - E(\mu|\mathbf{y}))^2$, as required.

The analytical posterior for the Normal-Normal model implies that we can visualize the posterior distribution and infer its characteristics (e.g., moments, credible intervals, etc) directly without using MCMC. For example, in the following R code, we simulated $n = 10$ observations from a normal distribution with mean 2 and variance 1:

BOX 7.1 NORMAL DATA SIMULATION AND POSTERIOR FOR μ

```
1  n=10
2  mu=2
3  s2=1
4  set.seed(1)
5  y=rnorm(n,mu,sqrt(s2))
6
7  mu0=0
8  s20=10
9
10 mu.var=1/(n/s2+1/s20)
11 mu.mean=mu.var*(sum(y)/s2+mu0/s20)
12 curve(dnorm(x,mu.mean,sqrt(mu.var)),from=-2,to=6)
13 curve(dnorm(x,mu0,sqrt(s20)),from=-2,to=6,lty=2,
14     add=TRUE)
15 rug(y,col=1)
```

---

and based on a normal prior for μ with mean $\mu_0 = 0$ and variance $\sigma_0^2 = 10$, this results in the posterior shown in Figure 7.1. Notice that the mildly informative prior does not appear to affect the posterior much because the posterior remains approximately centered on the data, even with only $n = 10$ observations. An often noted feature of the posterior distribution for μ is that, as the prior variance increases ($\sigma_0^2 \to \infty$), the posterior mean $E(\mu|\mathbf{y}) \to \sum_{i=1}^n y_i/n$. Thus, the Bayesian point estimate (which is also equal to the mode, because of symmetry in the posterior) approaches the MLE in the absence of prior information about μ.[3] Based on the data simulated in this example, the sample mean was $\bar{y} = 2.132$ and the posterior mean was $E(\mu|\mathbf{y}) = 2.111$. Thus, while the posterior mean is forced to be smaller than the sample mean because the prior mean was smaller than the sample mean, the posterior mean in this example was still quite similar to sample mean (Figure 7.1).

---

[3] We prefer not to use this as an excuse to always specify vague priors, but rather to assess the information content of the data.

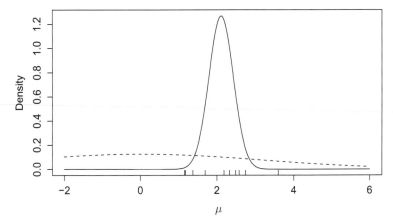

Figure 7.1 Posterior for μ (solid line) based on a model with normal likelihood and normal prior with hyperparameters $\mu_0 = 0$ and $\sigma_0^2 = 10$ (dashed line) and $n = 10$ data points illustrated as rug.

As with the Bernoulli-Beta model from previous chapters, we could also use MCMC to obtain a sample from the posterior associated with the Normal-Normal model. In fact, if we were to use M-H updates, a good proposal distribution is the normal random walk because the support matches and it is symmetric, simplifying the M-H ratio. However, we return to MCMC algorithms for this model in Chapter 9.

7.1   ADDITIONAL CONCEPTS AND READING

The univariate normal distribution is the most commonly specified prior for μ when the data model is also normal because of the conjugacy that results. However, a normal distribution may not accurately represent the prior information available about the parameter μ. For example, if we know that the population mean may assume an extreme value, then we could specify a prior distribution with heavier tails than the normal allows for. One example of a prior distribution with heavy tails is the non-central scaled t-distribution with location parameter μ_0, scale parameter σ_0, and degrees of freedom ν. As the degrees of freedom (ν) in the t-distribution decreases, the kurtosis increases, implying a heavier-tailed prior distribution.

Alternatively, if our prior information about μ suggests there is a high probability that μ occurs at a particular location and tapering off rapidly for smaller and larger values, then we may seek to specify a distribution with a completely different shape than the normal or t-distributions. The Laplace (or double-exponential) distribution has a PDF that tapers exponentially on both sides of the mean and contains only two parameters. Thus, if we specified the prior as $\mu \sim [\mu] \equiv \mathrm{Laplace}(\mu_0, \sigma_0)$, then

$$[\mu] = \frac{1}{2\sigma_0} \exp\left(-\frac{|\mu - \mu_0|}{\sigma_0}\right), \tag{7.12}$$

where μ_0 is the prior mean and $2\sigma_0^2$ is the prior variance. When the Laplace distribution is used as a prior, it induces a certain constraint on the parameter μ that can be helpful for model selection purposes when there are several mean parameters to be estimated in a single model (Park and Casella 2008; Hooten and Hobbs 2015). For example, in linear regression models (Chapter 11) with several coefficients, we can use Laplace priors (and normal priors) to shrink the parameters toward their prior means in such a way that the resulting model has improved predictive ability. This concept is known as "regularization" and we discuss its implementation in Chapter 14.

Finally, in cases where the population mean is known to be positive (or bounded in any way), a truncated prior provides an easy way to induce such known constraints on the estimation of μ. For example, suppose we know $\mu > 0$ *a priori*, then we could use a truncated normal prior such that $\mu \sim \text{TN}(\mu_0, \sigma_0^2)_0^\infty$ (where the normal distribution is truncated below at zero, but not above). When we fit the normal data model with truncated normal prior, the posterior distribution also will be appropriately bounded below by zero. Therefore, constrained inference is straightforward in the Bayesian setting. However, depending on the choice of prior, the resulting posterior distribution may not be conjugate nor analytically tractable. In those cases where the posterior is analytically intractable, we must use numerical integration or a stochastic sampling approach to characterize the posterior distribution.

8 Normal-Inverse Gamma

We also may be interested in estimating a population variance. Understanding the natural variation in data can help us account for measurement error, stochastic environmental events, and individual heterogeneity when making inference.

Assuming we observe the same data that we described in the previous chapter (y_i for $i = 1, \ldots, n$), we can use the same data model ($y_i \sim N(\mu, \sigma^2)$) but now consider the mean μ to be fixed and known. This method leaves σ^2 as the remaining unknown parameter in the model; thus we need to specify a prior for it.

Many different approaches have been used to provide prior information for σ^2, including log-normal, truncated normal, Cauchy, gamma, and inverse gamma ("IG" when used in an equation). The inverse gamma prior for σ^2 is conjugate, as we show next, but that does not mean that it is the most intuitive, nor does it have the best properties. As with the mean, the more data we have, the less the population variance prior affects the inference.

For the sake of completeness, we use the inverse gamma prior for the population variance in this chapter such that $\sigma^2 \sim IG(q, r) \equiv \frac{1}{r^q \Gamma(q)} (\sigma^2)^{-(q+1)} \exp(-\frac{1}{r\sigma^2})$. As before, using the independence assumption in the data model, we can write the posterior distribution for σ^2 as $[\sigma^2 | \mathbf{y}] \propto \prod_{i=1}^{n} [y_i | \sigma^2][\sigma^2]$. Again, the parameter μ does not appear in the posterior or joint distribution in our notation because we assume it is known in this chapter.

As mentioned, the inverse gamma prior for σ^2 combined with the normal likelihood results in an inverse gamma posterior because

$$[\sigma^2 | \mathbf{y}] \propto \prod_{i=1}^{n} [y_i | \sigma^2][\sigma^2] , \tag{8.1}$$

$$\propto \left(\prod_{i=1}^{n} \frac{1}{\sqrt{2\pi\sigma^2}} \exp\left(-\frac{1}{2\sigma^2} (y_i - \mu)^2 \right) \right) \frac{1}{r^q \Gamma(q)} (\sigma^2)^{-(q+1)} \times$$

$$\exp\left(-\frac{1}{r\sigma^2} \right) , \tag{8.2}$$

$$\propto (\sigma^2)^{-\frac{n}{2}} \exp\left(-\frac{1}{2\sigma^2} \sum_{i=1}^{n} (y_i - \mu)^2 \right) (\sigma^2)^{-(q+1)} \exp\left(-\frac{1}{r\sigma^2} \right) , \tag{8.3}$$

$$\propto (\sigma^2)^{-(\frac{n}{2}+q+1)} \exp\left(-\frac{1}{\sigma^2} \left(\frac{\sum_{i=1}^{n} (y_i - \mu)^2}{2} + \frac{1}{r} \right) \right) . \tag{8.4}$$

Thus, letting $\tilde{q} = \frac{n}{2} + q$ and $\tilde{r} = \left(\frac{\sum_{i=1}^{n} (y_i - \mu)^2}{2} + \frac{1}{r} \right)^{-1}$, the posterior distribution is

$[\sigma^2 | \mathbf{y}] = IG(\tilde{q}, \tilde{r})$. As with the Normal-Normal model, certain hyperparameters can reduce the influence of the prior on the posterior for σ^2. Using the properties of the inverse gamma distribution, as $q \to 0$ and $r \to \infty$, the posterior of σ^2 is less affected by the prior. In fact, setting q at a value close to zero and r at a large value has become a default choice in many Bayesian models. This choice could be due, in part, to the non-intuitiveness of the inverse gamma distribution and the difficulty in quantifying prior information for it.

One approach to specify hyperparameters for inverse gamma priors is to use moment matching (Hobbs and Hooten 2015). If we know the prior mean and variance of σ^2, then we can use the relationship between the parameters and the moments to solve for q and r. Recall that the prior mean of an inverse gamma random variable σ^2 is $E(\sigma^2) = \frac{1}{r(q-1)}$ and the prior variance is $\text{Var}(\sigma^2) = \frac{1}{r^2(q-1)^2(q-2)}$. Thus, solving the first equation for r, we have $r = \frac{1}{E(\sigma^2)(q-1)}$ and substituting r into the second equation and solving for q results in $q = \frac{E(\sigma^2)^2}{\text{Var}(\sigma^2)} + 2$. Thus, we can specify the prior based on the prior moments. This method makes it substantially easier to quantify prior knowledge, but variances can be difficult to interpret regardless.

As with the Normal-Normal model, the analytical posterior for the Normal-Inverse Gamma model implies that we can visualize the posterior distribution directly, without using MCMC. In the following R code, we simulated the same $n = 10$ observations from a normal distribution with mean 2 and variance 1:

BOX 8.1 NORMAL DATA SIMULATION AND POSTERIOR FOR σ^2

```
n=10
mu=2
s2=1
set.seed(1)
y=rnorm(n,mu,sqrt(s2))

s2.mn=2
s2.var=100
q=(s2.mn^2)/s2.var+2
r=1/(s2.mn*(q-1))

q.tilda=n/2+q
r.tilda=1/(sum((y-mu)^2)/2+1/r)

dIG <- function(x,r,q){x^(-(q+1))*exp(-1/r/x)/
    (r^q)/gamma(q)}

curve(dIG(x,r.tilda,q.tilda),from=0,to=6)
curve(dIG(x,r,q),from=0,to=6,lty=2,add=TRUE)
```

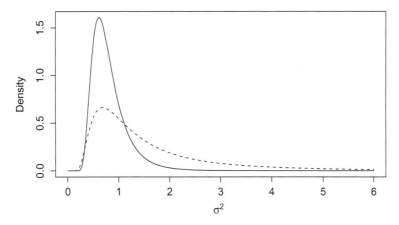

Figure 8.1 Posterior for σ^2 (solid line) based on a model with normal likelihood and inverse gamma prior (dashed line) with hyperparameters $q = 2.04$ and $r = 0.481$ and $n = 10$ observations.

We used a prior mean of 2 and prior variance of 100 to fit this model to the simulated data for illustration (these moments imply $q = 2.04$ and $r = 0.481$). The resulting posterior distribution for σ^2 is shown in Figure 8.1 with the prior for comparison. The resulting posterior distribution based on the prior and simulated data (Figure 8.1) captures the true value of $\sigma^2 = 1$ used to simulate the data, but it remains relatively wide because of the limited observations. The sample variance for these data is 0.61, as compared to the posterior mean of 0.81, which is influenced slightly by the larger prior mean in this case. Yet the dispersion in the posterior is reduced as compared to the prior, so the data reduce uncertainty in our understanding of σ^2.

8.1 ADDITIONAL CONCEPTS AND READING

As with the normal prior for the mean parameter μ, the inverse gamma prior for the variance parameter σ^2 is only one option. In fact, although the inverse gamma prior is conjugate, it may not represent our prior knowledge well. Standard deviation often is easier to interpret than variance; thus, we may wish to model σ directly instead of σ^2. For example, a uniform prior for the standard deviation (i.e., $\sigma \sim \text{Unif}(0, u_\sigma)$) may be reasonable when we can identify a sensible upper bound for the standard deviation *a priori*. In the case where a uniform prior is used, the posterior for σ is

$$[\sigma|\mathbf{y}] \propto \left(\prod_{i=1}^{n}[y_i|\mu,\sigma^2]\right)[\sigma] , \tag{8.5}$$

$$\propto \prod_{i=1}^{n}[y_i|\mu,\sigma^2] , \tag{8.6}$$

where the bounded support for σ must be enforced to ensure the prior is proper (i.e., integrates to one over the support of σ).

Alternatively, because $\log(\sigma^2) = 2\log(\sigma)$, we may choose to specify a prior for $\log(\sigma)$ with unbounded support. A natural prior to use for the log standard deviation is $\log(\sigma) \sim N(\mu_\sigma, \tau_\sigma^2)$. However, as with the prior for the variance, a prior for the log standard deviation may be difficult to interpret. On the other hand, a prior on the log standard deviation provides a way to allow the standard deviation of the data model to vary with the observations (e.g., when some observations are expected to be more dispersed than others around the mean μ). For example, if we let the standard deviation vary with observation i (implying the new data model $y_i \sim N(\mu, \sigma_i^2)$ for $i = 1, \ldots, n$) and relate it to a positive predictor variable x_i (for $x_i > 0$) by $\log(\sigma_i) = \alpha x_i$, then we only need a prior for the new parameter α because the variables x_i, for $i = 1, \ldots, n$, are known. In this case, we could specify a prior for α such as $\alpha \sim N(0, \sigma_\alpha^2)$. This new model parameterization and prior imply that the posterior distribution is

$$[\alpha | \mathbf{y}] \propto \left(\prod_{i=1}^{n} [y_i | \mu, \alpha] \right) [\alpha] . \tag{8.7}$$

Allowing for heterogeneity in the standard deviation of a data model as we just described is also a form of regression modeling, but for the dispersion rather than the location of a data model. We introduce the concept of regression modeling in Chapter 11.

The specification of priors for variance components may become more important in hierarchical models (Chapter 19); particularly when more vague priors are desired. In fact, Gelman (2006) recommended against using vague inverse gamma priors for latent variance parameters in hierarchical models, and preferred instead half-Cauchy or uniform prior distributions for standard deviations.

9 Normal-Normal-Inverse Gamma

In the previous chapters, we explored how to construct MCMC algorithms for single-parameter Bayesian models. Single-parameter models are useful to help understand how Bayesian computing works, but the real power of MCMC is how it facilitates fitting multi-parameter models.

Our first useful multiparameter model arises as a combination of the previous two chapters. That is, suppose that we observe continuously valued data that are well-modeled by a normal distribution with mean (μ) and variance (σ^2) unknown. Also, suppose we assume independence among the observations and the same prior distributions for the model parameters as previously used (i.e., *a priori* independent normal and inverse gamma) resulting in the following model specification

$$y_i \sim N(\mu, \sigma^2) , \text{ for } i = 1, \ldots, n, \tag{9.1}$$

$$\mu \sim N(\mu_0, \sigma_0^2), \tag{9.2}$$

$$\sigma^2 \sim IG(q, r). \tag{9.3}$$

For this model, the posterior distribution can be written as

$$[\mu, \sigma^2 | \mathbf{y}] \propto \prod_{i=1}^{n} [y_i | \mu, \sigma^2][\mu][\sigma^2] . \tag{9.4}$$

Note that alternative priors could be used. For example, instead of assuming conditional independence in the priors, we could specify the joint prior as $[\mu, \sigma^2] = [\mu|\sigma^2][\sigma^2]$. We find that it makes little practical difference in this simple model to use a joint prior distribution, but there will be other situations in multivariate models where it is important to allow for parameters to be correlated *a priori*.

To implement the model in (9.1–9.3), we can use MCMC. Recall that a MCMC algorithm for multiparameter models typically involves deriving full-conditional distributions for the parameters and then sampling from them recursively. Therefore, to find the full-conditional distribution for μ, we start with the joint distribution in (9.4) and omit all multiplicative terms that do not involve μ. Thus, the full-conditional distribution for μ in our model is

$$[\mu|\cdot] \propto \prod_{i=1}^{n} [y_i | \mu, \sigma^2][\mu], \tag{9.5}$$

which is equivalent to the posterior for μ in Chapter 7. Thus, we have already derived the full-conditional distribution for μ in our current model and it is

$$[\mu|\cdot] = N\left(\left(\frac{n}{\sigma^2} + \frac{1}{\sigma_0^2}\right)^{-1}\left(\frac{\sum_{i=1}^n y_i}{\sigma^2} + \frac{\mu_0}{\sigma_0^2}\right), \left(\frac{n}{\sigma^2} + \frac{1}{\sigma_0^2}\right)^{-1}\right), \qquad (9.6)$$

which is a conjugate full-conditional (like the posterior we derived in Chapter 7). Conjugacy means that we can sample from the full-conditional directly in our MCMC algorithm providing Gibbs updates for μ.

Similarly, we derive the full-conditional distribution for the variance parameter σ^2 by retaining all terms involving σ^2 in the joint distribution in (9.4) which results in

$$[\sigma^2|\cdot] \propto \prod_{i=1}^n [y_i|\mu, \sigma^2][\sigma^2] . \qquad (9.7)$$

Like the full-conditional distribution for μ, this full-conditional distribution for σ^2 is conjugate and is equivalent to the posterior we found for σ^2 in Chapter 8. Taking the result directly from Chapter 8, we have

$$[\sigma^2|\cdot] = IG\left(\frac{n}{2} + q, \left(\frac{\sum_{i=1}^n (y_i - \mu)^2}{2} + \frac{1}{r}\right)^{-1}\right) \qquad (9.8)$$

as the full-conditional distribution for σ^2, which can be sampled from in a MCMC algorithm using Gibbs updates.

With the full-conditional distributions for both parameters being conjugate, we can use a fully Gibbs MCMC algorithm to obtain samples from the multiparameter posterior distribution. The MCMC algorithm to fit this model to data can be described as follows:

1. Assign an initial value for $\mu^{(0)}$
2. Set $k = 1$
3. Sample $(\sigma^2)^{(k)} \sim [\sigma^2|\cdot] \equiv [\sigma^2|\mathbf{y}, \mu^{(k-1)}]$
4. Sample $\mu^{(k)} \sim [\mu|\cdot] \equiv [\mu|\mathbf{y}, (\sigma^2)^{(k)}]$
5. Set $k = k + 1$, go to step 3; repeat until enough MCMC samples have been obtained.

We sampled σ^2 first in the algorithm because it is often easier to find a good starting value (i.e., close to the posterior) for μ than for σ^2. Thus, if we begin by specifying a starting value for μ, we condition on it when sampling σ^2 and thus do not need a starting value for σ^2.[1]

In R, the MCMC algorithm to fit the Normal-Normal-Inverse Gamma model to data is

---

[1] In models with more parameters, we need to specify starting values for all but one parameter. Thus, we must pay careful attention to the order of updates in multiparameter MCMC algorithms.

BOX 9.1   MCMC ALGORITHM FOR NORMAL-NORMAL-INVERSE GAMMA MODEL

```
 1  NNIG.mcmc <- function(y,mu0,s20,q,r,mu.strt,
 2      n.mcmc){
 3
 4  n=length(y)
 5  mu.save=rep(0,n.mcmc)
 6  s2.save=rep(0,n.mcmc)
 7  mu=mu.strt
 8
 9  for(k in 1:n.mcmc){
10
11    tmp.r=1/(sum((y-mu)^2)/2+1/r)
12    tmp.q=n/2+q
13    s2=1/rgamma(1,tmp.q,,tmp.r)
14
15    tmp.var=1/(n/s2+1/s20)
16    tmp.mn=tmp.var*(sum(y)/s2+mu0/s20)
17    mu=rnorm(1,tmp.mn,sqrt(tmp.var))
18
19    mu.save[k]=mu
20    s2.save[k]=s2
21
22  }
23
24  list(mu.save=mu.save,s2.save=s2.save,
25      n.mcmc=n.mcmc)
26  }
```

Notice that we stored this algorithm as an R function so that we can use it more than once without having to highlight and run in the R console.

In our R code above, we note a few things that we use hereinafter. First, the two actual Gibbs updates occur at lines 15 and 17. To sample σ^2 from an Inverse Gamma distribution, we can first sample from a Gamma distribution and then take the reciprocal as our sample.[2] Also note that the parameterization we use for the Inverse Gamma corresponds to q as the shape parameter and r as the scale parameter; hence, the two commas in the `rgamma(1,tmp.q,,tmp.r)` function.

The second thing to note in this R code is that we only compute the full-conditional variance once and then use it to calculate the full-conditional mean. That reuse saves

---

[2] Be careful, this does not mean that the PDF for an Inverse Gamma is the reciprocal of the Gamma PDF!

us from having to calculate it again and will speed up the MCMC algorithm. Finally, note that we use the variable n.mcmc as the number of MCMC iterations (K) to make it more clear.

We applied this MCMC algorithm to the same simulated data we used in the previous two chapters. In this example, we simulated data ($n = 10$) from a normal distribution with mean 2 and variance 1 and used hyperparameters $\mu_0 = 0$, $\sigma_0^2 = 10$, $q = 2.04$, and $r = 0.481$. The R code to fit the model and display the results shown in Figure 9.1 is

---

BOX 9.2   NORMAL DATA SIMULATION AND MODEL FITNORMAL DATA SIMULATION AND MODEL FIT

```
n=10
mu=2
s2=1
set.seed(1)
y=rnorm(n,mu,sqrt(s2))

s2.mn=2
s2.var=100
q=(s2.mn^2)/s2.var+2
r=1/(s2.mn*(q-1))
mu0=0
s20=10

source("NNIG.mcmc.R")
mcmc.out=NNIG.mcmc(y=y,mu0=mu0,s20=s20,q=q,r=r,
    mu.strt=0,n.mcmc=1000)

dIG <- function(x,r,q){x^(-(q+1))*exp(-1/r/x)/
    (r^q)/gamma(q)}

layout(matrix(c(1,2,1,2,3,4),2,3))
plot(mcmc.out$mu.save,type="l",xlab="k",ylab
    =bquote(mu))
plot(mcmc.out$s2.save,type="l",xlab="k",
    ylab=bquote(sigma^2))
hist(mcmc.out$mu.save,col=8,prob=TRUE)
curve(dnorm(x,mu0,sqrt(s20)),from=0,to=6,lty=2,
    add=TRUE)
hist(mcmc.out$s2.save,col=8,prob=TRUE)
curve(dIG(x,r,q),from=0,to=6,lty=2,add=TRUE)
```

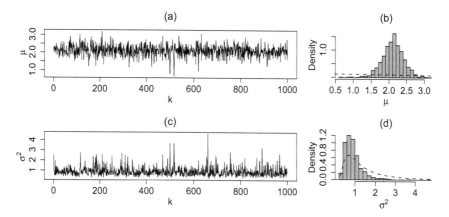

Figure 9.1 Trace plots (a, c) and posterior histograms (b, d) for μ and σ^2 and prior PDFs (dashed line).

Notice in Figure 9.1 that the Markov chains converge quickly, they are both well-mixed, and no burn-in period is necessary. As before, we only obtained $K = 1000$ MCMC iterations to illustrate the behavior of the Markov chains, but in practice we would increase this to a number that provides accurate approximations to our posterior quantities of interest. In a model like this, a reasonable number of MCMC iterations in practice is probably closer to $K = 10000$. Models with highly correlated parameters will require larger MCMC samples.

The posterior inference in Figure 9.1 is similar to what we reported in Chapters 7 and 8, but in the multiparameter model in this chapter, we account for the uncertainty associated with both parameters in the estimation of the other one. Recall that, in Chapters 7 and 8, we had to assume we knew a fixed value for either μ or σ^2. That assumption would not be realistic in practice, so the multiparameter Normal-Normal-Inverse Gamma model is useful for properly estimating a population mean and variance given normally distributed data.

A critical advantage of Bayesian inference is that it scales with sample size. Whereas traditional inference requires adjustments for small samples,[3] the Bayesian posterior automatically accounts for the both small and large samples. When small data sets are used, the uncertainty about parameters will increase appropriately, and when large samples are used, the inference is similar to that resulting from non-Bayesian models. However, for Bayesian models, the MCMC technique to obtain the inference is the same regardless of sample size.

---

[3] For example, to calculate confidence intervals for the sample mean as an estimate of the population mean, we can use the normal distribution when the sample size is large (i.e., $n > 30$), but must use the t-distribution for inference resulting from smaller samples.

The Normal-Normal-Inverse Gamma model also serves as a basis for Bayesian regression and analysis of variance. Thus, we return to these same ideas, approaches to derive full-conditional distributions, and algorithm components for fitting models in the chapters that follow.

9.1   ADDITIONAL CONCEPTS AND READING

As we discussed in the two preceding chapters, we are not restricted to using independent conjugate priors for the parameters in this model (or any model). Also, there is not necessarily a single set of conjugate priors for a given model. For example, in this case, because the inverse gamma distribution is a generalized version of the inverse χ-squared distribution, we could use the χ-squared distribution as an alternative prior for σ^2 and thus we may specify it more easily. As a prior for σ^2, we can write the inverse χ-squared PDF with hyperparameter v as

$$[\sigma^2] \equiv \frac{2^{-\frac{v}{2}}}{\Gamma\left(\frac{v}{2}\right)} (\sigma^2)^{-\left(\frac{v}{2}+1\right)} \exp\left(-\frac{1}{2\sigma^2}\right), \tag{9.9}$$

where the equivalent inverse gamma specification from the previous chapter has $q = v/2$ and $r = 2$.

Gelman et al. (2013) described how to use the inverse χ-squared distribution as a prior for σ^2 while allowing the prior for μ to depend on σ^2. Specifically, if we specify the conditional prior $[\mu|\sigma^2] \equiv N(\mu_0, \sigma^2/\tau_0)$ and marginal prior for σ^2 as inverse χ-squared, then the marginal posterior distribution for σ^2 will be inverse scaled χ-squared and the marginal posterior distribution for μ will be noncentral scaled t distribution (see Gelman et al. 2013 for details). Thus, the marginal posterior distributions for both model parameters are known and can be characterized analytically, without the use of MCMC. However, if we change the forms of the priors or generalize the data model, we will no longer have analytically tractable marginal posterior distributions and will need an algorithm like that presented in this chapter to fit the model.

Section III

Intermediate Models and Concepts

10 Mixture Models

In the previous chapters, we focused on building familiarity with basic models and MCMC algorithms. In the chapters in this part of the book, we build on those fundamental concepts to construct models and algorithms for more realistic and useful inferential situations. For example, one of the most useful classes of models in ecology are mixture models. In this chapter, we set the stage for this class of ecological models known as mixture models.

Mixture models are based on mixture distributions. We can write a mixture distribution $[\boldsymbol{\theta}]$, for the random vector $\boldsymbol{\theta}$, generically as

$$[\boldsymbol{\theta}] \equiv \sum_{l=1}^{L} p_l [\boldsymbol{\theta}]_l, \tag{10.1}$$

where $[\boldsymbol{\theta}]_l$ represents the component distributions that make up the mixture for $l = 1, \ldots, L$ components and p_l represents the mixture probabilities. The resulting mixture distribution $[\boldsymbol{\theta}]$ will be valid if the mixture probabilities are all positive ($p_l > 0$ for all $l = 1, \ldots, L$) and sum to one ($\sum_{l=1}^{L} p_l = 1$) and the components $[\boldsymbol{\theta}]_l$ are valid distributions. Although we have written a general mixture distribution in terms of a random vector $\boldsymbol{\theta}$, we can use mixture models for data, process, and parameter distributions.[1]

For example, consider a two component ($L = 2$) mixture of normal distributions with mixture components $[\boldsymbol{\theta}]_1 \equiv N(\mu_1, \sigma_1^2)$ and $[\boldsymbol{\theta}]_2 \equiv N(\mu_2, \sigma_2^2)$. The resulting mixture model for $\boldsymbol{\theta}$ is

$$[\boldsymbol{\theta}] \equiv p N(\mu_1, \sigma_1^2) + (1-p) N(\mu_2, \sigma_2^2), \tag{10.2}$$

where only a single p is required because $p + (1-p) = 1$. We used the following R code to create Figure 10.1 and Box 10.1:

BOX 10.1   EXAMPLE MIXTURE DISTRIBUTIONS

```
layout(matrix(1:3,1,3))
p=.25
curve(dnorm(x,-2,1),col=8,lty=1,main="a")
mtext("p = 0.25",,,.25)
```

(Continued)

---

[1] We briefly introduced the concept of hierarchical statistical models in Chapter 1, but return to it in Chapter 19. Hierarchical models typically comprise at least three conditionally specified probability distributions, which we often refer to as "data, process, and parameters" (Berliner 1996).

BOX 10.1 (Continued)   EXAMPLE MIXTURE DISTRIBUTIONS

```
5  curve(dnorm(x,1,2),col=8,lty=2,add=TRUE)
6  curve(p*dnorm(x,-2,1)+(1-p)*dnorm(x,1,2),col=1,
7         add=TRUE)
8
9  p=.5
10 curve(dnorm(x,-2,1),col=8,lty=1,main="b")
11 mtext("p = 0.5",,.25)
12 curve(dnorm(x,1,2),col=8,lty=2,add=TRUE)
13 curve(p*dnorm(x,-2,1)+(1-p)*dnorm(x,1,2),col=1,
14         add=TRUE)
15
16 p=.9
17 curve(dnorm(x,-2,1),col=8,lty=1,main="c")
18 mtext("p = 0.9",,.25)
19 curve(dnorm(x,1,2),col=8,lty=2,add=TRUE)
20 curve(p*dnorm(x,-2,1)+(1-p)*dnorm(x,1,2),col=1,
21         add=TRUE)
```

Based on values of $\mu_1 = -2$, $\sigma_1^2 = 1$, $\mu_2 = 1$, and $\sigma_2^2 = 4$ and three different values for p, Figure 10.1 illustrates that a wide range of PDF shapes are possible depending on the mixture specification. For example, Figure 10.1a shows a distinctly bimodal distribution, while Figure 10.1b shows a strongly right-skewed, but unimodal distribution. The mixture distribution in Figure 10.1c resembles the first mixture component distribution because $p = 0.9$ and the first mixture component is more precise.

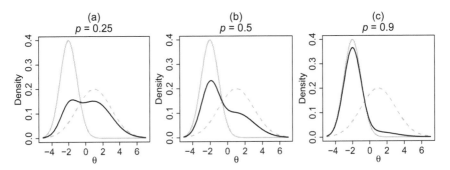

Figure 10.1 Mixture distributions (solid black line) for three values of mixture probability: (a) $p = 0.25$, (b) $p = 0.5$, and (c) $p = 0.9$. Mixture component distributions are $N(-2,1)$ (solid gray line) and $N(1,4)$ (dashed gray line).

Turning our attention to mixture models for data, we can formulate a simple two-component mixture model for a data set $\mathbf{y} \equiv (y_1,\ldots,y_n)'$ in two ways. The first formulation resembles that of the general mixture model in (10.1) where

$$y_i \sim [y_i|\boldsymbol{\theta}_1,\boldsymbol{\theta}_2,p] \equiv p[y_i|\boldsymbol{\theta}_1]_1 + (1-p)[y_i|\boldsymbol{\theta}_2]_2, \qquad (10.3)$$

for $i = 1,\ldots,n$. In (10.3), $[y_i|\boldsymbol{\theta}_1]_j$ corresponds to the jth mixture distribution, for $j = 1,2$ in this particular model. We often write the second formulation using cases notation, where

$$y_i \sim \begin{cases} [y_i|\boldsymbol{\theta}_1]_1, & \text{with probability } p \\ [y_i|\boldsymbol{\theta}_2]_2, & \text{with probability } 1-p \end{cases}. \qquad (10.4)$$

We can use either of the formulations in a Bayesian model to form components of a likelihood. In either case, the likelihood is

$$[\mathbf{y}|\boldsymbol{\theta}] = \prod_{i=1}^{n} p[y_i|\boldsymbol{\theta}_1]_1 + (1-p)[y_i|\boldsymbol{\theta}_2]_2, \qquad (10.5)$$

as long as we make the usual conditional independence assumption for y_i. The model for $\mathbf{y}$ implies that observations arise from a mixture of two conditional distributions. The mixture probability for the first component is p, implying that the probability for the second mixture component is $1-p$.

We can also use a hierarchical representation of the model

$$y_i \sim \begin{cases} [y_i|\boldsymbol{\theta}_1]_1, & z_i = 1 \\ [y_i|\boldsymbol{\theta}_2]_2, & z_i = 0 \end{cases}, \qquad (10.6)$$

with latent indicator variables z_i for $i = 1,\ldots,n$ that act as binary "switches," turning on and off the mixture components as needed, where $z_i \sim \text{Bern}(p)$ and the mixture probability p play the same role in the model as in (10.3). These types of mixture model specifications are fundamental to many ecological models that we describe in this third part of the book.

The conditional likelihood corresponding to the mixture model formulation with latent binary variables can be written in two different ways. The first way is similar to the former likelihood in (10.5), but with z_i appearing multiplicatively as

$$[\mathbf{y}|\boldsymbol{\theta},\mathbf{z}] = \prod_{i=1}^{n} (z_i[y_i|\boldsymbol{\theta}_1]_1 + (1-z_i)[y_i|\boldsymbol{\theta}_2]_2). \qquad (10.7)$$

The second form for the conditional likelihood is equivalent but has the latent binary variables z_i in the exponent such that

$$[\mathbf{y}|\boldsymbol{\theta},\mathbf{z}] = \prod_{i=1}^{n} [y_i|\boldsymbol{\theta}_1]_1^{z_i} [y_i|\boldsymbol{\theta}_2]_2^{1-z_i}. \qquad (10.8)$$

To complete the Bayesian model, we specify a prior for the mixture probability p and parameter vectors $\boldsymbol{\theta}_1$ and $\boldsymbol{\theta}_2$. For now, we assume $\boldsymbol{\theta}_1$ and $\boldsymbol{\theta}_2$ are independent *a priori*, but leave their priors generic ($\theta_1 \sim [\theta_1]$ and $\theta_2 \sim [\theta_2]$). As we described in Chapter 6, there are many options for priors for probability parameters, but we use a beta distribution for the prior in this case: $p \sim \text{Beta}(\alpha, \beta)$.

The second form of likelihood in (10.8) makes finding full-conditional distributions for this mixture model with latent variables very straightforward. To see this, consider the posterior distribution expression

$$[\theta_1, \theta_2, \mathbf{z}, p|\mathbf{y}] \propto \left(\prod_{i=1}^{n} [y_i|\theta_1]_1^{z_i} [y_i|\theta_2]_2^{1-z_i} [z_i|p] \right) [p][\theta_1][\theta_2]. \tag{10.9}$$

To derive the full-conditional distributions for this model, we begin with z_i. As we will see, the full-conditional distribution for the mixture indicator variable z_i is conjugate Bernoulli, regardless of the data model mixture component distributions. We can find the full-conditional distribution for z_i as before, by retaining only the terms involving z_i from the joint distribution–the right-hand side of (10.9)–and finding the normalizing constant that makes it a valid PMF (because z_i is discrete). In our model, the full-conditional distribution for z_i is

$$[z_i|\cdot] \propto [y_i|\theta_1]_1^{z_i} [y_i|\theta_2]_2^{1-z_i} [z_i|p], \tag{10.10}$$

$$\propto [y_i|\theta_1]_1^{z_i} [y_i|\theta_2]_2^{1-z_i} p^{z_i} (1-p)^{1-z_i}, \tag{10.11}$$

$$\propto (p[y_i|\theta_1]_1)^{z_i} ((1-p)[y_i|\theta_2]_2)^{1-z_i}, \tag{10.12}$$

by combining terms with the same exponents. Thus, we can see that (10.12) results in a Bernoulli distribution for z_i because it has a term raised to the z_i power in a product with another term raised to the $1 - z_i$ power. However, these two terms ($p[y_i|\theta_1]_1$ and $(1-p)[y_i|\theta_2]_2$) are not guaranteed to sum to one, as required in the Bernoulli distribution. Thus, we normalize them to yield $[z_i|\cdot] \equiv \text{Bern}(\tilde{p}_i)$ for $i = 1, \ldots, n$, where

$$\tilde{p}_i = \frac{p[y_i|\theta_1]_1}{p[y_i|\theta_1]_1 + (1-p)[y_i|\theta_2]_2}. \tag{10.13}$$

In a MCMC algorithm, we can perform n independent Gibbs updates to sample z_i simultaneously because the full-conditional distribution for z_i does not depend on the other latent indicator variables, only on the parameter vectors θ_1, θ_2, and p.

The full-conditional distribution for p is trivial because we already derived it in Chapter 6. Because we specified a beta prior for p, the full-conditional is also a conjugate beta distribution. To see this, recall that we can write the full-conditional distribution for p as

$$[p|\cdot] \propto \left(\prod_{i=1}^{n} [z_i|p] \right) [p], \tag{10.14}$$

$$\propto \left(\prod_{i=1}^{n} p_i^{z}(1-p)^{1-z_i} \right) p^{\alpha-1}(1-p)^{\beta-1}, \tag{10.15}$$

$$\propto p^{\left(\Sigma_{i=1}^{n} z_i \right) + \alpha - 1} (1-p)^{\left(\Sigma_{i=1}^{n} (1-z_i) \right) + \beta - 1}, \tag{10.16}$$

which implies that $[p|\cdot] \equiv \text{Beta}((\Sigma_{i=1}^{n} z_i) + \alpha, (\Sigma_{i=1}^{n}(1 - z_i)) + \beta)$, as we saw in Chapter 6. Thus, in a MCMC algorithm, we can perform Gibbs updates for both z_i (for $i = 1, \ldots, n$) and p.

We cannot derive the full-conditional distributions for the parameter vectors $\boldsymbol{\theta}_1$ and $\boldsymbol{\theta}_2$ completely because we have not yet specified parametric models for the mixture component distributions $[y_i|\boldsymbol{\theta}_2]_1$ and $[y_i|\boldsymbol{\theta}_2]_2$, nor priors. However, we can write the general form for these full-conditional distributions as

$$[\boldsymbol{\theta}_1|\cdot] \propto \left(\prod_{i=1}^{n} [y_i|\boldsymbol{\theta}_1]_1^{z_i} \right) [\boldsymbol{\theta}_1], \tag{10.17}$$

$$\propto \left(\prod_{\forall z_i=1} [y_i|\boldsymbol{\theta}_1]_1 \right) [\boldsymbol{\theta}_1], \tag{10.18}$$

and

$$[\boldsymbol{\theta}_2|\cdot] \propto \left(\prod_{i=1}^{n} [y_i|\boldsymbol{\theta}_2]_2^{1-z_i} \right) [\boldsymbol{\theta}_2], \tag{10.19}$$

$$\propto \left(\prod_{\forall z_i=0} [y_i|\boldsymbol{\theta}_2]_2 \right) [\boldsymbol{\theta}_2]. \tag{10.20}$$

The full-conditional distributions for $\boldsymbol{\theta}_1$ and $\boldsymbol{\theta}_2$ may or may not be conjugate depending our specification of the mixture components and priors. In the case where they are non-conjugate, we could use M-H updates in the MCMC algorithm for those parameters.

Mixture distributions arise in nature. A classic example of mixture distributions in nature involves Galapagos finches. The mixture of traits in Galapagos finches is well-known in evolutionary biology (e.g., Hendry et al. 2006, 2009) and several scientific studies have characterized these trait distributions with mixture models. We use data from Bierema and Rudge (2014) to motivate the construction of a two-component mixture model for observed beak depths (i.e., millimeters from top to bottom of beak) of *Geospiza fortis* on Daphne and James Islands in the Galapagos (Figure 10.2).

We specified a two-component mixture model for finch beak depths (y_i for $i = 1, \ldots, n$) as

$$y_i \sim \begin{cases} N(\mu_1, \sigma_1^2), & z_i = 1 \\ N(\mu_2, \sigma_2^2), & z_i = 0 \end{cases}, \tag{10.21}$$

$$z_i \sim \text{Bern}(p), \tag{10.22}$$

$$p \sim \text{Beta}(\alpha, \beta), \tag{10.23}$$

$$\mu_1 \sim N(\mu_{0,1}, \sigma_{0,1}^2), \tag{10.24}$$

$$\mu_2 \sim N(\mu_{0,2}, \sigma_{0,2}^2), \tag{10.25}$$

$$\sigma_1 \sim \text{Unif}(0, u), \tag{10.26}$$

$$\sigma_2 \sim \text{Unif}(0, u). \tag{10.27}$$

Note the departure from our conjugate inverse gamma prior for σ_1^2 and σ_2^2 to a uniform prior for σ_1 and σ_2 instead. This change allows us to restrict the upper bound (u) of the component standard deviation according to our prior knowledge. Based on this mixture model for finch beak depths, we can write the following posterior distribution:

$$[\mathbf{z}, \mu_1, \mu_2, \sigma_1, \sigma_2, p | \mathbf{y}] \propto \left(\prod_{i=1}^{n} [y_i | \mu_1, \sigma_1^2]^{z_i} [y_i | \mu_2, \sigma_2^2]^{1-z_i} [z_i | p] \right) [\mu_1][\mu_2][\sigma_1][\sigma_2][p]. \tag{10.28}$$

To construct a MCMC algorithm to fit the two-component mixture model for finch beak depths, we need to find the full-conditional distributions for the model parameters. Using the results we derived earlier for the full-conditional distribution of z_i, we have $[z_i | \cdot] = \text{Bern}(\tilde{p}_i)$, where

$$\tilde{p}_i = \frac{p N(y_i | \mu_1, \sigma_1^2)}{p N(y_i | \mu_1, \sigma_1^2) + (1 - p) N(y_i | \mu_2, \sigma_2^2)}, \tag{10.29}$$

for $i = 1, \ldots, n$. Similarly, for p, we showed the full-conditional is

$$[p | \cdot] \equiv \text{Beta} \left(\sum_{i=1}^{n} z_i + \alpha, \sum_{i=1}^{n} (1 - z_i) + \beta \right), \tag{10.30}$$

resulting in another Gibbs update in our MCMC algorithm.

Earlier in this chapter, we could not derive the exact full-conditional distributions for remaining model parameters because we had not fully specified the model. Given the normal priors for μ_1 and μ_2 as well as the uniform priors for σ_1 and σ_2, we can now derive their full-conditional distributions. To find the full-conditional distribution for μ_1, we use the same technique that we have in the past, retaining the components of the joint distribution in (10.28) that involve μ_1 and then find the normalizing constant that makes it a proper probability distribution, or recognize it as a kernel of a known distribution. Using our generic result for the full-conditional distribution for μ_1 and our results from the normal full-conditional distributions in Chapters 7 and 9, we have

$$[\mu_1|\cdot] \propto \left(\prod_{\forall z_i=1} [y_i|\mu_1,\sigma_1^2] \right) [\mu_1], \tag{10.31}$$

$$\propto \exp\left(-\frac{1}{2\sigma_1^2} \sum_{\forall z_i=1} (y_i-\mu_1)^2 \right) \exp\left(-\frac{1}{2\sigma_{0,1}^2} (\mu_1-\mu_{0,1})^2 \right), \tag{10.32}$$

$$\propto \exp\left(-\frac{1}{2} \left(-\frac{2\sum_{\forall z_i=1} y_i}{\sigma_1^2}\mu_1 + \frac{n_1}{\sigma_1^2}\mu_1^2 - \frac{2\mu_{0,1}}{\sigma_{0,1}^2}\mu_1 + \frac{1}{\sigma_{0,1}^2}\mu_1^2 \right) \right), \tag{10.33}$$

$$\propto \exp\left(-\frac{1}{2} \left(-2\left(\frac{\sum_{\forall z_i=1} y_i}{\sigma_1^2} + \frac{\mu_{0,1}}{\sigma_{0,1}^2} \right)\mu_1 + \left(\frac{n_1}{\sigma_1^2} + \frac{1}{\sigma_{0,1}^2} \right)\mu_1^2 \right) \right), \tag{10.34}$$

which has the form of a Gaussian kernel as we showed in Chapter 7. Thus, the full-conditional distribution for μ_1 is conjugate normal such that $[\mu_1|\cdot] \equiv N(a^{-1}b, a^{-1})$ with $a \equiv \frac{n_1}{\sigma_1^2} + \frac{1}{\sigma_{0,1}^2}$ and $b \equiv \frac{\sum_{\forall z_i=1} y_i}{\sigma_1^2} + \frac{\mu_{0,1}}{\sigma_{0,1}^2}$, and where $n_1 \equiv \sum_{i=1}^n z_i$.

The full-conditional distribution for μ_2 can be derived in the same way we derived the full-conditional distribution for μ_1, which results in $[\mu_2|\cdot] \equiv N(a^{-1}b, a^{-1})$ with $a \equiv \frac{n_0}{\sigma_2^2} + \frac{1}{\sigma_{0,2}^2}$ and $b \equiv \frac{\sum_{\forall z_i=0} y_i}{\sigma_2^2} + \frac{\mu_{0,2}}{\sigma_{0,2}^2}$, and where $n_0 \equiv \sum_{i=1}^n (1-z_i)$.

In contrast to μ_1 and μ_2, we use M-H to update σ_1 and σ_2, but we still need to know what to put in the numerator and denominator of the M-H ratio. The full-conditional distribution for σ_1 is

$$[\sigma_1|\cdot] \propto \left(\prod_{\forall z_i=1} [y_i|\mu_1,\sigma_1] \right) [\sigma_1], \tag{10.35}$$

$$\propto \left(\prod_{\forall z_i=1} N(y_i|\mu_1,\sigma_1^2) \right) 1_{\{0<\sigma_1<u\}}. \tag{10.36}$$

Note that we could omit additional terms (e.g., $\sqrt{2\pi}$), but they will cancel in the M-H ratio, so as long as they are not computationally expensive to compute, we can leave them in.

Similarly, the full-conditional distribution for σ_2 is

$$[\sigma_2|\cdot] \propto \left(\prod_{\forall z_i=0} N(y_i|\mu_2,\sigma_2^2) \right) 1_{\{0<\sigma_2<u\}}. \tag{10.37}$$

For a proposal distribution for σ_1 and σ_2, we could use the prior, but it may be inefficient if u is large. Alternatively, we could use the random walk normal proposal with a rejection sampler for proposals outside the support $(0,u)$ as described in Chapter 6.

We coded a MCMC algorithm called `mixnorm.mcmc.R` in R based on the Gibbs and M-H updates just described. This algorithm follows (Box 10.2):

BOX 10.2   MCMC ALGORITHM FOR MIXTURE NORMAL MODEL

```
mixnorm.mcmc <- function(y,n.mcmc,s.tune=0.5){

#####
#####  Setup Variables
#####

n=length(y)
mu.mat=matrix(0,n.mcmc,2)
s.mat=matrix(0,n.mcmc,2)
p.vec=rep(0,n.mcmc)
z.mean=rep(0,n)
n.burn=round(.2*n.mcmc)
y.sim=rep(0,n.mcmc)

#####
#####  Priors
#####

alpha=1
beta=1
mu01=14
mu02=18
s01=2
s02=2
u=2 # upper bound on sigma

#####
#####  Starting Values
#####

p=.5
tmp.mn=mean(y)
z=rep(0,n)
z[y<tmp.mn]=1
mu1=mu01
mu2=mu02
s1=1
s2=1
```

(Continued)

BOX 10.2 (Continued)   MCMC ALGORITHM FOR MIXTURE NORMAL MODEL

```
39 s12=s1^2
40 s22=s2^2
41
42 #####
43 #####  Begin MCMC Loop
44 #####
45
46 for(k in 1:n.mcmc){
47
48   #####
49   #####  Sample z
50   #####
51
52   p.tmp=p*dnorm(y,mu1,s1)/(p*dnorm(y,mu1,s1)+(1-p)
53         *dnorm(y,mu2,s2))
54   z=rbinom(n,1,p.tmp)
55
56   #####
57   #####  Sample mu1
58   #####
59
60   s2.tmp=(sum(z==1)/s12 + 1/s01)^(-1)
61   mu.tmp=s2.tmp*(sum(y[z==1])/s12 + mu01/s01)
62   mu1=rnorm(1,mu.tmp,sqrt(s2.tmp))
63
64   #####
65   #####  Sample mu2
66   #####
67
68   s2.tmp=(sum(z==0)/s22 + 1/s02)^(-1)
69   mu.tmp=s2.tmp*(sum(y[z==0])/s22+mu02/s02)
70   mu2=rnorm(1,mu.tmp,sqrt(s2.tmp))
71
72   #####
73   #####  Sample s1
74   #####
```

(Continued)

BOX 10.2 (Continued)   MCMC ALGORITHM FOR MIXTURE NORMAL MODEL

```
75  s1.star=rnorm(1,s1,s.tune)
76  if((s1.star > 0) & (s1.star < u)){
77    mh1=sum(z*dnorm(y,mu1,s1.star,log=TRUE))
78    mh2=sum(z*dnorm(y,mu1,s1,log=TRUE))
79    mh=exp(mh1-mh2)
80    if(mh > runif(1)){
81      s1=s1.star
82      s12=s1^2
83    }
84  }
85
86  #####
87  ##### Sample s2
88  #####
89
90  s2.star=rnorm(1,s2,s.tune)
91  if((s2.star > 0) & (s2.star < u)){
92    mh1=sum((1-z)*dnorm(y,mu2,s2.star,log=TRUE))
93    mh2=sum((1-z)*dnorm(y,mu2,s2,log=TRUE))
94    mh=exp(mh1-mh2)
95    if(mh > runif(1)){
96      s2=s2.star
97      s22=s2^2
98    }
99  }
100
101 #####
102 ##### Sample p
103 #####
104
105 p=rbeta(1,sum(z)+alpha,sum(1-z)+beta)
106
107 #####
108 ##### Save Samples
109 #####
110
111 mu.mat[k,]=c(mu1,mu2)
112 s.mat[k,]=c(s1,s2)
113 p.vec[k]=p
```

(Continued)

BOX 10.2 (Continued)   MCMC ALGORITHM FOR MIXTURE NORMAL MODEL

```
114    if(k>n.burn){
115       z.mean=z.mean+z/(n.mcmc-n.burn)
116    }
117 }
118
119 #####
120 #####   Write Output
121 #####
122
123 list(mu.mat=mu.mat,p.vec=p.vec,s.mat=s.mat,
124       z.mean=z.mean,n.burn=n.burn)
125
126 }
```

One benefit of including the latent indicator variables (z_i) in the mixture model is that we can make inference using them. On line 115 of this algorithm, notice that we calculate a running average of the z_i latent variables for $i = 1,\ldots,n$ (for the post burn-in portion of the MCMC sample).[2] This running average (i.e., z.mean) represents the posterior mean $E(z_i|\mathbf{y})$ and is a posterior probability of membership of each observation belonging to mixture component 1.

We specified hyperparameters for our two-component mixture model as $\mu_{0,1} = 14$, $\mu_{0,2} = 18$, $\sigma_{0,1}^2 = 2$, $\sigma_{0,2}^2 = 2$, $u = 2$, $\alpha = 1$, and $\beta = 1$. To fit the two-component mixture model to the finch beak depth data, we used the following R code (Box 10.3):

BOX 10.3   FIT MIXTURE MODEL TO FINCH BEAK DEPTH DATA

```
1 y=c(12.3,12.8,13.5,13.6,13.7,13.9,13.9,13.9,14.3,
2    14.4,14.5,14.7,14.7,15.5,15.8,15.8,15.8,16.7,
3    16.7,17.1,17.4,18.0,18.3,18.7)
4 n=length(y)
```

(Continued)

---

[2] In this case, and for many MCMC algorithms hereafter, we hard code a burn-in period based on a trial run of the algorithm to automatically calculate posterior quantities of interest such as posterior means of latent processes or derived quantities such as deviance and DIC. In practice, we may need to adjust this burn-in period for certain models and data sets.

> **BOX 10.3 (Continued)   FIT MIXTURE MODEL TO FINCH BEAK**
> **DEPTH DATA**

```
5  source("mixnorm.mcmc.R")
6  n.mcmc=20000
7  set.seed(1)
8  mcmc.out=mixnorm.mcmc(y=y,n.mcmc=n.mcmc)
9
10 n.burn=mcmc.out$n.burn
11 mu.mn=apply(mcmc.out$mu.mat[n.burn:n.mcmc,],
12     2,mean)
13 s.mn=apply(mcmc.out$s.mat[n.burn:n.mcmc,],2,mean)
14 p.mn=mean(mcmc.out$p.vec[n.burn:n.mcmc])
15
16 layout(matrix(1:2,2,1))
17 hist(y,breaks=20,border=0,prob=TRUE)
18 lines(density(mcmc.out$mu.mat[n.burn:n.mcmc,1],
19     from=10,to=20,adjust=5),col=8,lty=1)
20 lines(density(mcmc.out$mu.mat[n.burn:n.mcmc,2],
21     from=10,to=20,adjust=5),col=8,lty=2)
22 curve(p.mn*dnorm(x,mu.mn[1],s.mn[1])+(1-p.mn)
23     *dnorm(x,mu.mn[2],s.mn[2]),col=1,from=10,
24     to=20,add=TRUE)
25 rug(y,lwd=3)
26 plot(y,mcmc.out$z.mean,type="o",ylab="E(z|y)",
27     main="b")
```

The R code above results in Figure 10.2a displaying the two mixture components (gray lines) evaluated using a kernel density estimate based on their MCMC sample and the overall mixture distribution (black line) evaluated at the posterior mean of the parameters. We show the observed finch beak depths as black dashes at the bottom of the plot in Figure 10.2a.

In Figure 10.2a, the mixture components appear to be mostly separate, but any bi-modality in the overall mixture distribution is less pronounced. However, the mixture distribution is skewed right and covers the observed data well.

Returning now to $E(z_i|\mathbf{y})$, based on the fit of the model to the finch beak depth data, we show the posterior membership probabilities in Figure 10.2b as they correspond to the observed data (y_i). Notice the sigmoidal shape to the relationship in Figure 10.2b, which indicates that the smaller bird beak depths belong to group 1 (the solid gray line in Figure 10.2a) with higher posterior probability while the larger bird beak depths correspond group 2 (the dashed gray line in Figure 10.2a) with higher posterior probability.

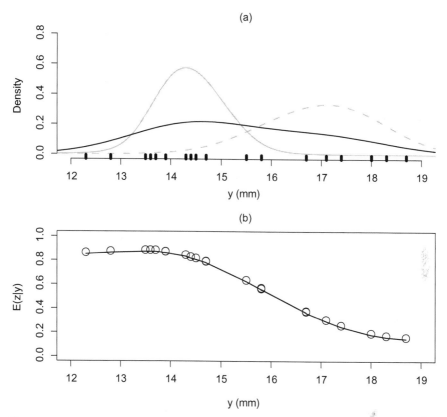

Figure 10.2 (a) Posterior mixture distribution (solid black line) for *G. fortis* based on component distributions shown in the gray solid (group 1) and dashed (group 2) lines. Observed beak depths shown as black dashes at the bottom. (b) Posterior membership probabilities ($E(z_i|\mathbf{y})$) associated with each of the observed data points (y_i).

10.1 ADDITIONAL CONCEPTS AND READING

Mixture models are a parametric form of clustering. We build mixture models to cluster data and processes into groups, or at least to account for the group structure that results in similar observations. As we mentioned at the beginning of this chapter, mixture models serve as the foundation for a large suite of models used in ecology and biology (especially wildlife biology). These include, zero-inflated models (Chapter 22), occupancy models (Chapter 23), and capture-recapture models (Chapter 24), among others. In these types of models, we specify the mixture aspect of the model to capture clustering in the data that is due to measurement error.

In Chapter 16, we describe temporally explicit models for animal movement trajectories based on telemetry data (i.e., observed locations of individual animals in geographic space). Depending on the type of sensor used to collect the telemetry data, they may be subject to measurement error that is a result of a mixture process due to the position of the sensor relative to the individual. Thus, Brost et al. (2015) and Buderman et al. (2016) developed mixture models to accommodate the different types of measurement error present in certain types of observed telemetry data.

Another type of mixture model is an infinite mixture model induced by a Dirichlet process (e.g., Neal 2000; Teh et al. 2006). These types of models consider the number of mixture components as infinite and are often used to represent complicated distributions for data that have no other well-known parametric form. Thus, we often refer to the field associated with Dirichlet process modeling as nonparametric Bayes (even though that is a misnomer). Gelman et al. (2013) provided a good overview of Dirichlet processes and how we can use them in Bayesian statistical models.

Dorazio et al. (2008), Johnson et al. (2013c), Brost et al. (2017), and Johnson and Sinclair (2017) provided good examples of applications of Dirichlet process models to environmental and ecological data. Specifically, Dorazio et al. (2008) used a Dirichlet process to represent the distribution of a latent set of population intensities (and therefore abundances) resulting in a flexible model that allows for natural clustering in the set of population abundances. Similarly, Johnson et al. (2013c) used a Dirichlet process to account for similarities among Northern fur seal (*Callorhinus ursinus*) population trends at a set of sites in the Pribilof Islands, Alaska, whereas Johnson and Sinclair (2017) showed how to cluster species into guilds more generally. Brost et al. (2017) presented a method that uses Dirichlet processes to identify the location of central places (e.g., nests, dens, haul outs, and rookeries) that are used repetitively by animals based on telemetry data.

11 Linear Regression

In the preceding chapters, we focused on single parameter models and multiparameter models with univariate independent priors. Linear regression presents a useful opportunity to learn about both multivariate priors and full-conditional distributions as well as multivariate (or "block") updates in MCMC.

In a first university course on regression, the multiple linear regression model is often presented as

$$y_i = \beta_0 + \beta_1 x_{1,i} + \cdots + \beta_{p-1} x_{p-1,i} + \varepsilon_i, \tag{11.1}$$

for $i = 1, \ldots, n$ and where ε_i represents independent, mean zero error with variance σ^2. Here, the $x_{j,i}$ represent $p - 1$ predictor variables (often refer to as "covariates") and $\boldsymbol{\beta} \equiv (\beta_0, \beta_1, \ldots, \beta_{p-1})'$ represents an intercept (β_0) and set of regression coefficients that linearly relate the predictor variables to the observed response data y_i.

Although it is common to write linear regression models as (11.1), we usually use the additional parametric assumption of normality for the errors in the Bayesian setting. Thus, we prefer to write the model such that the data arise directly from a named probability distribution,

$$y_i \sim N(\beta_0 + \beta_1 x_{1,i} + \cdots + \beta_{p-1} x_{p-1,i}, \sigma^2), \tag{11.2}$$

which allows us to avoid using the ε error notation altogether. The error notation may confuse some readers who are learning about statistical modeling for the first time because they may think that all statistical models have independent and additive error. While, in reality, those types of errors usually only pertain to Gaussian parametric models. Using the explicit named probability distribution as the data model reduces confusion.

It is also a good idea to use vector notation for regression models because it simplifies mathematical expressions. Thus, using linear algebra, the regression model

$$y_i \sim N(\mathbf{x}_i' \boldsymbol{\beta}, \sigma^2), \tag{11.3}$$

for $i = 1, \ldots, n$, where $\mathbf{x}_i \equiv (x_{0,i}, \ldots, x_{p-1,i})'$ and $\boldsymbol{\beta}$ are $p \times 1$ vectors (with $x_{0,i} = 1$ for $i = 1, \ldots, n$). Furthermore, we can write the full model jointly over all observations using vector/matrix notation as

$$\mathbf{y} \sim N(\mathbf{X}\boldsymbol{\beta}, \sigma^2 \mathbf{I}). \tag{11.4}$$

We refer to the $n \times p$ matrix $\mathbf{X}$ as the "design" matrix and it has $\mathbf{x}_i'$, for $i = 1, \ldots, n$, as rows. We refer to the $n \times n$ matrix $\mathbf{I}$ as the identity matrix and it has ones on the diagonal and zeros elsewhere. The data model represented in (11.4) is the multivariate normal distribution[1]; its expected value is the mean vector $\mathbf{X}\boldsymbol{\beta}$ and covariance matrix $\sigma^2\mathbf{I}$. In this particular case, with our assumption of conditional independence in the data model (resulting in the diagonal covariance matrix), the model expressed jointly as (11.4) has the same likelihood as implied by (11.2) and (11.3):

$$[\mathbf{y}|\boldsymbol{\beta},\sigma^2] \equiv \prod_{i=1}^{n} \frac{1}{\sqrt{2\pi\sigma^2}} \exp\left(-\frac{1}{2\sigma^2}(y_i - \mathbf{x}_i'\boldsymbol{\beta})^2\right). \tag{11.5}$$

We will generalize the model in (11.4) in later chapters to accommodate additional dependence in the data beyond that provided by the mean. In that case, the likelihood will take the form of a general multivariate normal distribution.[2]

To complete the Bayesian version of the linear regression model, we need priors for the model parameters. For the variance parameter σ^2, we can use any of the priors we presented for σ^2 in earlier chapters. In fact, if we use an inverse gamma prior for σ^2, it results in a conjugate inverse gamma full-conditional distribution for σ^2. To see this, we first write the posterior distribution for $\boldsymbol{\beta}$ and σ^2 as

$$[\boldsymbol{\beta},\sigma^2|\mathbf{y}] \propto [\mathbf{y}|\boldsymbol{\beta},\sigma^2][\boldsymbol{\beta}][\sigma^2]. \tag{11.6}$$

The full-conditional distribution for σ^2 then is comprised of the terms in the joint distribution (i.e., the right-hand side of (11.6)) that involve σ^2. Thus, similar to our derivation in Chapter 8, the full-conditional distribution for σ^2 in our multiple linear regression model is

$$[\sigma^2|\cdot] \propto [\mathbf{y}|\boldsymbol{\beta},\sigma^2][\sigma^2], \tag{11.7}$$

$$\propto \left(\prod_{i=1}^{n} \frac{1}{\sqrt{2\pi\sigma^2}} \exp\left(-\frac{1}{2\sigma^2}(y_i - \mathbf{x}_i'\boldsymbol{\beta})^2\right)\right) \frac{1}{r^q\Gamma(q)}(\sigma^2)^{-(q+1)} \times$$

$$\exp\left(-\frac{1}{r\sigma^2}\right), \tag{11.8}$$

$$\propto (\sigma^2)^{-\frac{n}{2}} \exp\left(-\frac{1}{2\sigma^2} \sum_{i=1}^{n}(y_i - \mathbf{x}_i'\boldsymbol{\beta})^2\right)(\sigma^2)^{-(q+1)} \exp\left(-\frac{1}{r\sigma^2}\right), \tag{11.9}$$

$$\propto (\sigma^2)^{-(\frac{n}{2}+q+1)} \exp\left(-\frac{1}{\sigma^2}\left(\frac{\sum_{i=1}^{n}(y_i - \mathbf{x}_i'\boldsymbol{\beta})^2}{2} + \frac{1}{r}\right)\right). \tag{11.10}$$

Thus, letting $\tilde{q} = \frac{n}{2} + q$ and $\tilde{r} = \left(\frac{\sum_{i=1}^{n}(y_i-\mathbf{x}_i'\boldsymbol{\beta})^2}{2} + \frac{1}{r}\right)^{-1}$, the full-conditional distribution is: $[\sigma^2|\cdot] \equiv \text{IG}(\tilde{q},\tilde{r})$.

---

[1] Also sometimes written as MVN.

[2] The general multivariate normal likelihood for a regression model specified such that $\mathbf{y} \sim \text{N}(\mathbf{X}\boldsymbol{\beta}, \boldsymbol{\Sigma})$ is $|2\pi\boldsymbol{\Sigma}|^{-\frac{1}{2}} \exp\left(-\frac{1}{2}(\mathbf{y} - \mathbf{X}\boldsymbol{\beta})'\boldsymbol{\Sigma}^{-1}(\mathbf{y} - \mathbf{X}\boldsymbol{\beta})\right)$.

In terms of a prior for the regression coefficients $\boldsymbol{\beta}$, we could use independent univariate normal distributions for each β_j for $j = 0, \ldots, p - 1$ or we could use a multivariate normal prior distribution for all the coefficients in $\boldsymbol{\beta}$ simultaneously: $\boldsymbol{\beta} \sim N(\boldsymbol{\mu}_\beta, \boldsymbol{\Sigma}_\beta)$, where $\boldsymbol{\mu}_\beta$ is the $p \times 1$ prior mean vector and $\boldsymbol{\Sigma}_\beta$ is the $p \times p$ prior covariance matrix. If we use a diagonal prior covariance matrix, then the multivariate joint prior is equivalent to individually specified independent normal priors.

The most common joint prior specification for $\boldsymbol{\beta}$ involves mean vector $\boldsymbol{\mu}_\beta \equiv \mathbf{0}$ and covariance matrix $\boldsymbol{\Sigma}_\beta \equiv \sigma_\beta^2 \mathbf{I}$. In fact, this particular prior has strong ties to a regularization approach referred to as ridge regression. We can use ridge regression to induce shrinkage (to zero) in the parameters, in effect, performing a type of model selection while also controlling for collinearity (Hoerl and Kennard 1970; Hooten and Hobbs 2015). We discuss this concept of regularization in Chapter 14.

Another specification for the prior covariance matrix of $\boldsymbol{\beta}$ could be $\boldsymbol{\Sigma}_\beta \equiv \sigma_\beta^2 (\mathbf{X}'\mathbf{X})^{-1}$ (Zellner 1986). This covariance matrix induces *a priori* dependence among the regression coefficients that is similar to what we would expect to result from a non-Bayesian model fit. To see this, consider the least squares estimate of the coefficient vector $\hat{\boldsymbol{\beta}} = (\mathbf{X}'\mathbf{X})^{-1}\mathbf{X}'\mathbf{y}$. The covariance of $\hat{\boldsymbol{\beta}}$ is $\sigma^2(\mathbf{X}'\mathbf{X})^{-1}$. Thus, the traditional covariance of the regression coefficient estimate is proportional to $(\mathbf{X}'\mathbf{X})^{-1}$, and does not include the response data so we can use it to help inform the dependence structure of $\boldsymbol{\beta}$ in our Bayesian model.

Regardless of how we specify the prior covariance for $\boldsymbol{\beta}$, we can find the full-conditional distribution for it using a similar derivation as we used in Chapter 7, but now we use for a multivariate random vector. The full-conditional distribution for $\boldsymbol{\beta}$ is

$$[\boldsymbol{\beta}|\cdot] \propto [\mathbf{y}|\boldsymbol{\beta}, \sigma^2][\boldsymbol{\beta}], \tag{11.11}$$

$$\propto \left(\prod_{i=1}^{n} \frac{1}{\sqrt{2\pi\sigma^2}} \exp\left(-\frac{1}{2\sigma^2}(y_i - \mathbf{x}_i'\boldsymbol{\beta})^2 \right) \right) |2\pi\boldsymbol{\Sigma}_\beta|^{-\frac{1}{2}} \times$$

$$\exp\left(-\frac{1}{2}(\boldsymbol{\beta} - \boldsymbol{\mu}_\beta)'\boldsymbol{\Sigma}_\beta^{-1}(\boldsymbol{\beta} - \boldsymbol{\mu}_\beta) \right), \tag{11.12}$$

$$\propto \exp\left(-\frac{1}{2\sigma^2}\sum_{i=1}^{n}(y_i - \mathbf{x}_i'\boldsymbol{\beta})^2 \right) \exp\left(-\frac{1}{2}(\boldsymbol{\beta} - \boldsymbol{\mu}_\beta)'\boldsymbol{\Sigma}_\beta^{-1}(\boldsymbol{\beta} - \boldsymbol{\mu}_\beta) \right), \tag{11.13}$$

$$\propto \exp\left(-\frac{1}{2}\left(-2\left(\frac{\sum_{i=1}^{n} y_i\mathbf{x}_i'}{\sigma^2} \right)\boldsymbol{\beta} + \boldsymbol{\beta}'\left(\frac{\sum_{i=1}^{n} \mathbf{x}_i\mathbf{x}_i'}{\sigma^2} \right)\boldsymbol{\beta} \right) \right) \times$$

$$\exp\left(-\frac{1}{2}\left(-2(\boldsymbol{\mu}_\beta'\boldsymbol{\Sigma}_\beta^{-1})\boldsymbol{\beta} + \boldsymbol{\beta}'\boldsymbol{\Sigma}_\beta^{-1}\boldsymbol{\beta} \right) \right), \tag{11.14}$$

$$\propto \exp\left(-\frac{1}{2}\left(-2\left(\frac{\sum_{i=1}^{n} y_i\mathbf{x}_i'}{\sigma^2} + \boldsymbol{\mu}_\beta'\boldsymbol{\Sigma}_\beta^{-1} \right)\boldsymbol{\beta} + \boldsymbol{\beta}'\left(\frac{\sum_{i=1}^{n} \mathbf{x}_i\mathbf{x}_i'}{\sigma^2} + \boldsymbol{\Sigma}_\beta^{-1} \right)\boldsymbol{\beta} \right) \right), \tag{11.15}$$

$$\propto \exp\left(-\frac{1}{2}\left(-2\mathbf{b}'\boldsymbol{\beta} + \boldsymbol{\beta}'\mathbf{A}\boldsymbol{\beta} \right) \right), \tag{11.16}$$

which has a similar form as the univariate full-conditional distribution for the mean in the Normal-Normal model from Chapter 7. In this case, using full vector/matrix notation, we can write

$$\mathbf{b} \equiv \mathbf{X}'(\sigma^2 \mathbf{I})^{-1} \mathbf{y} + \boldsymbol{\Sigma}_{\beta}^{-1} \boldsymbol{\mu}_{\beta}, \tag{11.17}$$

$$\mathbf{A} \equiv \mathbf{X}'(\sigma^2 \mathbf{I})^{-1} \mathbf{X} + \boldsymbol{\Sigma}_{\beta}^{-1}. \tag{11.18}$$

Thus, the full-conditional distribution for $\boldsymbol{\beta}$ is $[\boldsymbol{\beta}|\cdot] \equiv N(\mathbf{A}^{-1}\mathbf{b}, \mathbf{A}^{-1})$ using a similar trick as shown in Chapter 7 to complete the square (except with linear algebra).

The fact that the full-conditional distributions for both σ^2 and $\boldsymbol{\beta}$ are conjugate implies that we can easily construct a fully Gibbs MCMC algorithm (meaning Gibbs updates for all parameters). We just have to be sure to sample multivariate normal random vectors for $\boldsymbol{\beta}$ because the regression coefficients may be correlated due to possible collinearity among the covariates.

We wrote the following MCMC algorithm to fit the Bayesian multiple linear regression model in R based on the same basic structure of sampling sequentially from the full-conditional distributions that we introduced in Chapter 4. We named the algorithm norm.reg.mcmc.R as a function in R:

BOX 11.1   MCMC ALGORITHM FOR LINEAR REGRESSION

```
norm.reg.mcmc <- function(y,X,beta.mn,beta.var,
    s2.mn,s2.sd,n.mcmc){

###
### Subroutines
###

library(mvtnorm)

invgammastrt <- function(igmn,igvar){
  q <- 2+(igmn^2)/igvar
  r <- 1/(igmn*(q-1))
  list(r=r,q=q)
}

###
### Variables and Priors
###

n=dim(X)[1]
p=dim(X)[2]
```

(Continued)

BOX 11.1 (Continued)   MCMC ALGORITHM FOR LINEAR REGRESSION

```
22  r=invgammastrt(s2.mn,s2.sd^2)$r
23  q=invgammastrt(s2.mn,s2.sd^2)$q
24  Sig.beta.inv=diag(p)/beta.var
25
26  beta.save=matrix(0,p,n.mcmc)
27  s2.save=rep(0,n.mcmc)
28
29  ###
30  ### Starting Values
31  ###
32
33  beta=solve(t(X)%*%X)%*%t(X)%*%y
34
35  ###
36  ### MCMC Loop
37  ###
38
39  for(k in 1:n.mcmc){
40
41    ###
42    ### Sample s2
43    ###
44
45    tmp.r=(1/r+.5*t(y-X%*%beta)%*%(y-X%*%beta))^(-1)
46    tmp.q=n/2+q
47    s2=1/rgamma(1,tmp.q,,tmp.r)
48
49    ###
50    ### Sample beta
51    ###
52
53    tmp.var=solve(t(X)%*%X/s2 + Sig.beta.inv)
54    tmp.mn=tmp.var%*%(t(X)%*%y/s2
55       + Sig.beta.inv%*%beta.mn)
56    beta=as.vector(rmvnorm(1,tmp.mn,tmp.var,
57       method="chol"))
```

(*Continued*)

BOX 11.1 (Continued)   MCMC ALGORITHM FOR LINEAR REGRESSION

```
58  ###
59  ### Save Samples
60  ###
61
62      beta.save[,k]=beta
63      s2.save[k]=s2
64  }
65
66  ###
67  ###  Write Output
68  ###
69
70  list(beta.save=beta.save,s2.save=s2.save,y=y,X=X,
71      n.mcmc=n.mcmc,n=n,r=r,q=q,p=p)
72
73  }
```

There are several important features of this MCMC algorithm. First, on line 8 of the algorithm, we call the package `mvtnorm`. The `mvtnorm` package contains the function `rmvnorm` which allows us to sample multivariate normal random vectors. There are many other ways to sample multivariate normal random vectors, and some are much faster than `rmvnorm`, but for now, this is the most straightforward. Second, note that we define the `inversegammastrt` function inside the algorithm so that we can moment match the parameters of the inverse gamma distribution with hyperparameters specified as mean and variance (Hobbs and Hooten 2015). Third, we perform the Gibbs update for σ^2 as in Chapters 8 and 9, but the Gibbs update for $\boldsymbol{\beta}$ is slightly different than the univariate case we presented in Chapters 7 and 9. In this case, we can use linear algebra functions (e.g., `%*%` and `solve()`) to calculate the appropriate full-conditional mean (`tmp.mn`) and covariance matrix (`tmp.var`). Linear algebra functions in R allow us to avoid using loops and are very efficient for multivariate calculations, especially when a set of fast "basic linear algebra subprograms" (BLAS) are installed on your machine that is optimized to perform vector and matrix calculations.[3]

To demonstrate the MCMC algorithm for Bayesian multiple linear regression, we consider the white-tailed deer (*Odocoileus virginianus*) data set described by Hefley et al. (2013). Using a subset of hunter-harvest data ($n = 200$) on white-tailed deer from 2006, we used linear regression to infer the effects of an intercept $x_{0,i} = 1$ and

---

[3] We can install a fast BLAS with relative ease on most computers. For example, 'openBLAS' works well with Intel processors.

several covariates (i.e., $x_{1,i}$ = indicator of large cropland (large: 614 ha vs. small: 599 ha) in the sample, $x_{2,i}$ = sex (male = 1), $x_{3,i}$ = age in years, and $x_{4,i}$ = age squared on the body mass of individual deer (y_i, in kilograms). For hyperparameters, we used $\boldsymbol{\mu}_\beta = \mathbf{0}$, $\boldsymbol{\Sigma}_\beta = 10000\mathbf{I}$, $E(\sigma^2) = 50$, and $\text{Var}(\sigma^2) = 1000$. Using the data in deer.RData, we ran the algorithm for $K = 10000$ MCMC iterations using the following R code:

BOX 11.2   FIT LINEAR REGRESSION MODEL TO DEER DATA

```
 1 load("deer.RData")
 2 n=dim(deer.df)[1]
 3 y=deer.df$y
 4 X=cbind(rep(1,n),as.matrix(deer.df[,-1]))
 5 X[,2]=ifelse(X[,2]==599,0,1)
 6 p=dim(X)[2]
 7
 8 source("norm.reg.mcmc.R")
 9 set.seed(1)
10 mcmc.out=norm.reg.mcmc(y=y,X=X,beta.mn=rep(0,p),
11     beta.var=10000,s2.mn=50,s2.sd=1000,n.mcmc
        =10000)
12
13
14 dIG <- function(x,igmn,igvar){
15   q <- 2+(igmn^2)/igvar
16   r <- 1/(igmn*(q-1))
17   x^(-q-1) * exp(-1/r/x) / (r^q) / gamma(q)
18 }
19
20 layout(matrix(1:6,3,2))
21 hist(mcmc.out$beta.save[1,],prob=TRUE,main="a",
22     xlab=bquote(beta[0]))
23 curve(dnorm(x,0,100),add=TRUE)
24 hist(mcmc.out$beta.save[2,],prob=TRUE,main="b",
25     xlab=bquote(beta[1]))
26 curve(dnorm(x,0,100),add=TRUE)
27 hist(mcmc.out$beta.save[3,],prob=TRUE,main="c",
28     xlab=bquote(beta[2]))
29 curve(dnorm(x,0,100),add=TRUE)
30 hist(mcmc.out$beta.save[4,],prob=TRUE,main="d",
31     xlab=bquote(beta[3]))
```

(Continued)

BOX 11.2 (Continued)   FIT LINEAR REGRESSION MODEL TO DEER DATA

```
32  curve(dnorm(x,0,100),add=TRUE)
33  hist(mcmc.out$beta.save[5,],prob=TRUE,main="e",
34       xlab=bquote(beta[4]))
35  curve(dnorm(x,0,100),add=TRUE)
36  hist(mcmc.out$s2.save,prob=TRUE,main="f",xlab
37       =bquote(sigma^2))
38  curve(dIG(x,50,1000),add=TRUE)
```

The R code above produces the marginal (i.e., parameter-wise) posterior histograms in Figure 11.1. Note that, in the R code, we defined the function dIG, which is the density function of the inverse gamma distribution based on the mean and variance.

From Figure 11.1, we can immediately assess the magnitude of each coefficient and to what extent the posterior distribution may cover zero. In the regression context, coefficients that substantially overlap zero may not be useful in the model to explain variation in the response. For example, in Figure 11.1b, we can see that β_1 overlaps zero (95% credible interval: $(-0.65, 2.15)$) which implies that we were not able to detect an influence of the large cropland indicator on white-tailed deer body mass, given the other covariates in the model. However, the coefficient posterior distributions associated with sex, age, and age squared indicate a strong relationship between the covariates and response because they do not overlap zero (Figure 11.1c–e). In this case, the data and model provide evidence that male white-tailed deer and older deer have greater body mass. However, the posterior for β_4 (Figure 11.1e) is strongly negative, implying a quadratic shape in the relationship between age and body mass where body mass is greatest near middle ages and least for young and very old deer.

The fact that we are able to make inference based on marginal posterior distributions so easily using MCMC output is a feature known as the marginalization property of sampling-based methods (Hobbs and Hooten 2015). That is, to learn about the posterior distribution of a given parameter (or subset of parameters), we simply use the MCMC sample associated with that parameter to calculate posterior means, variances, credible intervals, histograms, and so on, while ignoring the MCMC sample for the other parameters.

Similarly, as we showed in Chapter 3, we can also use the equivariance property of MCMC to make inference on functions of model parameters or "derived quantities" (Hobbs and Hooten 2015). In analysis of variance (ANOVA) settings using the linear model framework, one type of derived quantity involving multiple parameters is called a "contrast." Following examples of contrasts related to our model are the

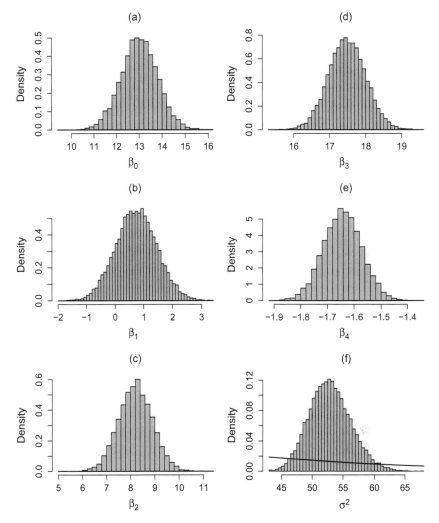

Figure 11.1 Marginal posterior histograms for (a) β_0 (intercept), (b) β_1 (large cropland), (c) β_2 (sex), (d) β_3 (age), (e) β_4 (age^2), and (f) σ^2. Prior PDFs shown as black lines.

derived quantities $\bar{\mathbf{x}}_0'\boldsymbol{\beta}$ and $\bar{\mathbf{x}}_1'\boldsymbol{\beta}$ (where $\bar{\mathbf{x}}_0' = (1,1,0,\bar{x}_3,\bar{x}_4)'$ and $\bar{\mathbf{x}}_1' = (1,1,1,\bar{x}_3,\bar{x}_4)'$). In this case, the derived quantities are sometimes also referred to as linear predictors. The interpretation of these two quantities is the average body mass in kilograms (kg) of white-tailed deer females and males in large cropland regions, holding all other covariates at their mean. To obtain the 95% credible interval for the two derived quantities we used the R function:

BOX 11.3   POSTERIOR INFERENCE FOR DERIVED QUANTITIES

```
1 > x.mean.1=apply(X,2,mean)
2 > x.mean.1[2]=1
3 > x.mean.1[3]=1
4 > x.mean.0=apply(X,2,mean)
5 > x.mean.0[3]=1
6 > x.mean.0[3]=0
7 > quantile(x.mean.1%*%mcmc.out$beta.save,
8       c(0.025,0.975))
9      2.5%      97.5%
10 43.49163  46.47422
11 > quantile(x.mean.0%*%mcmc.out$beta.save,
12       c(0.025,0.975))
13      2.5%      97.5%
14 35.41261  37.08845
```

---

which yields (43.5, 46.5) kg for male deer and (35.4, 37.1) kg for female deer. Thus, for average aged individual deer in large cropland regions, males weigh substantially more than females.

11.1   ADDITIONAL CONCEPTS AND READING

We are not limited to using a multivariate normal prior for the regression coefficients $\boldsymbol{\beta}$ or error variance parameter σ^2. We could use a number of other priors that suit our needs. For example, if the regression coefficients need to be non-negative (e.g., Gelfand et al. 1992), then we could specify independent truncated normal priors for them as $\beta_j \sim \text{TN}(0, \sigma_\beta^2)_0^\infty$ for $j = 0, \ldots, p - 1$. Alternatively, if our prior information about the regression coefficients is more concentrated around their means, then we could use a Laplace prior as discussed in Chapter 7. We also can use the Laplace prior to induce a specific type of model selection called regularization that we discuss more in Chapter 14 (Park and Casella 2008). Finally, there are several asymmetric distributions we could use as priors for $\boldsymbol{\beta}$. Asymmetric priors will place more emphasis at certain quantiles of the parameter distribution than others and may better account for skewness in our prior knowledge. We return to asymmetric distributions in Chapter 18 when we discuss quantile regression.

As we hinted at in this chapter, we can generalize the basic linear regression model in a number of ways. For example, it serves as a basis for spatially explicit regression models when we allow the covariance matrix of the data distribution to account for spatial structure among the observations beyond what the mean structure accounts for. We describe spatially explicit regression models in Chapter 17. Similarly, we also can extend the linear regression framework to accommodate temporal structure in data and it has been used to model continuous-time animal movement trajectories (Hooten and Johnson 2017), for example.

Most regression models focus on describing heterogeneity in the location (e.g., mean, median, etc) of the data. However, we also can model heterogeneity in the dispersion of the data by allowing the variance parameter to vary with observation and relating it to covariates. For example, a simple extension of the linear regression model in (11.3) is

$$y_i \sim N(\mathbf{x}_i'\boldsymbol{\beta}, \sigma_i^2), \qquad (11.19)$$

for $i = 1, \ldots, n$, where $\log(\sigma_i) = \mathbf{w}_i'\boldsymbol{\alpha}$ (Cook and Weisberg 1983; Aitkin 1987). In this regression model with heterogeneous mean and variance components, the covariates $\mathbf{w}_i$ associated with the variance could include those used to characterize the mean ($\mathbf{x}_i$) but we do not need them. In the Bayesian setting, the set of regression coefficients associated with the variance ($\boldsymbol{\alpha}$) also need a prior. For example, we could use a multivariate normal prior to characterize the knowledge about $\boldsymbol{\alpha}$ (i.e., $\boldsymbol{\alpha} \sim N(\boldsymbol{\mu}_\alpha, \boldsymbol{\Sigma}_\alpha)$), but it will not result in a conjugate relationship between the parameter model and data model as it did with $\boldsymbol{\beta}$. However, modifying the MCMC algorithm to include a M-H update for $\boldsymbol{\alpha}$ is straightforward. This model may better capture heteroskedasticity in data, but could also become overparameterized easily if we use too many covariates in the mean or variance components.

12 Posterior Prediction

Now that we have described how to implement a Bayesian regression model using MCMC, we may desire additional forms of inference. For example, in addition to making inference based on posterior distributions for parameters, we also may be interested in prediction. That is, for new covariate values $\tilde{\mathbf{X}}$, what is our prediction of the associated data $\tilde{\mathbf{y}}$?

For Bayesian prediction, we need to find the posterior predictive distribution (PPD) $[\tilde{\mathbf{y}}|\mathbf{y}]$, which, for a general Bayesian model with parameters $\boldsymbol{\theta}$, is

$$[\tilde{\mathbf{y}}|\mathbf{y}] = \int [\tilde{\mathbf{y}},\boldsymbol{\theta}|\mathbf{y}]d\boldsymbol{\theta}, \qquad (12.1)$$

$$= \int [\tilde{\mathbf{y}}|\boldsymbol{\theta},\mathbf{y}][\boldsymbol{\theta}|\mathbf{y}]d\boldsymbol{\theta}. \qquad (12.2)$$

Thus, to derive the PPD, we start with the joint distribution of the parameters and unobserved data conditioned on the observed data and then integrate out the parameters. The integrand in the PPD shown in (12.2) is the product of two distributions the second ($[\boldsymbol{\theta}|\mathbf{y}]$) is the usual posterior distribution and the first ($[\tilde{\mathbf{y}}|\boldsymbol{\theta},\mathbf{y}]$) is the full-conditional distribution of the unobserved data given the parameters and observed data. Using a procedure known as "composition sampling" (Tanner 1991) we can obtain a sample from the PPD by fitting the model first to obtain a sample of the parameters ($\boldsymbol{\theta}^{(k)}$, for $k=1,\ldots,K$) as usual, then sample $\tilde{\mathbf{y}}^{(k)} \sim [\tilde{\mathbf{y}}|\boldsymbol{\theta}^{(k)},\mathbf{y}]$. We can do this subsequent step of sampling $\tilde{\mathbf{y}}^{(k)}$ inside the main model fitting algorithm (but it will not affect the model fit) or in a second algorithm that uses the output (i.e., $\boldsymbol{\theta}^{(k)}$, for $k=1,\ldots,K$) from the first algorithm. We can parallelize this second approach easily because each draw $\tilde{\mathbf{y}}^{(k)}$ from the predictive full-conditional distribution is independent after conditioning on $\boldsymbol{\theta}^{(k)}$ and $\mathbf{y}$. Thus, when it is computationally expensive to sample the predictions, it may be advantageous to acquire the posterior predictive sample in pieces using multiple processors.

Although the procedure to obtain a sample from the PPD is straightforward when we know the predictive full-conditional distribution, the potential crux is in deriving the predictive full-conditional distribution. An important result is that the predictive full-conditional distribution is equal to the data model if the data model is conditionally independent for $\mathbf{y}$. Thus, for the linear regression model we specified in Chapter 11, we can obtain a sample from the PPD by drawing $\tilde{\mathbf{y}}^{(k)} \sim \text{N}(\tilde{\mathbf{X}}\boldsymbol{\beta}^{(k)}, \sigma^{2(k)}\mathbf{I})$ for $k=1,\ldots,K$, where $\tilde{\mathbf{X}}$ is a design matrix of prediction covariates and $\boldsymbol{\beta}^{(k)}$ and $\sigma^{2(k)}$ comprise the MCMC output from the model fit to data.

To obtain point predictions, we can simply compute the posterior predictive mean using MC integration as $E(\tilde{\mathbf{y}}|\mathbf{y}) \approx \sum_{k=1}^{K} \tilde{\mathbf{y}}^{(k)}/K$. Similarly, a key advantage of the Bayesian approach to prediction is how easy it is to estimate uncertainty in predictions; we just compute $\text{Var}(\tilde{\mathbf{y}}|\mathbf{y}) \approx \sum_{k=1}^{K} (\tilde{\mathbf{y}}^{(k)} - E(\tilde{\mathbf{y}}|\mathbf{y}))^2/K$. We also can compute prediction intervals using quantiles of the posterior predictive sample.

Consider the white-tailed deer example from the Chapter 11. In that example, we were interested in explaining variation in individual body mass given covariates related to the large cropland indicator, sex of the individual, age of the individual, and age squared. The white-tailed deer data set contains covariates for age ranging from 0.5 to 9.5 years, but there are no individuals at 6 years of age in the data set. Thus, suppose we wish to predict the body mass for female deer at 6 years of age in a location with large cropland. To characterize the appropriate posterior predictive distribution for this demographic, we define the predictive covariate vector $\tilde{\mathbf{x}} = (1,1,0,6,6^2)'$ and sample predictive realizations from the posterior as

$$\tilde{y}^{(k)} \sim \text{N}\left(\tilde{\mathbf{x}}'\boldsymbol{\beta}^{(k)}, \sigma^{2(k)}\right), \tag{12.3}$$

for $k = 1,\ldots,K$ MCMC iterations. Based on our model output from the previous chapter, we can compute the posterior predictive mean and 95% credible interval for our desired demographic of female deer at age 6 using R as

---

BOX 12.1   POSTERIOR PREDICTIVE INFERENCE

```
 1 > x.m=c(1,1,0,6,6^2)
 2 > y.m.pred=rep(0,mcmc.out$n.mcmc)
 3 > set.seed(1)
 4 > for(k in 1:mcmc.out$n.mcmc){
 5 +     y.m.pred[k]=rnorm(1,x.m%*%mcmc.out$beta.
 6         save[,k],sqrt(mcmc.out$s2.save[k]))
 7 + }
 8 > mean(y.m.pred)
 9 [1] 59.42572
10 > quantile(y.m.pred,c(0.025,0.975))
11     2.5%     97.5%
12 44.63422  74.07582
```

---

which yields a posterior predictive mean of 59.4 kg and 95% credible interval of $(44.6, 74.1)$. Note that we calculated the posterior predictive mean and credible interval using the MCMC sample and MC integration in the code just as we would for model parameters.

Another common form of prediction involves interpolating the data over a range of values for the covariates. Thus, building on our previous example, suppose we are interested in predicting the body mass of female deer for a range of ages spanning those present in the data set in a region with large cropland. In this case, we extend the previous R code to create a predictive design matrix $\tilde{\mathbf{X}}$ that contains a first column of ones, the second column with elements equal to one (i.e., indicating large cropland), a third column of zeros, a fourth column comprised of a regular sequence of ages from 0.5 to 9.5, and fifth column that is the element-wise square of the fourth column. We then sample a vector of posterior predictive realizations similar to before as

$$\tilde{\mathbf{y}}^{(k)} \sim \mathrm{N}(\tilde{\mathbf{X}}'\boldsymbol{\beta}^{(k)}, \sigma^{2(k)}\mathbf{I}) , \tag{12.4}$$

which can be summarized in a plot of a response curve with shaded predictive credible region (Figure 12.1) using the following R code that is based on the model output from the previous chapter

---

BOX 12.2 POSTERIOR PREDICTIVE CREDIBLE REGION

```
 1 n.pred=100
 2 x.mean=apply(X,2,mean)
 3 X.pred=matrix(0,n.pred,p)
 4 X.pred[,1]=1
 5 X.pred[,2]=mean(X[,2])
 6 X.pred[,4]=seq(min(X[,4]),max(X[,4]),,n.pred)
 7 X.pred[,5]=X.pred[,4]^2
 8 Y.pred=matrix(0,n.pred,mcmc.out$n.mcmc)
 9 set.seed(1)
10 for(k in 1:mcmc.out$n.mcmc){
11    Y.pred[,k]=rnorm(n.pred,X.pred%*%mcmc.out$beta.
12       save[,k],sqrt(mcmc.out$s2.save[k]))
13 }
14 y.pred.mean=apply(Y.pred,1,mean)
15 y.pred.l=apply(Y.pred,1,quantile,0.025)
16 y.pred.u=apply(Y.pred,1,quantile,0.975)
17
18 plot(X[,4],y,xlab="age")
19 polygon(c(X.pred[,4],rev(X.pred[,4])),c(y.pred.u,
20    rev(y.pred.l)))
21 lines(X.pred[,4],y.pred.mean)
22 points(X[,4],y)
```

We can use the same basic procedure to obtain a posterior predictive sample for each of the models we have presented thus far. Thus, this procedure provides an incredibly useful technique for interpolating data at places where we have not observed them. However, all of the models we have presented thus far were specified under the assumption that the data are conditionally independent, so the predictive full-conditional distribution $[\tilde{\mathbf{y}}|\boldsymbol{\theta},\mathbf{y}]$ is equal to the data model $[\tilde{\mathbf{y}}|\boldsymbol{\theta}]$ because, given the parameters, the unobserved data ($\tilde{\mathbf{y}}$) are independent of the observed data ($\mathbf{y}$). In cases where data are dependent, even when given the parameters, we need to derive the predictive full-conditional distribution as we did for the regular full-conditional distributions in previous chapters (e.g., in geostatistical models;

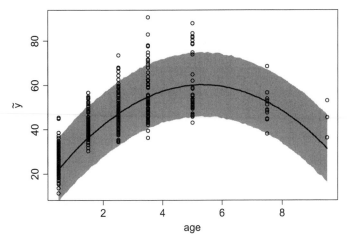

Figure 12.1 Posterior predictive mean (black line) and 95% credible interval (gray region) for female deer body mass (kg) in large cropland at a range of ages (years). Full data set shown as black circles.

Chapter 17). If we cannot derive it analytically, we can still sample from it using M-H, as before. We can perform these updates sequentially in the same algorithm that we use to fit the model. However, unlike the M-H updates for model parameters, these posterior predictive updates do not affect the model fit. We return to the topic of predictive full-conditional distributions for cases where conditional independence is not assumed in chapters that follow.

12.1 ADDITIONAL CONCEPTS AND READING

Prediction is a broad concept that includes both interpolation (i.e., predicting within the region of observed data) and extrapolation (i.e., predicting outside the region of observed data). We often frown upon extrapolation in many types of statistical analyses because we have little ability to learn the process outside the realm of observed data when using purely phenomenological models. However, when we use more mechanistic models and make certain assumptions about stationarity (i.e., the process does not change in different regions of the data), extrapolation is a more common form of prediction. For example, for temporally dynamic models (Chapter 16), we often refer to extrapolation beyond the data in time as forecasting. Forecasting is one of the main goals in time series analysis and we can use the same concept of posterior prediction for forecasting in the Bayesian setting. We formalize these ideas in Chapters 16 and 28.

Clark et al. (2001) and Dietze et al. (2018) noted many advantages associated with regular and frequent forecasting in ecological science. Forecasting is part of an operational program in some environmental sciences. For example, the science and industry focus on weather prediction has led to steady improvements in predictive ability because of the emphasis placed on forecasting in that field. Thus, we also can expect improvements in ecological forecasting over time if similar emphasis is placed on its importance (Dietze 2017).

Raftery (2016) suggested that the best way to use and communicate probabilistic forecasts depends on the users of the information and their utility of the information. Depending on how we value predictions, we can use a decision theory to optimize the amount of information communicated about probabilistic forecasts. We also can apply similar decision theoretic concepts to the management and conservation of natural systems (Williams and Hooten 2016). For example, Hobbs et al. (2015) were able to compare the effect of management strategies on wildlife disease outcomes using posterior predictions. The ability to obtain honest and reliable statistical forecasts also can help identify optimal monitoring strategies for the future (e.g., Hooten et al. 2009; Williams et al. 2018).

In addition to providing inference about unobserved data directly, statistical prediction and our ability to obtain a sample from the PPD is also useful for other reasons. For example, in the following chapter (Chapter 13) we discuss model comparison. Predictive ability serves as a primary way to score and compare statistical models (Hooten and Hobbs 2015). Thus, an understanding of how to obtain Bayesian predictions is imperative for quantifying the predictive performance of statistical models.

Posterior prediction also plays a role in model checking, as we discuss in Chapter 21. Model checking involves an assessment of model assumptions to determine if a model adequately represents the observed data (Gelman et al. 1996; Conn et al. 2018). The ability to obtain draws from the posterior predictive distribution allows us to compare features (such as the dispersion) of the predicted data with those of the observed data. If our model predicts data that fail to resemble the observed data, we have evidence for lack of fit which implies that our model assumptions may not be valid. This procedure allows us to determine which statistical models that we can consider further as a means to provide valid statistical inference.

13 Model Comparison

We can use predictive distributions to make inference on unobserved data and also to illustrate the shape of relationships in statistical models. Prediction also enables us to characterize how well our model represents data-generating mechanisms. If our model results in predictions that are close to real-world situations, perhaps better than alternative models, we may value the resulting inference more. Thus, the concept of scoring models based on their predictive ability provides a good way to better understand their value.

There are many ways to score statistical models (see Gelman et al. 2014, Hooten and Hobbs 2015, and Hooten and Cooch 2019 for reviews). In what follows, we describe how to assess models based on proper and local scores. For parametric statistical models (e.g., all Bayesian models), we can score them using a logarithmic score that is a function of the model. In this chapter, we focus on the deviance $D(\mathbf{y})$ as a logarithmic score. The deviance is defined as

$$D(\mathbf{y}, \boldsymbol{\theta}) = -2\log[\mathbf{y}|\boldsymbol{\theta}], \tag{13.1}$$

for a generic parametric statistical model and parameter vector $\boldsymbol{\theta}$ and data $\mathbf{y}$. Note that the term $[\mathbf{y}|\boldsymbol{\theta}]$ in (13.1) is merely the data model.[1] Lower values of deviance indicate better predicting models because the score is equivalent to the negative log likelihood. Thus, if we know the form of the model as well as the data and parameters, we can calculate the deviance. The deviance is a proper score because the expectation (over $\mathbf{y}$) of $D(\mathbf{y}, \boldsymbol{\theta})$ is guaranteed to be at a minimum when $[\mathbf{y}|\boldsymbol{\theta}]$ is the correct model (Gneiting and Raftery 2007). The deviance is local because it depends on observable data, rather than some other events (e.g., auxiliary data measuring a different part of the process than the data used to fit the model).

As a predictive score for a statistical model, the deviance will be optimistic if we calculate it using the same data that we used to fit the model. Also, the deviance depends on the parameters $\boldsymbol{\theta}$, which are unknown in real-world situations. Thus, to address these issues, we consider the following well-known deviance information criterion (DIC; Spiegelhalter et al. 2002),

$$\text{DIC} = \hat{D} + 2p_D, \tag{13.2}$$

where the term $\hat{D}$ is the deviance based on a plug-in estimate of the parameters

$$\hat{D} = D(\mathbf{y}, \hat{\boldsymbol{\theta}}), \tag{13.3}$$

$$= -2\log[\mathbf{y}|\hat{\boldsymbol{\theta}}], \tag{13.4}$$

---

[1] Recall that $[\mathbf{y}|\boldsymbol{\theta}]$ is a conditional PDF or PMF, also referred to as the likelihood.

where $\hat{\boldsymbol{\theta}}$ is the posterior mean $E(\boldsymbol{\theta}|\mathbf{y})$. The DIC score works well to characterize predictive ability and is usually easy to calculate, but some have argued (Gelman et al. 2014) that it is not a true Bayesian scoring function.

A better score would be based on the posterior predictive distribution $[\tilde{\mathbf{y}}|\mathbf{y}]$, where $\tilde{\mathbf{y}}$ represents new data, because that is the official Bayesian mechanism to obtain predictions. Because we usually want a single quantity to represent the score, we can take the expectation of the score over the marginal distribution of the new data $[\tilde{\mathbf{y}}]$ resulting in a score with the form

$$E_{[\tilde{\mathbf{y}}]}(-2\log[\tilde{\mathbf{y}}|\mathbf{y}]), \tag{13.5}$$

if we knew $[\tilde{\mathbf{y}}]$. Thus, we can estimate the score with $\hat{D}$, but we need to correct for its optimism. The amount of optimism can be estimated by

$$2(E_{[\boldsymbol{\theta}|\mathbf{y}]}(D(\mathbf{y},\boldsymbol{\theta})) - D(\mathbf{y},\hat{\boldsymbol{\theta}})) = 2(\bar{D} - \hat{D}), \tag{13.6}$$

$$= 2p_D, \tag{13.7}$$

where p_D appears in the equation for DIC (13.2), the optimism corrected score. DIC is also sometimes written as $DIC = \bar{D} + p_D$, which is equivalent to (13.2). The p_D is referred to as the "effective number of parameters" and has the same basic interpretation as a penalty for model complexity in other well-known information criteria. In the case of DIC, p_D naturally arises as a result of accounting for optimism in the predictive score. For simple models, a parameter counts as a unit of one in p_D if all the information in the model comes from the likelihood. Conversely, a parameter counts as zero in p_D if all the information about it comes from the prior. This illustrates why we cannot use an automatic penalty like the number of parameters in a Bayesian model to estimate the optimism. Some parameter information arises from the prior, thus we may not estimate the parameter entirely using the data alone. If we knew the parameters exactly (i.e., specified a perfectly precise prior distribution—a point mass distribution or Dirac delta function), then we could calculate the score directly using (13.1).

The expectation in (13.6) corresponds to the mean of the deviance over the posterior distribution

$$\bar{D} = E_{[\boldsymbol{\theta}|\mathbf{y}]}(D(\mathbf{y},\boldsymbol{\theta})), \tag{13.8}$$

$$= \int D(\mathbf{y},\boldsymbol{\theta})[\boldsymbol{\theta}|\mathbf{y}]d\boldsymbol{\theta}. \tag{13.9}$$

Thus, we can treat the deviance as a derived quantity and approximate its posterior mean using MC integration based on our MCMC sample just like any other posterior mean involving data and/or parameters. A MC integral approximation of $\bar{D}$ is

$$\bar{D} = \frac{\sum_{k=1}^{K} D\left(\mathbf{y}, \boldsymbol{\theta}^{(k)}\right)}{K}, \tag{13.10}$$

based on a MCMC sample of size K comprised of $\boldsymbol{\theta}^{(k)}$.

In general, studies have shown DIC to be valid for non-hierarchical and non-mixture models, cases where effective number of parameters is much less than the sample size of the data ($p_D << n$), and when the posterior mean of the parameters is a good summary of the central tendency of the posterior distribution.

Thus, we can use DIC to compare Bayesian linear regression models. For the regression model we introduced in the last two chapters, we can compute DIC using

$$\hat{D} = -2\log \prod_{i=1}^{n} N\left(y_i | \mathbf{x}_i'\hat{\boldsymbol{\beta}}, \hat{\sigma}^2\right), \tag{13.11}$$

$$= \sum_{i=1}^{n} -2\log N\left(y_i | \mathbf{x}_i'\hat{\boldsymbol{\beta}}, \hat{\sigma}^2\right), \tag{13.12}$$

and

$$\bar{D} = \frac{\sum_{k=1}^{K} -2\log \prod_{i=1}^{n} N\left(y_i | \mathbf{x}_i'\boldsymbol{\beta}^{(k)}, \sigma^{2(k)}\right)}{K}, \tag{13.13}$$

$$= \frac{\sum_{k=1}^{K} \sum_{i=1}^{n} -2\log N\left(y_i | \mathbf{x}_i'\boldsymbol{\beta}^{(k)}, \sigma^{2(k)}\right)}{K}, \tag{13.14}$$

where the $N(y_i | \cdots)$ notation refers to the PDF of a normal distribution evaluated at y_i.

We modified the MCMC algorithm for fitting the Bayesian linear regression model in Chapter 11 to calculate DIC (and $\hat{D}$, $\bar{D}$, p_D) resulting in the new algorithm norm.reg.mcmc.DIC:

BOX 13.1   MCMC ALGORITHM FOR LINEAR REGRESSION WITH DIC

```
norm.reg.mcmc.DIC <- function(y,X,beta.mn,
    beta.var,s2.mn,s2.sd,n.mcmc){

###
### Subroutines
###

library(mvtnorm)

invgammastrt <- function(igmn,igvar){
  q <- 2+(igmn^2)/igvar
  r <- 1/(igmn*(q-1))
  list(r=r,q=q)
}
```

(*Continued*)

BOX 13.1 (Continued)   MCMC ALGORITHM FOR LINEAR REGRESSION WITH DIC

```
15  ###
16  ### Set up Variables and Priors
17  ###
18
19  n=dim(X)[1]
20  p=dim(X)[2]
21  n.burn=round(.1*n.mcmc)
22  r=invgammastrt(s2.mn,s2.sd^2)$r
23  q=invgammastrt(s2.mn,s2.sd^2)$q
24  Sig.beta=beta.var*diag(p)
25
26  beta.save=matrix(0,p,n.mcmc)
27  s2.save=rep(0,n.mcmc)
28  Dbar=0
29
30  ###
31  ### Starting Values
32  ###
33
34  beta=solve(t(X)%*%X)%*%t(X)%*%y
35
36  ###
37  ### MCMC Loop
38  ###
39
40  for(k in 1:n.mcmc){
41
42     ###
43     ### Sample s2
44     ###
45
46     tmp.r=(1/r+.5*t(y-X%*%beta)%*%(y-X%*%beta))^(-1)
47     tmp.q=n/2+q
48     s2=1/rgamma(1,tmp.q,,tmp.r)
49
50     ###
51     ### Sample beta
52     ###
```

(Continued)

BOX 13.1 (Continued)   MCMC ALGORITHM FOR LINEAR
REGRESSION WITH DIC

```
53  tmp.var=solve(t(X)%*%X/s2 + solve(Sig.beta))
54  tmp.mn=tmp.var%*%(t(X)%*%y/s2 + solve(Sig.beta)%
       *%beta.mn)
55  beta=as.vector(rmvnorm(1,tmp.mn,tmp.var,
56    method="chol"))
57
58  ###
59  ### DIC Calculations
60  ###
61
62  if(k > n.burn){
63    Dbar=Dbar-2*sum(dnorm(y,X%*%beta,sqrt(s2),
64        log=TRUE))/(n.mcmc-n.burn)
65  }
66
67  ###
68  ### Save Samples
69  ###
70
71  beta.save[,k]=beta
72  s2.save[k]=s2
73
74  }
75
76  ###
77  ###   Calculate DIC
78  ###
79
80  if(dim(X)[2]==1){
81    postbetamn=mean(beta.save[,-(1:n.burn)])
82  }
83  if(dim(X)[2]>1){
84    postbetamn=apply(beta.save[,-(1:n.burn)],1,mean)
85  }
86  posts2mn=mean(s2.save[-(1:n.burn)])
87  Dhat=-2*(sum(dnorm(y,X%*%postbetamn,
88      sqrt(posts2mn),log=TRUE)))
```

(Continued)

BOX 13.1 (Continued)   MCMC ALGORITHM FOR LINEAR REGRESSION WITH DIC

```
89 pD=Dbar-Dhat
90 DIC=Dhat+2*pD
91
92 ###
93 ###   Write Output
94 ###
95
96 list(beta.save=beta.save,s2.save=s2.save,
97     Dhat=Dhat,Dbar=Dbar,pD=pD,DIC=DIC)
98
99 }
```

Note the new sections on lines 63 and 80–90 that we need to compute DIC. On line 63, we compute a running average for $\bar{D}$, and on lines 80–90, we compute the posterior mean of $\boldsymbol{\beta}$, then $\hat{D}$, then p_D, then DIC. We do not have to do these calculations concurrently with the model fit, we can do them using the output as we did in the previous chapter to obtain posterior predictive realizations. However, if the calculations are not too computationally intensive, then it will not slow down the MCMC algorithm noticeably to include them.

We used `norm.reg.mcmc.DIC` to fit $L = 11$ models with different combinations of parameters in them to the same set of deer body mass data we used in Chapters 11 and 12. Table 13.1 showing which covariates we included in each model as well as the resulting number of effective parameters (p_D) and DIC. The R code we used to fit the set of $L = 11$ models to the data was

---

BOX 13.2   COMPARE DIC FOR LINEAR REGRESSION MODELS

```
1 load("deer.RData")
2 n=dim(deer.df)[1]
3 y=deer.df$y
4 n=length(y)
5
6 L=11
7 X.list=vector("list",L)
8 X.list[[11]]=cbind(rep(1,n),as.matrix(deer.df[,
9     c(2:5)]))
10 X.list[[11]][,2]=ifelse(X.list[[11]][,2]==599,0,1)
```

(Continued)

BOX 13.2 (Continued)   COMPARE DIC FOR LINEAR REGRESSION MODELS

```
11 X.list[[1]]=X.list[[11]][,c(1,2)]
12 X.list[[2]]=X.list[[11]][,c(1,3)]
13 X.list[[3]]=X.list[[11]][,c(1,4)]
14 X.list[[4]]=X.list[[11]][,c(1,4:5)]
15 X.list[[5]]=X.list[[11]][,c(1,2:3)]
16 X.list[[6]]=X.list[[11]][,c(1,2,4)]
17 X.list[[7]]=X.list[[11]][,c(1,2,4,5)]
18 X.list[[8]]=X.list[[11]][,c(1,2,3,4)]
19 X.list[[9]]=X.list[[11]][,c(1,3,4)]
20 X.list[[10]]=X.list[[11]][,c(1,3,4,5)]
21
22 mcmc.out.list=vector("list",L)
23 DIC.vec=rep(0,L)
24 pD.vec=rep(0,L)
25 source("norm.reg.mcmc.DIC.R")
26 set.seed(1)
27 for(l in 1:L){
28   p=dim(X.list[[l]])[2]
29   mcmc.out.list[[l]]=norm.reg.mcmc.DIC(y=y,
30     X=X.list[[l]],beta.mn=rep(0,p),beta.var=10000,
31     s2.mn=50,s2.sd=1000,n.mcmc=10000)
32   DIC.vec[l]=mcmc.out.list[[l]]$DIC
33   pD.vec[l]=mcmc.out.list[[l]]$pD
34 }
35 cbind(1:L,pD.vec,DIC.vec)
```

The model with the smallest DIC (indicating better predictive ability) is model 10 (DIC = 3416.0) which includes the covariates sex, age, and age.[2] However, the next best DIC is associated with model 11 (DIC = 3416.8), which includes all covariates. Thus, models 10 and 11 differ by only parameter but have nearly the same DIC. It is customary to use the more parsimonious model in this situation—the one with fewer parameters—for prediction. Recall from Chapter 11 that the posterior distribution for β_1 (large cropland) overlapped zero substantially, thus we would not expect it to be a very important parameter in the model and DIC bears that out.

Predictive scores like DIC that depend on estimated corrections for optimism (e.g., $2p_D$) are useful because we can calculate them easily using within-sample data (i.e., the same data that are used to fit the model). The optimism corrections are optimal under certain, but not all, circumstances (Link and Barker 2006). Also, there is a huge variety of different types of information criteria, with some being more desired depending on the situation and model. We presented DIC here because it is the most

Table 13.1

The Resulting Effective Number of Parameters and DIC from $L = 11$ Linear Regression Models Fit to the White-Tailed Deer Body Mass Data

	Covariate					
Model	Crop	Sex	Age	Age^2	p_D	DIC
1	x				3.0	4158.9
2		x			3.0	4165.0
3			x		3.0	3809.3
4			x	x	4.0	3536.7
5	x	x			4.0	4160.8
6	x		x		4.0	3811.0
7	x		x	x	5.0	3538.8
8	x	x	x		5.0	3780.1
9		x	x		3.9	3779.3
10		x	x	x	5.0	3416.0
11	x	x	x	x	6.0	3416.8

Note: An intercept was included in all models and "x" indicates which covariates were included in each model.

popular and it provides an instructive example for calculating similar scores using MCMC output. DIC works for many Bayesian models and the basic formulation in (13.2) is general and we can apply it for any parametric likelihood. However, alternatives to within-sample scoring are numerous and we discuss some of those concepts (e.g., cross-validation and regularization) in the next chapter.

13.1 ADDITIONAL CONCEPTS AND READING

Although DIC is probably the most commonly used score to compare Bayesian models, in part because we can calculate it easily in automated Bayesian software, there are several other alternatives. Hooten and Hobbs (2015) provided an overview of several other approaches for comparing Bayesian models that we summarize in what follows. The so-called widely applicable information criterion (WAIC) is one such alternative (Gelman et al. 2014). The most common calculation of WAIC uses a point-wise score based on the PPD (Richardson 2002; Celeux et al. 2006; Watanabe 2010). Also, the point-wise WAIC calculation makes the assumption of conditional independence in the predictive distribution and may not be appropriate for models with correlation structure induced via covariance (e.g., spatial statistical models; Chapter 17).

Another commonly used proper score for comparing Bayesian models is referred to as the continuous ranked probability score (CRPS; Gneiting and Raftery 2007). The CRPS is used commonly in environmental science to score probabilistic weather and climate models (Zamo and Naveau 2018). The CRPS is used typically in forecasting situations as a way to score models based on the discrepancy between a probabilistic forecast distribution and the empirical cumulative distribution function of the observation(s) of interest. Forecasting is typically performed using dynamic temporal models, which we describe in Chapter 16.

Laud and Ibrahim (1995) and Gelfand and Ghosh (1998) described an approach for model comparison that balances the predictive ability of new data versus the data used to fit the statistical model. Their approach, often referred to as "posterior predictive loss," is based on decision theory and is very general. However, it depends on the choice of loss functions that quantify the loss associated with predictive error and the weight associated with the predictive ability for new data versus data used to fit the model.

Other forms of Bayesian model comparison include Bayes factors (Kass and Raftery 1995) and posterior model probabilities (Dawid 2011), which relate to Bayesian model averaging (Chapter 15). Similarly, several approaches for model-based comparison (Hooten and Hobbs 2015) also exist; many of which relate to Bayesian model averaging. Model-based model comparison methods such as indicator variable selection, Gibbs variable selection (Carlin and Chib 1995), and stochastic search variable selection (George and McCulloch 1993) explicitly augment the model to include parameters that can "select" which components should occur in the model. These binary indicator variables are stochastic and we need to estimate them jointly with the rest of the model parameters. Thus, implementing these methods requires us to fit more complicated models than we may have originally specified, but we only need to fit the augmented model a single time to perform both inference and parameter selection.

14 Regularization

Scoring a discrete set of nested models based on predictive ability as we did in the previous chapter is actually a specific type of multimodel inference called regularization (Bickel et al. 2006; Hooten and Hobbs 2015). The general concept of regularization involves putting constraints on models that result in better predictions. Other names for regularization include penalization and shrinkage, and specific types include ridge regression and lasso. The word regularization arises because of an original intent to create "regular" models out of irregular ones. In the context of linear regression, a simple way to think of regular regression models is as models with fewer unknown parameters than data. Irregularity in this context relates to the concept of redundancy or overparameterization, where there may be more parameters in a model than we can estimate using the available data and specified model structure.

To regularize a model using maximum likelihood methods, we add a penalty function to the likelihood that puts constraints on parameters to invoke a form of parsimony in the model. As we discussed in the previous chapter, parsimonious models tend to predict well. This is very similar to the information criterion concept. With information criteria, we added a penalty term to the deviance that gets stronger as the model complexity increases (e.g., two times the effective number of parameters in the case of DIC). However, most information criteria rely on penalties that fit into a larger class.

For example, consider the additive penalty $\gamma_1 \sum_{j=1}^{p} |\theta_j|^{\gamma_2}$, where $\boldsymbol{\theta} \equiv (\theta_1, \ldots, \theta_p)'$ represents a generic set of parameters. Then, if we let $\gamma_1 = 2$ and $\gamma_2 = 0$, the penalty equals $2p$. With less prior information, the effective number of parameters in DIC approaches p ($p_D \to p$), thus, this more general penalty accommodates DIC (also akaike information criterion [AIC] and bayer information criterion [BIC] in the non-Bayesian context).

However, we can go a step further with the regularization concept. Notice that the regularized score function for linear regression with respect to $\boldsymbol{\beta}$ can be written as the deviance plus this new general penalty term,

$$-2\left(-n\log(\sqrt{2\pi\sigma^2}) - \frac{1}{2\sigma^2}\sum_{i=1}^{n}(y_i - \beta_0 - \mathbf{x}_i'\boldsymbol{\beta})^2\right) + \gamma_1 \sum_{j=1}^{p} |\beta_j|^{\gamma_2}, \qquad (14.1)$$

where we separated the intercept (β_0) out of the coefficient vector for this example. If we let $\gamma_2 = 2$, then this regularized score function is linearly related to the log of the likelihood times a prior. In fact, the implied prior is normal with mean zero and variance σ^2/γ_1. Thus, as the regularization parameter γ_1 increases with respect to σ^2, the implied prior variance shrinks toward zero. When we use a prior with mean

119

zero and small variance, we force the coefficients $\boldsymbol{\beta}$ toward zero, effectively removing them from the model. However, because some coefficients vary in their magnitude, they will shrink toward zero at varying rates as γ_1 increases.

It is customary to first standardize the covariates $\mathbf{x}_i$ to have mean zero and variance one before fitting the model so that the coefficients are on the same scale. This standardization allows the single regularization parameter γ_1 to have the same influence across all coefficients (except the β_0, which we usually do not shrink toward zero).

This penalized regression example demonstrates that Bayesian models already contain a natural mechanism to induce regularization (i.e., the prior) that can improve predictive ability. We simply need to use more precise (i.e., stronger) priors. A more precise prior with mean zero for slope coefficients in Bayesian regression model will force them toward zero and perform a type of continuous model selection automatically.[1] In non-Bayesian statistics, this idea is known as ridge regression and we have used it traditionally to ameliorate the effects of multicollinearity among covariates (Hoerl and Kennard 1970). Multicollinearity can affect estimation because, as covariates become more linearly related, it is difficult to distinguish one coefficient from another, and the coefficient estimates will offset (one large positive and one small negative). Design matrices that contain covariates that are too similar are not regular. Thus, we can penalize the likelihood to induce regularity and force the offsetting coefficients toward zero. This also helps with model selection because the shrinkage reduces model complexity and more parsimonious models tend to predict better.

To perform Bayesian regularization for our regression model, we use a similar MCMC algorithm as in previous chapters, but with a prior that has mean zero and very small variance for the slope coefficients. It does not help us to shrink the intercept toward zero because we need it in the model to account for the location of the data when all the other parameters are shrunk to zero. In principle, the predictive score will always be worse for the null model if we have strong relationships between the covariates and response variable.

The big question is: How small do we need to make the prior variance (which is proportional to the reciprocal of γ_1) to yield an optimal predictive model? Traditionally, we tune the shrinkage parameter in regularization settings using cross-validation. In the regression setting, cross-validation involves holding out portions of response ($\mathbf{y}_u$) and covariate ($\mathbf{X}_u$) data and predicting them using only the remainder of the data (i.e., $\mathbf{y}_o$ and $\mathbf{X}_o$). We can perform this procedure either sequentially for each "fold" of data that we hold out or using parallel computing which can reduce computational time. Then we can compare the resulting predictions to the known hold-out data and compute a predictive score to measure how well the model predicts observable data. We can repeat this procedure (again, in parallel if desired) for a set of potential values for the regularization parameter γ_1

---

[1] In fact, in the models for binary data we discuss in Chapters 20 and 23, more precise priors are very reasonable (Hobbs and Hooten 2015; Hooten and Hobbs 2015; Northrup and Gerber 2018).

(or prior variance parameter for $\boldsymbol{\beta}$) to find the optimal amount of regularization that improves prediction.

To assess predictive ability, we can use a variety of methods, but a local and proper score like the deviance is desirable (Chapter 13). For linear regression, a proper score is the mean squared prediction error (MSPE) which is

$$\text{MSPE} = \frac{\sum_{i=1}^{n}(y_i - \hat{y}_i)^2}{n}, \tag{14.2}$$

where, in the Bayesian setting, we define a prediction as the posterior mean of the predicted hold-out data, $\hat{y}_i = E(y_{u,i}|\mathbf{y}_o)$. Because we are only interested in the posterior mean to represent predictions, we can approximate $\hat{y}_i$ using the MCMC output to calculate the derived quantity,

$$\hat{y}_i \approx \frac{\sum_{k=1}^{K} \mathbf{x}'_{u,i}\boldsymbol{\beta}^{(k)}}{K}. \tag{14.3}$$

This method is slightly more efficient than using composition sampling to first sample the predictive realizations $y_i^{(k)}$ and then compute the MCMC mean of them (because the conditional mean of $y_i^{(k)}$ is $\mathbf{x}'_{u,i}\boldsymbol{\beta}^{(k)}$).

We modified the MCMC algorithm from Chapter 11 to compute the predictive mean for hold-out data within the existing MCMC loop. This new algorithm (norm.reg.ridge.cv.mcmc) is

---

BOX 14.1   MCMC ALGORITHM FOR LINEAR REGRESSION WITH PREDICTION

```
1  norm.reg.ridge.cv.mcmc <- function(y,X,beta0.mn,
2      beta0.var,beta.mn,beta.var,s2.mn,s2.sd,
3      n.mcmc,X.pred){
4
5  ###
6  ### Subroutines
7  ###
8
9  library(mvtnorm)
10
11 invgammastrt <- function(igmn,igvar){
12   q <- 2+(igmn^2)/igvar
13   r <- 1/(igmn*(q-1))
14   list(r=r,q=q)
15 }
```

(*Continued*)

BOX 14.1 (Continued)   MCMC ALGORITHM FOR LINEAR
REGRESSION WITH PREDICTION

```
16 ###
17 ### Set up Variables and Priors
18 ###
19
20 n=dim(X)[1]
21 n.pred=dim(X.pred)[1]
22 p=dim(X)[2]
23 n.burn=round(.1*n.mcmc)
24 r=invgammastrt(s2.mn,s2.sd^2)$r
25 q=invgammastrt(s2.mn,s2.sd^2)$q
26 beta.mn.all=c(beta0.mn,beta.mn)
27 Sig.beta=diag(p)
28 diag(Sig.beta)=c(beta0.var,rep(beta.var,p-1))
29 Sig.beta.inv=solve(Sig.beta)
30
31 beta.save=matrix(0,p,n.mcmc)
32 s2.save=rep(0,n.mcmc)
33 y.pred.mn=rep(0,n.pred)
34
35 ###
36 ### Starting Values
37 ###
38
39 beta=solve(t(X)%*%X)%*%t(X)%*%y
40
41 ###
42 ### MCMC Loop
43 ###
44
45 for(k in 1:n.mcmc){
46
47   ###
48   ### Sample s2
49   ###
50
51   tmp.r=(1/r+.5*t(y-X%*%beta)%*%(y-X%*%beta))^(-1)
52   tmp.q=n/2+q
```

(Continued)

BOX 14.1 (Continued)   MCMC ALGORITHM FOR LINEAR REGRESSION WITH PREDICTION

```
53  s2=1/rgamma(1,tmp.q,,tmp.r)
54  s=sqrt(s2)
55
56  ###
57  ### Sample beta
58  ###
59
60  tmp.chol=chol(t(X)%*%X/s2 + Sig.beta.inv)
61  beta=backsolve(tmp.chol,backsolve(tmp.chol,
62      t(X)%*%y/s2 + Sig.beta.inv%*%beta.mn.all,
63      transpose=TRUE)+rnorm(p))
64
65  ###
66  ### Posterior Predictive Calculations
67  ###
68
69  if(k > n.burn){
70    y.pred.mn=y.pred.mn+X.pred%*%beta/
71        (n.mcmc-n.burn)
72  }
73
74  ###
75  ### Save Samples
76  ###
77
78  beta.save[,k]=beta
79  s2.save[k]=s2
80
81  }
82
83  ###
84  ###   Write Output
85  ###
86
87  list(beta.save=beta.save,s2.save=s2.save,
88      y=y,X=X,n.mcmc=n.mcmc,n=n,r=r,q=q,
89      p=p,y.pred.mn=y.pred.mn)
90
91  }
```

Recall that the MCMC algorithm does not need any major changes to do regularization because it already depends on the prior specification for β. However, we construct the prior covariance matrix for the entire vector β, including the intercept, so that it is diagonal as before, but has the first diagonal element which is the prior variance for the intercept, and the rest of the diagonal elements equal to the regularization variance parameter. It is convention to standardize (i.e., center by subtracting the mean and scale by the standard deviation) the covariates (X) before performing regularization because it allows us to use a single prior variance parameter and let it affect the shrinkage for the coefficients equally rather than have a separate variance parameter for each coefficient.

Notice that we also used a new method for sampling multivariate normal random variables in the portion of the algorithm where we update β. This method avoids calculating the inverse of the full-conditional precision matrix (11.18) directly and instead only depends on a Cholesky decomposition (i.e., a matrix square root) and two iterative "backsolves" which are much more computationally efficient than a direct matrix inverse using the base R functions. For example, the following R code demonstrates the potential speed up using the "backsolve" approach:

BOX 14.2   COMPARE SPEED OF MULTIVARIATE NORMAL SIMULATION METHODS

```
1 > library(mvtnorm)
2 > p=1000
3 > X=matrix(rnorm(p^2),p,p)
4 > A=t(X)%*%X
5 > b=rnorm(p)
6 >
7 > init.time=proc.time()
8 > tmp.var=solve(A)
9 > tmp.mn=tmp.var%*%b
10 > beta=as.vector(rmvnorm(1,tmp.mn,tmp.var,
11     method="chol"))
12 > (proc.time()-init.time)[3]
13 elapsed
14    0.202
15 >
16 > init.time=proc.time()
17 > tmp.chol=chol(A)
18 > beta=backsolve(tmp.chol,backsolve(tmp.chol,b,
19     transpose=TRUE)+rnorm(p))
20 > (proc.time()-init.time)[3]
21 elapsed
22    0.017
```

Thus, whereas using `rmvnorm` requires 0.2 seconds to obtain a p dimensional multivariate normal random vector, the "backsolve" approach requires only 0.017 seconds, an order of magnitude reduction in computing time. In MCMC, where we need to obtain multivariate normal samples thousands of times sequentially, this approach can lead to more efficient algorithms.

Using the regression model for the white-tailed deer data we presented in Chapter 11, but with standardized covariates,[2] we performed 10-fold cross-validation to obtain the optimal amount of regularization using the following R code:

BOX 14.3   CROSS-VALIDATION FOR LINEAR REGRESSION WITH REGULARIZATION

```
 1  load("deer.RData")
 2  n=dim(deer.df)[1]
 3  y=deer.df$y
 4  n=length(y)
 5  X=cbind(rep(1,n),scale(as.matrix(deer.df[,c(2:5)])
      ))
 6  p=dim(X)[2]
 7
 8  K=10
 9  set.seed(1)
10  rand.idx=sample(1:n,n) # randomize data for CV
11  X.rand=X[rand.idx,]
12  y.rand=y[rand.idx]
13  n.u=floor(n/K)
14  n.o=n.u*K-n.u
15  L=30
16  sig.beta.vec=seq(7,20,,L)
17  beta.mat=matrix(0,L,p)
18  spe.vec=rep(0,L)
19  source("norm.reg.ridge.cv.mcmc.R")
20  for(l in 1:L){
21    for(k in 1:K){
22      idx.u=(k-1)*n.u+(1:n.u)
23      X.o=X.rand[-idx.u,]
24      X.u=X.rand[idx.u,]
25      y.o=y.rand[-idx.u]
26      y.u=y.rand[idx.u]
```

(Continued)

BOX 14.3 (Continued)   CROSS-VALIDATION FOR LINEAR REGRESSION WITH REGULARIZATION

```
27    p=dim(X)[2]
28    mcmc.out=norm.reg.ridge.cv.mcmc(y=y.o,X=X.o,
29        beta0.mn=0,beta0.var=10000,beta.mn=rep
30        (0,p-1),beta.var=sig.beta.vec[l]^2,
31        s2.mn=50,s2.sd=1000,n.mcmc=100000,
32        X.pred=X.u)
33    spe.vec[l]=spe.vec[l]+
34        sum((y.u-mcmc.out$y.pred.mn)^2)
35    }
36    beta.mat[1,]=apply(norm.reg.ridge.cv.mcmc
37        (y=y.rand,X=X.rand,beta0.mn=0,beta0.var=10000,
38        beta.mn=rep(0,p-1),beta.var=sig.beta.vec[l]^2,
39        s2.mn=50,s2.sd=1000,n.mcmc=100000,
40        X.pred=matrix(0,2,p))$beta.save,1,mean)
41
42  }
43
44  sig.beta.opt=sig.beta.vec[spe.vec==min(spe.vec)]
45
46  layout(matrix(1:5,5,1))
47  plot(sig.beta.vec,beta.mat[,2],type="l",main="a")
48  abline(v=sig.beta.opt,col=8)
49  plot(sig.beta.vec,beta.mat[,3],type="l",main="b")
50  abline(v=sig.beta.opt,col=8)
51  plot(sig.beta.vec,beta.mat[,4],type="l",main="c")
52  abline(v=sig.beta.opt,col=8)
53  plot(sig.beta.vec,beta.mat[,5],type="l",main="d")
54  abline(v=sig.beta.opt,col=8)
55  plot(sig.beta.vec,spe.vec/n.u/K,type="l",main="e")
56  abline(v=sig.beta.opt,col=8)
```

This cross-validation and regularization algorithm results in the coefficient trajectories and predictive score shown in Figure 14.1. Coded in sequence, the procedure we just described, which includes cross-validation and a search for optimal regularization, may require substantial computing time, but we can reduce that using parallel computing.

Recall from the white-tailed deer body mass example from Chapter 11, we are interested in fitting a model that regresses individual body mass on a suite of (standardized) predictor variables (i.e., cropland, sex, age, and age^2). To interpret

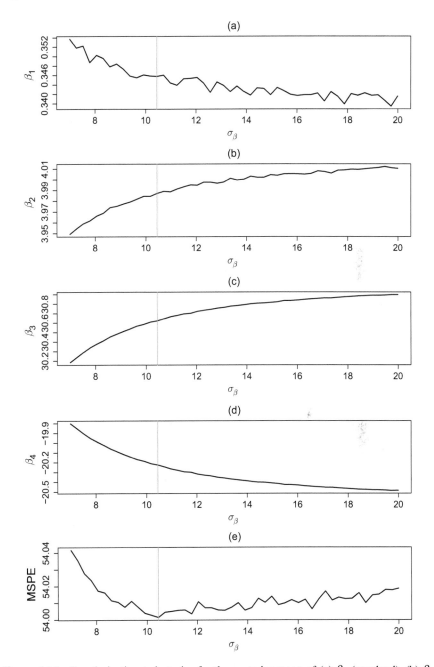

Figure 14.1 Regularization trajectories for the posterior mean of (a) β_1 (cropland), (b) β_2 (sex), (c) β_3 (age), and (d) β_4 (age^2) as well as (e) the predictive score (MSPE; lower is better) for a range of prior standard deviations (regularization parameter: σ_β). The gray vertical line represents the regularization parameter that yields the best predictive score.

the posterior mean coefficient trajectories in Figure 14.1a–d, we can think of those values at the far right as the unpenalized coefficient values, but as the prior variance is decreased, those coefficients with any non-negligible effect shrink toward zero. The vertical gray line in Figure 14.1 represents the optimal amount of shrinkage in the parameters that leads to the best predictive model. Note that the optimal value for σ_β for this example was 10.5, which results in a prior variance for the slope coefficients of approximately 109. This result implies that we needed a prior with variance two orders of magnitude smaller than what we used as a default to result in the best predictive model.

Overall, this form of regularization is easy to perform, allows us to use all the covariates in the model, ameliorates multicollinearity, and leads to a better overall predictive model. The algorithms we provided in this chapter perform the regularization parameter search and cross-validation in sequence, but recall that we can perform these tasks in parallel if desired.

From a model selection perspective, regularization has become popular among statisticians and ecologists (e.g., Gerber et al. 2015) and is a natural solution in the Bayesian setting where shrinkage is induced through the prior anyway. It can be applied to any type of regression model in the same way that we presented here, with a precise prior that shrinks β toward zero. When we have actual scientifically derived prior information, it will also perform a type of regularization and may improve predictive ability as well.

14.1   ADDITIONAL CONCEPTS AND READING

We only described the ridge regression approach to Bayesian regularization for linear models in this chapter and that relies on a normal prior for β_j. However, we can explore other forms of regularization by modifying the prior for β in our linear model. For example, we can induce the least absolute shrinkage and selection operator (lasso) approach to regularization by using a double exponential prior for β_j instead of a normal prior (Park and Casella 2008).

In our example, we identified the optimal amount of shrinkage by tuning the prior variance to yield the best predictive score for the regularized model. However, this approach is sometimes referred to as empirical Bayes because it uses the data twice to obtain the final inference: once to identify the optimal shrinkage and then again to fit the regularized model to the full data set. The fully Bayesian approach to regularization allows the prior variance to be a random parameter that has its own prior distribution (e.g., Park and Casella 2008). This extends the model by an additional unknown parameter, but only uses the data once. We must specify also the prior for the regularization parameter and there are a variety of ways to do that, some of which may yield a better predictive model than others.

On the topic of cross-validation, there are many perspectives on how best to do it for model selection (Arlot and Celisse 2010). Cross-validation requires us to fit the model several times to the folds of hold-out data and it can be computationally intensive for large models. Also, there may not be a straightforward way to partition the data for models that have complicated dependence structures and/or measurement error. However, we can approximate some scoring functions developed based on cross-validation without actually performing cross-validation. For example, Geisser (1993) described the conditional predictive ordinate (CPO), which is the predictive distribution for an observation in the data set obtained by holding that observation out. Although CPO was originally used for identifying outliers (Pettit 1990), it also provides a leave-one-out predictive score based on cross-validation. We can approximate the CPO using the harmonic posterior mean of the predictive distribution obtained evaluated at the data by fitting the model to the full data set (Held et al. 2010). For more details on this approach and others, see Hooten and Hobbs (2015).

We apply most uses of regularization in regression model settings and aim to shrink the regression coefficients toward zero. However, in principle, we can apply regularization in any model by penalizing parameters in ways that reduce model complexity.

15 Bayesian Model Averaging

We usually focus model selection procedures on identifying the model that best captures the data generating process.[1] After identifying the best model, we make a statistical inference on that model. Although this model selection procedure seems like it would provide inference based only on a single model, it actually results in multimodel inference because the entire statistical procedure is conditional on the set of models assessed. The final inference depends on the set of models considered because we do not know the selection of the model *a priori*. Bayesian model averaging (BMA) also considers a set of models, but allows us to weight all of the models explicitly when making final inference. Thus, BMA also provides a form of multimodel inference, but the final inference is not limited to the results of any one model in the set. The concept of BMA is similar to that of mixture models (Chapter 10), but it is usually presented in the slightly different context of multimodel inference.

Suppose that we are interested in a mixture distribution of the derived quantity $g \equiv g(\boldsymbol{\theta}, \tilde{\mathbf{y}})$, where the function g can depend on either the parameters $\boldsymbol{\theta}$ or data $\tilde{\mathbf{y}}$ (observed or unobserved) or both. Then we write the mixture distribution as a weighted average,

$$[g|\mathbf{y}] = \sum_{l=1}^{L} [g|\mathbf{y}, \mathcal{M}_l] P(\mathcal{M}_l|\mathbf{y}), \tag{15.1}$$

where $[g|\mathbf{y}, \mathcal{M}_l]$ represents the posterior distribution of g for model $\mathcal{M}_l$ and $P(\mathcal{M}_l|\mathbf{y})$ is the posterior probability of the model $\mathcal{M}_l$. The BMA method uses Bayes rule to find the posterior model probability $P(\mathcal{M}_l|\mathbf{y})$ based on the data, the set of models we wish to average, and their prior probabilities $P(\mathcal{M}_l)$, for $l = 1, \ldots, L$.

We use conditional probability in the form of Bayes rule to write the posterior model probability as

$$P(\mathcal{M}_l|\mathbf{y}) = \frac{[\mathbf{y}|\mathcal{M}_l]P(\mathcal{M}_l)}{\sum_{v=1}^{L}[\mathbf{y}|\mathcal{M}_v]P(\mathcal{M}_v)}, \tag{15.2}$$

where $[\mathbf{y}|\mathcal{M}_l]$ is the marginal distribution of the observed data for model $\mathcal{M}_l$. As we eluded to in Chapter 1, the marginal distribution of the data is the crux in implementing Bayesian models. If we could find the marginal distribution, then we could apply Bayes rule directly and avoid MCMC altogether. However, deriving the marginal distribution for a given model and data set often involves calculus and

---

[1] This method often is quantified based on predictive ability as discussed in the previous chapters.

may not be analytically tractable. We need the marginal distribution of the data to perform model averaging. Thus, we need a way to calculate it given MCMC output.

Recall from the Bayes rule that the posterior distribution is the joint distribution divided by the marginal distribution of the data. Thus, rearranging the Bayes rule we have

$$[\mathbf{y}|\mathcal{M}_l] = \frac{[\mathbf{y}|\boldsymbol{\theta},\mathcal{M}_l][\boldsymbol{\theta}]}{[\boldsymbol{\theta}|\mathbf{y},\mathcal{M}_l]}. \tag{15.3}$$

Therefore, if we have observed data and the parameters, then we can calculate the marginal distribution using the model components. The problem with this approach is that we do not know $\boldsymbol{\theta}$.

A second approach is to write the marginal distribution as an integrated version of the joint distribution

$$[\mathbf{y}|\mathcal{M}_l] = \int [\mathbf{y},\boldsymbol{\theta}|\mathcal{M}_l]d\boldsymbol{\theta}, \tag{15.4}$$

$$= \int [\mathbf{y}|\boldsymbol{\theta},\mathcal{M}_l][\boldsymbol{\theta}|\mathcal{M}_l]d\boldsymbol{\theta}, \tag{15.5}$$

which we can approximate using composition sampling by simulating $\boldsymbol{\theta}^{(k)}$ for $k = 1, \ldots, K$, from the prior $[\boldsymbol{\theta}|\mathcal{M}_l]$ and then averaging $[\mathbf{y}|\boldsymbol{\theta}^{(k)},\mathcal{M}_l]$ as a derived quantity. This approach may work well if the prior is informative and resembles the posterior distribution. Otherwise, it will be highly unstable due to overly diffuse prior samples.

Gelfand and Dey (1994) and Carlin and Louis (2009) described yet another approach for approximating the marginal data distribution. Their approach uses a form of harmonic averaging. Working from the typical posterior distribution expression, they rearrange the Bayes rule as

$$\frac{1}{[\mathbf{y}|\mathcal{M}_l]} = \frac{[\boldsymbol{\theta}|\mathbf{y},\mathcal{M}_l]}{[\mathbf{y}|\boldsymbol{\theta},\mathcal{M}_l][\boldsymbol{\theta}]}. \tag{15.6}$$

Then they multiply an integral that equals 1 to the right-hand side of (15.6) so that

$$\frac{1}{[\mathbf{y}|\mathcal{M}_l]} = \frac{[\boldsymbol{\theta}|\mathbf{y},\mathcal{M}_l]}{[\mathbf{y}|\boldsymbol{\theta},\mathcal{M}_l][\boldsymbol{\theta}]} \int f(\boldsymbol{\theta})d\boldsymbol{\theta}, \tag{15.7}$$

for any function $f(\boldsymbol{\theta})$ that integrates to 1. Because the right-hand side of (15.6) is constant with respect to observed $\mathbf{y}$, it can be placed inside the integral resulting in the posterior mean of a derived quantity

$$\frac{1}{[\mathbf{y}|\mathcal{M}_l]} = \int \frac{f(\boldsymbol{\theta})}{[\mathbf{y}|\boldsymbol{\theta},\mathcal{M}_l][\boldsymbol{\theta}]} [\boldsymbol{\theta}|\mathbf{y},\mathcal{M}_l] d\boldsymbol{\theta}, \tag{15.8}$$

$$\approx \frac{\sum_{k=1}^{K} \frac{f(\boldsymbol{\theta}^{(k)})}{[\mathbf{y}|\boldsymbol{\theta}^{(k)},\mathcal{M}_l][\boldsymbol{\theta}^{(k)}]}}{K}, \tag{15.9}$$

where $\boldsymbol{\theta}^{(k)} \sim [\boldsymbol{\theta}|\mathbf{y},\mathcal{M}_l]$, for $k = 1,\dots,K$, comprise a MCMC sample from the posterior. After we calculate (15.9) using MCMC output, we take the reciprocal of it to get the marginal data distribution $[\mathbf{y}|\mathcal{M}_l]$.

In practice, we still need to choose the function $f(\boldsymbol{\theta})$. The form for $f(\boldsymbol{\theta})$ will be most efficient if it is similar to the joint distribution in shape. Thus, we could use $f(\boldsymbol{\theta}) \equiv \mathrm{N}(E(\boldsymbol{\theta}|\mathbf{y}), \mathrm{Var}(\boldsymbol{\theta}|\mathbf{y}))$. In this case, the function $f(\boldsymbol{\theta})$ is a multivariate normal distribution that approximates the posterior.

To perform BMA, we specify the set of models we wish to average using good scientific judgement about their form and components. Then, we fit each model to the data set $\mathbf{y}$, using MCMC. Next, we use the MCMC samples to compute the marginal distributions $[\mathbf{y}|\mathcal{M}_l]$ for all $l = 1,\dots,L$ models. We then substitute the values for these marginal distributions evaluated at the observed data into (15.2) to get the posterior model probabilities for each model. Finally, we use the posterior model probabilities to calculate or summarize the BMA posterior distribution of the quantity g using (15.1).

In addition to selecting an appropriate function $f(\boldsymbol{\theta})$, this procedure relies on our ability to fit a set of models individually. For a relatively small or moderate set of models ($1 < L < 100$), it may be reasonable to fit each individually, but for larger model sets the procedure becomes computationally prohibitive. For example, if we had a regression model with 20 covariates and we wanted to use BMA over all possible subsets, that would require us to fit $2^{20} = 1,048,576$ models.

Thus, alternative methods were designed to perform model averaging and calculate posterior model probabilities without needing to fit each model individually. These methods are known as reversible-jump MCMC (RJMCMC) algorithms (Green 1995). RJMCMC approaches rely on model statements that treat the model index l (or a suitable reparameterization) as an unknown parameter in the model and then the form of the data and process models depend on that model index yielding a posterior of the form

$$[\boldsymbol{\theta},l|\mathbf{y}] \propto [\mathbf{y}|\boldsymbol{\theta},l][\boldsymbol{\theta}|l][l]. \tag{15.10}$$

When sampling the parameters associated with model $\mathcal{M}_l$, the dimensionality of the parameter space will likely change because some models have more parameters than others. In this case, the usual M-H ratio does not properly account for the transdimensionality and we must adjust it so that the Markov chain does not get stuck on models with larger spaces just because they are larger. The necessary MCMC algorithm adjustment is "reversible" to allow the chain to move among models with differing dimensionality.

   RJMCMC algorithms are model specific and somewhat complicated, thus we describe another approach to obtain posterior model probabilities and model averaging introduced by Barker and Link (2013). The benefit of the approach presented by Barker and Link (2013) is that it allows us to perform RJMCMC in two stages. In the first stage, we fit the models independently.[2] We can do this step in parallel, which can save substantial computational time. In the second stage, we post-process the output from the first stage, resampling the model index and parameters.

   The Barker and Link (2013) RJMCMC approach applied to our regression model is described as follows. Assuming we have a K-dimensional MCMC sample ($k = 1, \ldots, K$) for $\boldsymbol{\beta}^{(k)}$ and $\sigma^{2(k)}$ for each model $\mathcal{M}_l$ for $l = 1, \ldots, L$, we follow these steps:

1. Set the MCMC index $k = 1$.
2. Choose starting model $\mathcal{M}_l$ (the full model is a reasonable choice).
3. Select $\sigma^{2(k)}$ and $\beta_j^{(k)}$ for $j = 1, \ldots, p_l$, where p_l is the number of coefficients in model $\mathcal{M}_l$.
4. If the current model $\mathcal{M}_l$ is not the full model, then sample the remaining coefficients from a known distribution (e.g., a standard normal distribution); call these coefficients $\mathbf{u}^{(k)}$.
5. Compute the full-conditional model probability

$$P(\mathcal{M}_l|\cdot) = \frac{[\mathbf{y}|\boldsymbol{\beta}^{(k)}, \mathbf{u}^{(k)}, \sigma^{2(k)}, \mathcal{M}_l][\boldsymbol{\beta}^{(k)}|\mathcal{M}_l][\mathbf{u}^{(k)}|\mathcal{M}_l]P(\mathcal{M}_l)}{\sum_{v=1}^{L}[\mathbf{y}|\boldsymbol{\beta}^{(k)}, \mathbf{u}^{(k)}, \sigma^{2(k)}, \mathcal{M}_v][\boldsymbol{\beta}^{(k)}|\mathcal{M}_v][\mathbf{u}^{(k)}|\mathcal{M}_v]P(\mathcal{M}_v)}.$$

(15.11)

6. Sample the next model $\mathcal{M}_l$ from a categorical distribution with probabilities equal to the full-conditional model probabilities.
7. Let $k = k + 1$ and go to step 3 until $k = K$.[3]

We can summarize the output of this second stage algorithm for BMA inference. For example, the RJMCMC approximation of the posterior model probability $P(\mathcal{M}_l|\mathbf{y})$ is the number of times we selected model $\mathcal{M}_l$ at step 6 in this algorithm divided by K.

   We fit each of the same L models described in Chapter 13 (using the same priors) to the white-tailed deer data independently using the same MCMC algorithm that we used in Chapter 11. The R code we used to fit the models and store the MCMC output is

---

[2] Because this method requires us to fit each model individually, we cannot use it in cases with massive model sets. However, it allows us to avoid calculating the marginal data distribution directly, so it is a computational trade off.

[3] Barker and Link (2013) state that we can choose k at random from the initial set of MCMC iteration indices (which we do in our R code); this may reduce the autocorrelation in the initial MCMC output for use in the second stage RJMCMC algorithm.

BOX 15.1   FIT LINEAR REGRESSION MODELS TO DEER DATA

```
1  load("deer.RData")
2  n=dim(deer.df)[1]
3  y=deer.df$y
4  n=length(y)
5
6  L=11
7  X.list=vector("list",L)
8  X.list[[11]]=cbind(rep(1,n),as.matrix
9    (deer.df[,c(2:5)]))
10 X.list[[11]][,2]=ifelse(X.list[[11]][,2]==599,0,1)
11
12 X.list[[1]]=X.list[[11]][,c(1,2)]
13 X.list[[2]]=X.list[[11]][,c(1,3)]
14 X.list[[3]]=X.list[[11]][,c(1,4)]
15 X.list[[4]]=X.list[[11]][,c(1,4:5)]
16 X.list[[5]]=X.list[[11]][,c(1,2:3)]
17 X.list[[6]]=X.list[[11]][,c(1,2,4)]
18 X.list[[7]]=X.list[[11]][,c(1,2,4,5)]
19 X.list[[8]]=X.list[[11]][,c(1,2,3,4)]
20 X.list[[9]]=X.list[[11]][,c(1,3,4)]
21 X.list[[10]]=X.list[[11]][,c(1,3,4,5)]
22
23 mcmc.out.list=vector("list",L)
24 p.vec=rep(0,L)
25 n.mcmc=100000
26 source("norm.reg.mcmc.R")
27 set.seed(1)
28 for(l in 1:L){
29   p.vec[l]=dim(X.list[[l]])[2]
30   mcmc.out.list[[l]]=norm.reg.mcmc(y=y,
31     X=X.list[[l]],beta.mn=rep(0,p.vec[l]),
32     beta.var=10000,s2.mn=50,s2.sd=1000,
33     n.mcmc=n.mcmc)
34 }
35 p.max=max(p.vec)
```

Assuming equal prior model probabilities (i.e., $P(\mathcal{M}_l) = 1/L$), we processed the resulting MCMC output using the second stage RJMCMC algorithm described by Barker and Link (2013) that we just presented using the following R code:

BOX 15.2   SECOND STAGE RJMCMC ALGORITHM

```
 1 mod.mat=matrix(0,L,5)
 2 mod.mat[1,]=c(1,1,0,0,0)
 3 mod.mat[2,]=c(1,0,1,0,0)
 4 mod.mat[3,]=c(1,0,0,1,0)
 5 mod.mat[4,]=c(1,0,0,1,1)
 6 mod.mat[5,]=c(1,1,1,0,0)
 7 mod.mat[6,]=c(1,1,0,1,0)
 8 mod.mat[7,]=c(1,1,0,1,1)
 9 mod.mat[8,]=c(1,1,1,1,0)
10 mod.mat[9,]=c(1,0,1,1,0)
11 mod.mat[10,]=c(1,0,1,1,1)
12 mod.mat[11,]=c(1,1,1,1,1)
13
14 p.M=rep(1/L,L) # prior model probs
15 M=L # initial model index
16 M.save=rep(M,n.mcmc)
17 q=mcmc.out.list[[1]]$q
18 r=mcmc.out.list[[1]]$r
19 set.seed(1)
20 k.rand.seq=sample(2:n.mcmc,n.mcmc-1)
21 for(k in k.rand.seq){
22   beta=mcmc.out.list[[M]]$beta.save[,k]
23   s=sqrt(mcmc.out.list[[M]]$s2.save[k])
24   p.u=p.max-p.vec[M]
25   u=rnorm(p.u)
26   theta=rep(0,5)
27   theta[mod.mat[M,]==1]=beta
28   theta[mod.mat[M,]==0]=u
29   tmp.sum=rep(0,L)
30   for(l in 1:L){
31     tmp.sum[l]=sum(dnorm(y,X.list[[l]]%*%theta
32        [mod.mat[l,]==1],s,log=TRUE))+sum(dnorm
33        (theta[mod.mat[l,]==1],0,100,log=TRUE))
34        +sum(dnorm(theta[mod.mat[l,]==0],log=TRUE))
35        +dgamma(s^(-2),q,,r,log=TRUE)+log(p.M[M])
36   }
```

(Continued)

BOX 15.2 (Continued)   SECOND STAGE RJMCMC ALGORITHM

```
37  P.M=exp(tmp.sum-max(tmp.sum)-log(sum(exp
38    (tmp.sum-max(tmp.sum)))))
39  M=sample(1:L,1,prob=P.M)
40  M.save[k]=M
41  }
42  table(M.save)/n.mcmc
```

Note that, on line 35 of the RJMCMC algorithm, our use of dgamma($\cdots$, log=TRUE) corresponds to the natural logarithm of the inverse gamma prior PDF, $\log[\sigma^2]$, using the same parameterization from previous chapters. The resulting posterior model probabilities for each model are shown in Table 15.1. We can see that all models except 10 and 11 resulted in negligible posterior model probabilities and nearly all the probability mass in the model average was attributed to model 10, the model without the large cropland covariate (Table 15.1). This result agrees with our previous model selection analysis where we found that model 10 had the best predictive score (DIC) with model 11 a close second. The posterior model probabilities for models 10 and 11

Table 15.1

The Resulting Posterior Model Probabilities $P(\mathcal{M}_l|y)$ from $L = 11$ Linear Regression Models Fit to the White-Tailed Deer Body Mass Data

	Covariate				
Model	Crop	Sex	Age	Age^2	$P(\mathcal{M}_l|y)$
1	x				0.0000
2		x			0.0000
3			x		0.0000
4			x	x	0.0000
5	x	x			0.0000
6	x		x		0.0000
7	x		x	x	0.0000
8	x	x	x		0.0000
9		x	x		0.0000
10		x	x	x	0.9881
11	x	x	x	x	0.0119

Note: An intercept was included in all models and 'x' indicates which covariates were included in each model.

do not appear to be as close as the DIC results, but they are still the two best scoring models nonetheless.

Notice that we used a special technique in our second stage RJMCMC algorithm (line 37) that results in a much more stable way to compute ratios like that in the full-conditional model probability. For example, to compute a ratio $c_l/\sum_{v=1}^{L} c_v$, we can use

$$\frac{c_l}{\sum_{v=1}^{L} c_v} = \exp\left(\log(c_l) - a - \log\left(\sum_{v=1}^{L} \exp(\log(c_v) - a)\right)\right), \quad (15.12)$$

which takes advantage of the stability associated with calculating the natural logarithm of density functions in R (using the "log=TRUE" option) and minimizes numerical underflow/overflow issues by strategically subtracting a which is the maximum value of $\log(c_l)$ over all $l = 1,\ldots,L$. This technique (which is sometimes called the "log–sum–exp trick" works because

$$\sum_{v=1}^{L} c_v = \sum_{v=1}^{L} e^{\log(c_v)}, \quad (15.13)$$

$$= \sum_{v=1}^{L} e^a e^{-a} e^{\log(c_v)}, \quad (15.14)$$

$$= e^a \sum_{v=1}^{L} e^{-a} e^{\log(c_v)}, \quad (15.15)$$

$$= e^a \sum_{v=1}^{L} e^{\log(c_v)-a}. \quad (15.16)$$

Thus, taking the log of (15.16) results in

$$\log\left(\sum_{v=1}^{L} c_v\right) = a + \log\left(\sum_{v=1}^{L} e^{\log(c_v)-a}\right), \quad (15.17)$$

and setting a equal to the maximum of the $\log(c_v)$ values appropriately attenuates $e^{\log(c_v)-a}$ so that it is computationally stable. Because the technique involves multiplying the product $e^a e^{-a} = 1$, it will not change the result in theory, but should cause fewer problems than trying to compute a large sum of a product of several densities. We can also use this approach to fit mixture models with many mixture components (Chapter 10) because the full-conditional distribution for the latent mixture indicator involves a similar ratio.

To perform BMA for predicted body mass in the linear regression setting, we could either augment the second stage RJMCMC algorithm with a line that simulates a predictive realization using the predictive full-conditional or we could use a third algorithm based on the output of the second algorithm. We opted for a third algorithm to obtain BMA predictions for two individual deer (the first and last in the data set) using the output from the previous algorithms:

BOX 15.3   MODEL AVERAGED PREDICTIONS

```
1  y.pred.save=matrix(0,n.mcmc,2)
2  for(k in 1:n.mcmc){
3    beta=mcmc.out.list[[M.save[k]]]$beta.save[,k]
4    s=sqrt(mcmc.out.list[[M.save[k]]]$s2.save[k])
5    y.pred.save[k,]=rnorm(2,X.list[[M.save[k]]]
6      [c(1,n),]%*%beta,s)
7  }
8  boxplot(data.frame(y.pred.save),ylab="y",
9      names=c("a","b"))
```

This R code results in a RJMCMC sample from the PPDs summarized as boxplots in Figure 15.1 for body mass (kg) of two white-tailed deer. Notice that the separation in the posterior boxplots indicates that the body mass for individuals with these two different sets of covariates will be significantly different (Figure 15.1). This outcome is a result of individual (a) being female and individual (b) being male, as well as individual (b) being 4.5 years older than individual (a).

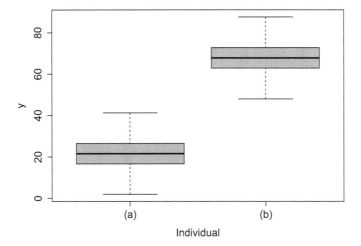

Figure 15.1 Posterior predictive distributions summarized as boxplots for individual deer with two different sets of covariates. Predictions are for covariates (a) $\mathbf{x}_a = (1,1,0,0.5,$ $0.25)'$ and (b) $\mathbf{x}_b = (1,0,1,5,25)'$.

15.1   ADDITIONAL CONCEPTS AND READING

A concept related to BMA is the Bayes factor (Kass and Raftery 1995). The Bayes factor is defined as the ratio of marginal data distributions associated with two models that we seek to compare. For example, for models $\mathcal{M}_1$ and $\mathcal{M}_2$, the Bayes factor is

$$B_{1,2} = \frac{[\mathbf{y}|\mathcal{M}_1]}{[\mathbf{y}|\mathcal{M}_2]} . \tag{15.18}$$

Studies have used Bayes factors to compare models directly because they are a ratio of the "evidence" for each model. Thus, conventional interpretations of Bayes factors used rules of thumb such as $B_{1,2} > 10$ implies strong evidence for model $\mathcal{M}_1$ versus $\mathcal{M}_2$ given the data (Jeffreys 1961). However, we also can use Bayes factors to perform model averaging because we can show that the posterior odds equals the product of the Bayes factor and the prior odds. For example, the posterior odds for models $\mathcal{M}_1$ and $\mathcal{M}_2$ are

$$\frac{P(\mathcal{M}_1|\mathbf{y})}{P(\mathcal{M}_2|\mathbf{y})} = B_{1,2} \frac{P(\mathcal{M}_1)}{P(\mathcal{M}_2)} . \tag{15.19}$$

Thus, the Bayes factor relates the prior model probabilities to the posterior model probabilities, which we need for Bayesian model averaging (Link and Barker 2006). Bayes factors provide a useful tool for comparing models and performing BMA, but Spiegelhalter and Smith (1982) noted that Bayes factors are not well-defined when improper priors are specified for model parameters. We also may find Bayes factors challenging to compute because they rely on a knowledge of the marginal data distributions.

   In terms of the range of applications for which BMA is useful, certain types of model averaging have come under scrutiny after decades of use in the fields of wildlife biology and ecology. The main caveat about model averaging is associated with the interpretation of parameters from different models that have been averaged. Cade (2015) and Banner and Higgs (2017) summarize the main issues associated with model averaging regression coefficients. In particular, it is tempting to model average coefficients associated with the same covariates used in a suite of different regression models. However, because the inference associated with a regression coefficient is conditional on the other parameters in the model (unless the covariates are all perfectly uncorrelated), its interpretation varies across models. This variation in interpretation implies that we should not model average regression coefficients in most cases (unless there is negligible collinearity among covariates). On the other hand, the response variables have the same interpretation regardless of model structure. Therefore, model averaged predictions of the data are valid and have improved predictive ability (Madigan and Raftery 1994; Raftery et al. 1997).

Overall, BMA and model averaging in general provide another set of tools for ecologists and environmental scientists to consider in their research. However, as we might expect, recent literature suggests matching the type of inference with the goals of the specific study (Fieberg and Johnson 2015; Ver Hoef and Boveng 2015).

Finally, we applied the RJMCMC method presented by Barker and Link (2013) in our BMA example. However, many other methods exist for performing BMA. For example, see Clyde et al. (2011) and the references therein for alternative sampling approaches.

16 Time Series Models

16.1 UNIVARIATE AUTOREGRESSIVE MODELS

We turn our attention to autoregression in this chapter, having covered regression in the preceding chapters. Autoregressive models are a class of models that explain temporal structure in data using dynamics (or covariance, depending on how we look at it). That is, instead of regressing a response variable on a set of covariates, in autoregressive models, we regress the response variable on itself. Although a huge suite of statistical models exist for temporal data, including both discrete-time and continuous-time models, we focus on first-order discrete-time autoregressive models in this chapter because they are commonly used in ecological and environmental analyses.

In this chapter, we denote response data as $\mathbf{y} \equiv (y_1, \ldots, y_t, \ldots, y_T)'$ and we assume that each observation y_t is continuously valued on real support ($y_t \in \Re$). Although many types of observed temporal data are actually positive-valued (or discrete-valued), we can transform them to have real support in some cases, or if discrete-valued, we can develop a hierarchical model with a continuous-valued latent temporal process. Hooten et al. (2017) provide an accessible overview of statistical models for temporal data, including traditional and Bayesian approaches for specifying and fitting them. In this chapter, we focus on the implementation of dynamic autoregressive models and point out a few important aspects of constructing MCMC algorithms to fit these models to time series data.

Traditionally, autoregressive time series models are specified conditionally so that the data at time t depend on the previous data at time $t-1$ (and perhaps earlier, depending on the order of the model). For example, we can specify a Gaussian autoregressive time series model as

$$y_t \sim \text{N}(\alpha y_{t-1}, \sigma^2), \tag{16.1}$$

for $t = 2, \ldots, T$. This conditional specification appears to imply that y_t only depends on y_{t-1}, but its appearance can be misleading. In fact, the entire set of observations are dependent according to the model (16.1), but y_t is independent of the rest of the data when conditioned on y_{t-1}.[1] This type of dependence is referred to as Markov dependence. Thus, the autoregressive model of order 1 (AR(1)) in (16.1) is a type of Markov model. The AR(1) model has two parameters, a variance parameter σ^2 and an autoregressive parameter α that controls the amount of dependence in the temporal process. Increasing dependence can be interpreted visually as smoothness in the

---

[1] We can show the joint distribution associated with the conditional model in (16.1) as multivariate normal with mean zero and covariance matrix that has $\frac{\sigma^2}{1-\alpha^2}$ for the diagonal elements and $\frac{\sigma^2}{1-\alpha^2}\alpha^{|t-\tau|}$ for the off-diagonal elements corresponding to times t and τ, when $|\alpha| < 1$.

time series. Smoother time series will have larger values for α. Based on the specification in (16.1), the process will be stationary[2] around zero when $-1 < \alpha < 1$. Thus, most model specifications constrain α based on an assumption that **y** is stationary (i.e., will return to zero eventually). Furthermore, Figure 16.1 illustrates that positive values for α lead to positive dependence (i.e., smoothness in the temporal process) and negative values lead to negative dependence in the time series (i.e., a saw-like shape to the temporal process). The following R code simulates three separate AR(1) processes with the same variance ($\sigma^2 = 1$) but with varying amounts and types of dependence ($\alpha = 0.99$, $\alpha = 0$, and $\alpha = -0.8$):

BOX 16.1   SIMULATE AR(1) PROCESSES

```
T=100
y.1=rep(0,T)
y.2=rep(0,T)
y.3=rep(0,T)
alpha.1=.99
alpha.2=0
alpha.3=-.8
sig=1

set.seed(1)
for(t in 2:T){
   y.1[t]=rnorm(1,alpha.1*y.1[t-1],sig)
   y.2[t]=rnorm(1,alpha.2*y.2[t-1],sig)
   y.3[t]=rnorm(1,alpha.3*y.3[t-1],sig)
}

layout(matrix(1:3,3,1))
plot(1:T,y.1,type="l",main="a")
abline(h=0,col=8)
plot(1:T,y.2,type="l",main="b")
abline(h=0,col=8)
plot(1:T,y.3,type="l",main="c")
abline(h=0,col=8)
```

[2] In this time series setting, stationarity implies that the properties of the joint distribution of all y_t, for $t = 1,\ldots,T$, do not change if the process is shifted earlier or later in time. Also, stationarity in these models implies that the process is "mean-reverting" and stochasticity early in the time series decays over time. A non-stationary AR(1) time series (i.e., when $|\alpha| \geq 1$) has a variance that diverges to infinity as the length of the time series increases.

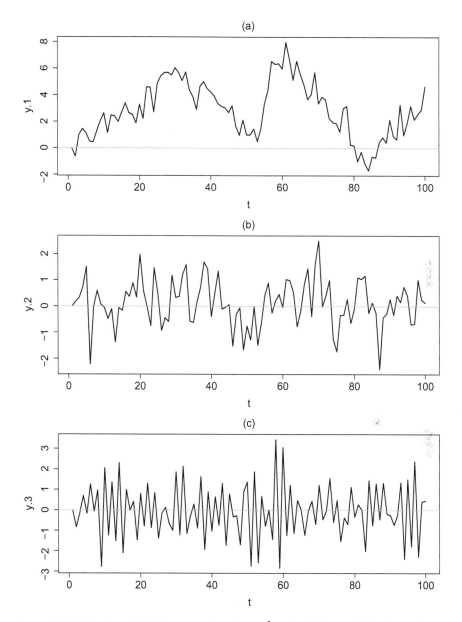

Figure 16.1 Simulated AR(1) processes based on $\sigma^2 = 1$ and (a) $\alpha = 0.99$, (b) $\alpha = 0$, and (c) $\alpha = -0.8$. The horizontal gray line indicates zero, the mean of the AR(1) processes shown.

Notice in this R code for creating Figure 16.1, we set all the initial values equal to zero, the expected value of the process. We could use any value we prefer for a starting value, but it does raise an interesting question about how to estimate the initial state. When existing data are available, and there is no latent process in the model (i.e., in a hierarchical model), we can condition on the first observation as the initial state. However, when we have an AR(1) model for a latent temporal process, we may wish to estimate the initial state, y_0 in this case. We do that by specifying a prior for it and estimating it with the other model parameters. For now, though, we condition on y_1 as the initial state.

From a Bayesian perspective, the stability constraint on α is easy to enforce with a prior. For example, we could specify a prior as $\alpha \sim \text{Unif}(0,1)$ to constrain the process to have positive dependence only.

We can specify the prior for σ^2 as in standard regression models. For example, we could specify an inverse gamma prior distribution for σ^2 such that $\sigma^2 \sim \text{IG}(q,r)$. Although the process (16.1) will have mean zero if $-1 < \alpha < 1$, the parameter σ^2 controls the magnitude of the process. Larger values of σ^2 allow the y_t to venture farther from the expected value of zero.

Consider the time series of average daily soil moisture data (in volumetric water content from August 11 to November 6, 2017) at the Kenai Site in the U.S. Regional Climate Reference Network (USRCRN) shown in Figure 16.2. We first centered the data by subtracting the mean so that the transformed data have mean zero and then created Figure 16.2 using the following R code:

BOX 16.2   PLOT SOIL MOISTURE DATA

```
1 load("soil.RData")
2 T=length(y)
3 y.c=y-mean(y)
4
5 plot(y,type="l",main="")
6 abline(h=0,col=8)
```

Then we specified the following model for these soil moisture measurements as

$$y_t \sim \text{N}(\alpha y_{t-1}, \sigma^2), \tag{16.2}$$

$$\alpha \sim \text{N}(\mu_\alpha, \sigma_\alpha^2), \tag{16.3}$$

$$\sigma^2 \sim \text{IG}(q,r), \tag{16.4}$$

for $t = 2, \ldots, T$. The posterior distribution for this model is

$$[\alpha, \sigma^2|\mathbf{y}] \propto \left(\prod_{t=2}^{T} [y_t|y_{t-1}, \alpha, \sigma^2] \right) [\alpha][\sigma^2], \tag{16.5}$$

where $[y_t|y_{t-1}, \alpha, \sigma^2]$ is the normal data model (16.1) and $[\alpha]$ and $[\sigma^2]$ are the priors.

To construct a MCMC algorithm for this model, we need to derive the full-conditional distributions for the parameters α and σ^2. Notice that we did not use

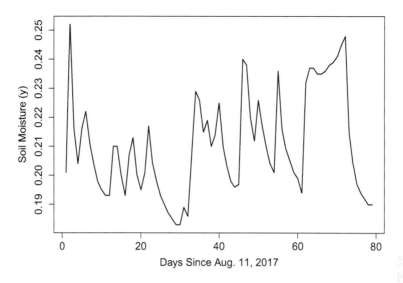

Figure 16.2 Soil moisture data for 79 days at the USRCRN station, Kenai Site.

a constrained prior for α, instead we used a normal prior. Based on the normal prior, the full-conditional distribution for α is conjugate and results in a Gibbs update in our MCMC algorithm because

$$[\alpha|\cdot] \propto \left(\prod_{t=2}^{T} [y_t|y_{t-1}, \alpha, \sigma^2] \right) [\alpha], \tag{16.6}$$

$$\propto \exp\left(-\frac{1}{2} \frac{\sum_{t=2}^{T}(y_t - \alpha y_{t-1})^2}{\sigma^2} \right) \exp\left(-\frac{1}{2} \frac{(\alpha - \mu_\alpha)^2}{\sigma_\alpha^2} \right), \tag{16.7}$$

$$\propto \exp\left(-\frac{1}{2} \left(-\frac{2(\sum_{t=2}^{T} y_t y_{t-1})\alpha}{\sigma^2} + \frac{(\sum_{t=2}^{T} y_{t-1}^2)\alpha^2}{\sigma^2} - \frac{2\mu_\alpha \alpha}{\sigma_\alpha^2} + \frac{\alpha^2}{\sigma_\alpha^2} \right) \right), \tag{16.8}$$

$$\propto \exp\left(-\frac{1}{2} \left(-2\left(\frac{\sum_{t=2}^{T} y_t y_{t-1}}{\sigma^2} + \frac{\mu_\alpha}{\sigma_\alpha^2} \right) \alpha + \left(\frac{\sum_{t=2}^{T} y_{t-1}^2}{\sigma^2} + \frac{1}{\sigma_\alpha^2} \right) \alpha^2 \right) \right), \tag{16.9}$$

$$\propto \exp\left(-\frac{1}{2} \left(-2b\alpha + a\alpha^2 \right) \right), \tag{16.10}$$

which implies that $[\alpha|\cdot] \equiv N(a^{-1}b, a^{-1})$, where

$$a \equiv \frac{\sum_{t=2}^{T} y_{t-1}^2}{\sigma^2} + \frac{1}{\sigma_\alpha^2}, \tag{16.11}$$

$$b \equiv \frac{\sum_{t=2}^{T} y_t y_{t-1}}{\sigma^2} + \frac{\mu_\alpha}{\sigma_\alpha^2}, \tag{16.12}$$

by completing the square as shown in Chapter 7.

The full-conditional distribution for σ^2 is also conjugate because we used an inverse gamma prior. We derive the full-conditional distribution for σ^2 using the same technique we presented in Chapter 8

$$[\sigma^2|\cdot] \propto \left(\prod_{t=2}^{T} [y_t|y_{t-1}, \alpha, \sigma^2] \right) [\sigma^2], \tag{16.13}$$

$$\propto (\sigma^2)^{-\frac{T-1}{2}} \exp\left(-\frac{1}{2} \frac{\sum_{t=2}^{T} (y_t - \alpha y_{t-1})^2}{\sigma^2} \right) (\sigma^2)^{-(q+1)} \exp\left(-\frac{1}{\sigma^2 r} \right), \tag{16.14}$$

$$\propto (\sigma^2)^{-\left(\frac{T-1}{2}+q+1\right)} \exp\left(-\frac{1}{\sigma^2} \left(\frac{\sum_{t=2}^{T} (y_t - \alpha y_{t-1})^2}{2} + \frac{1}{r} \right) \right), \tag{16.15}$$

$$\propto (\sigma^2)^{-(\tilde{q}+1)} \exp\left(-\frac{1}{\sigma^2 \tilde{r}} \right), \tag{16.16}$$

which implies that $[\sigma^2|\cdot] \equiv \text{IG}(\tilde{q}, \tilde{r})$, where

$$\tilde{q} \equiv \frac{T-1}{2} + q, \tag{16.17}$$

$$\tilde{r} \equiv \left(\frac{\sum_{t=2}^{T} (y_t - \alpha y_{t-1})^2}{2} + \frac{1}{r} \right)^{-1}. \tag{16.18}$$

Thus, we can construct a MCMC algorithm by iteratively sampling $\sigma^{2(k)}$ and $\alpha^{(k)}$ from their full-conditional distributions for $k = 1, \ldots, K$ MCMC iterations. Our MCMC algorithm is shown in the following R code and we used it to fit the AR(1) model in (16.2)–(16.4) to the centered soil moisture data (Figure 16.2):

BOX 16.3   MCMC ALGORITHM FOR THE AR(1) MODEL

```
1  ar1.mcmc <- function(y,n.mcmc){
2
3  ###
4  ###  Variables, Priors, Starting Values
5  ###
```

(Continued)

BOX 16.3 (Continued)   MCMC ALGORITHM FOR THE AR(1) MODEL

```
6  T=length(y)
7  s2.save=rep(1,n.mcmc)
8  alpha.save=rep(0,n.mcmc)
9
10 alpha=0.1
11 mu.alpha=0
12 s2.alpha=1
13 r=1000
14 q=0.001
15
16 ###
17 ###   MCMC Loop
18 ###
19 for(k in 1:n.mcmc){
20
21   ###
22   ###   Sample s2
23   ###
24
25   r.tmp=1/(sum((y[-1]-alpha*y[-T])^2)/2 + 1/r)
26   q.tmp=(T-1)/2 + q
27   s2=1/rgamma(1,q.tmp,,r.tmp)
28
29   ###
30   ###   Sample alpha
31   ###
32
33   tmp.var=1/(sum(y[-T]^2)/s2 + 1/s2.alpha)
34   tmp.mn=tmp.var*(sum(y[-1]*y[-T])/s2
35     + mu.alpha/s2.alpha)
36   alpha=rnorm(1,tmp.mn,sqrt(tmp.var))
37
38   ###
39   ###   Save Samples
40   ###
41
42   alpha.save[k]=alpha
43   s2.save[k]=s2
44
45 }
```

(Continued)

**BOX 16.3 (Continued)   MCMC ALGORITHM FOR THE AR(1)
MODEL**

```
46 ###
47 ###   Write Output
48 ###
49
50 list(y=y,s2.save=s2.save,alpha.save=alpha.save,
51    n.mcmc=n.mcmc,mu.alpha=mu.alpha,s2.alpha=s2.
        alpha,r=r,q=q)
52
53 }
```

In this MCMC algorithm, we set the hyperparameters as $\mu_\alpha = 0$, $\sigma_\alpha^2 = 1$, $q = 0.001$, and $r = 1000$. These hyperparameter values result in a weakly informative prior for α and a relatively vague prior for σ^2. Using $K = 10000$ MCMC iterations, we fit the AR(1) model to the centered soil moisture data with the following R code:

BOX 16.4   FIT AR(1) MODEL TO SOIL MOISTURE DATA

```
1 source("ar1.mcmc.R")
2 set.seed(1)
3 mcmc.out=ar1.mcmc(y=y.c,n.mcmc=10000)
4
5 dIG <- function(x,r,q){
6   x^(-(q+1))*exp(-1/r/x)/(r^q)/gamma(q)
7 }
8
9 layout(matrix(1:2,1,2))
10 hist(mcmc.out$alpha.save,xlab=bquote(alpha),
11    main="a",xlim=c(-1,1),prob=TRUE)
12 curve(dnorm(x,mcmc.out$mu.alpha,sqrt(mcmc.out$s2.
13    alpha)),add=TRUE,lwd=1.5,from=-2,to=2)
14 hist(mcmc.out$s2.save,xlab=bquote(sigma^2),
15    main="b",prob=TRUE)
16 curve(dIG(x,mcmc.out$r,mcmc.out$q),add=TRUE,
17    lwd=1.5)
```

We summarized the MCMC output in terms of a marginal posterior density for α and σ^2 shown in Figure 16.3. The results of our AR(1) model fit (Figure 16.3) indicate that the soil moisture data show strong positive dependence in time because most of the posterior mass is between 0.5 and 1 in Figure 16.3a. This result is evidence

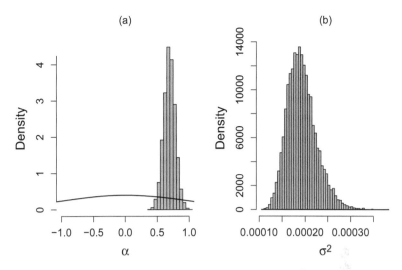

Figure 16.3 Marginal posterior distributions for (a) α and (b) σ^2 as a result of fitting the AR(1) model to the centered soil moisture data. Priors for each parameter shown as black lines.

against a hypothesis that the soil moisture values at this site and during this time frame (August 11 to November 6, 2017) are uncorrelated.

16.2 AUTOREGRESSIVE MODELS FOR POPULATIONS

The models we presented in the previous section are classical time series models in that they are purely phenomenological. Ecologists commonly do not use classical time series models. For example, ecologists may wish to characterize the dynamics in populations of organisms over time and there does not appear to be an obvious connection between AR(1) models and mechanistic population dynamics. However, we can develop a similar AR(1) time series model based on first principles of population growth. To do so, suppose that a population can increase or decrease from one period of time to the next by accumulating the individuals from the previous time period and then either adding or subtracting new individuals based on a per capita growth rate r.[3] This conceptual model for population growth is referred to as Malthusian (or exponential) growth and, in discrete time, we can formulate it stochastically as

$$y_t = (1+r)y_{t-1}e^{\varepsilon_t}, \tag{16.19}$$

---

[3] We use r in this example to denote a growth rate because this notation is common in mathematical population models. To avoid conflicting notation, we use α and β in the inverse gamma distribution for this example instead of q and r.

where $\varepsilon_t \sim N(0, \sigma^2)$. A reasonable minimum value for r is -1 because it is not realistic to lose more than the entire population in a single time period. Thus, the minimum for $1 + r$ is 0. The "error" (e^{ε_t}) in this case is multiplicative and positive only, which ensures that negative population sizes do not occur.

Even though this population model above looks different than the AR(1) time series we presented in the previous section, we can use similar methods to fit it to data. For example, if we apply the natural logarithm to both sides of (16.19), the following AR(1) model results

$$\log(y_t) \sim N(\log(1+r) + \log(y_{t-1}), \sigma^2), \tag{16.20}$$

because the multiplicative error becomes additive error when log transformed. Thus, this takes a similar form as the AR(1) model from the previous section, except that there is an "intercept" term ($\log(1+r)$) in (16.20) and the previous autoregression parameter is implicitly one ($\alpha = 1$). These conditions imply that the log transformed time series is nonstationary and that increasing populations will grow exponentially.

To specify a Bayesian population model for Malthusian growth, we let $z_t = \log(y_t)$ for $t = 1, \ldots, T$ and formulate the model components as

$$z_t \sim N(\theta + z_{t-1}, \sigma^2), \tag{16.21}$$

$$\theta \sim N(\mu_\theta, \sigma^2), \tag{16.22}$$

$$\sigma^2 \sim IG(\alpha, \beta), \tag{16.23}$$

where $\theta = \log(1+r)$. This model specification results in the posterior distribution

$$[\theta, \sigma^2 | \mathbf{z}] \propto \left(\prod_{t=2}^{T} [z_t | z_{t-1}, \theta, \sigma^2] \right) [\theta][\sigma^2]. \tag{16.24}$$

To fit this model to data using MCMC, we need to find the full-conditional distributions for θ and σ^2. In what follows, we show that the full-conditional distributions are conjugate and similar to that of the AR(1) model in the previous section.

Starting with θ, the full-conditional distribution involves all components of the joint distribution in (16.24) that include θ so that

$$[\theta | \cdot] \propto \left(\prod_{t=2}^{T} [z_t | z_{t-1}, \theta, \sigma^2] \right) [\theta], \tag{16.25}$$

$$\propto \exp\left(-\frac{1}{2} \frac{\sum_{t=2}^{T}(z_t - \theta - z_{t-1})^2}{\sigma^2} \right) \exp\left(-\frac{1}{2} \frac{(\theta - \mu_\theta)^2}{\sigma_\theta^2} \right), \tag{16.26}$$

$$\propto \exp\left(-\frac{1}{2} \left(\frac{-2\sum_{t=2}^{T}(z_t - z_{t-1})\theta}{\sigma^2} + \frac{(T-1)\theta^2}{\sigma^2} + \frac{-2\mu_\theta\theta}{\sigma_\theta^2} + \frac{\theta^2}{\sigma_\theta^2} \right) \right), \tag{16.27}$$

$$\propto \exp\left(-\frac{1}{2}\left(-2\frac{\sum_{t=2}^{T}(z_t - z_{t-1})}{\sigma^2} + \frac{\mu_\theta}{\sigma_\theta^2}\right)\theta + \left(\frac{T-1}{\sigma^2} + \frac{1}{\sigma_\theta^2}\right)\theta^2\right), \quad (16.28)$$

$$\propto \exp\left(-\frac{1}{2}\left(-2b\theta + a\theta^2\right)\right), \quad (16.29)$$

which implies that $[\theta|\cdot] \equiv N(a^{-1}b, a^{-1})$ with

$$a \equiv \frac{T-1}{\sigma^2} + \frac{1}{\sigma_\theta^2}, \quad (16.30)$$

$$b \equiv \frac{\sum_{t=2}^{T}(z_t - z_{t-1})}{\sigma^2} + \frac{\mu_\theta}{\sigma_\theta^2}. \quad (16.31)$$

Thus, we can sample from $[\theta|\cdot]$ using a Gibbs update in a MCMC algorithm.

We can derive the full-conditional distribution for σ^2 the same way we did in the previous section and the result is very similar. In fact, the main changes are the specific form of the likelihood and the change in hyperparameters for σ^2. Therefore, the full-conditional distribution for σ^2 in our Malthusian growth model is

$$[\sigma^2|\cdot] \propto \left(\prod_{t=2}^{T}[z_t|z_{t-1}, \theta, \sigma^2]\right)[\sigma^2], \quad (16.32)$$

$$\propto (\sigma^2)^{-\frac{T-1}{2}}\exp\left(-\frac{1}{2}\frac{\sum_{t=2}^{T}(z_t - \theta - z_{t-1})^2}{\sigma^2}\right)(\sigma^2)^{-(\alpha+1)}\exp\left(-\frac{1}{\sigma^2\beta}\right), \quad (16.33)$$

$$\propto (\sigma^2)^{-\left(\frac{T-1}{2}+\alpha+1\right)}\exp\left(-\frac{1}{\sigma^2}\left(\frac{\sum_{t=2}^{T}(z_t - \theta - z_{t-1})^2}{2} + \frac{1}{\beta}\right)\right), \quad (16.34)$$

$$\propto (\sigma^2)^{-(\tilde\alpha+1)}\exp\left(-\frac{1}{\sigma^2\tilde\beta}\right), \quad (16.35)$$

which implies that $[\sigma^2|\cdot] \equiv IG(\tilde\alpha, \tilde\beta)$, where

$$\tilde\alpha \equiv \frac{T-1}{2} + \alpha, \quad (16.36)$$

$$\tilde\beta \equiv \left(\frac{\sum_{t=2}^{T}(z_t - \theta - z_{t-1})^2}{2} + \frac{1}{\beta}\right)^{-1}. \quad (16.37)$$

Thus, the full-conditional distribution for σ^2 is conjugate inverse gamma and we can use a Gibbs update to sample it in the MCMC algorithm.

We constructed a MCMC algorithm for fitting the Malthusian growth time series model based on the full-conditional distributions for θ and σ^2 that we just derived. The MCMC algorithm, which we called `malthus.mcmc.R`, in R is

BOX 16.5 MCMC ALGORITHM FOR MALTHUSIAN GROWTH MODEL

```
1  malthus.mcmc <- function(z,n.mcmc){
2
3  ###
4  ###  Variables, Priors, Starting Values
5  ###
6
7  T=length(z)
8  s2.save=rep(1,n.mcmc)
9  theta.save=rep(0,n.mcmc)
10
11 theta=0.1
12 mu.theta=0
13 s2.theta=1
14 beta=1000
15 alpha=0.001
16
17 ###
18 ###  MCMC Loop
19 ###
20 for(k in 1:n.mcmc){
21
22   ###
23   ###  Sample s2
24   ###
25
26   beta.tmp=1/(sum((z[-1]-theta-z[-T])^2)/
27     2 + 1/beta)
28   alpha.tmp=(T-1)/2 + alpha
29   s2=1/rgamma(1,alpha.tmp,,beta.tmp)
30
31   ###
32   ###  Sample theta
33   ###
34
35   tmp.var=1/((T-1)/s2 + 1/s2.theta)
36   tmp.mn=tmp.var*(sum(z[-1]-z[-T])/s2
37          + mu.theta/s2.theta)
38   theta=rnorm(1,tmp.mn,sqrt(tmp.var))
```

(Continued)

BOX 16.5 (Continued)   MCMC ALGORITHM FOR MALTHUSIAN
GROWTH MODEL

```
39  ###
40  ###   Save Samples
41  ###
42
43  theta.save[k]=theta
44  s2.save[k]=s2
45
46 }
47
48 ###
49 ###   Write Output
50 ###
51
52 list(z=z,s2.save=s2.save,theta.save=theta.save,
53     n.mcmc=n.mcmc,mu.theta=mu.theta,s2.theta=s2.
           theta,alpha=alpha,beta=beta)
54
55 }
```

Using hyperparameters $\mu_\theta = 0$, $\sigma^2 = 1$, $\alpha = 0.001$, and $\beta = 1000$, we fit the Malthusian growth model to a set of log transformed annual observed population abundances of scaup (*Aythya spp.*) in a region of southeastern North Dakota from 1980 to 2009 (Ross et al. 2015). In the study region for these data, the scaup population has increased substantially over the past 50 years. We used the following R code to fit the Bayesian time series model based on Malthusian dynamics to the scaup data to estimate the per capita growth rate r:

BOX 16.6   FIT MALTHUSIAN GROWTH MODEL TO SCAUP DATA

```
1 y=c(13,14,40,84,113,53,71,42,80,57,45,20,47,24,49,
2   199,173,185,133,204,308,368,185,234,618,177,353,
3   193,206,152)
4 years=1980:2009
5 z=log(y)
6
7 source("malthus.mcmc.R")
8 set.seed(1)
9 mcmc.out=malthus.mcmc(z=z,n.mcmc=10000)
```

(*Continued*)

**BOX 16.6 (Continued)   FIT MALTHUSIAN GROWTH MODEL TO
SCAUP DATA**

```
10 dIG <- function(x,r,q){
11   x^(-(q+1))*exp(-1/r/x)/(r^q)/gamma(q)
12 }
13
14 layout(matrix(1:3,1,3))
15 hist(mcmc.out$theta.save,xlab=bquote(theta),
16     main="a",prob=TRUE)
17 curve(dnorm(x,mcmc.out$mu.theta,sqrt(mcmc.
18     out$s2.theta)),add=TRUE,from=-2,to=2)
19 abline(v=0,lty=2,lwd=2)
20 hist(exp(mcmc.out$theta.save)-1,xlab="r",
21     main="b",prob=TRUE)
22 lines(density(exp(rnorm(10000))-1,from=-2,to=2),
23     lwd=1.5)
24 abline(v=0,lwd=2,lty=2)
25 hist(mcmc.out$s2.save,xlab=bquote(sigma^2),
26     main="c",prob=TRUE)
27 curve(dIG(x,mcmc.out$beta,mcmc.out$alpha),
28     add=TRUE)
```

This R code creates the posterior density plots for θ, r, and σ^2 shown in Figure 16.4. Several features of the graphics in Figure 16.4 are noteworthy. First, notice that we treated the per capita growth rate r as a derived quantity in the model (a function of θ). Second, notice that we approximated the implied prior for r (recall the original prior was on θ) by simulating from the prior for θ and transforming those samples using the relationship between θ and r (Figure 16.4b). Third, notice that the symmetric standard normal prior for θ implies a right-skewed prior for r with most of the mass less than zero (indicating a shrinking population). In fact, this prior is not a very good representation of our actual prior knowledge for r because we expect the scaup population to be growing over time in this region. Finally, notice that the marginal posterior distributions for θ and r indicate some overlap of zero. In fact, using the MCMC output, we can calculate the amount that r overlaps zero as the probability of $r > 0$, given the data, using the R code:

BOX 16.7   POSTERIOR PROBABILITY OF $r > 0$

```
1 > mean(exp(mcmc.out$theta.save)-1>0)
2 [1] 0.7518
```

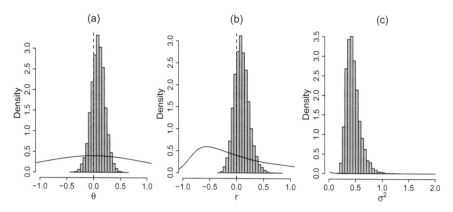

Figure 16.4 Marginal posterior distributions for (a) θ, (b) r, and (c) σ^2 as a result of fitting the Malthusian growth time series model to the scaup abundance data. Priors for each parameter shown as solid black lines. Vertical dashed black lines indicate zero.

Thus, our relative abundance data suggest that the scaup population in this region is growing with probability $P(r > 0|\mathbf{z}) = 0.75$, according to the Malthusian growth model.

16.3 PREDICTION WITH TIME SERIES MODELS

We discussed posterior predictive distributions in Chapter 11 within the context of regression models. However, our focus is on dynamic models in this chapter and we may want to use the model to interpolate temporal data or predict into the future (i.e., extrapolate or forecast in time). The concept of PPDs is the same as in Chapter 11, but to implement it using MCMC, we need to examine the predictive full-conditional more closely.

For example, consider the Malthusian population growth model from the previous section formulated in (16.21)–(16.23). Suppose we wish to make inference on the population at a given time, z_t. In this case, we can write the PPD as

$$[z_t|\mathbf{z}] = \int \int [z_t|\mathbf{z}, \theta, \sigma^2][\theta, \sigma^2|\mathbf{z}]d\theta d\sigma^2, \tag{16.38}$$

where $[z_t|\mathbf{z}, \theta, \sigma^2]$ is the predictive full-conditional distribution we need to derive. In the regression setting with independent errors, the predictive full-conditional distribution was trivial—it was the data model. In the case of models with dependence through dynamics or second-order structure (e.g., spatial models), the predictive full-conditional distribution involves more components. To make sure we correctly identify all the relevant components of the joint distribution to the predictive full-conditional, we need to examine carefully the likelihood $\prod_{t=2}^{T}[z_t|z_{t-1}, \theta, \sigma^2]$. Notice that the variable of interest z_t appears in two components of this likelihood: $[z_t|z_{t-1},\theta,\sigma^2]$ and $[z_{t+1}|z_t,\theta,\sigma^2]$, as long as $1 < t < T$. Thus, our predictive full-conditional distribution for z_t is

$$[z_t|\cdot] \propto [z_{t+1}|z_t,\theta,\sigma^2][z_t|z_{t-1},\theta,\sigma^2], \tag{16.39}$$

$$\propto \exp\left(-\frac{1}{2}\frac{(z_{t+1}-\theta-z_t)^2}{\sigma^2}\right)\exp\left(-\frac{1}{2}\frac{(z_t-\theta-z_{t-1})^2}{\sigma^2}\right), \tag{16.40}$$

$$\propto \exp\left(-\frac{1}{2}\left(\frac{-2(z_{t+1}-\theta)z_t}{\sigma^2}+\frac{z_t^2}{\sigma^2}+\frac{-2(\theta+z_{t-1})z_t}{\sigma^2}+\frac{z_t^2}{\sigma^2}\right)\right), \tag{16.41}$$

$$\propto \exp\left(-\frac{1}{2}\left(-2\left(\frac{z_{t+1}-\theta}{\sigma^2}+\frac{\theta+z_{t-1}}{\sigma^2}\right)z_t+\left(\frac{1}{\sigma^2}+\frac{1}{\sigma^2}\right)z_t^2\right)\right), \tag{16.42}$$

$$\propto \exp\left(-\frac{1}{2}\left(-2bz_t+az_t^2\right)\right), \tag{16.43}$$

which is the kernel of a normal distribution and implies $[z_t|\cdot] \equiv \mathrm{N}(a^{-1}b,a^{-1})$ where

$$a = \frac{2}{\sigma^2}, \tag{16.44}$$

$$b = \frac{z_{t+1}+z_{t-1}}{\sigma^2}. \tag{16.45}$$

Thus, the predictive full-conditional mean is $a^{-1}b = (z_{t+1}+z_{t-1})/2$, the mean of z_{t-1} and z_{t+1}, which is intuitive. In fact, for this particular Malthusian growth model, our predictive realizations for z_t when $1 < t < T$ do not depend on the parameter θ directly (only indirectly through the MCMC sample of σ^2)! We can sample a predictive full-conditional realization for any z_t of interest ($1 < t < T$) in the loop of a MCMC algorithm, or in a subsequent algorithm using the output from the model fit to data.

We can derive the predictive full-conditional distributions for the cases where $t = 1$ and $t = T$ in the same way and that procedure yields

$$[z_1|\cdot] = \mathrm{N}(z_2 - \theta, \sigma^2), \tag{16.46}$$

$$[z_T|\cdot] = \mathrm{N}(\theta + z_{T-1}, \sigma^2), \tag{16.47}$$

which we also can sample from easily in a MCMC algorithm, or using MCMC output.

To obtain the posterior forecast distribution for z_{T+1}, we still use the predictive full-conditional distribution, but now we need to sample the data at one step ahead using

$$[z_{T+1}|\cdot] = \mathrm{N}(\theta + z_T, \sigma^2). \tag{16.48}$$

However, to forecast two time steps ahead using an autoregressive model, we need to sample from

$$[z_{T+2}|\cdot] = \mathrm{N}(\theta + z_{T+1}, \sigma^2). \tag{16.49}$$

We do not have z_{T+1} in the data set, so we first need to sample z_{T+1} from (16.48) and then sample z_{T+2} from (16.49). We can sample data farther ahead in time as well using the same strategy to bridge up to the time of interest.

For example, using the MCMC output from fitting the Malthusian growth model in the previous section to the scaup data, we predicted scaup population sizes for all times where data exist (1980–2009) and then for three times in the future (2010–2012). The R code to obtain posterior predictive realizations and summarize them graphically in Figure 16.5 is

---

BOX 16.8   OBTAIN POSTERIOR PREDICTIONS

```
1  n.mcmc=10000
2  T=length(z)
3  z.pred.mat=matrix(0,n.mcmc,T+3)
4  set.seed(1)
5  z.pred.mat[,1]=rnorm(n.mcmc,z[2]-mcmc.out$theta.
6      save,sqrt(mcmc.out$s2.save))
7  for(t in 2:(T-1)){
8    z.pred.mat[,t]=rnorm(n.mcmc,(z[t-1]+z[t+1])/2,
9      sqrt(mcmc.out$s2.save))
10 }
11 z.pred.mat[,T]=rnorm(n.mcmc,mcmc.out$theta.save
12     +z[T-1],sqrt(mcmc.out$s2.save))
13 z.pred.mat[,T+1]=rnorm(n.mcmc,mcmc.out$theta.save
14     +z[T],sqrt(mcmc.out$s2.save))
15 z.pred.mat[,T+2]=rnorm(n.mcmc,mcmc.out$theta.save
16     +z.pred.mat[,T+1],sqrt(mcmc.out$s2.save))
17 z.pred.mat[,T+3]=rnorm(n.mcmc,mcmc.out$theta.save
18     +z.pred.mat[,T+2],sqrt(mcmc.out$s2.save))
19
20 z.pred.mn=apply(z.pred.mat,2,mean)
21 z.pred.l=apply(z.pred.mat,2,quantile,0.025)
22 z.pred.u=apply(z.pred.mat,2,quantile,0.975)
23
24 y.pred.mn=apply(exp(z.pred.mat),2,mean)
25 y.pred.l=apply(exp(z.pred.mat),2,quantile,0.025)
26 y.pred.u=apply(exp(z.pred.mat),2,quantile,0.975)
27
28 layout(matrix(1:2,2,1))
29 matplot(1980:2012,cbind(z.pred.l,z.pred.u),
30     type="n",ylab="z",xlab="year",main="a")
31 polygon(c(1980:2012,2012:1980),c(z.pred.u,
32     rev(z.pred.l)),col=rgb(0,0,0,.2),border=NA)
```

(Continued)

BOX 16.8 (Continued)   OBTAIN POSTERIOR PREDICTIONS

```
33 lines(1980:2012,z.pred.mn)
34 points(1980:2009,z)
35 abline(v=2009,lty=2)
36 matplot(1980:2012,cbind(y.pred.l,y.pred.u),
37     type="n",ylab="y",xlab="year",main="b")
38 polygon(c(1980:2012,2012:1980),c(y.pred.u,
39     rev(y.pred.l)),col=rgb(0,0,0,.2),border=NA)
40 lines(1980:2012,y.pred.mn)
41 points(1980:2009,y)
42 abline(v=2009,lty=2)
```

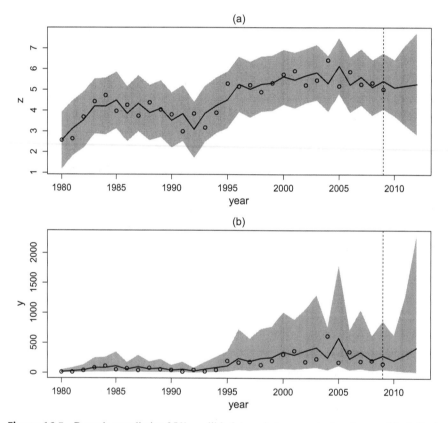

Figure 16.5 Posterior predictive 95% credible intervals (gray region) and mean (black line) for (a) the log transformed relative abundances z_t and (b) the original untransformed relative abundances y_t. Vertical dashed lines indicate the last time for which data exist. Data are shown as points.

Figure 16.5a shows the predictions for the log transformed data and because we can use the equivariance property of MCMC to make inference on transformations of random variables, Figure 16.5b shows the same predictions, but on the original scale of the data (relative abundance). Notice that our forecast distribution increases in variability as we forecast farther into the future. This increasing forecast uncertainty is a feature of dynamic autoregressive time series models.

16.4 MULTIVARIATE AUTOREGRESSIVE MODELS FOR ANIMAL MOVEMENT

Studies often focus conventional dynamic models on univariate time series as we described in the previous section. However, the real power of dynamic modeling comes into play with multivariate temporal processes. For example, many spatio-temporal models are specifically designed to account for dynamics of evolving spatial processes over time (see Chapter 28 as well as Wikle and Hooten 2010; Cressie and Wikle 2011). An example of a low-dimensional spatio-temporal process is a trajectory, or the movement of an object in two or more dimensions. An example of an ecological trajectory is an animal track in space over time. Animal movement is a critical element of ecological processes and has been the subject of statistical modeling for decades. Individual-based animal movement trajectories in two dimensions can be modeled in many different ways (Hooten et al. 2017). Perhaps the simplest way to model a movement trajectory based on a time series of position data $\mathbf{s}_t$ (often called telemetry data), for $t = 1, \ldots, T$, is as a generalized version of the AR(1) process we described in the previous section, but jointly as a vector. Thus, a simple spatio-temporal model is a two-dimensional vector autoregressive model of order one (VAR(1)), which we write as $\mathbf{s}_t \sim N(\mathbf{M}\mathbf{s}_{t-1}, \sigma^2 \mathbf{I})$ for $t = 2, \ldots, T$ (assuming the trajectory has mean zero vector). In this VAR(1) model, the matrix $\mathbf{M}$ is often called the propagator matrix or evolution matrix and is of dimension 2×2 if the position $\mathbf{s}_t$ is 2×1 (e.g., the position of an individual animal in geographic space at time t). We can parameterize the propagator matrix $\mathbf{M}$ in many ways, but a simple form is $\mathbf{M} \equiv \alpha \mathbf{I}$, which implies that the dynamics are the same for each dimension and α controls the smoothness in the same way it did in the univariate AR(1) we described in the previous section. However, we can also parameterize the propagator matrix such that each of the four elements of $\mathbf{M}$ are individual parameters, for example, $\alpha_{1,1}$, $\alpha_{1,2}$, $\alpha_{2,1}$, and $\alpha_{2,2}$. In a Bayesian VAR(1) model, we can specify priors for these parameters, as well as for σ^2, and fit the model using MCMC.

One issue with the VAR(1) we formulated is that it is quite noisy as a model for animal trajectories. Thus, rather than model the observed positions directly, we could model the velocities instead. The velocity vector of a multivariate temporal process is defined as $\boldsymbol{\delta}_t = \mathbf{s}_t - \mathbf{s}_{t-1}$, which is the temporal difference of the position process. The velocity process contains information about the speed of the trajectory as well as its direction. A VAR(1) model for the velocity process can be accumulated over time to result in a smooth position process because $\mathbf{s}_t = \mathbf{s}_{t-1} + \boldsymbol{\delta}_t$. Thus, if we know the starting position and the velocities $\boldsymbol{\delta}_t$, for $t = 2, \ldots, T$, we can recover the position process (i.e., the trajectories) $\mathbf{s}_t$. Also notice that the $\sqrt{\delta_t'\delta_t}$ for $t = 2, \ldots, T$ are the step lengths (or displacement distances between times $t - 1$ and t) and the associated speed per step is $\sqrt{\delta_t'\delta_t}/\Delta t$, where Δt is the time between $t - 1$ and t.

Jonsen et al. (2005) described a VAR(1) model for velocities that was specified as

$$\delta_t \sim N(\mathbf{M}\delta_{t-1}, \sigma^2 \mathbf{I}), \tag{16.50}$$

for $t = 2, \ldots, T$. We could parameterize the propagator matrix $\mathbf{M}$ in this model as we just described in the VAR(1), or we could use a more mechanistic specification such as the rotation matrix

$$\mathbf{M} \equiv \gamma \begin{pmatrix} \cos(\theta) & -\sin(\theta) \\ \sin(\theta) & \cos(\theta) \end{pmatrix}, \tag{16.51}$$

where, γ $(0 \leq \gamma \leq 1)$ controls the autocorrelation in the velocity process and θ is the angle in radians $(-\pi < \theta \leq \pi)$ between two consecutive velocities δ_{t-1} and δ_t. Thus, the propagator matrix in (16.51) rotates and dampens (i.e., smooths) the velocity process.

We often refer to the time series model in (16.50) as a correlated random walk (CRW) model for the position process (i.e., $\mathbf{s}_t$, the original trajectories) and it simplifies to a random walk model if $\gamma = 0$ because (16.50) reduces to $\mathbf{s}_t \sim N(\mathbf{s}_{t-1}, \sigma^2 \mathbf{I})$.

To complete the Bayesian CRW model, we specify priors for each of the parameters as

$$\gamma \sim \text{Unif}(0, 1), \tag{16.52}$$

$$\theta \sim \text{Unif}(-\pi, \pi), \tag{16.53}$$

$$\sigma^2 \sim \text{IG}(q, r), \tag{16.54}$$

which results in the posterior distribution

$$[\gamma, \theta, \sigma^2 | \delta_1, \ldots, \delta_T] \propto \left(\prod_{t=2}^{T} [\delta_t | \delta_{t-1}, \gamma, \theta, \sigma^2] \right) [\gamma][\theta][\sigma^2]. \tag{16.55}$$

To fit the model using MCMC, we take the usual approach of first deriving the full-conditional distributions, if they are analytically tractable, and then performing the updates in a MCMC algorithm in R. The first full-conditional is for the autocorrelation parameter γ, which appears in the likelihood component of (16.55) and its prior, yielding

$$[\gamma | \cdot] \propto \left(\prod_{t=2}^{T} [\delta_t | \delta_{t-1}, \gamma, \theta, \sigma^2] \right) [\gamma], \tag{16.56}$$

which is temptingly close to appearing conjugate because of the normal form of the likelihood and the linearity of the parameter in the mean term of that normal. Because the prior $[\gamma]$ is uniform on compact support $(0, 1)$, the full-conditional cannot be normal. However, because the full-conditional distribution for γ has a normal kernel and the prior restricts the support to between zero and one, the full-conditional is truncated normal. To show this, we write the full-conditional explicitly as

$$[\gamma|\cdot] \propto \left(\prod_{t=2}^{T} [\delta_t | \delta_{t-1}, \gamma, \theta, \sigma^2] \right) [\gamma], \tag{16.57}$$

$$\propto \exp\left(-\frac{1}{2} \sum_{t=2}^{T} (\delta_t - \gamma\Theta\delta_{t-1})'(\sigma^2 I)^{-1}(\delta_t - \gamma\Theta\delta_{t-1}) \right) 1_{\{0 \leq \gamma \leq 1\}}, \tag{16.58}$$

$$\propto \exp\left(-\frac{1}{2} \left(-2 \left(\sum_{t=2}^{T} \delta_t'(\sigma^2 I)^{-1}\Theta\delta_{t-1} \right) \gamma \right. \right.$$
$$\left. \left. + \gamma \left(\sum_{t=2}^{T} \delta_{t-1}'\Theta'(\sigma^2 I)^{-1}\Theta\delta_{t-1} \right) \gamma \right) \right) 1_{\{0 \leq \gamma \leq 1\}}, \tag{16.59}$$

$$\propto \exp\left(-\frac{1}{2} \left(-2b\gamma + a\gamma^2 \right) \right) 1_{\{0 \leq \gamma \leq 1\}}, \tag{16.60}$$

where the matrix $\Theta \equiv M/\gamma$ is 2×2 with $\cos(\theta)$ on the diagonal, $\sin(\theta)$ on the lower left off-diagonal, and $-\sin(\theta)$ on the upper right off-diagonal. Following the same procedure we showed for the autocorrelation parameter in the previous section, we define a and b as

$$a \equiv \sum_{t=2}^{T} \delta_{t-1}'\Theta'(\sigma^2 I)^{-1}\Theta\delta_{t-1}, \tag{16.61}$$

$$b \equiv \sum_{t=2}^{T} \delta_t'(\sigma^2 I)^{-1}\Theta\delta_{t-1}, \tag{16.62}$$

which, together with the constraint $0 \leq \theta \leq 1$, implies $[\gamma|\cdot] = \text{TN}(a^{-1}b, a^{-1})_0^1$ (a truncated normal with lower bound of zero and upper bound of one). To sample γ in a MCMC algorithm, we can use a direct update from an existing R function that provides truncated normal realizations. In the following algorithm, we use a direct update from the truncated normal distribution, which results in a Gibbs update.

The full-conditional distribution for θ is similar in that it involves the likelihood and a uniform prior on compact support $(-\pi < \theta < \pi)$. However, because θ appears inside a set of nonlinear functions (i.e., sine and cosine) in the mean of the normal, we cannot easily derive a closed form full-conditional distribution for θ. Thus, we write the full-conditional distribution for θ generically as

$$[\theta|\cdot] \propto \left(\prod_{t=2}^{T} [\delta_t | \delta_{t-1}, \gamma, \theta, \sigma^2] \right) [\theta], \tag{16.63}$$

and use a M-H update to sample θ in our MCMC algorithm. We can use a uniform proposal distribution on $(-\pi, \pi)$, which is also the prior, as we described in Chapter 6 so that all of our proposed values for $\theta^{(k)}$ meet the constraints on the parameter.

If the posterior distribution is very precise, our MCMC algorithm may not be very efficient, prohibiting us from accepting many proposals. However, using the prior as the proposal distribution will be helpful in this case because of the circularity in the relationship between the parameter θ and the resulting turning angle in the observed data. For example, if $\theta = \pi$ or $\theta = -\pi$ the resulting turning angle is 180 degrees and this would result in a Markov chain that has to traverse back and forth in the parameter space. If we used a normal random walk proposal for this parameter, rejecting any proposed values for θ that fell outside the support, it may move very slowly from one end of the parameter space to the other, or not be able to transition to the other side at all if the posterior is very precise and the tuning parameter in the random walk proposal is small. Using the prior as a proposal eliminates the need for one tuning parameter, is easy to implement, and allows the Markov chain to quickly transition back and forth in the parameter space as needed.

The variance parameter σ^2 for the CRW model has a similar full-conditional as we derived in the previous section for the AR(1) model and it is inverse gamma. However, because the response data are multivariate, we now use a combination of summation and vector notation. Also, there is one important difference with the q parameter based on the determinant of $\sigma^2\mathbf{I}$. The full-conditional distribution for σ^2 is

$$[\sigma^2|\cdot] \propto \left(\prod_{t=2}^{T}[\delta_t|\delta_{t-1}, \gamma, \theta, \sigma^2]\right)[\sigma^2], \tag{16.64}$$

$$\propto |\sigma^2\mathbf{I}|^{-\frac{T-1}{2}}\exp\left(-\frac{1}{2}\sum_{t=2}^{T}(\delta_t - \mathbf{M}\delta_{t-1})'(\sigma^2\mathbf{I})^{-1}(\delta_t - \mathbf{M}\delta_{t-1})\right) \times \tag{16.65}$$

$$(\sigma^2)^{-(q+1)}\exp\left(-\frac{1}{\sigma^2 r}\right), \tag{16.66}$$

$$\propto (\sigma^2)^{-(T-1+q+1)}\exp\left(-\frac{1}{\sigma^2}\left(\frac{\sum_{t=2}^{T}(\delta_t - \mathbf{M}\delta_{t-1})'(\delta_t - \mathbf{M}\delta_{t-1})}{2} + \frac{1}{r}\right)\right), \tag{16.67}$$

$$\propto (\sigma^2)^{-(\tilde{q}+1)}\exp\left(-\frac{1}{\sigma^2 \tilde{r}}\right), \tag{16.68}$$

which implies that $[\sigma^2|\cdot] \equiv \text{IG}(\tilde{q}, \tilde{r})$, where

$$\tilde{q} \equiv T - 1 + q, \tag{16.69}$$

$$\tilde{r} \equiv \left(\frac{\sum_{t=2}^{T}(\delta_t - \mathbf{M}\delta_{t-1})'(\delta_t - \mathbf{M}\delta_{t-1})}{2} + \frac{1}{r}\right)^{-1}. \tag{16.70}$$

Note that the determinant of the product of a scalar and an identity matrix is equal to the scalar raised to the dimension of the matrix. Thus, $|\sigma^2\mathbf{I}|^{-\frac{T-1}{2}} = (\sigma^2)^{-(T-1)}$

because the identity matrix $\mathbf{I}$ in the conditional covariance for δ_t is 2×2 for a trajectory in two dimensions. The full-conditional distribution for σ^2 implies that we can use a Gibbs update to sample σ^2 in the MCMC algorithm to fit the CRW model to data.

We constructed the MCMC algorithm in R using the mixture of Gibbs and M-H updates just described. We named the R function for this MCMC algorithm crw.mcmc.R:

BOX 16.9 MCMC ALGORITHM TO FIT THE CRW MOVEMENT MODEL

```
1  crw.mcmc <- function(S,n.mcmc){
2
3  ####
4  ####   Libraries and Subroutines
5  ####
6
7  library(msm)
8
9  make.Theta <- function(theta){
10    Theta=matrix(0,2,2)
11    Theta[1,1]=cos(theta)
12    Theta[1,2]=-sin(theta)
13    Theta[2,1]=sin(theta)
14    Theta[2,2]=cos(theta)
15    Theta
16  }
17
18  make.M <- function(gamma,theta){
19    Theta=make.Theta(theta)
20    M=gamma*Theta
21    M
22  }
23
24  ####
25  ####   Set Up Variables
26  ####
27
28  Delta=apply(S,2,diff)
29  T=dim(Delta)[1]
```

(Continued)

BOX 16.9 (Continued)   MCMC ALGORITHM TO FIT THE CRW MOVEMENT MODEL

```
30  theta.save=rep(0,n.mcmc)
31  gamma.save=rep(0,n.mcmc)
32  s2.save=rep(0,n.mcmc)
33
34  ####
35  ####  Priors and Starting Values
36  ####
37
38  theta=0
39  gamma=.9
40  Theta=make.Theta(theta)
41  M=make.M(gamma,theta)
42  s2=.01
43  Sig=s2*diag(2)
44  Sig.inv=solve(Sig)
45
46  q=.0001
47  r=1000
48
49  ####
50  ####  Begin MCMC Loop
51  ####
52
53  for(k in 1:n.mcmc){
54
55    ####
56    ####  Sample theta
57    ####
58
59    theta.star=runif(1,-pi,pi)
60    M.star=make.M(gamma,theta.star)
61    tmp.theta.sum.1=0
62    tmp.theta.sum.2=0
63    for(t in 2:T){
64      delta.star=M.star%*%Delta[t-1,]
65      delta=M%*%Delta[t-1,]
```

(Continued)

**BOX 16.9 (Continued)   MCMC ALGORITHM TO FIT THE CRW
MOVEMENT MODEL**

```
66   tmp.theta.sum.1=tmp.theta.sum.1-t(Delta[t,]
67       -delta.star)%*%Sig.inv%*%(Delta[t,]
68       -delta.star)/2
69   tmp.theta.sum.2=tmp.theta.sum.2-t(Delta[t,]
70       -delta)%*%Sig.inv%*%(Delta[t,]-delta)/2
71   }
72
73   mh1=tmp.theta.sum.1
74   mh2=tmp.theta.sum.2
75   mh=exp(mh1-mh2)
76   if(mh > runif(1)){
77     theta=theta.star
78     Theta=make.Theta(theta)
79   }
80
81   ####
82   ####   Sample gamma
83   ####
84
85   tmp.b=0
86   tmp.a=0
87   for(t in 2:T){
88     delta=Theta%*%Delta[t-1,]
89     tmp.b=tmp.b+t(Delta[t,])%*%Sig.inv%*%delta
90     tmp.a=tmp.a+t(delta)%*%Sig.inv%*%delta
91   }
92   tmp.var=1/tmp.a
93   tmp.mn=tmp.var*tmp.b
94   gamma=rtnorm(1,tmp.mn,sqrt(tmp.var),0,1)
95   M=make.M(gamma,theta)
96
97   ####
98   ####   Sample s2
99   ####
100
101  tmp.diff=Delta[2:T,]-t(M%*%t(Delta[1:(T-1),]))
102  tmp.sum=sum(apply(tmp.diff^2,1,sum))
103  s2=1/rgamma(1,T-1+q,,1/(tmp.sum/2+1/r))
```

(*Continued*)

BOX 16.9 (Continued)   MCMC ALGORITHM TO FIT THE CRW MOVEMENT MODEL

```
104  Sig=s2*diag(2)
105  Sig.inv=diag(2)/s2
106
107  ####
108  ####  Save Samples
109  ####
110
111  theta.save[k]=theta
112  gamma.save[k]=gamma
113  s2.save[k]=s2
114
115  }
116
117  ####
118  #### Write Output
119  ####
120
121  list(s2.save=s2.save,n.mcmc=n.mcmc,S=S,
122      Delta=Delta,theta.save=theta.save,
123      gamma.save=gamma.save,r=r,q=q)
124
125  }
```

In this MCMC algorithm, notice that we used the rtnorm function from the msm R package to sample from a truncated normal to update γ. Also, notice that we sampled θ first, followed by γ, and finally σ^2. The order of updates is not important in standard MCMC algorithms as long as the order is consistent throughout iterations. On lines 66–70 of the MCMC algorithm, we calculated the required sums in the full-conditional directly instead of using dmvnorm (or similar) because it results in a more efficient algorithm when we only have to compute the inverse of the covariance matrix once and we can avoid calculating the determinant of the covariance matrix completely because the update for θ does not require it.

To demonstrate the VAR(1) model for animal trajectories, we used the MCMC algorithm just described to fit the CRW model to global positioning system (GPS) data associated with the migration of a sandhill crane (SACR; *Grus canadensis*) in North America between September 29 and October 10, 2013 (Hooten et al. 2018). Figure 16.6 shows the set of $T = 47$ observed telemetry positions for an individual

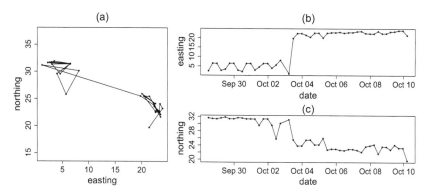

Figure 16.6 (a) A subset of $T = 47$ GPS observations $\mathbf{s}_t$ in geographic space using a centered coordinate system in kilometers and (b) the time series of observed data in easting (east/west) and (c) northing (north/south).

SACR collected approximately every 6 hours during this period. We converted the GPS observations $\mathbf{s}_t$ for $t = 1, \ldots, T$ to velocities using the relationship $\boldsymbol{\delta}_t = \mathbf{s}_t - \mathbf{s}_{t-1}$ and fit the CRW model using the MCMC algorithm above and the following R code:

BOX 16.10   FIT CRW MODEL TO SACR DATA

```
load("sacr.RDdata")
set.seed(1)
source("crw.mcmc.R")
mcmc.out=crw.mcmc(S=S.sm,n.mcmc=100000)

layout(matrix(1:3,1,3))
hist(mcmc.out$gamma.save,xlab=bquote(gamma),
    main="a",prob=TRUE)
curve(dunif(x,0,1),add=TRUE)
hist(mcmc.out$theta.save,xlab=bquote(theta),
    main="b",prob=TRUE)
curve(dunif(x,-pi,pi),add=TRUE)
abline(v=0,lty=2,col=1)
hist(mcmc.out$s2.save,xlab=bquote(sigma^2),
    main="c",prob=TRUE)
curve(dIG(x,mcmc.out$r,mcmc.out$q),add=TRUE)
```

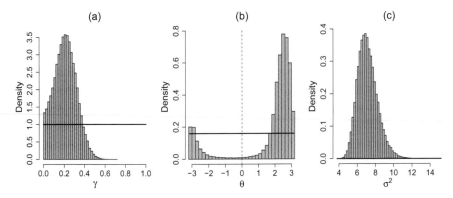

Figure 16.7 Marginal posterior histograms based on the MCMC output for (a) γ, (b) θ, and (c) σ^2. Priors are shown at solid black lines and the vertical dashed line in (b) indicates zero.

This R code produces the marginal posterior histograms in Figure 16.7 for γ, θ, and σ^2.[4] As we mentioned previously, the bimodality in Figure 16.7b is a function of the circularity of the relationship between θ and the turning angle in the data. Recall that, $\theta = -\pi$ results in the same behavior as $\theta = \pi$ in terms of the trajectory and both imply a rapid oscillation in the time series. This bimodality is also evident in Figure 16.6 where it appears that the individual SACR is flying back and forth every 6 hours. In fact, that is likely what the individual is doing because they are transitioning between different water bodies a few times per day in the prairie pothole region of North America (Hooten et al. 2018).

As we discussed, the parameter θ directly affects the turning angle. Values of θ less than zero result in right turns and values of θ greater than zero result in left turns. The closer θ is to zero the more straight the resulting trajectory will be and the closer θ is to π or $-\pi$ the more the trajectory will oscillate back and forth (180 degree turns). Thus, we can infer the individual SACR propensity to turn one direction or the other by calculating $P(\theta > 0 | \delta_1, \ldots, \delta_T)$ using the MCMC output:

BOX 16.11   POSTERIOR PROBABILITY OF $\theta > 0$

```
> mean(mcmc.out$theta.save>0)
[1] 0.84882
```

---

which results in the $P(\theta > 0 | \delta_1, \ldots, \delta_T) \approx 0.85$. Thus, based on the SACR data we analyzed, the probability of the individual making a left turn in its discrete trajectory every 6 hours is 0.85. However, the posterior mean of the parameter γ was

---

[4] Bimodality like that shown in Figure 16.7b may indicate non-identifiability in parameters in other types of models if the switches between modes appear concurrently in the Markov chains, so it is good to check.

$E(\gamma|\boldsymbol{\delta}_1,\ldots,\boldsymbol{\delta}_T) = 0.21$ which indicates a fairly low level of autocorrelation and thus characteristics in the trajectory due to turning angle induced by θ are minimal relative to random directions.

16.5 ADDITIONAL CONCEPTS AND READING

We only scratched the surface of temporal modeling in this chapter. Our focus was mainly on autoregressive models because they are used most commonly in ecological studies; however, moving average models are very common in the time series literature. Various combinations of autoregressive and moving average models make up a large class of methods in econometrics, for example. Also, autoregressive and moving average models have important connections to spatial statistics (Chapter 17). Finally, we have only presented models for discrete temporal processes; however, we model many temporal processes with continuous-time statistical models based on stochastic differential equations. In particular, continuous-time approaches provide an important class of models for studying animal movement (e.g., Johnson et al. 2008; McClintock et al. 2014; Hooten and Johnson 2017; Hooten et al. 2017).

There is an important connection between conditionally specified dynamic time series models and jointly specified time series models. In fact, recall that we mentioned the simple AR(1) model can also be written jointly as

$$\mathbf{y} \sim \mathrm{N}(\mathbf{0}, \sigma^2(\mathrm{diag}(\mathbf{W1}) - \rho\mathbf{W})^{-1}), \tag{16.71}$$

where σ^2 is the variance parameter as before, $\mathbf{1}$ is a $T \times 1$ vector of ones, $\mathbf{W}$ is an adjacency or neighborhood matrix with diagonal elements equal to zero and ones indicating neighbors in time, and ρ is a temporal dependence parameter similar to α. Ver Hoef et al. (2018a) discussed these same ideas and the details associated with them for spatial autoregressive models. Similar connections between first- and second-order model specifications exist in the continuous-time and continuous-space context (Hefley et al. 2017a).

Autoregressive models are not limited to lag 1 dependencies in time (e.g., AR(1), VAR(1)). In fact, longer range dependence can be included using lag 2 (or higher) specifications. For example, we can write a parametric AR(2) model as $y_t \sim \mathrm{N}(\alpha_1 y_{t-1} + \alpha_2 y_{t-2}, \sigma^2)$ for $t = 2, \ldots, T$. Higher order model specifications can result in smoother overall time series. Also, there are connections between higher order autoregressive models and the "integrated" time series models like the one we presented for animal movement. In fact, the VAR(1) CRW model we described for velocities can be reformulated as a VAR(2) model for position. To see this, recall that the CRW model was specified as $\boldsymbol{\delta}_t \sim \mathrm{N}(\mathbf{M}\boldsymbol{\delta}_{t-1}, \sigma^2\mathbf{I})$, where the position at time t is $\mathbf{s}_t = \mathbf{s}_{t-1} + \boldsymbol{\delta}_t$. Substituting the vector difference of positions into the CRW model for $\boldsymbol{\delta}_t$ we have

$$\mathbf{s}_t - \mathbf{s}_{t-1} \sim \mathrm{N}(\mathbf{M}(\mathbf{s}_{t-1} - \mathbf{s}_{t-1}), \sigma^2\mathbf{I}), \tag{16.72}$$

$$\sim \mathrm{N}(\mathbf{Ms}_{t-1} - \mathbf{Ms}_{t-2}, \sigma^2\mathbf{I}), \tag{16.73}$$

which implies that $\mathbf{s}_t \sim \mathrm{N}((\mathbf{I}+\mathbf{M})\mathbf{s}_{t-1} - \mathbf{M}\mathbf{s}_{t-2}, \sigma^2 \mathbf{I})$. Thus, there are two effective propagator matrices: $\mathbf{I}+\mathbf{M}$ for lag one dynamics and $-\mathbf{M}$ for lag two dynamics. This substitution means that a VAR(1) model for velocity is equivalent to a VAR(2) model for position and is the reason why the CRW is a more appropriate model for animal movement, because it accommodates smoother trajectories than a VAR(1) model would have on position directly.

In this chapter, we focused solely on continuous-valued time series because that is what the classical methods are designed for. However, many data sets in wildlife ecology are comprised of discrete-valued observations (e.g., zeros and ones or counts). Thus, we return to models for discrete time series data in later chapters of this book. Also, there are a variety of methods for analyzing discrete-valued series from a variety of perspectives that are outside the conventional suite of models used in wildlife ecology (e.g., Davis et al. 2016).

Many ecologists are interested in nonlinear models for temporal processes. The autoregressive models we presented in this chapter have homogeneous dynamics, but we can relax them by allowing the dynamics to vary as a function of time. For example, we can generalize the AR(1) time series model so that $y_t \sim \mathrm{N}(\alpha_t y_{t-1}, \sigma^2)$ where $\mathrm{logit}(\alpha_t) = \mathbf{x}_t' \boldsymbol{\beta}$ (see Hefley et al. (2016) for details on these types of specifications). This model relates the autocorrelation parameter α_t to a set of temporally varying covariates $\mathbf{x}_t$ through a nonlinear link function involving a new set of coefficients $\boldsymbol{\beta}$. This is the same type of expression we present in later chapters, but applied to continuously valued time series here. Of course, we can specify other types of nonlinear time series models as well. For example, in contrast to the Malthusian growth model we presented in this chapter, consider the nonlinear population model $y_t = y_{t-1} \exp\left(r\left(1 - \frac{y_{t-1}}{K}\right) + \varepsilon_t\right)$, where r is the population growth rate, K (in this case) is the carrying capacity or density dependence parameter, and $\varepsilon_t \sim \mathrm{N}(0, \sigma^2)$. This type of model is a stochastic version of a Ricker model (Turchin 2003).

Finally, an often neglected aspect of time series data in classical models is measurement error. Measurement error can range from very simple (e.g., additive, normally distributed, and independent) to very complicated (e.g., including missingness, mixture distributed, and dependent). All types of measurement error can potentially affect statistical inference, so it is good to incorporate mechanisms that give rise to the observed data when possible. In time series analyses, measurement error is often accommodated using hierarchical models (Chapter 19), which are sometimes called state-space models in the temporal context (e.g., de Valpine and Hastings 2002; Dennis et al. 2006; Patterson et al. 2008). We return to hierarchical models in the final part of this book, but in brief, hierarchical models allow us to partition uncertainty into components associated with the different mechanisms that give rise to our data. Hierarchical models are often specified conditionally (like most Bayesian models), where the data arise from a distribution that depends on a latent process, and the latent process arises from a distribution that depends on process-level parameters that control the dynamics. Berliner (1996) presented the quintessential hierarchical model structure for temporal processes[5] in terms of data, process, and parameter models, structured as

$$\mathbf{y} \sim [\mathbf{y}|\mathbf{z},\boldsymbol{\phi}], \tag{16.74}$$

$$\mathbf{z} \sim [\mathbf{z}|\boldsymbol{\theta}], \tag{16.75}$$

$$\boldsymbol{\phi} \sim [\boldsymbol{\phi}], \tag{16.76}$$

$$\boldsymbol{\theta} \sim [\boldsymbol{\theta}], \tag{16.77}$$

where $\mathbf{y}$ are the data, $\mathbf{z}$ is the process, $\boldsymbol{\phi}$ are parameters affecting the observation process, and $\boldsymbol{\theta}$ are parameters affecting the latent physical/ecological process. That is the basic framework behind all hierarchical models and we show how to formulate them more specifically as well as how to implement them in later chapters.

---

[5] See Wikle et al. (2019) pg. 13 for a helpful explanation of the paper by Berliner (1996).

17 Spatial Models

17.1 GEOSTATISTICAL MODELS

The three most commonly modeled types of spatial processes are point processes, continuous spatial processes, and discrete spatial processes. We return to point processes in Chapter 21, but cover continuous and discrete spatial processes in this chapter. Continuous spatial processes can be thought of as maps of spatial gradients of a response variable of interest (which does not have to be continuous), for example, topography maps and temperature maps. Discrete spatial processes (or areal, or lattice processes) can be thought of as maps comprised of distinct regions, for example, maps of aggregated data within political boundaries, or points such as small alpine lakes.

We begin with an introduction to continuous spatial models. The field associated with the development and implementation of continuous spatial models is referred to as geostatistics because of its history in geology and mining (Cressie 1990). Associated concepts are optimal spatial prediction of continuous processes, also called kriging. The goals for inference with continuous spatial process models often focus on (1) accounting for dependence in data beyond that accommodated by the predictor variables, (2) interpolation (i.e., spatial prediction), and (3) mimicking an ecological process in nature that leads to spatial structure in the response variable of interest. In all these cases, we can develop spatially explicit statistical models that we can use to make the desired form of inference.

Parametric statistical models for continuous and discrete spatial processes are different from most other statistical models used in ecological studies because they explicitly account for dependence in data, in part, through covariance (or second-order structure).[1] By far, the most common type of parametric statistical model for continuous spatial processes is the Gaussian process. That is, we assume that, out of the infinite locations that exist in a study area, the process of interest is measured at a finite subset of those, giving rise to actual data. However, the probability model is based on the fact that the process exists continuously in the spatial domain. The explicit Gaussian process[2] notation is not particularly helpful for introducing the concept. Thus, we introduce a model for the response variable $\mathbf{y}$ in the form of a linear regression model, but we can generalize that.

---

[1] Not to be confused with the term "second-order" used in time series models (e.g., AR(2)) which refers to the temporal lag of the dependence in the model. Used in this context, second-order refers to second-moment properties like covariance as opposed to first-moment or mean properties.

[2] It is called a "process" because it is a model for an infinite dimensional process or field (in space in this setting).

Suppose that we could model the observed process as normally distributed, with potentially heterogeneous mean and spatially explicit covariance such that $\mathbf{y} \sim \mathrm{N}(\mathbf{X}\boldsymbol{\beta}, \boldsymbol{\Sigma})$. This model is similar to what we presented in Chapter 11, except for the potentially non-diagonal covariance matrix $\boldsymbol{\Sigma}$. The key in spatially explicit models is that we parameterize this covariance matrix $\boldsymbol{\Sigma}$ so that it is a function of spatial locations associated with the process. For example, we specify most geostatistical models such that the second-order correlation in the process between locations should decay toward zero as the distance between them increases. Thus, we specify a function that relates distance between locations to their covariance in a way that results in valid covariance for the process (mainly that the function is non-negative definite). One of the most commonly used spatial covariance functions is the exponential function $\Sigma_{i,j} \equiv \sigma^2 \exp(-d_{i,j}/\phi)$, where $d_{i,j} = \sqrt{(\mathbf{s}_i - \mathbf{s}_j)'(\mathbf{s}_i - \mathbf{s}_j)}$ is the distance between locations $\mathbf{s}_i$ and $\mathbf{s}_j$ and ϕ is a parameter that controls the range of spatial structure in the process. Critical to this type of covariance formulation is that we assume the spatial process exists everywhere (for any $\mathbf{s}$) in the spatial domain of interest. As formulated, large values of ϕ correspond to smoother spatial processes. The variance parameter σ^2 adjusts the overall variation or magnitude of the spatial process.

Our geostatistical model only involves one additional parameter (ϕ) beyond the regression model we described in Chapter 11. Thus, we need a prior for ϕ as well as the other model parameters ($\boldsymbol{\beta}$ and σ^2). The parameter has positive support, and there are many options for priors including gamma, inverse gamma, log-normal, truncated normal, uniform, and discrete uniform.[3] However, in what follows, we use a gamma prior distribution for ϕ.

To complete the Bayesian geostatistical model specification, we write the complete model statement as

$$\mathbf{y} \sim \mathrm{N}(\mathbf{X}\boldsymbol{\beta}, \sigma^2 \mathbf{R}(\phi)), \tag{17.1}$$

$$\boldsymbol{\beta} \sim \mathrm{N}(\boldsymbol{\mu}_\beta, \boldsymbol{\Sigma}_\beta), \tag{17.2}$$

$$\sigma^2 \sim \mathrm{IG}(q, r), \tag{17.3}$$

$$\phi \sim \mathrm{Gamma}(\gamma_1, \gamma_2), \tag{17.4}$$

where we have specified a gamma distribution (with $E(\phi) = \gamma_1/\gamma_2$) as the prior for ϕ because the data set in the example that follows is relatively small ($n = 115$) and $\boldsymbol{\Sigma} \equiv \sigma^2 \mathbf{R}(\phi)$. The posterior distribution for this Bayesian geostatistical model is

$$[\boldsymbol{\beta}, \sigma^2, \phi | \mathbf{y}] \propto [\mathbf{y} | \boldsymbol{\beta}, \sigma^2, \phi][\boldsymbol{\beta}][\sigma^2][\phi]. \tag{17.5}$$

---

[3] The discrete uniform distribution is a commonly selected prior for ϕ because it allows us to precalculate the correlation portion of the covariance matrix before fitting the model to data so that the MCMC algorithm is more computationally efficient. In practice, a reasonable prior support for ϕ into a fine enough grid so that it provides a close approximation to a continuous uniform prior in terms of posterior inference.

Notice in (17.5) that we cannot write the likelihood as a product of conditionally independent components because the data are correlated even after accounting for the first-order mean structure in $\mathbf{X}\boldsymbol{\beta}$.

To construct a MCMC algorithm for fitting this model to data, we first need to find the full-conditional distributions for the parameters as we have for other models. In this case, we can borrow linear algebra results from Chapter 11 because the geostatistical model is very similar to the multiple linear regression model, except that, in the independent error regression model, the covariance in the data model was $\sigma^2\mathbf{I}$ and now it is $\sigma^2\mathbf{R}(\phi)$.

Starting with the full-conditional distribution for $\boldsymbol{\beta}$, we can write it the same way as before, where $[\boldsymbol{\beta}|\cdot]$ is proportional to the product of the data model and the prior, using the multivariate normal PDF for the data model:

$$[\boldsymbol{\beta}|\cdot] \propto [\mathbf{y}|\boldsymbol{\beta},\sigma^2,\phi][\boldsymbol{\beta}], \tag{17.6}$$

$$\propto \exp\left(-\frac{1}{2}(\mathbf{y}-\mathbf{X}\boldsymbol{\beta})'\boldsymbol{\Sigma}^{-1}(\mathbf{y}-\mathbf{X}\boldsymbol{\beta})\right)\exp\left(-\frac{1}{2}(\boldsymbol{\beta}-\boldsymbol{\mu}_\beta)'\boldsymbol{\Sigma}_\beta^{-1}(\boldsymbol{\beta}-\boldsymbol{\mu}_\beta)\right), \tag{17.7}$$

$$\propto \exp\left(-\frac{1}{2}(-2\mathbf{y}'\boldsymbol{\Sigma}^{-1}\mathbf{X}\boldsymbol{\beta} + \boldsymbol{\beta}'\mathbf{X}'\boldsymbol{\Sigma}^{-1}\mathbf{X}\boldsymbol{\beta} - 2\boldsymbol{\mu}_\beta'\boldsymbol{\Sigma}_\beta^{-1}\boldsymbol{\mu}_\beta)\right), \tag{17.8}$$

$$\propto \exp\left(-\frac{1}{2}(-2(\mathbf{y}'\boldsymbol{\Sigma}^{-1}\mathbf{X} + \boldsymbol{\mu}_\beta'\boldsymbol{\Sigma}_\beta^{-1})\boldsymbol{\beta} + \boldsymbol{\beta}'(\mathbf{X}'\boldsymbol{\Sigma}^{-1}\mathbf{X} + \boldsymbol{\Sigma}_\beta^{-1})\boldsymbol{\beta})\right), \tag{17.9}$$

$$\propto \exp\left(-\frac{1}{2}(-2\mathbf{b}'\boldsymbol{\beta} + \boldsymbol{\beta}'\mathbf{A}\boldsymbol{\beta})\right), \tag{17.10}$$

where, as before, we dropped terms that do not affect the proportionality as we work through the derivation. Thus, $[\boldsymbol{\beta}|\cdot] = N(\mathbf{A}^{-1}\mathbf{b}, \mathbf{A}^{-1})$, where $\mathbf{A}$ and $\mathbf{b}$ are defined as

$$\mathbf{A} \equiv \mathbf{X}'\boldsymbol{\Sigma}^{-1}\mathbf{X} + \boldsymbol{\Sigma}_\beta^{-1}, \tag{17.11}$$

$$\mathbf{b} \equiv \mathbf{y}'\boldsymbol{\Sigma}^{-1}\mathbf{X} + \boldsymbol{\mu}_\beta'\boldsymbol{\Sigma}_\beta^{-1}. \tag{17.12}$$

In deriving the full-conditional distribution for σ^2, we also can use a similar approach to that of Chapter 11 where it is proportional to the product of the data model and prior for σ^2:

$$[\sigma^2|\cdot] \propto [\mathbf{y}|\boldsymbol{\beta},\sigma^2,\phi][\sigma^2], \tag{17.13}$$

$$\propto |\sigma^2\mathbf{R}(\phi)|^{-\frac{1}{2}}\exp\left(-\frac{1}{2}(\mathbf{y}-\mathbf{X}\boldsymbol{\beta})'(\sigma^2\mathbf{R}(\phi))^{-1}(\mathbf{y}-\mathbf{X}\boldsymbol{\beta})\right)(\sigma^2)^{-(q+1)}\times$$

$$\exp\left(-\frac{1}{r\sigma^2}\right), \tag{17.14}$$

$$\propto (\sigma^2)^{-\frac{n}{2}} \exp\left(-\frac{1}{\sigma^2}\frac{(\mathbf{y}-\mathbf{X}\boldsymbol{\beta})'\mathbf{R}(\phi)^{-1}(\mathbf{y}-\mathbf{X}\boldsymbol{\beta})}{2}\right)(\sigma^2)^{-(q+1)}\times$$

$$\exp\left(-\frac{1}{r\sigma^2}\right), \tag{17.15}$$

$$\propto (\sigma^2)^{-(\frac{n}{2}+q+1)} \exp\left(-\frac{1}{\sigma^2}\left(\frac{(\mathbf{y}-\mathbf{X}\boldsymbol{\beta})'\mathbf{R}(\phi)^{-1}(\mathbf{y}-\mathbf{X}\boldsymbol{\beta})}{2}+\frac{1}{r}\right)\right), \tag{17.16}$$

$$\propto (\sigma^2)^{-(\tilde{q}+1)} \exp\left(-\frac{1}{\tilde{r}\sigma^2}\right). \tag{17.17}$$

Thus, $[\sigma^2|\cdot] = \text{IG}(\tilde{q},\tilde{r})$, where

$$\tilde{q} \equiv \frac{n}{2}+q, \tag{17.18}$$

$$\tilde{r} \equiv \left(\frac{(\mathbf{y}-\mathbf{X}\boldsymbol{\beta})'\mathbf{R}(\phi)^{-1}(\mathbf{y}-\mathbf{X}\boldsymbol{\beta})}{2}+\frac{1}{r}\right)^{-1}. \tag{17.19}$$

The full-conditional distribution for the spatial range parameter ϕ will not be conjugate, but we can sample it using a M-H update. Generically, we can write the full-conditional distribution for ϕ as

$$[\phi|\cdot] \propto [\mathbf{y}|\boldsymbol{\beta},\sigma^2,\phi][\phi]. \tag{17.20}$$

If we use the random walk method with rejection sampling for proposing values of $\phi^{(*)}$ (where we reject the update when $\phi^{(*)} \leq 0$) as we described in Chapter 6, the resulting M-H ratio is

$$mh = \frac{\text{N}(\mathbf{y}|\mathbf{X}\boldsymbol{\beta}^{(k)},\boldsymbol{\Sigma}(\sigma^{2(k)},\phi^{(*)}))\text{Gamma}(\phi^{(*)}|\gamma_1,\gamma_2)}{\text{N}(\mathbf{y}|\mathbf{X}\boldsymbol{\beta}^{(k)},\boldsymbol{\Sigma}(\sigma^{2(k)},\phi^{(k-1)}))\text{Gamma}(\phi^{(k-1)}|\gamma_1,\gamma_2)}, \tag{17.21}$$

where the named distributions in (17.21) are PDFs pertaining to the random variables shown on the left-hand side of the conditional bar. We also note that we can omit terms that cancel in the numerator and denominator (e.g., normalizing constants) in the M-H ratio. As a proposal distribution for $\phi^{(*)}$ we use $\phi^{(*)} \sim \text{N}(\phi^{(k-1)},\sigma_{\text{tune}}^2)$, where we can tune the algorithm to be efficient by adjusting σ_{tune}^2 and rejecting proposed values of the range parameters for which $\phi^{(*)} \leq 0$.

We constructed the MCMC algorithm (norm.geostat.mcmc.R) based on the full-conditional distributional distribution just described and hyperparameters $\mu_\beta = 0$, $\boldsymbol{\Sigma}_\beta = 1000 \cdot \mathbf{I}$, $r = 1000$, $q = 0.0001$, $\gamma_1 = 0.95$, and $\gamma_2 = 1.78$, and we specified the tuning parameter as $\sigma_{\text{tune}}^2 = 0.13$. The R code for the MCMC algorithm is

BOX 17.1   MCMC ALGORITHM FOR GEOSTATISTICAL MODEL

```
norm.geostat.mcmc <- function(y,X,locs,beta.mn,
    beta.var,n.mcmc){

###
### Subroutines
###

ldmvnorm<-function(y,mean,Sig.inv,logdet){
    logretval=-(logdet+t(y-mean)%*%Sig.inv%
        *%(y-mean))/2
    logretval
}

###
### Set up Variables and Priors
###

n=dim(X)[1]
p=dim(X)[2]
r=1000
q=0.0001
Sig.beta=beta.var*diag(p)
Sig.beta.inv=solve(Sig.beta)

D=rdist.earth(locs,locs)
max.d=max(D)
phi.mn=max.d/4/4
phi.var=10*phi.mn
gamma.1=(phi.mn^2)/phi.var
gamma.2=phi.mn/phi.var
phi=phi.mn
phi.tune=phi/4

beta.save=matrix(0,p,n.mcmc)
s2.save=rep(0,n.mcmc)
phi.save=rep(0,n.mcmc)

###
### Starting Values
###
```

(Continued)

BOX 17.1 (Continued)   MCMC ALGORITHM FOR GEOSTATISTICAL MODEL

```
41 beta=solve(t(X)%*%X)%*%t(X)%*%y
42 Xbeta=X%*%beta
43
44 R=exp(-D/phi)
45 R.inv=solve(R)
46 logdet=determinant(R)$modulus
47
48 ###
49 ### MCMC Loop
50 ###
51
52 for(k in 1:n.mcmc){
53
54   ###
55   ### Sample s2
56   ###
57
58   tmp.r=(1/r+.5*t(y-Xbeta)%*%R.inv%*%
59     (y-Xbeta))^(-1)
60   tmp.q=n/2+q
61   s2=1/rgamma(1,tmp.q,,tmp.r)
62   Sig.inv=R.inv/s2
63
64   ###
65   ### Sample beta
66   ###
67
68   tmp.chol=chol(t(X)%*%Sig.inv%*%X + Sig.beta.inv)
69   beta=backsolve(tmp.chol,backsolve(tmp.chol,
70     t(X)%*%Sig.inv%*%y + Sig.beta.inv%*%beta.mn,
71     transpose=TRUE)+rnorm(p))
72   Xbeta=X%*%beta
73
74   ###
75   ### Sample phi
76   ###
```

(*Continued*)

BOX 17.1 (Continued)   MCMC ALGORITHM FOR GEOSTATISTICAL MODEL

```
77   phi.star=rnorm(1,phi,phi.tune)
78   if(phi.star>0){
79     R.star=exp(-D/phi.star)
80     R.inv.star=solve(R.star)
81     logdet.star=determinant(R.star)$modulus
82     mh.1=ldmvnorm(y,Xbeta,R.inv.star/s2,
83         logdet.star)+dgamma(phi.star,gamma.1,
84         gamma.2,log=TRUE)
85     mh.2=ldmvnorm(y,Xbeta,R.inv/s2,logdet)
86         +dgamma(phi,gamma.1,gamma.2,log=TRUE)
87     if(exp(mh.1-mh.2)>runif(1)){
88       phi=phi.star
89       R.inv=R.inv.star
90       logdet=logdet.star
91     }
92   }
93
94   ###
95   ### Save Samples
96   ###
97
98   beta.save[,k]=beta
99   s2.save[k]=s2
100  phi.save[k]=phi
101
102 }
103
104 ###
105 ###   Write Output
106 ###
107
108 list(beta.save=beta.save,s2.save=s2.save,
109     phi.save=phi.save,y=y,X=X,n.mcmc=n.mcmc,n=n,
110     r=r,q=q,p=p,gamma.1=gamma.1,
111     gamma.2=gamma.2,phi.tune=phi.tune)
112
113 }
```

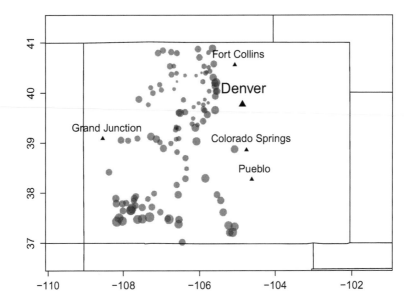

Figure 17.1 Map showing the SNOTEL data locations (circles) in Colorado, USA, with cities identified as triangles. Smallest circles indicate 10°F minimum temperature on October 14, 2017, and largest circles indicate 34°F minimum temperature.

Notice that we defined our own function to calculate the log of the multivariate normal density function (`ldmvnorm`) that includes only the elements that are necessary for the M-H update for ϕ. We computed the distances d_{ij} as great circle distances (line 25) in kilometers. Also, we used the same backsolve method to perform the Gibbs update for β that we described in Chapter 14 because MCMC algorithms for models with second-order dependence require many more matrix calculations and can be computationally intensive. Finally, for this MCMC algorithm and the next, we used input arguments for the prior mean and variance of the regression coefficients to allow for more generality in their use.[4]

We fit our Bayesian geostatistical model to minimum daily temperatures recorded at a set of $n = 115$ weather stations throughout Colorado, USA, as part of the Snow Telemetry (SNOTEL) Data Collection Network (`www.wcc.nrcs.usda.gov`). The set of SNOTEL data we analyze were recorded on October 14, 2017, at the locations shown in Figure 17.1. Both latitude and longitude may be useful predictor variables for these data to help characterize the spatial trend (i.e., the first-order or mean structure). In fact, a quadratic transformation of longitude as a predictor variable may be important as well. Thus, we fit a geostatistical regression model with three

---

[4] When building more general functions, it is usually a good idea to allow the user to specify variables related to priors and tuning parameters rather than hard coding them in the MCMC algorithm. In our previous algorithms, we often hard coded variables to streamline the R scripts.

standardized covariates: longitude, latitude, and longitude squared. However, because we desire predictions of temperature throughout the region, we standardized the covariates jointly for the observed data and prediction locations (i.e., a fine grid of points equally spaced throughout the region). The code to read the data into R and perform the standardization is

---

BOX 17.2   PREPARE SNOTEL DATA

```
temp.df=read.csv("20171014_Temp.csv")
y=temp.df$Temp
n=length(y)

locs=as.matrix(temp.df[,2:1])
n.lat=75  # Controls the number of points in
    latitude dimension
n.lon=round(n.lat*diff(range(locs[,1])))/
    diff(range(locs[,2])))
n.pred=n.lat*n.lon
lon.vals=seq(min(locs[,1]),max(locs[,1]),,n.lon)
lat.vals=seq(min(locs[,2]),max(locs[,2]),,n.lat)
locs.pred=as.matrix(expand.grid(lon.vals,
    lat.vals))
locs.full=rbind(locs,locs.pred)
locs.full.sc=locs.full
locs.full.sc=scale(locs.full)

X.full=cbind(rep(1,n.pred+n),locs.full.sc,
    locs.full.sc[,1]^2)
X=X.full[1:n,]
X.pred=X.full[-(1:n),]
p=dim(X)[2]
```

To fit the Bayesian geostatistical model to the SNOTEL data and create the posterior histograms in Figure 17.2, we used the R code:

BOX 17.3   FIT GEOSTATISTICAL MODEL TO SNOTEL DATA

```
source("norm.geostat.mcmc.R")
n.mcmc=10000
set.seed(1)
```

(Continued)

BOX 17.3 (Continued)   FIT GEOSTATISTICAL MODEL TO SNOTEL DATA

```
4  mcmc.out=norm.geostat.mcmc(y=y,X=X,locs=locs,
5      beta.mn=rep(0,p),beta.var=1000,n.mcmc=n.mcmc)
6
7  dIG <- function(x,r,q){
8    x^(-(q+1))*exp(-1/r/x)/(r^q)/gamma(q)
9  }
10
11  layout(matrix(1:6,3,2,byrow=TRUE))
12  hist(mcmc.out$beta.save[1,],prob=TRUE,xlab=bquote
13      (beta[1]),main="a")
14  curve(dnorm(x,0,sqrt(1000)),add=TRUE)
15  hist(mcmc.out$beta.save[2,],prob=TRUE,
16      xlab=bquote(beta[2]),main="b")
17  curve(dnorm(x,0,sqrt(1000)),add=TRUE)
18  hist(mcmc.out$beta.save[3,],prob=TRUE,
19      xlab=bquote(beta[3]),main="c")
20  curve(dnorm(x,0,sqrt(1000)),add=TRUE)
21  hist(mcmc.out$beta.save[4,],prob=TRUE,
22      xlab=bquote(beta[4]),main="d")
23  curve(dnorm(x,0,sqrt(1000)),add=TRUE)
24  hist(mcmc.out$s2.save,prob=TRUE,
25      xlab=bquote(sigma^2),main="e")
26  curve(dIG(x,mcmc.out$r,mcmc.out$q),add=TRUE)
27  hist(mcmc.out$phi.save,prob=TRUE,xlab=bquote(phi),
28      main="f")
29  curve(dgamma(x,mcmc.out$gamma.1,mcmc.out$gamma.2),
30      from=0.001,add=TRUE)
```

From Figure 17.2, we can see that all the regression coefficients differed from zero except β_1 (longitude; Figure 17.2b) but that it was left in the model because the posterior distribution for the quadratic term associated with it was different from zero (Figure 17.2d). We can interpret the range parameter ϕ on the scale of the distances $d_{i,j}$ (in km) such that 3ϕ indicates the effective range of spatial structure in the process after we account for the covariate effects. The maximum distance between observation locations was 274.28 km and the posterior mean of ϕ was $E(\phi|\mathbf{y}) = 9.59$; thus the range of spatial structure in the SNOTEL data was approximately 28.8 km (after accounting for longitude and latitude as covariates).

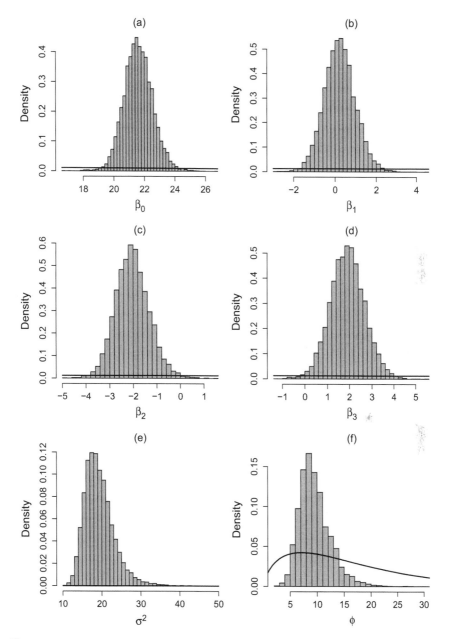

Figure 17.2 Marginal posterior histograms and associated priors (black lines) for (a) β_0 (intercept), (b) β_1 (longitude), (c) β_2 (latitude), (d) β_3 (longitude^2), (e) σ^2, and (f) ϕ (range). All covariates were scaled before fitting the model.

17.2 BAYESIAN KRIGING

As we discussed in the previous section, spatially explicit statistical models are useful for (a) properly accounting for second-order dependence when making inference on first-order effects, (b) mimicking mechanisms in the systems we study, and (c) obtaining predictions of the process of interest at locations other than those where data were collected. In the geostatistical literature, optimal spatial prediction is referred to as "kriging" (often with a lowercase "k") after D.L. Krige a mining engineer (Cressie 1990). Bayesian kriging is based on the same concept for prediction we presented in Chapter 12, the PPD.

In Chapter 12, we showed that we can write the PPD generically as

$$[\mathbf{y}_u|\mathbf{y}_o] = \int [\mathbf{y}_u|\boldsymbol{\theta},\mathbf{y}_o][\boldsymbol{\theta}|\mathbf{y}_o]d\boldsymbol{\theta}, \tag{17.22}$$

where $\mathbf{y}_u$ and $\mathbf{y}_o$ represent the response variable at a set of unobserved and observed locations and the parameter vector $\boldsymbol{\theta}$ serves as a place holder for $\boldsymbol{\beta}$, σ^2, and ϕ. In Chapter 12, we referred to the distribution $[\tilde{\mathbf{y}}|\boldsymbol{\theta},\mathbf{y}]$ (which is the same as $[\mathbf{y}_u|\boldsymbol{\theta},\mathbf{y}_o]$ in (17.22)) as the predictive full-conditional distribution and saw that it was trivial to find for the regression model with independent errors in Chapter 12.

For the Bayesian geostatistical model, we can find the predictive full-conditional distribution using existing results for conditional multivariate normal distributions. Geostatistical models are based on Gaussian processes and have continuous spatial support. Therefore, we could specify a model for the response variable anywhere in the spatial domain. Typically, we specify the geostatistical model only for the observed data (a subset of the infinite process), but we can also specify the model jointly for the observation and prediction locations as

$$\begin{pmatrix} \mathbf{y}_o \\ \mathbf{y}_u \end{pmatrix} \sim \mathrm{N}\left(\begin{pmatrix} \mathbf{X}_o\boldsymbol{\beta} \\ \mathbf{X}_u\boldsymbol{\beta} \end{pmatrix}, \begin{pmatrix} \boldsymbol{\Sigma}_{oo} \boldsymbol{\Sigma}_{ou} \\ \boldsymbol{\Sigma}_{uo} \boldsymbol{\Sigma}_{uu} \end{pmatrix} \right). \tag{17.23}$$

Using properties of the multivariate normal distribution, the resulting marginal model for the observed data is $[\mathbf{y}_o|\boldsymbol{\beta},\sigma^2,\phi] = \mathrm{N}(\mathbf{X}_o\boldsymbol{\beta},\boldsymbol{\Sigma}_{oo})$, which matches the specification we used in the previous section for the geostatistical model. Also, and critically, we can show the conditional distribution of the unobserved data $\mathbf{y}_u$ given $\mathbf{y}_o$ as $[\mathbf{y}_u|\boldsymbol{\beta},\sigma^2,\phi,\mathbf{y}_o] \equiv \mathrm{N}(\tilde{\boldsymbol{\mu}},\tilde{\boldsymbol{\Sigma}})$, where

$$\tilde{\boldsymbol{\mu}} = \mathbf{X}_u\boldsymbol{\beta} + \boldsymbol{\Sigma}_{uo}\boldsymbol{\Sigma}_{oo}^{-1}(\mathbf{y}_o - \mathbf{X}_o\boldsymbol{\beta}), \tag{17.24}$$

$$\tilde{\boldsymbol{\Sigma}} = \boldsymbol{\Sigma}_{uu} - \boldsymbol{\Sigma}_{uo}\boldsymbol{\Sigma}_{oo}^{-1}\boldsymbol{\Sigma}_{ou}, \tag{17.25}$$

which is the predictive full-conditional distribution for $\mathbf{y}_u$. Thus, to obtain a MCMC sample from the PPD to use for predictive inference, we can sample $\mathbf{y}_u^{(k)} \sim \mathrm{N}(\tilde{\boldsymbol{\mu}}^{(k)},\tilde{\boldsymbol{\Sigma}}^{(k)})$, where $\tilde{\boldsymbol{\mu}}^{(k)}$ and $\tilde{\boldsymbol{\Sigma}}^{(k)}$ are formed based on the current values for $\boldsymbol{\beta}^{(k)}$, $\sigma^{2(k)}$, and $\phi^{(k)}$ in the MCMC sample. This composition sampling technique allows us to avoid calculating the integral in (17.22) analytically. Then we can use the resulting PPD sample with MC integration methods to find the posterior predictive mean and variance, for example.

As discussed in Chapter 12, we can perform composition sampling to obtain $\mathbf{y}_u^{(k)}$ for $k = 1, \ldots, K$ in the main MCMC algorithm or as a separate second-stage algorithm using the output from the main MCMC algorithm. For large prediction problems where we are interested in predicting at a fine grid of locations throughout the study area, we may want to use two algorithms, the first for fitting the model to data and second for obtaining predictions. Although it is straightforward to use composition sampling to obtain samples from the predictive full-conditional distribution, there are still several computationally intensive matrix calculations that we require. For example, if the size of the observed data set is small, then calculating inverse $\boldsymbol{\Sigma}_{oo}^{-1}$ is no problem, but for large data sets, this calculation can be a bottleneck. Likewise, when using standard sequential computing (a "for" loop in most programming languages) and the set of prediction locations is large, we need to sample from a large multivariate normal distribution, which can be time consuming. However, if parallel computing resources are available, we can always reduce the size of the prediction problem and sample the predictive realizations in smaller groups because the predictive full-conditional distribution is general for any set of prediction locations regardless of size. When the size of the observed data set is large, it is more challenging to improve computational efficiency because we need to calculate the inverse $\boldsymbol{\Sigma}_{oo}^{-1}$ repeatedly to fit the model.

Using the MCMC output from the previous section where we fit the Bayesian geostatistical model to the SNOTEL temperature data, we obtained posterior predictive means and standard deviations for a grid of $n_u = 4800$ locations in western Colorado, USA, shown in Figure 17.3) using the following R code:

BOX 17.4   OBTAIN POSTERIOR PREDICTIONS

```
library(mvtnorm)
library(fields)
library(maps)

D=rdist.earth(locs,locs)
D.full=rdist.earth(locs.full,locs.full)
D.pred=rdist.earth(locs.pred,locs.pred)

D.uo=D.full[-(1:n),1:n]
n.pred=n.lat*n.lon
y.pred.save=matrix(0,n.pred,n.mcmc)
set.seed(1)
for(k in 1:n.mcmc){
    s2.tmp=mcmc.out$s2.save[k]
    phi.tmp=mcmc.out$phi.save[k]
    beta.tmp=mcmc.out$beta.save[,k]
```

(*Continued*)

BOX 17.4 (Continued)   OBTAIN POSTERIOR PREDICTIONS

```
17  Sig.inv.tmp=solve(s2.tmp*exp(-D/phi.tmp))
18  tmp.mn=X.pred%*%beta.tmp+s2.tmp*exp
19  (-D.uo/phi.tmp)%*%Sig.inv.tmp%*%(y-X%*%beta.tmp)
20  tmp.var=s2.tmp*exp(-D.pred/phi.tmp)-s2.tmp
21  *exp(-D.uo/phi.tmp)%*%solve(s2.tmp*exp
22  (-D/phi.tmp))%*%t(s2.tmp*exp(-D.uo/phi.tmp))
23  y.pred.save[,k]=as.vector(rmvnorm(1,tmp.mn,
24  tmp.var,method="chol"))
25  }
26  y.pred.mn=apply(y.pred.save,1,mean)
27  y.pred.sd=apply(y.pred.save,1,sd)
28
29  gray.colors.rev <- function (n, start = .2,end=1,
30  gamma = 1){gray(seq.int(to= start^gamma,
31  from = end^gamma,length.out = n)^(1/gamma))
32  }
33
34  data(us.cities)
35  CO.cities=us.cities[c(322,247,360,728,192),]
36  CO.cities$name=c("Fort Collins","Denver",
37  "Grand Junction","Pueblo","Colorado Springs")
38  layout(matrix(1:2,2,1))
39  map("state","Colorado",xlim=c(-110,-101),
40  ylim=c(36.5,41.5))
41  title("a")
42  image(matrix(y.pred.mn,n.lon,n.lat),y=lat.vals,
43  x=lon.vals,col=gray.colors.rev(100),asp=TRUE,
44  add=TRUE)
45  points(locs,pch=20,cex=.5)
46  map("state",add=TRUE)
47  map.axes()
48  map("state","Colorado",xlim=c(-110,-101),
49  ylim=c(36.5,41.5))
50  title("b")
51  image(matrix(y.pred.sd,n.lon,n.lat),y=lat.vals,
52  x=lon.vals,col=gray.colors.rev(100),asp=TRUE,
53  add=TRUE)
54  points(locs,pch=20,cex=.5)
55  map("state",add=TRUE)
56  map.axes()
```

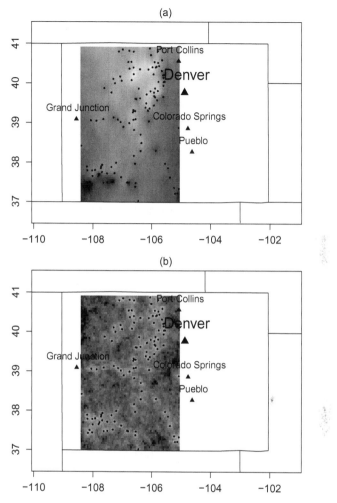

Figure 17.3 Posterior predictive (a) mean and (b) standard deviation maps for minimum temperature on Oct. 14, 2017 in western Colorado, USA. Darkest and lightest image colors in (a) correspond to 11°F and 34°F and in (b) correspond to 0.63°F and 6.75°F. Black points correspond to SNOTEL locations.

In Figure 17.3a we can see that, despite the dominant north-south trend with warmer temperatures in the south (due to the negative regression coefficient β_2 for latitude in Figure 17.2c), a distinctly nonlinear spatial pattern appears in temperature that we would not be able to characterize with standard regression modeling alone. Similarly, notice that the pattern for predictive standard deviation is strongly tied to the position of the observation locations and predictions far from the observed data are more uncertain (by as much as 6.75°F on the map in Figure 17.3b in northwestern Colorado, USA).

17.3   AREAL DATA MODELS

In the previous section, we focused on models for continuous spatial processes; however, often data and/or processes only exist on discrete spatial support. Examples of types of data on discrete spatial support include measured abundance on a set of islands, timber harvest data in a set of management regions, and public health data in census tracts. Thus, a popular class of spatial statistical models are designed for use with these data types explicitly (Ver Hoef et al. 2018a). We can express these models, also known as simultaneous autoregressive (SAR) models and conditional autoregressive (CAR) models, in a regression framework in the same way that we write geostatistical models. For example, for an $n \times 1$ vector of continuously valued response data $\mathbf{y}$, we can specify the data distribution for a SAR or CAR model as

$$\mathbf{y} \sim \text{N}(\mathbf{X}\boldsymbol{\beta}, \boldsymbol{\Sigma}), \qquad (17.26)$$

where we use the linear regression specification ($\mathbf{X}\boldsymbol{\beta}$) to account for the first-order structure and a covariance matrix $\boldsymbol{\Sigma}$ accounts for second-order spatial structure in the discrete domain.

The key difference between areal data models and geostatistical models is in the covariance specification ($\boldsymbol{\Sigma}$). Whereas covariance models for continuous spatial processes typically involve a decaying covariance function that depends on distance between locations, CAR and SAR covariance models depend on a network structure expressed in terms of an $n \times n$ matrix $\mathbf{W}$. The elements of $\mathbf{W}$, w_{ij} for $i \neq j$, indicate the proximity between the ith and jth spatial region; thus $\mathbf{W}$ is often referred to as a proximity or neighborhood matrix. Most commonly, we express the elements as $w_{ij} = 1$ if area i and j are neighbors and zero otherwise (with $w_{ii} = 0, \forall i$). The definition of the neighborhood matrix is part of the model specification. A neighbor could be the nearest area, it could be an area within a certain distance (based on the centroids of the areas, or other point location), or it could be an area sharing a boundary, and so on. However, proximity matrices do not have to be comprised of binary elements, they could be non-negative weights that indicate the strength of the association. The resulting statistical models are often referred to as weighted graph (or network) models. They are called graph or network models because the diagonal elements of $\mathbf{W}$ correspond to the nodes of a network, and the off-diagonal elements of $\mathbf{W}$ correpond to the potential edges between nodes. In this interpretation, if $w_{ij} = 0$, then there is no edge between areas i and j.

In this chapter, we focus on CAR specifications because they are used commonly in ecological models. SAR models also have some utility, especially from a model construction perspective. SAR and CAR models are very similar, but different in a few critical ways. Ver Hoef et al. (2018b) compare SAR and CAR models from a technical perspective and examine their similarities and differences, some of which are important but not described previously.

We described the CAR model covariance structure in Chapter 16 when we presented autoregressive models for time series that we can specify using second-order covariance structures. We noted that we could parameterize the covariance matrix for autoregressive models as $\boldsymbol{\Sigma} \equiv \sigma^2 (\text{diag}(\mathbf{W1}) - \rho\mathbf{W})^{-1}$, where $\text{diag}(\mathbf{W1})$ is a diagonal matrix with the row sums of $\mathbf{W}$ as diagonal elements. This specification allows us to restrict ρ to fall between 0 and 1. Note that there are many other ways to specify CAR models (Ver Hoef et al. 2018a, 2018b), but the specification we use here is a good place to start.

The covariance structure we presented in the AR(1) time series context is actually a one-dimensional CAR model. To see this, sort the times from earliest to latest and let the first off-diagonals of $\mathbf{W}$ equal 1 and all others equal zero. The parameter ρ in the covariance ($0 < \rho < 1$; which we called α in Chapter 16) controls the type and amount of spatial dependence. Values of ρ close to 1 imply strong positive autocorrelation in the process $\mathbf{y}$ and negative values for ρ imply negative dependence. However, in real data sets, there is almost always positive dependence which implies that $\rho > 0$ and, in fact, only values of ρ that are quite close to 1 will result in noticeable spatial structure in $\mathbf{y}$. Thus, it is common to use priors for ρ that favor larger values.

To specify the complete CAR model we use the formulation in (17.26) for the data model and priors for the remaining parameters $\boldsymbol{\beta}$, σ^2, and ρ as

$$\mathbf{y} \sim \mathrm{N}(\mathbf{X}\boldsymbol{\beta}, \boldsymbol{\Sigma}(\sigma^2, \rho)), \tag{17.27}$$

$$\boldsymbol{\beta} \sim \mathrm{N}(\boldsymbol{\mu}_\beta, \boldsymbol{\Sigma}_\beta), \tag{17.28}$$

$$\sigma^2 \sim \mathrm{IG}(q, r), \tag{17.29}$$

$$\rho \sim \mathrm{Beta}(\alpha_1, \alpha_2), \tag{17.30}$$

where we write the CAR covariance matrix as $\boldsymbol{\Sigma}(\sigma^2, \rho) \equiv \sigma^2 \mathbf{R}(\rho)$ and $\mathbf{R}(\rho) \equiv (\mathrm{diag}(\mathbf{W1}) - \rho\mathbf{W})^{-1}$.

The joint posterior distribution associated with this CAR model is

$$[\boldsymbol{\beta}, \sigma^2, \rho | \mathbf{y}] \propto [\mathbf{y} | \boldsymbol{\beta}, \sigma^2, \rho][\boldsymbol{\beta}][\sigma^2][\rho], \tag{17.31}$$

which results in the same full-conditional distributions for $\boldsymbol{\beta}$ and σ^2 as we derived for the geostatistical model in the previous section. The fact that we can use the same Gibbs updates for $\boldsymbol{\beta}$ and σ^2 as we used in the MCMC algorithm to fit the geostatistical model implies that we only need a minor modification of the geostatistical MCMC algorithm to fit the Bayesian CAR model. The only part of the MCMC algorithm that changes is that we need to replace the update for ϕ (the spatial range parameter) with the update for ρ (the spatial autocorrelation parameter).

As with the parameter ϕ from the previous section, the full-conditional distribution for ρ is

$$[\rho | \cdot] \propto [\mathbf{y} | \boldsymbol{\beta}, \sigma^2, \rho][\rho], \tag{17.32}$$

which is not conjugate, but we can use a M-H update by sampling $\rho^{(*)}$ from the prior as a proposal and using the M-H ratio

$$mh = \frac{[\mathbf{y} | \boldsymbol{\beta}^{(k)}, \sigma^{2(k)}, \rho^{(*)}]}{[\mathbf{y} | \boldsymbol{\beta}^{(k)}, \sigma^{2(k)}, \rho^{(k-1)}]}, \tag{17.33}$$

to accept or reject the proposal for MCMC iterations $k = 1, \ldots, K$. Notice that the proposal and prior cancel out in (17.33) as we described in Chapter 6.

We modified the MCMC algorithm for the geostatistical model in the previous section to fit the Bayesian CAR model to an areal spatial process. The R code for this algorithm, which we called norm.car.mcmc.R, is as follows:

BOX 17.5   MCMC ALGORITHM FOR CAR MODEL

```
1  norm.car.mcmc <- function(y,X,W,beta.mn,beta.var,
2      n.mcmc){
3
4  ###
5  ### Subroutines
6  ###
7
8  ldmvnorm<-function(y,mean,Sig.inv,logdet){
9      logretval=-(logdet+t(y-mean)%*%Sig.inv%
10         *%(y-mean))/2
11     logretval
12  }
13
14  ###
15  ### Set up Variables and Priors
16  ###
17
18  n=dim(X)[1]
19  p=dim(X)[2]
20  r=1000
21  q=0.0001
22  Sig.beta=beta.var*diag(p)
23  Sig.beta.inv=solve(Sig.beta)
24
25  alpha.1=1
26  alpha.2=1
27  rho=.8
28
29  beta.save=matrix(0,p,n.mcmc)
30  s2.save=rep(0,n.mcmc)
31  rho.save=rep(0,n.mcmc)
```

(Continued)

BOX 17.5 (Continued)   MCMC ALGORITHM FOR CAR MODEL

```
32 ###
33 ### Starting Values
34 ###
35
36 beta=solve(t(X)%*%X)%*%t(X)%*%y
37 Xbeta=X%*%beta
38 s2=mean((y-Xbeta)^2)
39
40 W.rowsum=diag(apply(W,1,sum))
41 R.inv=W.rowsum-rho*W
42 R=solve(R.inv)
43 logdet=determinant(R)$modulus
44
45 ###
46 ### MCMC Loop
47 ###
48
49 for(k in 1:n.mcmc){
50
51   ###
52   ### Sample s2
53   ###
54
55   tmp.r=(1/r+.5*t(y-Xbeta)%*%R.inv%*%(y-Xbeta))
          ^(-1)
56   tmp.q=n/2+q
57   s2=1/rgamma(1,tmp.q,,tmp.r)
58   Sig.inv=R.inv/s2
59
60   ###
61   ### Sample beta
62   ###
63
64  tmp.chol=chol(t(X)%*%Sig.inv%*%X + Sig.beta.inv)
65   beta=backsolve(tmp.chol,backsolve(tmp.chol,
66     t(X)%*%Sig.inv%*%y + Sig.beta.inv%*%beta.mn,
67     transpose=TRUE)+rnorm(p))
68   Xbeta=X%*%beta
```

(Continued)

BOX 17.5 (Continued) MCMC ALGORITHM FOR CAR MODEL

```
69  ###
70  ### Sample rho
71  ###
72
73  rho.star=rbeta(1,alpha.1,alpha.2)
74  R.inv.star=W.rowsum-rho.star*W
75  R.star=solve(R.inv.star)
76  logdet.star=determinant(R.star)$modulus
77  mh.1=ldmvnorm(y,Xbeta,R.inv.star/s2,logdet.star)
78  mh.2=ldmvnorm(y,Xbeta,R.inv/s2,logdet)
79  if(exp(mh.1-mh.2)>runif(1)){
80    rho=rho.star
81    R.inv=R.inv.star
82    logdet=logdet.star
83  }
84
85  ###
86  ### Save Samples
87  ###
88
89  beta.save[,k]=beta
90  s2.save[k]=s2
91  rho.save[k]=rho
92
93  }
94
95  ###
96  ###  Write Output
97  ###
98
99  list(beta.save=beta.save,s2.save=s2.save,
100     rho.save=rho.save,y=y,X=X,n.mcmc=n.mcmc,
101     n=n,r=r,q=q,p=p)
102
103  }
```

To demonstrate the use of this MCMC algorithm, we fit the Bayesian CAR model to log species richness for birds observed in counties throughout Colorado, USA (Figure 17.4). We used hyperparameters $\mu_\beta = 0$, $\Sigma_\beta = 1000\mathbf{I}$, $r = 1000$, $q = 0.0001$, $\alpha_1 = 1$, $\alpha_2 = 1$, and $K = 10000$ MCMC iterations to fit the Bayesian CAR model to the log species richness data using the following R code:

BOX 17.6   FIT CAR MODEL TO LOG SPECIES RICHNESS DATA

```
 1  library(maps)
 2  library(rgeos)
 3  library(rgdal)
 4
 5  co=readOGR("Colorado/","COUNTIES")
 6  co.cents=gCentroid(co,byid=TRUE)
 7  counties=as.character(co$COUNTY)
 8  locs=co.cents@coords
 9  n=dim(locs)[1]
10
11  birds.df=read.csv("co_birds.csv")
12  tmp=birds.df
13  for(i in 1:n){
14     tmp.idx=(1:n)[birds.df$county==counties[i]]
15     tmp[i,]=birds.df[tmp.idx,]
16  }
17  birds.df=tmp
18
19  D=as.matrix(dist(locs))
20
21  y=log(birds.df$richness)
22  x1=log(birds.df$minelev)
23  x2=log(birds.df$pop)
24  x3=birds.df$total
25  X=cbind(rep(1,n),scale(cbind(x1,x2,x3)))
26  p=dim(X)[2]
27
28  thresh=.15*max(D)
29  W=ifelse(D>thresh,0,1)
30  diag(W)=0
31
32  source("norm.car.mcmc.R")
33  n.mcmc=10000
34  set.seed(1)
35  mcmc.out=norm.car.mcmc(y=y,X=X,W=W,beta.mn=rep
36      (0,p),beta.var=1000,n.mcmc=n.mcmc)
```

(Continued)

**BOX 17.6 (Continued)   FIT CAR MODEL TO LOG SPECIES
RICHNESS DATA**

```
37 dIG <- function(x,r,q){
38   x^(-(q+1))*exp(-1/r/x)/(r^q)/gamma(q)
39 }
40
41
42 layout(matrix(1:6,3,2,byrow=TRUE))
43 hist(mcmc.out$beta.save[1,],prob=TRUE,
44     xlab=bquote(beta[0]),main="a")
45 curve(dnorm(x,0,sqrt(1000)),add=TRUE)
46 hist(mcmc.out$beta.save[2,],prob=TRUE,
47     xlab=bquote(beta[1]),main="b")
48 curve(dnorm(x,0,sqrt(1000)),add=TRUE)
49 hist(mcmc.out$beta.save[3,],prob=TRUE,
50     xlab=bquote(beta[3]),main="c")
51 curve(dnorm(x,0,sqrt(1000)),add=TRUE)
52 hist(mcmc.out$beta.save[4,],prob=TRUE,
53     xlab=bquote(beta[4]),main="d")
54 curve(dnorm(x,0,sqrt(1000)),add=TRUE)
55 hist(mcmc.out$s2.save,prob=TRUE,
56     xlab=bquote(sigma^2),main="e")
57 curve(dIG(x,mcmc.out$r,mcmc.out$q),add=TRUE)
58 hist(mcmc.out$rho.save,prob=TRUE,xlab=bquote(rho),
59     main="f")
60 curve(dbeta(x,mcmc.out$alpha.1,mcmc.out$alpha.2),
61     add=TRUE)
```

which results in the marginal posterior histograms shown in Figure 17.5. Notice that we created the proximity matrix $\mathbf{W}$ by defining counties within 0.15% of the maximum distance between all county centroids as neighbors (this distance is approximately 100 km). From Figure 17.5 we can infer that the marginal posterior distributions for all the regression coefficients ($\boldsymbol{\beta}$) indicate relationships between log species richness and the covariates except for β_1 which corresponds to minimum elevation. Also, the marginal posterior distribution for the autocorrelation parameter ρ tends toward larger values of ρ indicating the presence of residual spatial structure at the scale captured by $\mathbf{W}$ after accounting for the covariates in the first-order trend.

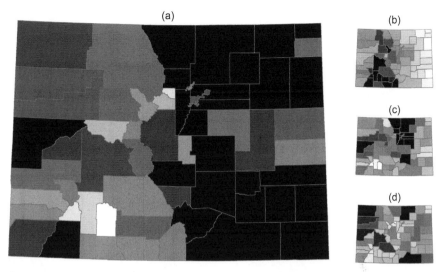

Figure 17.4   (a) log species richness for birds in counties within Colorado, USA, and a set of standardized covariates pertaining to (b) minimum elevation in the county, (c) log human population density per county, and (d) total area of the county. Dark shaded counties correspond to large values and light shaded counties correspond to small values.

17.4   ADDITIONAL CONCEPTS AND READING

One of the most common extensions to both the geostatistical and CAR models is to include a "nugget" effect. A nugget effect (a term arising from the geology and mining literature) refers to unstructured error in the process that may arise due to measurement error or small-scale variability. To incorporate a nugget effect in a spatial statistical model, we can simply add another variance component in the covariance matrix so that $\Sigma \equiv \tau^2 \mathbf{I} + \sigma^2 \mathbf{R}$, where τ^2 is referred to as the nugget variance and σ^2 is now referred to as the partial sill, where $\mathbf{R}$ is defined using an explicitly spatial specification as we presented earlier in this chapter. The new parameter τ^2 requires a prior in the Bayesian paradigm and we can sample it in a MCMC algorithm using its full-conditional distribution just like the other parameters in the model.

An alternative way to think about this generalized spatial statistical model with a nugget effect is as a hierarchical model (Chapter 19) where the data arise from a distribution $\mathbf{y} \sim N(\mathbf{X}\boldsymbol{\beta} + \boldsymbol{\eta}, \tau^2 \mathbf{I})$ that depends on the parameters $\boldsymbol{\beta}$ and unstructured error variance τ^2, as well as the spatially correlated random effect $\boldsymbol{\eta}$, where $\boldsymbol{\eta} \sim N(\mathbf{0}, \sigma^2 \mathbf{R})$. This hierarchical model formulation makes it clear that there is a latent correlated process that we wish to account for directly and make inference about (Arab et al. 2008a). Typically, this type of model is fit using the integrated likelihood rather than the conditional form with $\boldsymbol{\eta}$ appearing explicitly, for computational

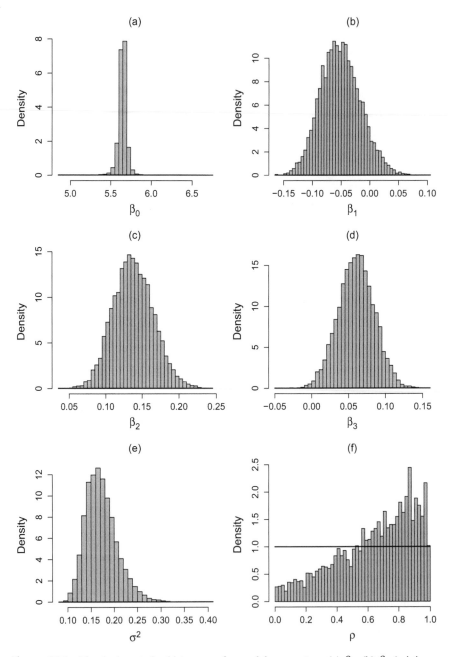

Figure 17.5 Marginal posterior histograms for model parameters: (a) β_0, (b) β_1 (minimum elevation), (c) β_2 (log human population density), (d) β_3 (county area), (e) σ^2, and (f) ρ. All covariates were standardized to have mean zero and variance one. Prior PDFs shown as black lines.

efficiency. In either case, if there is only weak evidence for spatial structure in $\mathbf{R}$, and without replication, the two variance components may not be fully identifiable. However, we can use standard exploratory spatial data analysis techniques to examine the residual structure before fitting the model. See Hooten et al. (2017) Chapter 2 for a primer and set of relevant references on the suite of spatial statistical data analysis approaches. We also cover hierarchical models in much more detail in the next part of this book.

Just as an integrated form of the hierarchical spatial statistical model can be more computationally efficient, we can also integrate the likelihood again with respect to the regression coefficients $\boldsymbol{\beta}$ which can also increase algorithm stability in some cases. To show this, consider the model presented in the first section of this chapter

$$\mathbf{y} \sim N(\mathbf{X}\boldsymbol{\beta}, \sigma^2 \mathbf{R}), \tag{17.34}$$

where $\mathbf{R}$ takes the geostatistical correlation matrix form and prior $\boldsymbol{\beta} \sim N(\boldsymbol{\mu}_\beta, \boldsymbol{\Sigma}_\beta)$. To arrive at a fully integrated likelihood, we integrate out $\boldsymbol{\beta}$ from the product of the original likelihood and prior for $\boldsymbol{\beta}$ which yields

$$[\mathbf{y}|\sigma^2, \phi] = \int [\mathbf{y}|\boldsymbol{\beta}, \sigma^2, \phi][\boldsymbol{\beta}]d\boldsymbol{\beta}, \tag{17.35}$$

$$\propto \int \exp\left(-\frac{1}{2}(\mathbf{y}-\mathbf{X}\boldsymbol{\beta})'\boldsymbol{\Sigma}^{-1}(\mathbf{y}-\mathbf{X}\boldsymbol{\beta})\right)$$

$$\times \exp\left(-\frac{1}{2}(\boldsymbol{\beta}-\boldsymbol{\mu}_\beta)'\boldsymbol{\Sigma}_\beta^{-1}(\boldsymbol{\beta}-\boldsymbol{\mu}_\beta)\right)d\boldsymbol{\beta}, \tag{17.36}$$

$$\propto \exp\left(-\frac{1}{2}\mathbf{y}'\boldsymbol{\Sigma}^{-1}\mathbf{y}\right)\int \exp\left(-\frac{1}{2}\left(-2(\mathbf{y}'\boldsymbol{\Sigma}^{-1}\mathbf{X}+\boldsymbol{\mu}_\beta'\boldsymbol{\Sigma}_\beta^{-1})\boldsymbol{\beta}\right.\right.$$

$$\left.\left.+\boldsymbol{\beta}'(\mathbf{X}'\boldsymbol{\Sigma}^{-1}\mathbf{X}+\boldsymbol{\Sigma}_\beta^{-1})\boldsymbol{\beta}\right)\right)d\boldsymbol{\beta}, \tag{17.37}$$

$$\propto \exp\left(-\frac{1}{2}\mathbf{y}'\boldsymbol{\Sigma}^{-1}\mathbf{y}\right)\int \exp\left(-\frac{1}{2}\left(-2\mathbf{b}'\boldsymbol{\beta}+\boldsymbol{\beta}'\mathbf{A}\boldsymbol{\beta}\right)\right)d\boldsymbol{\beta}. \tag{17.38}$$

Thus, the integrand has the form of a multivariate Gaussian kernel for $\boldsymbol{\beta}$ where

$$\mathbf{b}' \equiv \mathbf{y}'\boldsymbol{\Sigma}^{-1}\mathbf{X}+\boldsymbol{\mu}_\beta'\boldsymbol{\Sigma}_\beta^{-1}, \tag{17.39}$$

$$\mathbf{A} \equiv \mathbf{X}'\boldsymbol{\Sigma}^{-1}\mathbf{X}+\boldsymbol{\Sigma}_\beta^{-1}. \tag{17.40}$$

With the integrand in (17.38) taking the form of a multivariate Gaussian kernel, we can set the integral to one effectively by replacing it with the reciprocal of the normalizing constant associated with that multivariate Gaussian kernel. Then, we will associate the remaining terms in (17.38) with the data $\mathbf{y}$ and we can derive the integrated data model by recognizing the remaining expression as the kernel of another multivariate Gaussian, but this time with respect to $\mathbf{y}$:

$$[\mathbf{y}|\sigma^2, \phi] \propto \exp\left(-\frac{1}{2}\mathbf{y}'\boldsymbol{\Sigma}^{-1}\mathbf{y}\right)\exp\left(\frac{1}{2}\mathbf{b}'\mathbf{A}^{-1}\mathbf{b}\right), \tag{17.41}$$

$$\propto \exp\left(-\frac{1}{2}\left(\mathbf{y}'\boldsymbol{\Sigma}^{-1}\mathbf{y} - (\mathbf{y}'\boldsymbol{\Sigma}^{-1}\mathbf{X} + \boldsymbol{\mu}'_\beta\boldsymbol{\Sigma}_\beta^{-1})'(\mathbf{X}'\boldsymbol{\Sigma}^{-1}\mathbf{X} + \boldsymbol{\Sigma}_\beta^{-1})^{-1}\right.\right.$$
$$\left.\left.(\mathbf{y}'\boldsymbol{\Sigma}^{-1}\mathbf{X} + \boldsymbol{\mu}'_\beta\boldsymbol{\Sigma}_\beta^{-1})\right)\right), \tag{17.42}$$

$$\propto \exp\left(-\frac{1}{2}\left(\mathbf{y}'(\boldsymbol{\Sigma}^{-1} - \boldsymbol{\Sigma}^{-1}\mathbf{X}(\mathbf{X}'\boldsymbol{\Sigma}^{-1}\mathbf{X} + \boldsymbol{\Sigma}_\beta^{-1})^{-1}\mathbf{X}'\boldsymbol{\Sigma}^{-1})\mathbf{y} - \right.\right. \tag{17.43}$$

$$\left.\left.-2\boldsymbol{\mu}'_\beta\boldsymbol{\Sigma}_\beta^{-1}(\mathbf{X}'\boldsymbol{\Sigma}^{-1}\mathbf{X} + \boldsymbol{\Sigma}_\beta^{-1})^{-1}\mathbf{X}'\boldsymbol{\Sigma}^{-1}\mathbf{y}\right)\right), \tag{17.44}$$

$$\propto \exp\left(-\frac{1}{2}\left(\mathbf{y}'\tilde{\mathbf{A}}\mathbf{y} - 2\tilde{\mathbf{b}}'\mathbf{y}\right)\right), \tag{17.45}$$

which implies that $[\mathbf{y}|\sigma^2, \phi] \equiv \mathrm{N}(\tilde{\mathbf{A}}^{-1}\mathbf{b}, \tilde{\mathbf{A}}^{-1})$, where

$$\tilde{\mathbf{b}}' \equiv \boldsymbol{\mu}'_\beta\boldsymbol{\Sigma}_\beta^{-1}(\mathbf{X}'\boldsymbol{\Sigma}^{-1}\mathbf{X} + \boldsymbol{\Sigma}_\beta^{-1})^{-1}\mathbf{X}'\boldsymbol{\Sigma}^{-1}, \tag{17.46}$$

$$\tilde{\mathbf{A}} \equiv \boldsymbol{\Sigma}^{-1} - \boldsymbol{\Sigma}^{-1}\mathbf{X}(\mathbf{X}'\boldsymbol{\Sigma}^{-1}\mathbf{X} + \boldsymbol{\Sigma}_\beta^{-1})^{-1}\mathbf{X}'\boldsymbol{\Sigma}^{-1}. \tag{17.47}$$

Thus, we can use the integrated data model directly in a MCMC algorithm to fit the Bayesian spatial model to data and avoid sampling the coefficients $\boldsymbol{\beta}$. We can recover the regression coefficients using a second-stage algorithm that samples $\boldsymbol{\beta}$ from its full-conditional distribution (see first section in this chapter for this full-conditional distribution) using the output from the first-stage algorithm that contains a MCMC sample for σ^2 and ϕ.[5] Then, if we further desire predictive realizations, we can obtain those using a third MCMC composition sampling algorithm as described in this chapter. This technique of integrating out fixed effects to fit models is often referred to as Rao-Blackwellization and is used in the spBayes R package which fits Bayesian spatial models like those discussed in this chapter (Finley et al. 2015).

Another useful concept related to statistical models for dependence is the relationship between first- and second-order formulations of the models (Sampson 2010). Hefley et al. (2017a) describe the connections between these two different approaches to specifying and fitting models and how they can both lead to equivalent statistical inference. In particular, Hefley et al. (2017a) describe how to formulate models with spatial and/or temporal structure using first-order terms involving basis functions and associated basis function coefficients. To demonstrate how this works briefly, consider the hierarchical version of the spatial statistical model with a nugget effect described previously:

$$\mathbf{y} \sim \mathrm{N}(\mathbf{X}\boldsymbol{\beta} + \boldsymbol{\eta}, \tau^2\mathbf{I}), \tag{17.48}$$

---

[5] Also see Van Dyk and Park (2008) for details on a related approach called a "partially collapsed Gibbs sampler."

where $\eta \sim N(\mathbf{0}, \sigma^2 \mathbf{R})$. Now expand the spatial random vector η as a linear combination of basis vectors $\mathbf{H}$ and basis coefficients α such that $\eta \equiv \mathbf{H}\alpha$ (where $\mathbf{H}$ has certain properties; Wikle 2010; Hefley et al. 2017a). Then, if $\alpha \sim N(\mathbf{0}, \sigma^2 \mathbf{I})$ serve as the new random effects, the matrix of basis vectors $\mathbf{H}$ provide the necessary correlation structure because $\mathbf{H}\alpha \sim N(\mathbf{0}, \sigma^2 \mathbf{HH'})$ as a result of multivariate normal properties; thus, the correlation matrix for the spatial random vector η is $\mathbf{R} \equiv \mathbf{HH'}$.

The basis function approach to formulating spatial (and temporal) statistical models reconciles several perspectives on modeling dependence and may be useful from a model specification and implementation perspective. In particular, writing models from the first-order perspective can illuminate ways to reduce the dimensionality and increase the speed of model fitting algorithms. For example, various "reduced-rank" approaches for specifying models have become popular recently because they accommodate large data sets (Wikle 2010). See Heaton et al. (2018) for an excellent overview of approaches for specifying and implementing geostatistical models for large data sets. Basis function specifications of spatial statistical models also can help accommodate nonstationarity in the spatial process (Sampson 2010).

The first-order perspective also can indicate potentially problematic model structure issues. For example, consider the issue known as "spatial confounding" (Hodges and Reich 2010; Hanks et al. 2015b). Spatial confounding arises because of overlap in explained variation in the response variable by the two effective design matrices in models with both first- and second-order effects. Note that we can write the basis function formulation of the spatial model we described previously as $\mathbf{y} \sim N(\mathbf{X}\boldsymbol{\beta} + \mathbf{H}\alpha, \tau^2 \mathbf{I})$. In this formulation, we can see that $\mathbf{X}$ and $\mathbf{H}$ act as design matrices in the model and that, if their columns of predictor variables are correlated, the inference pertaining to $\boldsymbol{\beta}$ (and α) could be affected. Thus, remedial approaches have been proposed for orthogonalizing $\mathbf{X}$ and $\mathbf{H}$ so that they account for different types of dependence in the response variable; although there is debate regarding when that type of modification is appropriate in spatial models (Hanks et al. 2015b). Hooten et al. (2017) also describe this spatial confounding issue and its potential ramifications for inference and Hefley et al. (2017b) and Bose et al. (2018) provide alternative ways to check for confounding and account for it in Bayesian spatial models.

Finally, in our Gaussian CAR model example, we analyzed log transformed counts associated with species richness in Colorado, USA, counties. This begs the question, how can we account for the natural support of the data (e.g., binary or counts) more directly in our modeling? In the next part of this book, we describe generalized linear models and discuss how we can extend them to hierarchical settings that have latent Gaussian processes. These latent processes could be unstructured random effects or they could be spatial random effects. Thus, this chapter on spatial models has utility that extends beyond the typical Gaussian data assumptions and we can apply it in a variety of other ecological model, including binary regression (e.g., Hooten et al. 2003), occupancy models (e.g., Johnson et al. 2013a), N-mixture models (e.g., Hooten et al. 2007), and so on.

Section IV

Advanced Models and Concepts

18 Quantile Regression

18.1 QUANTILE MODELS FOR CONTINUOUS DATA

We covered a variety of approaches for modeling mean (and covariance) structure of data in the preceding chapters. For example, the classical linear regression model allowed us to infer how the heterogeneous mean of the data varies with predictor variables of interest. To do so, we parameterized a statistical model as $y_i \sim N(\mathbf{x}_i'\boldsymbol{\beta}, \sigma^2)$, for $i = 1, \ldots, n$, where $\mathbf{x}_i'\boldsymbol{\beta}$ represents the mean of the data distribution at covariate values $\mathbf{x}_i$. However, while this regression model tells us something about the average relationship between $\mathbf{x}_i$ and y_i, there could be other important aspects of the relationship that it does not provide inference about. For example, what if we are interested in learning about the relationship between $\mathbf{x}_i$ and y_i associated with the median of the data, or even the upper or lower tail of the data?

In a seminal paper, Koenker and Basset (1978) described a way to generalize the regression concept to model relationships between $\mathbf{x}_i$ and y_i at various quantiles of the data. The later expansions of these ideas have spawned a variety of generalizations of quantile regression approaches to study relationships in data. In ecology specifically, Cade and Noon (2003) advocated the use of quantile regression methods for ecological inference, but the details on how to implement and extend quantile regression models were beyond the scope of their paper. In what follows, we provide a short introduction to the concept of quantile regression and show how to extend this concept to the Bayesian setting. We also present two different and contrasting perspectives on quantile regression and how we can use it in ecological and environmental science.

In the classical linear regression model, we assumed that $E(y_i|\boldsymbol{\beta}, \sigma^2) = \mathbf{x}_i'\boldsymbol{\beta}$. By contrast, in quantile regression, we assume that $q_\tau(y_i|\boldsymbol{\beta}, \sigma^2) = \mathbf{x}_i'\boldsymbol{\beta}$, where q_τ is the τth ($0 < \tau < 1$) quantile function (instead of the mean function E). In the non-Bayesian setting we can find the optimal $\boldsymbol{\beta}$ by minimizing the function $\sum_{i=1}^{n} v_i (\tau - 1_{\{v_i < 0\}})$ with respect to $\boldsymbol{\beta}$, where $v_i = y_i - \mathbf{x}_i'\boldsymbol{\beta}$. Koenker and Machado (1999) showed that the asymmetric Laplace distribution (ALD) is the parametric equivalent to the least squares optimization for quantile regression and Yu and Moyeed (2001) demonstrated that we can use the ALD to form the likelihood in a Bayesian model.

Yu and Zhang (2005) proposed the following form for the PDF of an ALD based on the τth percentile:

$$[y_i|\mu_i, \sigma] \equiv \frac{\tau(1-\tau)}{\sigma} \exp\left(-z_i(\tau - 1_{\{z_i < 0\}})\right), \tag{18.1}$$

where $z_i \equiv (y_i - \mu_i)/\sigma$ and $\mu_i = \mathbf{x}_i'\boldsymbol{\beta}$. Thus, we can use the ALD definition in (18.1) as a data model in a Bayesian framework with priors for the model parameters specified as

$$\boldsymbol{\beta} \sim \text{N}(\boldsymbol{\mu}_\beta, \sigma_\beta^2 \mathbf{I}), \tag{18.2}$$

$$\sigma \sim \text{Unif}(0, u), \tag{18.3}$$

where u is a reasonable upper bound for σ and the user sets τ. Based on this Bayesian quantile regression model, the full-conditional distributions for $\boldsymbol{\beta}$ and σ will not be conjugate and thus we sample them using M-H updates in a MCMC algorithm.

We can write the posterior distribution for the quantile regression model as

$$[\boldsymbol{\beta}, \sigma | \mathbf{y}] \propto \left(\prod_{i=1}^{n} [y_i | \boldsymbol{\beta}, \sigma] \right) [\boldsymbol{\beta}][\sigma], \tag{18.4}$$

where $[y_i | \boldsymbol{\beta}, \sigma]$ is the ALD defined in (18.1). Thus, to identify the correct M-H ratio for updating $\boldsymbol{\beta}$, we use the portion of the joint distribution in (18.4) that is proportional to $\boldsymbol{\beta}$

$$[\boldsymbol{\beta} | \cdot] \propto \left(\prod_{i=1}^{n} [y_i | \boldsymbol{\beta}, \sigma] \right) [\boldsymbol{\beta}]. \tag{18.5}$$

If we use a multivariate normal random walk as a proposal for the regression coefficients where $\boldsymbol{\beta}^{(*)} \sim \text{N}(\boldsymbol{\beta}^{(k-1)}, \sigma_{\beta,\text{tune}}^2 \mathbf{I})$ for MCMC iteration k with $\sigma_{\beta,\text{tune}}^2$ as a tuning parameter, then the M-H ratio is

$$mh = \frac{\left(\prod_{i=1}^{n} [y_i | \boldsymbol{\beta}^{(*)}, \sigma^{(k)}] \right) [\boldsymbol{\beta}^{(*)}]}{\left(\prod_{i=1}^{n} [y_i | \boldsymbol{\beta}^{(k-1)}, \sigma^{(k)}] \right) [\boldsymbol{\beta}^{(k-1)}]}. \tag{18.6}$$

The M-H update for σ can be found similarly, by identifying the components in the joint distribution (18.4) that are proportional with respect to σ and using them to form the M-H ratio. There are several options for proposal distributions for σ. Because we used a prior that has compact support (i.e., uniform on $(0, u)$), we could use the prior as a proposal, but if it is diffuse relative to the posterior it will result in an inefficient MCMC algorithm. Similarly, we could use importance sampling resampling (Chapter 5) instead of M-H, but again, if u is chosen to be much larger than the posterior for σ it will require excess computational resources to obtain an accurate importance sample. Thus, we use the random walk with rejection sampling technique described in Chapter 6 (i.e., $\sigma^{(*)} \sim \text{N}(\sigma^{(k-1)}, \sigma_{\sigma,\text{tune}}^2)$) which allows us to tune the updates to be more efficient (i.e., select a tuning parameter $\sigma_{\sigma,\text{tune}}^2$ that does not yield too many or too few M-H proposal acceptances). With an unconstrained random walk proposal, we skip the update for proposed values outside the support $(0, u)$ of σ and only proceed with the usual M-H update if our proposed value $0 \leq \sigma^{(*)} \leq u$. In this case, the M-H ratio is

$$mh = \frac{\left(\prod_{i=1}^{n} [y_i | \boldsymbol{\beta}^{(k)}, \sigma^{(*)}] \right) [\sigma^{(*)}]}{\left(\prod_{i=1}^{n} [y_i | \boldsymbol{\beta}^{(k)}, \sigma^{(k-1)}] \right) [\sigma^{(k-1)}]}, \tag{18.7}$$

where the prior for σ evaluates to $[\sigma] = 1/u$ if $0 \leq \sigma \leq u$ and to $[\sigma] = 0$ if $\sigma < 0$ or $\sigma > u$.

We constructed the MCMC algorithm to fit the quantile regression model based on the full-conditional distributions and these M-H ratios as well as knowledge of fixed parameters τ (specifying the quantile of interest) and tuning parameters $\sigma^2_{\beta,\text{tune}}$ and $\sigma^2_{\sigma,\text{tune}}$. The R code for our MCMC algorithm (called reg.qr.mcmc.R) is as follows:

BOX 18.1   MCMC ALGORITHM FOR QUANTILE REGRESSION

```
1  reg.qr.mcmc <- function(y,X,tau,beta.mn,beta.var,
2      s.u,n.mcmc){
3
4  ###
5  ### Subroutines
6  ###
7
8  ldasymlap <- function(y,tau,mu,sig){
9    rho <- function(z,tau){z*(tau-(z<0))}
10   ld=log(tau)+log(1-tau)-log(sig)-rho((y-mu)/
11   sig,tau)
12   ld
13 }
14
15 ###
16 ### Setup Variables
17 ###
18
19 n=dim(X)[1]
20 p=dim(X)[2]
21 n.burn=round(.1*n.mcmc)
22
23 beta.save=matrix(0,p,n.mcmc)
24 s2.save=rep(0,n.mcmc)
25 D.bar=0
26
27 ###
28 ### Starting Values and Priors
29 ###
30
31 beta=solve(t(X)%*%X)%*%t(X)%*%y
32 XpX=t(X)%*%X
```

(Continued)

BOX 18.1 (Continued) MCMC ALGORITHM FOR QUANTILE REGRESSION

```
33 s.tune=.1
34 beta.tune=.1
35
36 s2=mean((y-X%*%beta)^2)
37 sig=sqrt(s2)
38
39 ###
40 ### MCMC Loop
41 ###
42
43 for(k in 1:n.mcmc){
44
45   ###
46   ### Sample s2
47   ###
48
49   s.star=rnorm(1,sig,s.tune)
50   if((s.star < s.u) & (s.star > 0)){
51     mh.1=sum(ldasymlap(y,tau,X%*%beta,s.star))
52     mh.2=sum(ldasymlap(y,tau,X%*%beta,sig))
53     mh=exp(mh.1-mh.2)
54
55     if(mh > runif(1)){
56       sig=s.star
57     }
58   }
59
60   ###
61   ### Sample beta
62   ###
63
64   beta.star=rnorm(p,beta,beta.tune)
65   mh.1=sum(ldasymlap(y,tau,X%*%beta.star,sig))
66       +sum(dnorm(beta.star,beta.mn,
67       sqrt(beta.var),log=TRUE))
68   mh.2=sum(ldasymlap(y,tau,X%*%beta,sig))+sum
69     (dnorm(beta,beta.mn,sqrt(beta.var),log=TRUE))
70   mh=exp(mh.1-mh.2)
```

(Continued)

BOX 18.1 (Continued)   MCMC ALGORITHM FOR QUANTILE REGRESSION

```
71    if(mh > runif(1)){
72       beta=beta.star
73    }
74
75    ###
76    ### Posterior Predictive Calculations
77    ###
78
79    if(k > n.burn){
80       D.bar=D.bar-2*sum(ldasymlap(y,tau,X%*%beta,
81          sig))/(n.mcmc-n.burn)
82    }
83
84    ###
85    ### Save Samples
86    ###
87
88    beta.save[,k]=beta
89    s2.save[k]=sig^2
90
91 }
92
93 ###
94 ###  DIC Calculations
95 ###
96
97 beta.mean=as.vector(apply(beta.save
98    [,n.burn:n.mcmc],1,mean))
99 sig.mean=mean(sqrt(s2.save[n.burn:n.mcmc]))
100 D.hat=-2*sum(ldasymlap(y,tau,X%*%beta.mean,
101     sig.mean))
102 pD=D.bar-D.hat
103 DIC=D.hat+2*pD
```

(Continued)

BOX 18.1 (Continued)   MCMC ALGORITHM FOR QUANTILE REGRESSION

```
104  ###
105  ###   Write Output
106  ###
107
108  list(beta.save=beta.save,s2.save=s2.save,y=y,
109      X=X,n.mcmc=n.mcmc,n.burn=n.burn,n=n,p=p,
110      tau=tau,D.bar=D.bar,D.hat=D.hat,pD=pD,
111      DIC=DIC)
112
113  }
```

Note that we created our own R function that calculates the log density function of the ALD function (called `ldasymlap`).

There are two philosophical perspectives that we can assume when using quantile regression. The first takes an exploratory data analysis perspective where we are interested in exploring the relationship in the data at a particular quantile or at different quantiles, perhaps with the intention of comparing them. For example, we may be interested in the $\tau = 0.1$ versus $\tau = 0.9$ percentiles because they may represent different patterns in the response variables at lower levels than at higher levels.[1] The second philosophy is as a generalization of the regression model that can accommodate skewness in the data. Taking this perspective, we may be interested in identifying the skewness parameter that provides the best predictive model (which we can assess using model selection or estimate it formally).

For example, consider the white-tailed deer data we described in Chapter 11. Suppose that we are interested in inferring the relationship between body mass of individual deer and their age at one of several quantiles. Recall that there was a quadratic relationship with age in our previous regression models. Thus, ignoring the other covariates for illustrative purposes, we specify the design matrix in our quantile regression model as having three columns, the first with a vector of ones to represent the intercept, and the second and third as age and age.[2] We investigated a set of five quantile regression models with τ equal to 0.10, 0.25, 0.5, 0.75, and 0.90. The second through fourth quantiles in our set represent the two quartiles and the median.

---

[1] We note that, when the pattern in the data varies strongly with the quantile, the assumptions of the model may no longer be valid. In this case, we may be limited to an exploratory perspective only because the model is unlikely to have generated the observed data, although we may be able to use model checking techniques to examine this.

To fit the five quantile regression models to data, we used the following R code, which reads in the data, fits each of the models in a loop, and reports the posterior information in Table 18.1:

BOX 18.2   FIT QUANTILE REGRESSION MODEL TO DEER DATA

```
 1  load("deer.RData")
 2  n=dim(deer.df)[1]
 3  y=deer.df$y
 4  X=cbind(rep(1,n),as.matrix(deer.df[,4:5]))
 5  n=dim(X)[1]
 6  p=dim(X)[2]
 7
 8  tau.vec=c(.1,.25,.5,.75,.9)
 9  n.tau=length(tau.vec)
10  out.list=vector("list",n.tau)
11  DIC.vec=rep(0,n.tau)
12  beta.mat=matrix(0,p,n.tau)
13  s.vec=rep(0,n.tau)
14  source("reg.qr.mcmc.R")    # Code Box 18.1
15  n.mcmc=50000
16  set.seed(1)
17  for(j in 1:n.tau){
18    out.list[[j]]=reg.qr.mcmc(y=y,X=X,
19       tau=tau.vec[j],beta.mn=rep(0,p),
20       beta.var=1000,s.u=100,n.mcmc=n.mcmc)
21    n.burn=out.list[[j]]$n.burn
22    DIC.vec[j]=out.list[[j]]$DIC
23    beta.mat[,j]=apply(out.list[[j]]$beta.save
24       [,n.burn:n.mcmc],1,mean)
25    s.vec[j]=mean(sqrt(out.list[[j]]$s2.
26       save[n.burn:n.mcmc]))
27  }
28  cbind(tau.vec,DIC.vec,t(beta.mat),s.vec)
```

Thus, based on the quantiles we examined, the model with $\tau = 0.25$ appears to be the best predictive model, with the smallest DIC value of 3465.4. The fact that a model selection procedure indicates a smaller skewness parameter than 0.5 suggests that left-skewness in the data is important to account for to improve predictive performance (and likely goodness of fit).

We can examine the relationships at the set of five quantiles that we specified by plotting the posterior mean of the regression quantile $E(\tilde{\mathbf{X}}\boldsymbol{\beta}|\mathbf{y})$ for a range of

Table 18.1

Percentiles (τ), Posterior Means for Model Parameters β and σ, and the Resulting DIC from Each Model

		Covariate			
τ	Intercept	Age	Age^2	σ	DIC
0.10	13.72	13.16	-1.30	1.09	3505.5
0.25	15.33	14.19	-1.37	2.19	3465.4
0.50	18.33	15.19	-1.44	3.09	3524.6
0.75	19.45	19.06	-1.76	2.70	3673.9
0.90	20.78	22.03	-2.00	1.52	3834.3

covariate values spanning the space of our observed covariates. We used the following R code to obtain the estimated quantile regression lines and create Figure 18.1:

BOX 18.3   POSTERIOR QUANTILE FUNCTION

```
 1 n.pred=100
 2 X.pred=matrix(1,n.pred,p)
 3 X.pred[,2]=seq(min(X[,2]),max(X[,2]),,n.pred)
 4 X.pred[,3]=X.pred[,2]^2
 5
 6 plot(X[,2],y,xlab="age",ylab="body mass (kg)")
 7 for(j in 1:n.tau){
 8   n.burn=out.list[[j]]$n.burn
 9   y.q=apply(X.pred%*%out.list[[j]]$beta.save
10     [,n.burn:n.mcmc],1,mean)
11   lines(X.pred[,2],y.q,type="l")
12 }
```

The procedure we described thus far in this chapter is typical of most studies using quantile regression for ecological and environmental learning. As a review, we specify a quantile regression model that we wish to fit to data and then fit it using a known quantile to assess the pattern. For example, Figure 18.1 indicates that, at the higher quantiles, the parabolic shape in the relationship between individual body mass and age is slightly more accentuated than at lower quantiles. However, the second main approach is ALD regression modeling while considering the "quantile" to be an unknown skewness parameter in the model that we wish to learn about given the data. Based on the individual model fits for different quantiles (Table 18.1), we can see that

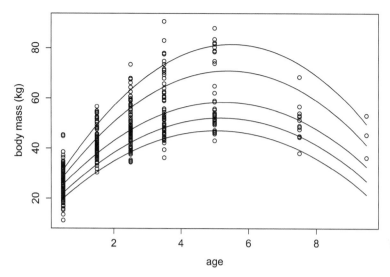

Figure 18.1 White-tailed deer scatterplot of body mass (kg) versus age (points) and quantile regression lines of $E(\tilde{\mathbf{X}}\boldsymbol{\beta}|\mathbf{y})$ for values of τ set at 0.10, 0.25, 0.50, 0.75, and 0.90, shown as parabolic lines from bottom to top.

some skewness parameter $\overset{S}{V}$ provide better predictive performance than others. Thus, treating the quantile as an unknown skewness parameter and allowing the data to inform it can lead to an optimal model that better characterizes the data-generating mechanisms.

To generalize the model so that τ is unknown, we retain the rest of the model components because we have specified them and include a prior distribution for τ in the posterior, which now becomes

$$[\boldsymbol{\beta}, \sigma, \tau | \mathbf{y}] \propto \left(\prod_{i=1}^{n} [y_i | \boldsymbol{\beta}, \sigma, \tau] \right) [\boldsymbol{\beta}][\sigma][\tau]. \tag{18.8}$$

If we specify a uniform prior for τ such that $\tau \sim \text{Unif}(0, 1)$, the posterior distribution leads to the following full-conditional distribution for τ

$$[\tau | \cdot] \propto \left(\prod_{i=1}^{n} [y_i | \boldsymbol{\beta}, \sigma, \tau] \right), \tag{18.9}$$

for $0 < \tau < 1$, which is not conjugate. Thus, we can perform a M-H update for τ in the MCMC algorithm using the M-H ratio

$$mh = \frac{\prod_{i=1}^{n} [y_i | \boldsymbol{\beta}^{(k)}, \sigma^{(k)}, \tau^{(*)}]}{\prod_{i=1}^{n} [y_i | \boldsymbol{\beta}^{(k)}, \sigma^{(k)}, \tau^{(k-1)}]}, \tag{18.10}$$

if we use a random walk with rejection sampling ($\tau^{(*)} \sim N(\tau^{(k-1)}, \sigma^2_{\tau,\text{tune}})$) as we did with the parameter σ in the quantile regression model except that, if τ is not between zero and one, we skip the M-H update.

To construct a MCMC algorithm to fit the quantile regression model treating τ as an unknown parameter to be estimated, we only have to modify the previous R code to include the new M-H update for τ. We called this new MCMC algorithm reg.qr.tau.mcmc.R:

BOX 18.4   MCMC ALGORITHM FOR ALD REGRESSION WITH UNKNOWN SKEWNESS PARAMETER

```
 1  reg.qr.tau.mcmc <- function(y,X,beta.mn,
 2      beta.var,u,n.mcmc){
 3
 4  ###
 5  ### Subroutines
 6  ###
 7
 8  ldasymlap <- function(y,tau,mu,sig){
 9    rho <- function(z,tau){z*(tau-(z<0))}
10    ld=log(tau)+log(1-tau)-log(sig)-rho((y-mu)/
11        sig,tau)
12    ld
13  }
14
15  ###
16  ### Setup Variables
17  ###
18
19  n=dim(X)[1]
20  p=dim(X)[2]
21  n.burn=round(.1*n.mcmc)
22
23  beta.save=matrix(0,p,n.mcmc)
24  s2.save=rep(0,n.mcmc)
25  tau.save=rep(0,n.mcmc)
26  D.bar=0
```

(Continued)

BOX 18.4 (Continued)   MCMC ALGORITHM FOR ALD REGRESSION WITH UNKNOWN SKEWNESS PARAMETER

```
27 ###
28 ### Starting Values and Priors
29 ###
30
31 beta=solve(t(X)%*%X)%*%t(X)%*%y
32 XpX=t(X)%*%X
33
34 s.tune=.1
35 beta.tune=.1
36 p.tune=.01
37
38 tau=.5
39 s2=mean((y-X%*%beta)^2)
40 sig=sqrt(s2)
41
42 ###
43 ### MCMC Loop
44 ###
45
46 for(k in 1:n.mcmc){
47
48   ###
49   ### Sample s2
50   ###
51
52   s.star=rnorm(1,sig,s.tune)
53   if((s.star < u) & (s.star > 0)){
54     mh.1=sum(ldasymlap(y,tau,X%*%beta,s.star))
55     mh.2=sum(ldasymlap(y,tau,X%*%beta,sig))
56     mh=exp(mh.1-mh.2)
57
58     if(mh > runif(1)){
59       sig=s.star
60     }
61   }
62
63   ###
64   ### Sample beta
65   ###
```

(Continued)

**BOX 18.4 (Continued)　MCMC ALGORITHM FOR ALD
REGRESSION WITH UNKNOWN SKEWNESS PARAMETER**

```
66    beta.star=rnorm(p,beta,beta.tune)
67    mh.1=sum(ldasymlap(y,tau,X%*%beta.star,sig))
68          +sum(dnorm(beta.star,beta.mn,
69          sqrt(beta.var),log=TRUE))
70    mh.2=sum(ldasymlap(y,tau,X%*%beta,sig))
71          +sum(dnorm(beta,beta.mn,sqrt(beta.var),
72          log=TRUE))
73    mh=exp(mh.1-mh.2)
74
75    if(mh > runif(1)){
76      beta=beta.star
77    }
78
79    ###
80    ### Sample tau (skewness parameter)
81    ###
82
83    tau.star=rnorm(1,tau,p.tune)
84    if((tau.star >0) & (tau.star<1)){
85      mh.1=sum(ldasymlap(y,tau.star,X%*%beta,sig))
86      mh.2=sum(ldasymlap(y,tau,X%*%beta,sig))
87      mh=exp(mh.1-mh.2)
88
89      if(mh > runif(1)){
90        tau=tau.star
91      }
92    }
93
94    ###
95    ### Posterior Predictive Calculations
96    ###
97
98    if(k > n.burn){
99      D.bar=D.bar-2*sum(ldasymlap(y,tau,X%*%beta,
100         sig))/(n.mcmc-n.burn)
101    }
```

(Continued)

BOX 18.4 (Continued)   MCMC ALGORITHM FOR ALD REGRESSION WITH UNKNOWN SKEWNESS PARAMETER

```
102   ###
103   ### Save Samples
104   ###
105
106   beta.save[,k]=beta
107   s2.save[k]=sig^2
108   tau.save[k]=tau
109
110   }
111
112   ###
113   ###   DIC Calculations
114   ###
115
116   beta.mean=as.vector(apply(beta.save
117       [,n.burn:n.mcmc],1,mean))
118   sig.mean=mean(sqrt(s2.save[n.burn:n.mcmc]))
119   D.hat=-2*sum(ldasymlap(y,tau,X%*%beta.mean,
120       sig.mean))
121   pD=D.bar-D.hat
122   DIC=D.hat+2*pD
123
124   ###
125   ###   Write Output
126   ###
127
128   list(beta.save=beta.save,s2.save=s2.save,y=y,
129       X=X,n.mcmc=n.mcmc,n=n,p=p,D.bar=D.bar,
130       D.hat=D.hat,pD=pD,DIC=DIC,tau.save=tau.save,
131       n.burn=n.burn)
132
133   }
```

which can be used to fit the model to the white-tailed deer data using the R code and hyperparameters $\mu_\beta = 0$, $\sigma_\beta^2 = 1000$, and $u = 100$, with $K = 50000$ MCMC iterations:

BOX 18.5   POSTERIOR DISTRIBUTIONS FOR ALD REGRESSION PARAMETERS

```
source("reg.qr.tau.mcmc.R")
n.mcmc=50000
set.seed(1)
mcmc.out=reg.qr.tau.mcmc(y=y,X=X,beta.mn=rep
    (0,p),beta.var=1000,u=100,n.mcmc=n.mcmc)
keep.idx=mcmc.out$n.burn:n.mcmc

layout(matrix(1:4,2,2,byrow=TRUE))
hist(mcmc.out$beta.save[2,keep.idx],prob=TRUE,
    breaks=40,main="a",xlab=bquote(beta[1]))
curve(dnorm(x,0,sqrt(10000)),lwd=2,add=TRUE)
hist(mcmc.out$beta.save[3,keep.idx],prob=TRUE,
    breaks=40,main="b",xlab=bquote(beta[2]))
curve(dnorm(x,0,sqrt(10000)),lwd=2,add=TRUE)
hist(sqrt(mcmc.out$s2.save[keep.idx]),prob=TRUE,
    breaks=40,main="c",xlab=bquote(sigma))
abline(h=1/100,lwd=2)
hist(mcmc.out$tau.save[keep.idx],prob=TRUE,
    breaks=10,xlim=c(0,1),main="d",
    xlab=bquote(tau))
abline(h=1,lwd=2)
abline(v=0.25,lwd=2,lty=2)
```

This R code results in the marginal posterior distributions shown in Figure 18.2. Thus, as before, we can see in Figure 18.2b that the quantile regression coefficient β_2 is negative, indicating a downward concave parabolic relationship between body mass and age for white-tailed deer. Also, note that the marginal posterior distribution for τ indicates a mean of $E(\tau|\mathbf{y}) = 0.27$, which is very close to the first quartile (dashed line in Figure 18.2d) that we examined in the previous analysis where we held τ fixed at several values. Thus, this new model where we treat τ as unknown confirms the model selection procedure that we performed before; however, it also allows us to account for the uncertainty when estimating τ.[2]

---

[2] It also provides an example where the data may be informative about a parameter that is usually considered fixed when used to fit the model at a specific quantile.

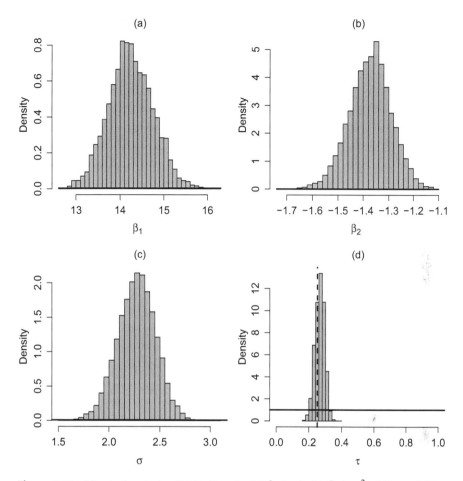

Figure 18.2 Marginal posterior distributions for (a) β_1 (age), (b) β_2 (age^2), (c) σ, and (d) τ. Prior PDFs are shown as horizontal black lines and the vertical dashed line represents the value of τ ($\tau = 0.25$) that had the best DIC score from the previous analysis for reference. Note that the intercept coefficient (β_0) is not shown, but it had posterior mean of $E(\beta_0|\mathbf{y}) = 15.5$.

18.2 ADDITIONAL CONCEPTS AND READING

Quantile regression arose as an alternative to testing for homoskedasticity and conditional symmetry in regression models (Newey and Powell 1987). Koenker and Basset (1978) introduced regression quantiles to extend the concept of order statistics to the regression setting. Since then, the expansion of quantile regression ideas has led to application in many different classes of models and data.

In Chapters 20 and 21, we describe ways to analyze binary and count data from a quantile regression perspective. For binary data, this amounts to using the ALD model for a latent continuous process (Benoit and Van Den Poel 2012; Benoit and

Van Den Poel 2017). In the case of count data, Machado and Santos Silva (2005) and Lee and Neocleous (2010) described a clever way to jitter the data such that we can use the original ALD model that we presented in this chapter to perform quantile regression. However, in both cases, issues with interpretation and implementation remain unsolved.

Quantile regression has been extended also to the spatial and spatio-temporal setting. For example, Reich et al. (2011) developed a method that allows the regression coefficients associated with a quantile regression to vary spatially. Similarly, Reich (2012) presented a slightly different approach that models all quantiles simultaneously and allows the quantile function to vary smoothly in space and time. Yang and Tokdar (2017) also describe approaches for jointly estimating quantile functions over a range of quantiles in such a way that the estimated functions do not intersect. Although these types of approaches are more challenging to implement, they may lead to novel inferences in the study of ecological and environmental processes.

There are many alternatives to quantile regression that relax the assumptions associated with conventional mean regression (Kneib 2013). For example, so-called expectile regression (Aigner et al. 1976; Waltrup et al. 2015) uses a squared term in the objective function and includes mean regression as a special case. Bayesian expectile regression results in the use of an asymmetric normal distribution (or generalized normal distribution) for a data model instead of the ALD that we presented in this chapter (Xing and Qian 2017). In fact, to accommodate skewness, we could use any asymmetric distribution to model the data and are not limited to the ALD or asymmetric normal. For example, an asymmetric t-distribution could accommodate skewness and kurtosis in the data and may be useful in some situations.

19 Hierarchical Models

19.1 HIERARCHICAL GAUSSIAN MODELS

Bayesian methods are incredibly powerful for building and fitting hierarchical models that can accommodate mechanisms in the underlying process, the parameters, and the measurement instrumentation used to observe the process (Cressie et al. 2009). Bayesian methods also are useful for reconciling multiple types of data (a phrase proposed by Hanks et al. (2011) for what are now often referred to as "integrated population models" in wildlife biology; see Chapter 25).

We briefly touched on hierarchical models in previous chapters, but did not expand on them. In this chapter, we present a quintessential hierarchical model to set the stage for specific forms of ecological data that appear in later chapters.[1]

Assume we observe continuous-valued data y_{ij} for replicate i ($i = 1, \ldots, n_j$) on individual j ($j = 1, \ldots, J$). An example might be a physiological measurement such as body temperature collected several (n_j) times independently for a set of J individual organisms in a study chosen randomly from a larger population. If we are interested in population-level inference based on these data, the population-level mean temperature μ, for example, we would not want to treat the individual-level measurements y_{ij} as the sample unit to avoid pseudo-replication (Hurlbert 1984) and because of potentially unbalanced sampling (i.e., n_j may not be equal for all J individuals). Thus, if we fit the model with a likelihood based on $y_{ij} \sim N(\mu, \sigma^2)$, we could obtain misleading inference for μ that is too precise.

The design of hierarchical models accounts for these so-called repeated measures on each individual in a way that explicitly addresses measurement uncertainty, pseudo-replication, and different amounts of individual-level uncertainty due to varying sample sizes. Hierarchical models provide a formal mechanism to incorporate random effects because, in this case, they treat the individual-level parameters as the random effect. To formulate a hierarchical model for this situation, we let the original data arise from a distribution with individual-level parameters and allow those parameters, in turn, to arise from a population-level process. We specify the hierarchical model as

$$y_{ij} \sim N(\mu_j, \sigma_j^2), \tag{19.1}$$

$$\mu_j \sim N(\mu, \sigma^2), \tag{19.2}$$

$$\sigma_j^2 \sim IG(q, r), \tag{19.3}$$

---

[1] Although we prefer the term "hierarchical" to describe nested, conditionally specified, statistical models, other terms such as "multi-level" model (e.g., Gelman and Hill 2006) and "state-space" model (see Chapter 28) are equivalent and more common in certain fields.

$$\mu \sim N(\mu_0, \sigma_0^2), \tag{19.4}$$

$$\sigma^2 \sim IG(q, r), \tag{19.5}$$

for $i = 1, \ldots, n_j$, and where, μ_j represent the individual-level random effects (for $j = 1, \ldots, J$) and σ_j^2 account for individual-level variation (perhaps due to measurement error or natural within-individual variability). The population-level parameters μ and σ^2 are usually of most interest scientifically and they each need priors, which we specified as normal and inverse gamma. In the hierarchical model framework, we also refer to the latent individual-level parameters μ_j for $j = 1, \ldots, J$, as the "process." Thus, the data arise conditionally from a model that depends on the latent process, which in turn arises from a distribution that depends on a set of parameters.

The joint posterior distribution for the hierarchical Gaussian model in (19.1)–(19.5) is

$$\left[\mu, \sigma^2, \mu, \sigma^2 | \{y_{ij}, \forall i, j\} \right] \propto \left(\prod_{j=1}^{J} \left(\prod_{i=1}^{n_j} \left[y_{ij} | \mu_j, \sigma_j^2 \right] \right) \left[\mu_j | \mu, \sigma^2 \right] \left[\sigma_j^2 \right] \right) [\mu] \left[\sigma^2 \right], \tag{19.6}$$

where $\mu \equiv (\mu_1, \ldots, \mu_J)'$, $\sigma^2 \equiv (\sigma_1^2, \ldots, \sigma_J^2)'$, and $\{y_{ij}, \forall i, j\}$ corresponds to the entire set of data y_{ij} for $i = 1, \ldots, n_j$ and $j = 1, \ldots, J$. Thus, to fit the hierarchical model in (19.1)–(19.5) using MCMC, we first find the full-conditional distributions for all the unknown model quantities. Working our way through the parameters in the posterior distribution (19.6), we first derive the full-conditional distribution for the individual-level μ_j by starting with the terms in the joint distribution that involve μ_j and simplifying to yield

$$[\mu_j | \cdot] \propto \left(\prod_{i=1}^{n_j} \left[y_{ij} | \mu_j, \sigma_j^2 \right] \right) \left[\mu_j | \mu, \sigma^2 \right], \tag{19.7}$$

$$\propto \exp\left(-\frac{1}{2} \frac{\sum_{i=1}^{n_j} (y_{ij} - \mu_j)^2}{\sigma_j^2} \right) \exp\left(-\frac{1}{2} \frac{(\mu_j - \mu)^2}{\sigma^2} \right), \tag{19.8}$$

$$\propto \exp\left(-\frac{1}{2} \left(-2 \frac{(\sum_{i=1}^{n_j} y_{ij}) \mu_j}{\sigma_j^2} + \frac{n_j \mu_j^2}{\sigma_j^2} - 2 \frac{\mu \mu_j}{\sigma^2} + \frac{\mu^2}{\sigma^2} \right) \right), \tag{19.9}$$

$$\propto \exp\left(-\frac{1}{2} \left(-2 \left(\frac{\sum_{i=1}^{n_j} y_{ij}}{\sigma_j^2} + \frac{\mu}{\sigma^2} \right) \mu_j + \left(\frac{n_j}{\sigma_j^2} + \frac{1}{\sigma^2} \right) \mu_j^2 \right) \right), \tag{19.10}$$

$$\propto \exp\left(-\frac{1}{2} \left(-2 b \mu_j + a \mu_j^2 \right) \right), \tag{19.11}$$

which has the form of a conjugate normal distribution such that $[\mu_j|\cdot] \equiv N(a^{-1}b, a^{-1})$, where

$$a \equiv \frac{n_j}{\sigma_j^2} + \frac{1}{\sigma^2}, \tag{19.12}$$

$$b \equiv \frac{\sum_{i=1}^{n_j} y_{ij}}{\sigma_j^2} + \frac{\mu}{\sigma^2}. \tag{19.13}$$

Thus, the full-conditional distribution for μ_j is conjugate and we can use a Gibbs update in our MCMC algorithm to sample it. Also, because the full-conditional distributions have the same form for all μ_j, we can sample them independently from each other, but they all depend on the latent random parameter μ.

We can find similarly the full-conditional distribution for σ_j^2, by retaining the components involving σ_j^2 in the joint distribution (19.6) and then simplifying to recognize the resulting form as a known distribution, which is

$$[\sigma_j^2|\cdot] \propto \left(\prod_{i=1}^{n_j} [y_{ij}|\mu_j, \sigma_j^2] \right) [\sigma_j^2], \tag{19.14}$$

$$\propto (\sigma_j^2)^{-\frac{n_j}{2}} \exp\left(-\frac{1}{\sigma_j^2} \frac{\sum_{i=1}^{n_j} (y_{ij} - \mu_j)^2}{2} \right) (\sigma_j^2)^{-(q+1)} \exp\left(-\frac{1}{\sigma_j^2} \frac{1}{r} \right), \tag{19.15}$$

$$\propto (\sigma_j^2)^{-\left(\frac{n_j}{2}+q+1\right)} \exp\left(-\frac{1}{\sigma_j^2} \left(\frac{\sum_{i=1}^{n_j} (y_{ij} - \mu_j)^2}{2} + \frac{1}{r} \right) \right), \tag{19.16}$$

$$\propto (\sigma_j^2)^{-(\tilde{q}+1)} \exp\left(-\frac{1}{\sigma_j^2} \frac{1}{\tilde{r}} \right), \tag{19.17}$$

where

$$\tilde{q} \equiv \frac{n_j}{2} + q, \tag{19.18}$$

$$\tilde{r} \equiv \left(\frac{\sum_{i=1}^{n_j} (y_{ij} - \mu_j)^2}{2} + \frac{1}{r} \right)^{-1}. \tag{19.19}$$

Thus, the full-conditional distribution is conjugate inverse gamma, $[\sigma_j^2|\cdot] \equiv IG(\tilde{q}, \tilde{r})$, which implies that we can use a Gibbs update for it in a MCMC algorithm for $j = 1, \ldots, J$.

The latent process, or population-level mean parameter, μ, also has a conjugate full-conditional distribution. To see this, we use the same procedure

$$[\mu|\cdot] \propto \left(\prod_{j=1}^{J}[\mu_j|\mu,\sigma^2]\right)[\mu], \tag{19.20}$$

$$\propto \exp\left(-\frac{1}{2}\frac{\sum_{j=1}^{J}(\mu_j-\mu)^2}{\sigma^2}\right)\exp\left(-\frac{1}{2}\frac{(\mu-\mu_0)^2}{\sigma_0^2}\right), \tag{19.21}$$

$$\propto \exp\left(-\frac{1}{2}\left(-2\frac{\left(\sum_{j=1}^{J}\mu_j\right)\mu}{\sigma^2}+\frac{J\mu^2}{\sigma^2}-2\frac{\mu_0\mu}{\sigma_0^2}+\frac{\mu_0^2}{\sigma_0^2}\right)\right), \tag{19.22}$$

$$\propto \exp\left(-\frac{1}{2}\left(-2\left(\frac{\sum_{j=1}^{J}\mu_j}{\sigma^2}+\frac{\mu_0}{\sigma_0^2}\right)\mu+\left(\frac{J}{\sigma^2}+\frac{1}{\sigma_0^2}\right)\mu^2\right)\right), \tag{19.23}$$

$$\propto \exp\left(-\frac{1}{2}\left(-2b\mu+a\mu^2\right)\right), \tag{19.24}$$

which implies that $[\mu|\cdot] \equiv N(a^{-1}b, a^{-1})$, where

$$a \equiv \frac{J}{\sigma^2}+\frac{1}{\sigma_0^2}, \tag{19.25}$$

$$b \equiv \frac{\sum_{j=1}^{J}\mu_j}{\sigma^2}+\frac{\mu_0}{\sigma_0^2}. \tag{19.26}$$

Thus, we can sample the latent process in our MCMC algorithm using a Gibbs update as well.

Finally, we can derive the full-conditional distribution for the latent variance parameter σ^2 in the same way, by retaining the components of the joint distribution in (19.6) that involve σ^2 and then simplify as

$$[\sigma^2|\cdot] \propto \left(\prod_{j=1}^{J}[\mu_j|\mu,\sigma^2]\right)[\sigma^2], \tag{19.27}$$

$$\propto (\sigma^2)^{-\frac{J}{2}}\exp\left(-\frac{1}{\sigma^2}\frac{\sum_{j=1}^{J}(\mu_j-\mu)^2}{2}\right)(\sigma^2)^{-(q+1)}\exp\left(-\frac{1}{\sigma^2}\frac{1}{r}\right), \tag{19.28}$$

$$\propto (\sigma^2)^{-(\frac{J}{2}+q+1)}\exp\left(-\frac{1}{\sigma^2}\left(\frac{\sum_{j=1}^{J}(\mu_j-\mu)^2}{2}+\frac{1}{r}\right)\right), \tag{19.29}$$

$$\propto (\sigma^2)^{-(\tilde{q}+1)}\exp\left(-\frac{1}{\sigma_j^2}\frac{1}{\tilde{r}}\right), \tag{19.30}$$

where

$$\tilde{q} \equiv \frac{J}{2} + q, \tag{19.31}$$

$$\tilde{r} \equiv \left(\frac{\sum_{j=1}^{J}(\mu_j - \mu)^2}{2} + \frac{1}{r} \right)^{-1}. \tag{19.32}$$

Thus, like the individual-level variance parameters, the population-level variance parameter is also conjugate inverse gamma such that $[\sigma^2|\cdot] \equiv \text{IG}(\tilde{q}, \tilde{r})$ and we can sample it in the MCMC algorithm using a Gibbs update.

To create a MCMC algorithm to fit the hierarchical model in (19.1)–(19.5) using the full-conditional distributions that we derived previously, we sampled from each full-conditional distribution sequentially starting with σ^2. We called the MCMC algorithm in R norm.hier.mcmc.R and it is as follows:

BOX 19.1   MCMC ALGORITHM FOR HIERARCHICAL MODEL

```
1  norm.hier.mcmc <- function(y.list,n.mcmc){
2
3  ####
4  ####   Setup Variables
5  ####
6
7  J=length(y.list)
8  nj=sapply(y.list,length)
9
10 mu.save=rep(0,n.mcmc)
11 s2.save=rep(0,n.mcmc)
12
13 muj.save=matrix(0,J,n.mcmc)
14 s2j.save=matrix(0,J,n.mcmc)
15
16 ####
17 ####   Priors and Starting Values
18 ####
19
20 q=0.001
21 r=1000
22
23 mu.0=0
24 s2.0=10000
```

(*Continued*)

BOX 19.1 (Continued)   MCMC ALGORITHM FOR HIERARCHICAL MODEL

```
25 mu=mean(unlist(y.list))
26 muj=sapply(y.list,mean)
27 s2j=sapply(y.list,var)
28 rl=as.relistable(y.list)
29
30 ####
31 ####  Begin MCMC Loop
32 ####
33
34 for(k in 1:n.mcmc){
35
36   ####
37   ####  Sample s2
38   ####
39
40   q.tmp=J/2+q
41   r.tmp=1/(sum((muj-mu)^2)/2+1/r)
42   s2=1/rgamma(1,q.tmp,,r.tmp)
43
44   ####
45   ####  Sample mu
46   ####
47
48   tmp.var=1/(J/s2+1/s2.0)
49   tmp.mn=tmp.var*(sum(muj)/s2+mu.0/s2.0)
50   mu=rnorm(1,tmp.mn,sqrt(tmp.var))
51
52   ####
53   ####  Sample mu_j
54   ####
55
56   tmp.var=1/(nj/s2j+1/s2)
57   tmp.mn=tmp.var*(sapply(y.list,sum)/s2j+mu/s2)
58   muj=rnorm(J,tmp.mn,sqrt(tmp.var))
```

(Continued)

BOX 19.1 (Continued)   MCMC ALGORITHM FOR HIERARCHICAL MODEL

```
59   ####
60   ####    Sample s2_j
61   ####
62
63   q.tmp=nj/2+q
64   tmp.sum=relist((unlist(y.list)-rep(muj,
65      nj))^2,rl)
66   r.tmp=1/(sapply(tmp.sum,sum)/2+1/r)
67   s2j=1/rgamma(J,q.tmp,,r.tmp)
68
69   ####
70   ####    Save Samples
71   ####
72
73   mu.save[k]=mu
74   s2.save[k]=s2
75   muj.save[,k]=muj
76   s2j.save[,k]=s2j
77
78   }
79
80   ####
81   ####    Write Output
82   ####
83
84   list(n.mcmc=n.mcmc,mu.save=mu.save,
85      s2.save=s2.save,muj.save=muj.save,s2j.
86      save=s2j.save)
87
88   }
```

To demonstrate this MCMC algorithm, we fit the hierarchical Gaussian model to a set of eye region temperature data taken on a sample of 14 blue tits (*Cyanistes caeruleus*). In a study of individual-level versus population-level variation in wild birds, Jerem et al. (2018) measured the eye region temperature (in Celsius) of a sample of blue tits using a non-invasive thermal imaging method. For each individual, eye region temperature was recorded a different number of times. We read in the data to R and fit our hierarchical Gaussian model to the blue tit temperature data using the following R code:

BOX 19.2   READ IN BLUE TIT DATA AND FIT HIERARCHICAL MODEL

```
et.df=read.table("eyetemp.txt")
names(et.df)=c("id","sex","date","time","event",
   "airtemp","relhum","sun","bodycond","eyetemp")
head(et.df)

id.unique=unique(et.df$id)
J=length(id.unique)
y.list=vector("list",J)
for(j in 1:J){
  y.list[[j]]=as.vector(et.df[et.df$id==id.
    unique[j],"eyetemp"])
}

source("norm.hier.mcmc.R")
n.mcmc=10000
set.seed(1)
mcmc.out=norm.hier.mcmc(y.list=y.list,n.mcmc=n.
   mcmc)

layout(matrix(1:2,2,1))
plot(1:(J+1),rep(0,J+1),ylim=range(apply(all.mu,1,
    quantile,c(0.025,0.975))),ylab=
    bquote(mu["j"]),xlab="individual",xaxt="n",
    type="n",main="a")
mtext(bquote(mu),4,.5,cex=1.25)
axis(1,1:(J+1),labels=c(1:J,"pop"),cex.axis=1.1)
segments(1:(J+1),apply(all.mu,1,quantile,0.025),
    1:(J+1),apply(all.mu,1,quantile,0.975),lwd=2,
    lty=c(rep(1,J),6))
points(1:(J+1),apply(all.mu,1,mean),pch=20,
    cex=1.5)
abline(h=mean(mcmc.out$mu.save),col=8)
points(1:J,sapply(y.list,mean),pch=4,col=8,
    cex=1.5)
plot(1:(J+1),rep(0,J+1),ylim=range(apply(all.s2,1,
    quantile,c(0.025,0.975))),ylab=bquote
    (sigma["j"]^2),xlab="individual",
    xaxt="n",type="n",main="b")
```

(*Continued*)

**BOX 19.2 (Continued)   READ IN BLUE TIT DATA AND FIT
HIERARCHICAL MODEL**

```
38 mtext(bquote(sigma^2),4,.5,cex=1.25)
39 axis(1,1:(J+1),labels=c(1:J,"pop"),cex.axis=1.1)
40 segments(1:(J+1),apply(all.s2,1,quantile,0.025),
41    1:(J+1),apply(all.s2,1,quantile,0.975),lwd=2,
42    lty=c(rep(1,J),6))
43 points(1:(J+1),apply(all.s2,1,mean),pch=20,
44    cex=1.5)
45 abline(h=mean(mcmc.out$s2.save),col=8)
```

This R code resulted in the posterior means and 95% credible intervals shown in Figure 19.1. Notice how the posterior means shrink toward the population mean for individuals 13 and 14 in Figure 19.1a, whereas the rest of the means are very similar to the sample means. This shrinkage occurs because those individuals (with only 2 temperature measurements each) are "borrowing strength" from the other individuals with more data. For comparison, we used the following R code to fit the nonhierarchical Normal-Normal-Inverse Gamma model (Chapter 9) to all the blue tit eye temperature data combined together without recognizing individuals explicitly:

BOX 19.3   FIT NONHIERARCHICAL MODEL TO BLUE TIT DATA

```
1 source("NNIG.mcmc.R")
2 set.seed(1)
3 NNIG.mcmc.out=NNIG.mcmc(y=unlist(y.list),mu0=0,
4    s20=10000,q=0.001,r=1000,mu.strt=30,
5    n.mcmc=n.mcmc)
6 mean(NNIG.mcmc.out$mu.save)
7 mean(mcmc.out$mu.save)
8 quantile(NNIG.mcmc.out$mu.save,c(0.025,0.975))
9 quantile(mcmc.out$mu.save,c(0.025,0.975))
```

which shows that the posterior means resulting from both models are nearly the same (29.6 versus 29.5 for the nonhierarchical and hierarchical models, respectively).[2] However, the resulting 95% credible intervals are substantially different

---

[2] Note that we write our NNIG.mcmc function slightly more generic, allowing for hyperparameter values as function inputs. Which values to include as function arguments and which are hard-coded is a user decision. It is more general to include hyperparameters in as function arguments, but also makes for more cumbersome scripts. Thus, we often hard coded hyperparameter values for examples throughout this book to simplify the R scripts, but the user can modify that easily.

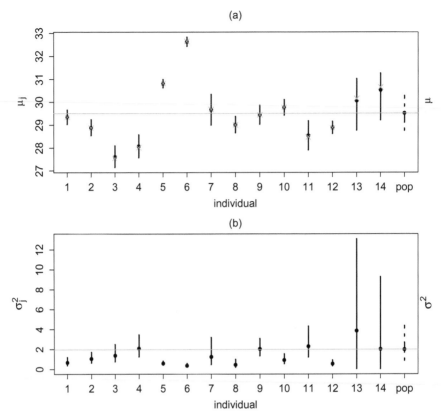

Figure 19.1 Marginal posterior means (black points) and 95% credible intervals (vertical lines) for the individual-level (denoted by individual ID number) and population-level (denoted by "pop") (a) means (μ_j and μ; in Celsius) and (b) variance parameters (σ_j^2 and σ^2). The $\times$ symbols in (a) indicate the sample means for each individual and the horizontal gray line is the population-level posterior mean of μ, shown for reference.

with $(29.5, 29.8)$ for the nonhierarchical model and $(28.8, 30.3)$ for the hierarchical model. Thus, by not recognizing the individual-level variability in temperature, we end up with uncertainty estimates that are overly confident.[3] Also, the nonhierarchical model does not take into account the imbalance in sample size across the different individuals explicitly.

19.2   TWO-STAGE MODEL FITTING ALGORITHMS

There is no doubt that hierarchical models are useful and have led to advancements in science across many disciplines. However, with increasing complexity in models comes an increasing computational burden. The simple hierarchical model that we

---

[3] Sometimes this is also referred to as pseudo-replication (Hurlbert 1984).

presented in the previous section is trivial to implement, but still requires more analytical and computational steps than the models that we presented previously in this
book.

We can add more levels in the hierarchical model, include more complicated data
models that may not be conditionally independent, and specify sophisticated process
models with spatial and/or temporal structure that can make implementing the model
very challenging. Then, even if we can construct MCMC algorithms for more complex models, they can result in Markov chains with poor mixing and instability due
to the various interdependencies among model parameters and processes. Also, conventional MCMC algorithms to fit models in automated software (e.g., BUGS, JAGS,
STAN) may be very slow to run due to not optimizing the algorithms for particular
model structures. We have found that the ability to implement models using custom software in R can improve computational efficiency substantially beyond many
existing Bayesian software packages.

There are many approaches to facilitate computation using custom algorithms
specific to certain types of model specifications; thus it is not feasible to review them
all in a single volume. However, there is one that is particularly useful for certain
classes of hierarchical models that we cover here. We refer to this computing strategy as the "Lunn method," named after David Lunn, who pioneered some of the
early efforts to develop ways to fit Bayesian models to data using automated MCMC
software (BUGS; Lunn et al. 2000, 2009).[4] The Lunn Method involves fitting hierarchical models with many data-level models in two stages (Lunn et al. 2013). The
first stage involves fitting the set of nonhierarchical data-level models independently
using placeholder priors for model parameters. The second stage involves a substantially simplified MCMC algorithm to fit the full hierarchical model using the output
from the first stage.

Critically, the Lunn method yields exact inference for the model parameters (in
the same sense that MCMC is exact[5]) and the second stage algorithm does not rely
on the data directly, only indirectly through the first-stage output. Hooten et al. (2016)
showed how to implement the Lunn method for hierarchical animal movement models to make population-level inference on common latent parameters or processes.
We review this approach here, but in the context of the hierarchical Gaussian model
from the previous section.

The main "trick" to the Lunn method is in the second-stage algorithm where we
use the first stage MCMC samples for the data-level parameters as proposals. The
use of first-stage MCMC samples as proposals allows us to dramatically simplify the
M-H update in the second-stage algorithm for the data-level parameters. The resulting M-H ratio does not rely on the data model and we can achieve the updates for

---

[4] Hooten et al. (2019) refers to this approach as "proposal recursive Bayes" to contrast it with other forms
of sequential model fitting referred to as "prior recursive Bayes."

[5] In this sense, "exact" refers to the properties of MCMC that guarantee we will be able to characterize the
posterior distribution exactly with infinite computing time (i.e., infinite MCMC sample size). Of course,
in practice our final inference based on MCMC output is approximate due to finite Markov chain length,
but we can get it as close as we want by extending the Markov chain length.

the rest of the model parameters as described in the previous section. To see how this works, consider the hierarchical Gaussian model from the previous section. First, we split up the hierarchical model to form a separate model for the sets of data $\mathbf{y}_j$ arising from each individual

$$y_{ij} \sim N(\mu_j, \sigma_j^2), \tag{19.33}$$

and fit each of those using a placeholder prior for model parameters. In this case, we could assume priors $[\mu_j] \equiv N(\mu_{\text{temp}}, \sigma_{\text{temp}}^2)$ and $[\sigma_j^2] \equiv IG(q_{\text{temp}}, r_{\text{temp}})$. Fitting each of these individual-level models yields a separate posterior distribution for μ_j and σ_j^2

$$\left[\mu_j, \sigma_j^2 | \mathbf{y}_j\right] \propto \prod_{i=1}^{n_j} \left[y_{ij} | \mu_j, \sigma_j^2\right] [\mu_j][\sigma_j^2]. \tag{19.34}$$

Thus, the first stage actually involves fitting J models separately. However, because these first-stage models are treated as independent, we can fit them in parallel across many processors. In many cases, depending on the size of the computing facility and data set, the parallelization implies that the first stage of the Lunn method only requires slightly more time than fitting a single model to a single individual.

To outline the second stage, recall that the MCMC algorithm to fit the full model that we presented in the previous section required us to sample from each full-conditional distribution for the parameters sequentially. Focusing on the update for μ_j in the full-model algorithm from the previous section, recall that we found the full-conditional distribution for μ_j analytically and could use a Gibbs update to sample it. However, we also could have used M-H updates to sample μ_j based on proposed values $\mu_j^{(*)}$. Also, because we are now considering M-H updates, we can propose values $\mu_j^{(*)}$ and $\sigma_j^{2(*)}$ jointly using the proposal distribution $[\mu_j, \sigma_j^2]^*$ and thus use block updates in the MCMC algorithm. The resulting M-H ratio for this proposal scenario is

$$mh = \frac{\prod_{i=1}^{n_j} \left[y_{ij} | \mu_j^{(*)}, \sigma_j^{2(*)}\right] \left[\mu_j^{(*)} | \mu^{(k)}, \sigma^{2(k)}\right] \left[\sigma_j^{2(*)}\right] \left[\mu_j^{(k-1)}, \sigma_j^{2(k-1)}\right]^*}{\prod_{i=1}^{n_j} \left[y_{ij} | \mu_j^{(k-1)}, \sigma_j^{2(k-1)}\right] \left[\mu_j^{(k-1)} | \mu^{(k)}, \sigma^{2(k)}\right] \left[\sigma_j^{2(k-1)}\right] \left[\mu_j^{(*)}, \sigma_j^{2(*)}\right]^*}, \tag{19.35}$$

where $[\mu_j^{(*)}, \sigma_j^{2(*)}]^*$ represents the joint proposal distribution evaluated at $\mu_j^{(*)}$ and $\sigma_j^{2(*)}$.

Instead of using the usual random walk proposal or prior proposal, we use thinned output from the first stage algorithms as the proposals $\mu_j^{(*)}$ and $\sigma_j^{2(*)}$. In this case, the form of the proposal is the posterior distribution from the first stage

$$[\mu_j, \sigma_j^2]^* \propto \prod_{i=1}^{n_j} [y_{ij} | \mu_j, \sigma_j^2][\mu_j][\sigma_j^2], \tag{19.36}$$

from (19.34).[6] Thus, the second stage M-H ratio for updating μ_j and σ_j^2 becomes

$$mh = \frac{\prod_{i=1}^{n_j}[y_{ij}|\mu_j^{(*)},\sigma_j^{2(*)}][\mu_j^{(*)}|\mu^{(k)},\sigma^{2(k)}][\sigma_j^{2(*)}]\prod_{i=1}^{n_j}[y_{ij}|\mu_j^{(k-1)},\sigma_j^{2(k-1)}][\mu_j^{(k-1)}][\sigma_j^{2(k-1)}]}{\prod_{i=1}^{n_j}[y_{ij}|\mu_j^{(k-1)},\sigma_j^{2(k-1)}][\mu_j^{(k-1)}|\mu^{(k)},\sigma^{2(k)}][\sigma_j^{2(k-1)}]\prod_{i=1}^{n_j}[y_{ij}|\mu_j^{(*)},\sigma_j^{2(*)}][\mu_j^{(*)}][\sigma_j^{2(*)}]},$$
(19.37)

$$= \frac{[\mu_j^{(*)}|\mu^{(k-1)},\sigma^{2(k)}][\mu_j^{(k-1)}]}{[\mu_j^{(k-1)}|\mu^{(k-1)},\sigma^{2(k)}][\mu_j^{(*)}]}.$$
(19.38)

The resulting M-H ratio in (19.38) is a substantially simplified version of original M-H ratio to fit the full-model because the likelihood components and priors for σ_j^2 cancel out.[7] In fact, if the first stage priors for μ_j are very diffuse, they will nearly cancel out in the second-stage M-H ratio (19.38) leaving only the ratio of process models. Further, because the resulting second-stage M-H ratio for μ_j does not involve σ_j^2 and neither do any of the other full-conditional distributions for the remaining model parameters μ or σ^2, we can ignore the first-stage MCMC samples for σ_j^2 completely.

Thus, to implement the Lunn method for fitting hierarchical models like that in the previous section, we first fit the Normal-Normal-Inverse Gamma model described in Chapter 9 to each of the individual data sets separately. We then use the MCMC output from the individual model fits for μ_j as proposals in a second stage algorithm with M-H updates for μ_j for $j = 1,\ldots,J$ and perform all other updates for model parameters exactly as described in the previous section, except for the updates for σ_j^2, which can safely be ignored in our second-stage MCMC algorithm (unless we want inference involving them).

We used the MCMC algorithm constructed in Chapter 9 called NNIG.mcmc.R to fit the first-stage models and then modified the MCMC algorithm from the previous section, which we renamed norm.hier.L.mcmc.R, as

---

BOX 19.4   SECOND-STAGE MCMC ALGORITHM TO FIT HIERARCHICAL MODEL

```
1  norm.hier.L.mcmc  <- function(muj.mat){
2
3  ####
4  ####   Setup Variables
5  ####
```

(Continued)

---

[6] This proposal is actually conditional on the data, but we suppress the conditional notation for simplicity.

[7] The cancellation only occurs when there is no autocorrelation in the sample from the first stage output. Thus, we thin or sample randomly with replacement the first stage MCMC sample to reduce the autocorrelation in the Markov chain.

BOX 19.4 (Continued)   SECOND-STAGE MCMC ALGORITHM TO FIT HIERARCHICAL MODEL

```
 6 n.mcmc=dim(muj.mat)[2]
 7 J=dim(muj.mat)[1]
 8
 9 mu.save=rep(0,n.mcmc)
10 s2.save=rep(0,n.mcmc)
11 muj.save=matrix(0,J,n.mcmc)
12
13 ####
14 ####   Priors and Starting Values
15 ####
16
17 q=0.001
18 r=1000
19
20 mu.0=0
21 s2.0=10000
22
23 mu=mean(unlist(y.list))
24 muj=sapply(y.list,mean)
25 s2j=sapply(y.list,var)
26 rl=as.relistable(y.list)
27
28 ####
29 ####   Begin MCMC Loop
30 ####
31
32 for(k in 1:n.mcmc){
33
34    ####
35    ####   Sample s2
36    ####
37
38    q.tmp=J/2+q
39    r.tmp=1/(sum((muj-mu)^2)/2+1/r)
40    s2=1/rgamma(1,q.tmp,,r.tmp)
```

(Continued)

BOX 19.4 (Continued)   SECOND-STAGE MCMC ALGORITHM TO FIT HIERARCHICAL MODEL

```
41   ####
42   ####   Sample mu
43   ####
44
45   tmp.var=1/(J/s2+1/s2.0)
46   tmp.mn=tmp.var*(sum(muj)/s2+mu.0/s2.0)
47   mu=rnorm(1,tmp.mn,sqrt(tmp.var))
48
49   ####
50   ####   Sample mu_j
51   ####
52
53   muj.star=muj.mat[,k]
54   mh.1=dnorm(muj.star,mu,sqrt(s2),log=TRUE)
55     +dnorm(muj,0,sqrt(10000),log=TRUE)
56   mh.2=dnorm(muj,mu,sqrt(s2),log=TRUE)
57     +dnorm(muj.star,0,sqrt(10000),log=TRUE)
58   keep.idx=exp(mh.1-mh.2)>runif(J)
59   muj[keep.idx]=muj.star[keep.idx]
60
61   ####
62   ####   Save Samples
63   ####
64
65   mu.save[k]=mu
66   s2.save[k]=s2
67   muj.save[,k]=muj
68 }
69
70 ####
71 ####   Write Output
72 ####
73
74 list(n.mcmc=n.mcmc,mu.save=mu.save,s2.save=
75     s2.save,muj.save=muj.save)
76
77 }
```

where mu.mat is a $J \times K$ matrix containing the MCMC output $(\mu_j^{(k)}$, for $j = 1,\ldots,J$ and $k = 1,\ldots,K)$ from the first stage model fits.

To fit the hierarchical model to the blue tit eye region temperature data using the two-stage method, we used the following R code, which first fits the individual-level models independently and then uses the second-stage algorithm to obtain the rest of the sample:

BOX 19.5   FIT HIERARCHICAL MODEL USING TWO-STAGE APPROACH

```
 1 n.mcmc.1=100000
 2 n.mcmc.2=10000
 3 mu.mat=matrix(0,J,n.mcmc.2)
 4 source("NNIG.mcmc.R")
 5 set.seed(1)
 6 for(j in 1:J){
 7   mu.mat[j,]=NNIG.mcmc(y=y.list[[j]],mu0=0,
 8     s20=10000,q=0.001,r=1000,mu.strt=30,
 9     n.mcmc=n.mcmc.1)$mu.save[round(seq(1,
10     n.mcmc.1,,n.mcmc.2))]
11 }
12
13 source("norm.hier.L.mcmc.R")
14 mcmc.L.out=norm.hier.L.mcmc(muj.mat=mu.mat)
15
16 all.mu=rbind(mcmc.out$muj.save,mcmc.out$mu.save)
17 all.mu.L=rbind(mcmc.L.out$muj.save,
18     mcmc.L.out$mu.save)
19 plot(1:(J+1),rep(0,J+1),ylim=range(apply(all.mu,1,
20     quantile,c(0.025,0.975))),ylab=bquote(mu["j"]),
21     xlab="individual",xaxt="n",type="n")
22 mtext(bquote(mu),4,.5,cex=1.25)
23 axis(1,1:(J+1),labels=c(1:J,"pop"),cex.axis=1.1)
24 segments((1:(J+1))-0.15,apply(all.mu,1,quantile,
25     0.025),(1:(J+1))-0.15,apply(all.mu,1,quantile,
26     0.975),lwd=2,lty=c(rep(1,J),6))
27 points((1:(J+1))-0.15,apply(all.mu,1,mean),pch=20,
28     cex=1.5)
29 segments((1:(J+1))+0.15,apply(all.mu.L,1,quantile,
30     0.025),(1:(J+1))+0.15,apply(all.mu.L,1,
31     quantile,0.975),lwd=2,lty=c(rep(1,J),6),col=8)
```

(Continued)

**BOX 19.5 (Continued)   FIT HIERARCHICAL MODEL USING
TWO-STAGE APPROACH**

```
32  points((1:(J+1))+0.15,apply(all.mu.L,1,mean),
33     pch=20,cex=1.5,col=8)
34  segments(1:J,apply(mu.mat,1,quantile,0.025),1:J,
35     apply(mu.mat,1,quantile,0.975),lwd=2,
36     col=gray(.5))
37  points(1:J,apply(mu.mat,1,mean),pch=20,cex=1.5,
38     col=gray(.5))
```

Combined with the algorithm output from the previous section (mcmc.out$muj.
save), this R code results in the comparison of posterior distributions resulting from
the single algorithm versus the Lunn method illustrated in Figure 19.2. The posterior means and credible intervals shown in Figure 19.2 indicate that, although the
independent nonhierarchical model fits (shown in dark gray) for individuals 13 and
14 (the two individuals with only two observations each) have very large uncertainty
and are slightly larger on average, the uncertainty shrinks toward the inference from
the simultaneous model fit to yield the corrected inference for μ_j (for $j = 1, \ldots, J$)
and μ.

In general, we can apply the two-stage method to fit any hierarchical model with
multiple data model components and independent data model nuisance parameters
(i.e., σ_j^2). Although the design of the Lunn method was mainly for meta-analysis,
Hooten et al. (2016) used it for hierarchical models more generally and found it

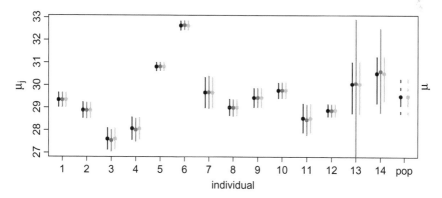

Figure 19.2 Marginal posterior means (points) and 95% credible intervals (vertical lines)
for the individual-level (denoted by individual ID number) and population-level (denoted by
"pop") means. The black intervals (left in each cluster) correspond to the results from the
simultaneous algorithm, the dark gray (middle in each cluster) intervals correspond to the
results from the first-stage algorithm, and the light gray intervals (right in each cluster) correspond to the results from the second-stage algorithm.

especially useful with large numbers of data models that can be fit in parallel for the first stage. Hooten et al. (2016) also found that the second stage algorithm was well-mixed, requiring a smaller MCMC sample to approximate posterior quantities than with the single-stage algorithm. Finally, when the likelihood component of a hierarchical model is particularly challenging to work with, the Lunn method proves useful because the likelihood does not appear in the second-stage algorithm.

19.3   ADDITIONAL CONCEPTS AND READING

In addition to more general books on hierarchical modeling (Gelman and Hill 2006), there are several good references for hierarchical Bayesian modeling in ecology and environmental science. Aside from the relevant chapters in this book, Hobbs and Hooten (2015) provided a more accessible introduction to Bayesian hierarchical modeling that strongly emphasizes the graph structure of the models. We can represent any Bayesian model as a directed acyclic graph (DAG), and while we do not visually display the DAGs associated with the models in this book, that perspective can be useful for those learning Bayesian methods the first time. In particular, DAGs help to emphasize the conditioning inherent to Bayesian models and are especially helpful for graphically displaying models that contain multiple data sources (Chapter 25). Clark (2007) also emphasized DAGs in his comprehensive treatment of Bayesian hierarchical modeling that contains a variety of examples spanning the field of ecology, including several temporal (Chapter 16 in this book) and spatio-temporally dynamic models (Chapter 28 in this book) for ecological processes.

Royle and Dorazio (2008) and Link and Barker (2010) provided an excellent overview of hierarchical models for animal ecologists. They go into substantially more detail about occupancy models (Chapter 23 in this book) and capture-recapture models (Chapter 24 in this book) than we do. In particular, they introduce the concepts of multispecies occupancy models and dynamic occupancy models as well as capture-recapture models that account for individual heterogeneity and open populations. Kery and Royle (2016) provided a more recent volume that includes substantial detail about animal ecological models for unmarked individuals (Chapter 24 in this book) including distance sampling methods and N-mixture models. They also translated their models to the BUGS language which provides automated inference using built-in MCMC algorithms.

The concept of multi-stage approaches for fitting Bayesian models is not new, in fact, Chopin (2002) presented ways to sequentially fit Bayesian models using multiple stages and importance sampling (Chapter 5 in this book). Hooten et al. (2019) extended these ideas to the MCMC setting to provide an accessible way to fit nearly any Bayesian model in a sequence of stages, using the MCMC output from previous stages. In particular, they demonstrated how to accelerate Bayesian model fitting for spatial statistical models (Chapter 17 in this book) and time series (Chapter 16 in this book).

20 Binary Regression

20.1 GENERALIZED LINEAR MODELS

The development of generalized linear models (GLMs) more explicitly deals with non-Gaussian data and has been tremendously useful in many areas of science. In particular, ecological and environmental science often involves discrete-valued data. Ecological and environmental data are often binary (e.g., presence or absence of a species) or integer-valued (e.g., counts of individual plants or animals; also called relative abundance). Thus, in the chapters that follow, we present the most common types of statistical models formulated as GLMs, many designed specifically for ecological and environmental data.

To specify a parametric GLM, we assume the data y_i, for $i = 1, \ldots, n$, arise from a probability distribution (e.g., Gaussian, exponential, binomial, gamma, etc) that depends on a set of parameters θ_i, such that $y_i \sim [y_i|\theta_i]$, where the conditional mean of y_i is a distribution-specific function of the natural parameters, $E(y_i|\theta_i)$. For example, in the Gaussian GLM (which is just the regular linear model from Chapter 11), the function $E(y_i|\mu_i, \sigma^2) = \mu_i$ is the identity function of the parameter μ_i (because it is the mean of the normal distribution).

To complete the GLM specification, we link the expected value of the data distribution to a linear combination of predictor variables as in the regular Gaussian regression model. Thus, we model a transformation of the mean function as $g(E(y_i|\theta_i)) \equiv \mathbf{x}_i'\boldsymbol{\beta}$, where $\boldsymbol{\beta}$ are the regression coefficients and the $p \times 1$ vector $\mathbf{x}_i$ typically has the first element equal to one and the remainder of the elements are covariates associated with the ith observation. The function $g(\cdot)$ is referred to as the "link" function and it is often seen inverted on the right-hand side of the relationship between the mean and the covariates such that $E(y_i|\theta_i) = g^{-1}(\mathbf{x}_i'\boldsymbol{\beta})$.

In what follows, we present GLM specifications for several different types of common ecological questions and data, beginning with binary data.

20.2 LOGISTIC REGRESSION

Logistic regression is a common approach to modeling heterogeneity in binary data; thus we present logistic regression from the same model-based perspective as we have done so far in this book. To do that, we consider the data and the parametric distribution they may arise from. For binary data, model building is straightforward. If we consider that there is a conditional probability associated with the event that $y_i = 1$ (and hence $y_i = 0$), then the conditional distribution of y_i must be Bernoulli. Thus, the data model for logistic regression is $y_i \sim \text{Bern}(p_i)$, where p_i is the probability that $y_i = 1$. To make this a GLM, we consider the conditional mean of y_i, which is p_i, and use a link function to connect the mean to the covariate effects. There are many link functions that have been used for the Bernoulli probability p_i, but the most common is the logistic link (and hence "logistic regression"), where $\text{logit}(p_i) = \mathbf{x}_i'\boldsymbol{\beta}$ and the 'logit' function is defined as $\text{logit}(p_i) = \log(p_i/(1-p_i))$.[1]

To complete the Bayesian logistic model specification, we select an appropriate prior for $\boldsymbol{\beta}$. In Chapter 11, we used the multivariate normal distribution as a prior for the regression coefficients such that $\boldsymbol{\beta} \sim \text{N}(\boldsymbol{\mu}_\beta, \boldsymbol{\Sigma}_\beta)$ and we can use the same prior here, but because the data model is not Gaussian, we will not be able to derive a conjugate full-conditional distribution.

For the Bayesian logistic regression model with multivariate normal prior for $\boldsymbol{\beta}$, there is only one (multivariate) full-conditional distribution required to construct a MCMC algorithm to fit the model. Thus, the full-conditional distribution for $\boldsymbol{\beta}$ is also the posterior distribution for this model

$$[\boldsymbol{\beta}|\mathbf{y}] \propto \prod_{i=1}^{n} [y_i|\boldsymbol{\beta}][\boldsymbol{\beta}], \tag{20.1}$$

because $\boldsymbol{\beta}$ are the only model parameters. It is possible to build a MCMC algorithm based on individual updates for β_j and, in some cases, that approach may lead to better mixing in the resulting Markov chains. However, we use the "block" updating approach throughout because it allows us to use the same MCMC algorithm to fit models with different numbers of coefficients while avoiding extra `for` loops.

The conventional implementation of the Bayesian logistic regression model using MCMC relies on M-H updates for $\boldsymbol{\beta}$ because of the lack of conjugacy. Thus, we use a multivariate normal proposal distribution to jointly sample $\boldsymbol{\beta}^{(*)} \sim \text{N}(\boldsymbol{\beta}^{(k-1)}, \sigma_{\text{tune}}^2\mathbf{I})$ in the following MCMC algorithm (`logit.reg.mcmc.R`) written in R:

BOX 20.1   MCMC ALGORITHM FOR LOGISTIC REGRESSION

```
1  logit.reg.mcmc <- function(y,X,beta.mn,beta.var,
2       beta.tune,n.mcmc){
3
4  ###
5  ###   Libraries and Subroutines
6  ###
7
8  logit <- function(p){
9    log(p/(1-p))
10 }
```

(Continued)

---

[1] The inverse logit function is sometimes called the "expit" function and, in this case, implies that $p_i = \frac{e^{\mathbf{x}_i'\beta}}{1+e^{\mathbf{x}_i'\beta}}$.

BOX 20.1 (Continued)   MCMC ALGORITHM FOR LOGISTIC REGRESSION

```
11 logit.inv <- function(z){
12   exp(z)/(1+exp(z))
13 }
14
15 ###
16 ###   Preliminary Variables
17 ###
18
19 n.burn=round(n.mcmc/10)
20 n=length(y)
21 p=dim(X)[2]
22
23 beta.save=matrix(0,p,n.mcmc)
24 Davg=0
25
26 ###
27 ###   Starting Values
28 ###
29
30 Sig.beta.inv=solve(beta.var*diag(p))
31 beta=beta.mn
32 pp=logit.inv(X%*%beta)
33
34 ###
35 ###   MCMC Loop
36 ###
37
38 for(k in 1:n.mcmc){
39
40   ###
41   ### Sample beta
42   ###
43
44   beta.star=rnorm(p,beta,beta.tune)
45   pp.star=logit.inv(X%*%beta.star)
46   mh1=sum(dbinom(y,1,pp.star,log=TRUE))-0.5*t
47     (beta.star-beta.mn)%*%Sig.beta.inv%*%
48     (beta.star-beta.mn)
```

(Continued)

BOX 20.1 (Continued)   MCMC ALGORITHM FOR LOGISTIC REGRESSION

```
49  mh2=sum(dbinom(y,1,pp,log=TRUE))-0.5*t
50    (beta-beta.mn)%*%Sig.beta.inv%*%(beta-beta.mn)
51  mh=exp(mh1-mh2)
52  if(mh > runif(1)){
53    beta=beta.star
54    pp=pp.star
55  }
56
57  ###
58  ### DIC Calculations
59  ###
60
61  if(k>n.burn){
62    Davg=Davg-2*(sum(dbinom(y,1,pp,log=TRUE)))/
63       (n.mcmc-n.burn)
64  }
65
66  ###
67  ### Save Samples
68  ###
69
70  beta.save[,k]=beta
71
72 }
73
74 ###
75 ###   Calculate DIC
76 ###
77
78 postbeta.mn=apply(beta.save[,-(1:n.burn)],1,mean)
79 Dhat=-2*(sum(dbinom(y,1,logit.inv(X%*%postbeta.
80    mn),log=TRUE)))
81 pD=Davg-Dhat
82 DIC=2*Davg-Dhat
```

(*Continued*)

BOX 20.1 (Continued)   MCMC ALGORITHM FOR LOGISTIC REGRESSION

```
83 ###
84 ###   Write output
85 ###
86
87 list(y=y,X=X,n.mcmc=n.mcmc,beta.save=beta.save,
88     DIC=DIC)
89
90 }
```

At the beginning of this MCMC algorithm, we defined two functions: `logit` and `logit.inv` because those are not included in the base R software. Also, notice that we added DIC calculations to our MCMC algorithm using the deviance equation that we presented in Chapter 13, adapted for the Bernoulli likelihood. Finally, in `logit.reg.mcmc`, we assumed that the prior covariance matrix for β was $\Sigma_\beta \equiv \sigma_\beta^2 \mathbf{I}$; we can generalize this specification easily to accommodate any prior covariance matrix.

We return to the binary data example involving observed presence or absence of the forest understory plant *Desmodium glutinosum* (Chapter 6) to demonstrate the MCMC algorithm. Hooten et al. (2003), Hefley et al. (2017a), and Hefley et al. (2017b) examined relationships between the presence or absence of *Desmodium glutinosum* and several landscape variables including "southwestness" (an aspect transform where the southwest direction is the largest value), relative elevation (elevation scaled to the study area), and variable depth soil type. Using a prior for β with mean zero and homogeneous variance of 2.25, we fit the Bayesian logistic regression model to the *Desmodium glutinosum* data using the following R code:

BOX 20.2   READ IN DATA AND FIT LOGISTIC REGRESSION MODEL

```
1 desglut.df=read.csv("d.glutinosum.survey.
2     data.csv",header=TRUE)
3 desglut.df=desglut.df[!is.na(desglut.df$present),]
4
5 y=desglut.df$present
6 n=length(y)
7 X=as.matrix(cbind(rep(1,n),desglut.df[,2:4]))
8 p=dim(X)[2]
```

(Continued)

BOX 20.2 (Continued)   READ IN DATA AND FIT LOGISTIC REGRESSION MODEL

```
 9 source("logit.reg.mcmc.R")
10 n.mcmc=100000
11 set.seed(1)
12 mcmc.out=logit.reg.mcmc(y=y,X=X,beta.mn=rep(0,p),
13     beta.var=2.25,beta.tune=.1,n.mcmc=n.mcmc)
14
15 layout(matrix(1:4,2,2,byrow=TRUE))
16 hist(mcmc.out$beta.save[1,],prob=TRUE,breaks=40,
17     main="a",xlab=bquote(beta[0]))
18 curve(dnorm(x,0,2.25),lwd=2,add=TRUE)
19 hist(mcmc.out$beta.save[2,],prob=TRUE,breaks=40,
20     main="b",xlab=bquote(beta[1]))
21 curve(dnorm(x,0,2.25),lwd=2,add=TRUE)
22 hist(mcmc.out$beta.save[3,],prob=TRUE,breaks=40,
23     main="c",xlab=bquote(beta[2]))
24 curve(dnorm(x,0,2.25),lwd=2,add=TRUE)
25 hist(mcmc.out$beta.save[4,],prob=TRUE,breaks=40,
26     main="d",xlab=bquote(beta[3]))
27 curve(dnorm(x,0,2.25),lwd=2,add=TRUE)
```

This R code produces the posterior histograms in Figure 20.1. From Figure 20.1, it is clear that all the covariates are important in explaining variation in *Desmodium glutinosum* presence or absence except for β_3, which corresponds to variable soil depth. Also notice that the prior for β appears to be more precise than those we have used in the past for regression coefficients. Hobbs and Hooten (2015) pointed out that vague priors for coefficients in GLMs can be misleading in that they do not necessarily imply a flat prior for the mean of the response variable. In particular, for the Bernoulli data model with an intercept only, a mean zero normal prior for β_0 with variance 2.25 is approximately flat for most of the support of p.

To interpret the posterior distributions for logistic regression coefficients, recall that the odds of an event occurring are $p/(1-p)$. Thus, the logit function, $\log(p/(1-p))$ is the log odds of $y_i = 1$. We can interpret the effect corresponding to a regression coefficient β_j as an increase in the odds by a factor of $\exp(\beta_j)$ for every 1-unit change in the covariate x_j. In Figure 20.1b, the posterior mean of β_1 is approximately -1; thus a 1-unit change in southwestness decreases the odds of *Desmodium glutinosum* by a multiplicative factor of $\exp(-1) \approx 0.37$.

In the case where we standardize the covariates before the model is fit to data, the logistic regression coefficients have a slightly different interpretation. Consider a standardized covariate, say $\tilde{x}_i = (x_i - \bar{x})/s$ where $\bar{x}$ and s are the sample mean and

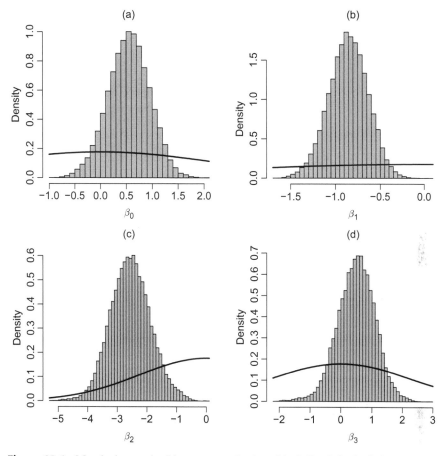

Figure 20.1 Marginal posterior histograms and priors (black lines) for logistic regression coefficients (a) β_0 (intercept), (b) β_1 (southwestness), (c) β_2 (relative elevation), and (d) β_3 (variable depth soil).

standard deviation of $\mathbf{x} \equiv (x_1, \ldots, x_n)'$. Then, in a simplified logistic regression model with only the single standardized covariate, the link function is

$$\text{logit}(p_i) = \beta_0 + \beta_1 \bar{x}_i, \tag{20.2}$$

$$= \beta_0 + \beta_1 \frac{x_i - \bar{x}}{s}, \tag{20.3}$$

$$= \beta_0 - \frac{\beta_1 \bar{x}}{s} + \frac{\beta_1}{s} x_i. \tag{20.4}$$

Thus, a 1-unit change in x_i increases the odds by a multiplicative factor of $\exp(\beta_1/s)$.

20.3   PROBIT REGRESSION

As we mentioned in the previous section on logistic regression, there are many different link functions that we can use to connect the data model mean to the linear combination of covariates. Although the logistic function that we described in the previous section is most commonly used, it leads to a non-conjugate posterior distribution and a MCMC algorithm with M-H updates that we need to tune. The so-called "probit" link function $\Phi^{-1}(\cdot)$, which is the inverse cumulative standard normal distribution, offers an alternative.[2] If we use the probit link function in our binary regression model instead of the logit link, we can develop a MCMC algorithm that has analytically tractable full-conditional distributions and thus we can use Gibbs updates for all parameters. The catch is that we need to introduce auxiliary variables into the model that will assist us with conjugacy. These auxiliary variables actually lead to a hierarchical binary regression model specification that yields the same inference as the non-hierarchical model.

Before we introduce the probit regression model, we review a few properties of the normal distribution. Recall that, for a random variable z arising from a normal distribution with mean μ and variance 1 (i.e., $z \sim N(\mu, 1)$), the difference between z and μ is distributed standard normal (i.e., $z - \mu \sim N(0, 1)$). Also, we can write the probability that $z > 0$ as

$$P(z > 0) = P(z > \mu - \mu), \tag{20.5}$$

$$= P(z - \mu > -\mu), \tag{20.6}$$

$$= P(z - \mu < \mu), \tag{20.7}$$

$$= \Phi(\mu), \tag{20.8}$$

where we used the properties of symmetry in the normal distribution to equate $P(z - \mu > -\mu)$ and $P(z - \mu < \mu)$ and $\Phi(\mu)$ is the standard normal cumulative distribution function (CDF) evaluated at μ. Thus, if $y \sim \text{Bern}(p)$ with link function $\Phi^{-1}(p) = \mu$, then $P(y = 1) = p = \Phi(\mu)$ is equivalent to

$$y = \begin{cases} 0 & , z \leq 0 \\ 1 & , z > 0 \end{cases}, \tag{20.9}$$

where $z \sim N(\mu, 1)$.

This connection between the standard normal CDF and the Bernoulli distribution implies that we can introduce the auxiliary variables z_i into our binary regression model that are deterministically linked to the response variables y_i, but conditionally normal with mean μ_i. We can specify the conditional mean in terms of the linear combination of covariates to form the regression component of the model such that

---

[2] Recall that the CDF of the standard normal distribution is $\Phi(z) = \int_{-\infty}^{z} (2\pi)^{-1/2} \exp(-\tilde{z}^2/2) d\tilde{z}$ and that we can calculate it in R using pnorm(z).

$\mu_i \equiv \mathbf{x}_i'\boldsymbol{\beta}$. Albert and Chib (1993) first described this technique and it has been used in many other binary regression models since, including the models for forest plants presented by Hooten et al. (2003).

A probit regression is still a GLM and could be alternatively fit using the same methods as the previous section where we specify the link function directly specified as $g(\cdot) = \Phi^{-1}(\cdot)$ however, we formulate the model with auxiliary variables in this section because it induces conjugacy in the full-conditional distributions and allows us to construct a fully Gibbs MCMC algorithm. The hierarchical probit regression model is

$$y_i = \begin{cases} 0 & , z_i \leq 0 \\ 1 & , z_i > 0 \end{cases}, \tag{20.10}$$

$$z_i \sim N(\mathbf{x}_i'\boldsymbol{\beta}, 1), \tag{20.11}$$

$$\boldsymbol{\beta} \sim N(\boldsymbol{\mu}_\beta, \boldsymbol{\Sigma}_\beta), \tag{20.12}$$

which implies that $y_i = 1$ with probability $p_i = \Phi(\mathbf{x}_i'\boldsymbol{\beta})$ for $i = 1, \ldots, n$.

The posterior distribution for the probit regression model with auxiliary variables is

$$[\boldsymbol{\beta}, \mathbf{z}|\mathbf{y}] \propto \left(\prod_{i=1}^{n} [y_i|z_i, \boldsymbol{\beta}][z_i|\boldsymbol{\beta}] \right) [\boldsymbol{\beta}], \tag{20.13}$$

$$\propto \left(\prod_{i=1}^{n} \left(1_{\{z_i>0\}} 1_{\{y_i=1\}} + 1_{\{z_i\leq0\}} 1_{\{y_i=0\}} \right) [z_i|\boldsymbol{\beta}] \right) [\boldsymbol{\beta}], \tag{20.14}$$

where, $[z_i|\boldsymbol{\beta}] \equiv N(z_i|\mathbf{x}_i'\boldsymbol{\beta}, 1)$ and the ones with subscripts are indicator variables that equal one when the condition in their subscript is true and zero otherwise.[3]

In contrast to the traditional logistic regression model we described in the previous section, the probit model we specified in (20.10)–(20.12) involves two sets of full-conditional distributions, one for the regression coefficients $\boldsymbol{\beta}$ as before, and one for the auxiliary variables $\mathbf{z}$. It may seem like the probit regression model specified as in (20.10)–(20.12) is overparameterized because there are n auxiliary variables z_i, for $i = 1, \ldots, n$, in addition to the regression coefficients $\boldsymbol{\beta}$. However, because of the deterministic relationship between y_i and z_i and the result we showed in (20.8), it has the same number of effective parameters as the logistic regression.

---

[3] The components $[y_i|z_i, \boldsymbol{\beta}]$ are actually deterministic when we use this model specification and serve to ensure we have all the conditions met in (20.10). However, we admit that the use of bracket notation may be misleading here.

Recall that the joint conditional distribution for the auxiliary variables is $\mathbf{z} \sim N(\mathbf{X}\boldsymbol{\beta}, \mathbf{I})$ because of the conditional independence in our model specification. Thus, the full-conditional distribution for $\boldsymbol{\beta}$ is derived as

$$[\boldsymbol{\beta}|\cdot] \propto [\mathbf{z}|\boldsymbol{\beta}][\boldsymbol{\beta}], \tag{20.15}$$

$$\propto \exp\left(-\frac{1}{2}(\mathbf{z} - \mathbf{X}\boldsymbol{\beta})'(\mathbf{z} - \mathbf{X}\boldsymbol{\beta})\right) \exp\left(-\frac{1}{2}(\boldsymbol{\beta} - \boldsymbol{\mu}_{\boldsymbol{\beta}})'\boldsymbol{\Sigma}_{\boldsymbol{\beta}}^{-1}(\boldsymbol{\beta} - \boldsymbol{\mu}_{\boldsymbol{\beta}})\right), \tag{20.16}$$

$$\propto \exp\left(-\frac{1}{2}(-2(\mathbf{z}'\mathbf{X} + \boldsymbol{\mu}_{\boldsymbol{\beta}}'\boldsymbol{\Sigma}_{\boldsymbol{\beta}}^{-1})\boldsymbol{\beta}) + \left(\boldsymbol{\beta}'\left(\mathbf{X}'\mathbf{X} + \boldsymbol{\Sigma}_{\boldsymbol{\beta}}^{-1}\right)\boldsymbol{\beta}\right)\right), \tag{20.17}$$

$$\propto \exp\left(-\frac{1}{2}(-2\mathbf{b}'\boldsymbol{\beta} + \boldsymbol{\beta}'\mathbf{A}\boldsymbol{\beta})\right), \tag{20.18}$$

where

$$\mathbf{b}' \equiv \mathbf{z}'\mathbf{X} + \boldsymbol{\mu}_{\boldsymbol{\beta}}'\boldsymbol{\Sigma}_{\boldsymbol{\beta}}^{-1}, \tag{20.19}$$

$$\mathbf{A} \equiv \mathbf{X}'\mathbf{X} + \boldsymbol{\Sigma}_{\boldsymbol{\beta}}^{-1}, \tag{20.20}$$

implies that $[\boldsymbol{\beta}|\cdot] \equiv N(\mathbf{A}^{-1}\mathbf{b}, \mathbf{A}^{-1})$ using the same multivariate normal properties that we presented in Chapter 11. Thus, the full-conditional distribution for the probit regression coefficients is conjugate and we can sample $\boldsymbol{\beta}$ using a Gibbs update in our MCMC algorithm.

The full-conditional distribution for z_i involves all components of the joint distribution (20.14) that include z_i, which is the set of indicator functions and the normal distribution for z_i:

$$[z_i|\cdot] \propto \left(1_{\{z_i>0\}}1_{\{y_i=1\}} + 1_{\{z_i\leq0\}}1_{\{y_i=0\}}\right)[z_i|\boldsymbol{\beta}], \tag{20.21}$$

$$\propto \begin{cases} \text{TN}(\mathbf{x}_i'\boldsymbol{\beta}, 1)_0^\infty & , y_i = 1 \\ \text{TN}(\mathbf{x}_i'\boldsymbol{\beta}, 1)_{-\infty}^0 & , y_i = 0 \end{cases}, \tag{20.22}$$

where $\text{TN}(\mu, \sigma^2)_a^b$ refers to a truncated normal distribution with parameters μ and σ^2 truncated below at a and above at b. A truncated normal distribution results as the full-conditional distribution for z_i because the conditions stated in the indicator functions truncate the normal in two different ways depending on whether y_i is one or zero. The truncated normal is renormalized version of the normal distribution after it has been truncated so that it is proper (i.e., integrates to one). Thus, the full-conditional distribution for z_i, for $i = 1, \ldots, n$, has a known form and we can sample it using a Gibbs update in a MCMC algorithm.

To construct the MCMC algorithm for fitting the probit regression model with auxiliary variables, we use the same procedure as before, sampling each set of the model parameters ($\boldsymbol{\beta}$ and $\mathbf{z}$) sequentially so that they depend on each other. We programmed the following MCMC algorithm called `probit.reg.mcmc.R` in R:

BOX 20.3   MCMC ALGORITHM FOR PROBIT REGRESSION

```
probit.reg.mcmc <- function(y,X,beta.mn,beta.var,
    n.mcmc){

###
###   Subroutines
###

rtn <- function(n,mu,sig2,low,high){
  flow=pnorm(low,mu,sqrt(sig2))
  fhigh=pnorm(high,mu,sqrt(sig2))
  u=runif(n)
  tmp=flow+u*(fhigh-flow)
  x=qnorm(tmp,mu,sqrt(sig2))
  x
}

###
###   Preliminary Variables
###

X=as.matrix(X)
y=as.vector(y)
n=length(y)
p=dim(X)[2]

beta.save=matrix(0,p,n.mcmc)
z.save=matrix(0,n,n.mcmc)

y1=(y==1)
y0=(y==0)

###
###   Priors and Starting Values
###

Sig.beta.inv=solve(beta.var*diag(p))
beta=beta.mn

###
###   MCMC Loop
###
```

(*Continued*)

BOX 20.3 (Continued)   MCMC ALGORITHM FOR PROBIT REGRESSION

```
42 for(k in 2:n.mcmc){
43
44   ###
45   ### Sample z
46   ###
47
48   z=matrix(0,n,1)
49   z1=rtn(sum(y),(X%*%beta)[y1],1,0,Inf)
50   z0=rtn(sum(1-y),(X%*%beta)[y0],1,-Inf,0)
51   z[y1,]=z1
52   z[y0,]=z0
53
54   ###
55   ### Sample beta
56   ###
57
58   tmp.chol=chol(t(X)%*%X + Sig.beta.inv)
59   beta=backsolve(tmp.chol,backsolve(tmp.chol,t(X)%
60       *%z + Sig.beta.inv%*%beta.mn,transpose=TRUE)
         +rnorm(p))
61
62   ###
63   ### Save Samples
64   ###
65
66   beta.save[,k]=beta
67   z.save[,k]=z
68
69 }
70
71 ###
72 ###   Write output
73 ###
74
75 list(y=y,X=X,n.mcmc=n.mcmc,z.save=z.save,
76     beta.save=beta.save)
77
78 }
```

Note that we defined the new function rtn to sample from the truncated normal distribution. There are alternative R functions that allow us to sample truncated normal random variables (e.g., in the msm package), but rtn does not require any additional packages. Also, as in the previous example, we assumed that the prior covariance matrix for $\boldsymbol{\beta}$ was $\boldsymbol{\Sigma}_\beta \equiv \sigma_\beta^2 \mathbf{I}$.

We fit the probit regression model with auxiliary variables to the *Desmodium glutinosum* presence or absence data that we described in the previous section using $\sigma_\beta^2 = 1$ as a prior variance for $\boldsymbol{\beta}$ and the following R code:

BOX 20.4 FIT PROBIT REGRESSION MODEL TO PRESENCE OR ABSENCE DATA

```
1  source("probit.reg.mcmc.R")
2  set.seed(1)
3  mcmc.out=probit.reg.mcmc(y=y,X=X,beta.mn=rep(0,p),
4      beta.var=1,n.mcmc=n.mcmc)
5
6  layout(matrix(1:4,2,2,byrow=TRUE))
7  hist(mcmc.out$beta.save[1,],prob=TRUE,breaks=40,
8      main="a",xlab=bquote(beta[0]))
9  curve(dnorm(x,0,1),add=TRUE)
10 hist(mcmc.out$beta.save[2,],prob=TRUE,breaks=40,
11     main="b",xlab=bquote(beta[1]))
12 curve(dnorm(x,0,1),add=TRUE)
13 hist(mcmc.out$beta.save[3,],prob=TRUE,breaks=40,
14     main="c",xlab=bquote(beta[2]))
15 curve(dnorm(x,0,1),add=TRUE)
16 hist(mcmc.out$beta.save[4,],prob=TRUE,breaks=40,
17     main="d",xlab=bquote(beta[3]))
18 curve(dnorm(x,0,1),add=TRUE)
```

This R code results in the marginal posterior histograms for the probit regression coefficients shown in Figure 20.2. Notice that the histograms in Figure 20.2 indicate the same basic relationships between the covariates and the response data as those we obtained in the logistic regression analysis (Figure 20.1), where the posteriors for β_1 and β_2 imply that southwestness and relative elevation are negatively related to *Desmodium glutinosum* presence.

Another aspect of the binary regression model that we described is that we also can make posterior predictive inference for a unobserved data point $\tilde{y}$ given the data $\mathbf{y}$. We can approximate the posterior predictive mean $E(\tilde{y}|\mathbf{y})$ using MC integration as

$$E(\tilde{y}|\mathbf{y}) \approx \frac{\sum_{k=1}^{K} \tilde{y}^{(k)}}{K}, \tag{20.23}$$

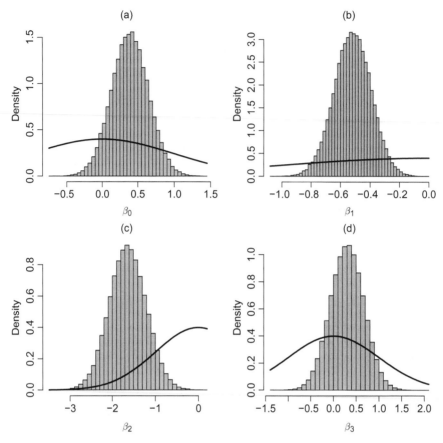

Figure 20.2 Marginal posterior histograms and priors (black lines) for probit regression coefficients (a) β_0 (intercept), (b) β_1 (southwestness), (c) β_2 (relative elevation), and (d) β_3 (variable depth soil).

where $\tilde{y}^{(k)}$ is sampled on the kth MCMC iteration from the predictive full-conditional distribution. The predictive full-conditional distribution for the probit regression model is $[\tilde{y}|\cdot] \equiv \text{Bern}(\Phi(\tilde{\mathbf{x}}'\boldsymbol{\beta}))$, where $\tilde{\mathbf{x}}$ is a $p \times 1$ vector containing the covariates associated with the desired prediction. Because the mean function is linear and the mean of a Bernoulli distribution is the Bernoulli probability, we can approximate the posterior predictive mean of $\tilde{y}$ with

$$E(\tilde{y}|\mathbf{y}) \approx \frac{\sum_{k=1}^{K} \Phi(\tilde{\mathbf{x}}'\boldsymbol{\beta}^{(k)})}{K}, \tag{20.24}$$

For example, we can calculate the posterior predictive probability associated with $\tilde{y}$ for the covariate vector $\tilde{\mathbf{x}}$ where the predictor variables are all equal to the sample mean of the covariates for the observed data in R using the MCMC output from the model fit as

---

BOX 20.5   CALCULATE POSTERIOR PREDICTIVE PROBABILITY

```
1 > mean(pnorm(apply(X,2,mean)%*%mcmc.out
2     $beta.save))
3 [1] 0.2966466
```

Thus, the posterior predictive probability is $P(\tilde{y} = 1 | \mathbf{y}) \approx 0.297$. Note also that the R function pnorm computes the probit function $\Phi(\cdot)$. The same approach to calculating the posterior predictive mean works for logistic regression as well, but in that case, we use the logit link function instead of the probit function. Thus, we can perform predictive classification using either logistic or probit Bayesian regression.

20.4   QUANTILE MODELS FOR BINARY DATA

As we have seen in this chapter, conventional GLMs allow us to make inference on regression parameters linked to the expected value of the data distribution. However, suppose we are interested in describing the relationship between binary data and the predictor variables in terms of quantiles as in Chapter 18. To develop a quantile regression framework for binary data, Benoit and Van Den Poel (2012) considered a model using auxiliary variables where

$$y_i = \begin{cases} 1 & , z_i > 0 \\ 0 & , z_i \leq 0 \end{cases},$$
(20.25)

$$z_i \sim \text{ALD}(\mathbf{x}_i'\boldsymbol{\beta}, 1, \tau),$$
(20.26)

where τ is the percentile of interest ($0 < \tau < 1$) and ALD refers to the same asymmetric Laplace distribution that we introduced in Chapter 18, but with $\sigma^2 = 1$. The form in (20.26) is very similar to the probit regression that we presented in the previous section except that the auxiliary variables (z_i for $i = 1, \ldots, n$) are conditionally ALD instead of normal. Thus, if we specify a multivariate normal prior for the regression coefficients $\boldsymbol{\beta}$ as we did in the probit regression case (i.e., $\boldsymbol{\beta} \sim \text{N}(\boldsymbol{\mu}_\beta, \sigma_\beta^2 \mathbf{I})$), we could fit the model to data using a similar MCMC algorithm that has two updates ($\mathbf{z}$ and $\boldsymbol{\beta}$). To construct a MCMC algorithm for the binary quantile regression model, we need to derive two full-conditional distributions. The full-conditional distribution for z_i will be truncated ALD (truncated above at 0 when $y_i = 0$ and below at zero when $y_i = 1$). Thus, we need a function in R to sample truncated ALD random variables efficiently, and given that, the update for z_i is trivial as in the probit regression model.

However, unlike for the probit regression model, the full-conditional distribution for $\boldsymbol{\beta}$ will not be analytically tractable in the binary quantile regression model. Thus, we need to sample $\boldsymbol{\beta}$ using a M-H update. Although Benoit and Van Den Poel (2017) proposed a model reparameterization that exploits the relationship between the ALD and a scale mixture of normals to induce a form of conjugacy that alleviates the need for tuning, we take an alternative approach here to simplify the model by eliminating the auxiliary variables.

Our approach simply relies on an appropriate link function that connects the probability $P(y_i = 1)$, and hence the parameter in the Bernoulli distribution, to the covariates based on the auxiliary variable model specification in (20.26). Based on the auxiliary quantile regression model, $P(y_i = 1|\boldsymbol{\beta}) = P(z_i > 0|\boldsymbol{\beta})$ and the conditional probability that $z_i > 0$ is

$$P(z_i > 0|\boldsymbol{\beta}) = 1 - \int_{-\infty}^{0} [z_i|\boldsymbol{\beta}]dz_i, \quad (20.27)$$

where the integral is the CDF of the ALD from (20.26) evaluated at zero. Thus, to construct the MCMC algorithm for the binary quantile regression model without auxiliary variables, we have the following posterior distribution for $\boldsymbol{\beta}$

$$[\boldsymbol{\beta}|\mathbf{y}] \propto \left(\prod_{i=1}^{n}[y_i|\boldsymbol{\beta}]\right)[\boldsymbol{\beta}], \quad (20.28)$$

where the data model is Bernoulli with probability as shown in (20.27). Then we can use the same random walk proposal for $\boldsymbol{\beta}$ that we used in the first section of this chapter where $\boldsymbol{\beta}^{(*)} \sim N(\boldsymbol{\beta}^{(k-1)}, \sigma_{\text{tune}}^2 \mathbf{I})$ and calculate the M-H ratio as

$$mh = \frac{\left(\prod_{i=1}^{n}[y_i|\boldsymbol{\beta}^{(*)}]\right)[\boldsymbol{\beta}^{(*)}]}{\left(\prod_{i=1}^{n}[y_i|\boldsymbol{\beta}^{(k-1)}]\right)[\boldsymbol{\beta}^{(k-1)}]}. \quad (20.29)$$

We constructed the following MCMC algorithm called `binary.qr.mcmc.R` in R to fit the Bayesian binary quantile regression model:

BOX 20.6   MCMC ALGORITHM FOR BINARY QUANTILE REGRESSION

```
binary.qr.mcmc <- function(y,X,tau,beta.mn,
    beta.var,beta.tune,n.mcmc){

###
###   Libraries and Subroutines
###

library(ald)

###
###   Preliminary Variables
###
```

(Continued)

BOX 20.6 (Continued)   MCMC ALGORITHM FOR BINARY QUANTILE REGRESSION

```
13 n.burn=round(n.mcmc/10)
14 n=length(y)
15 p=dim(X)[2]
16
17 beta.save=matrix(0,p,n.mcmc)
18 Davg=0
19
20 ###
21 ###   Starting Values
22 ###
23
24 Sig.beta.inv=solve(beta.var*diag(p))
25 beta=beta.mn
26 pp=1-pALD(-X%*%beta,0,1,tau)
27
28 ###
29 ###   MCMC Loop
30 ###
31
32 for(k in 1:n.mcmc){
33
34    ###
35    ### Sample beta
36    ###
37
38    beta.star=rnorm(p,beta,beta.tune)
39    Xbeta.star=X%*%beta.star
40    pp.star=1-pALD(-Xbeta.star,0,1,tau)
41    mh1=sum(dbinom(y,1,pp.star,log=TRUE))-0.5*t
42      (beta.star-beta.mn)%*%Sig.beta.inv%*%
43      (beta.star-beta.mn)
44    mh2=sum(dbinom(y,1,pp,log=TRUE))-0.5*t
45      (beta-beta.mn)%*%Sig.beta.inv%*%(beta-beta.mn)
46    mh=exp(mh1-mh2)
47    if(mh > runif(1)){
48      beta=beta.star
49      pp=pp.star
50    }
```

(Continued)

BOX 20.6 (Continued)   MCMC ALGORITHM FOR BINARY QUANTILE REGRESSION

```
51  ###
52  ### DIC Calculations
53  ###
54
55  if(k>n.burn){
56    Davg=Davg-2*(sum(dbinom(y,1,pp,log=TRUE)))/
57        (n.mcmc-n.burn)
58  }
59
60  ###
61  ### Save Samples
62  ###
63
64  beta.save[,k]=beta
65
66 }
67
68 ###
69 ###   Calculate DIC
70 ###
71
72 postbeta.mn=apply(beta.save[,-(1:n.burn)],1,mean)
73 Dhat=-2*(sum(dbinom(y,1,1-pALD(-X%*%postbeta.mn,
74     0,1,tau),log=TRUE)))
75 pD=Davg-Dhat
76 DIC=2*Davg-Dhat
77
78 ###
79 ###   Write output
80 ###
81
82 list(y=y,X=X,n.mcmc=n.mcmc,beta.save=beta.save,
83     DIC=DIC)
84
85 }
```

Note that we defined $P(y_i = 1|\boldsymbol{\beta})$ as pp to distinguish it from the number of regression coefficients in the R code. Also, note that pALD(-X%*%beta,0,1,tau) is an alternative way to write the ALD CDF evaluated at zero. We used that particular syntax because the pALD function (i.e., the ALD CDF) does not accept vectorized location parameters.

We fit the Bayesian quantile regression model to the same *Desmodium glutinosum* data that we analyzed previously in this chapter using a sequence of quantiles (τ): 0.10, 0.25, 0.50, 0.75, and 0.90, using the following R code:

BOX 20.7   FIT BINARY QUANTILE REGRESSION MODELS

```
1  tau.vec=c(.1,.25,.5,.75,.9)
2  n.tau=length(tau.vec)
3  out.list=vector("list",n.tau)
4  DIC.vec=rep(0,n.tau)
5  beta.mat=matrix(0,p,n.tau)
6  s.vec=rep(0,n.tau)
7  source("binary.qr.mcmc.R") # Code Box 20.6
8  n.mcmc=50000
9  set.seed(1)
10 for(j in 1:n.tau){
11   out.list[[j]]=binary.qr.mcmc(y=y,X=X,tau=tau.
12     vec[j],beta.mn=rep(0,p), beta.var=1000,
13     beta.tune=1,n.mcmc=n.mcmc)
14   DIC.vec[j]=out.list[[j]]$DIC
15   beta.mat[,j]=apply(out.list[[j]]$beta.save,
16     1,mean)
17 }
18 cbind(tau.vec,DIC.vec,t(beta.mat))
```

The posterior means for $\boldsymbol{\beta}$ and DIC values for each of the five Bayesian quantile regression models are shown in Table 20.1. The posterior means for $\boldsymbol{\beta}$ in Table 20.1 indicate a distinct pattern as the quantile increases from 0.10 to 0.50 (the median) to 0.90 where they shrink toward zero as the quantile approaches the median. This result indicates stronger relationships between the *Desmodium glutinosum* presence or absence and the covariates in the tails of the distribution associated with the model. In fact, the DIC indicates that the Bayesian quantile model may predict best when $\tau = 0.90$ (out of the set of quantiles we fit the model with).

Table 20.1

Percentiles, Posterior Means for Model Parameters β, and the Resulting DIC from Each Model

		Covariate			
τ	Intercept	Southwestness	Relative Elevation	Variable Depth	DIC
0.10	-4.35	-4.90	-13.65	3.64	251.5
0.25	-0.51	-2.14	-6.34	1.25	251.2
0.50	1.50	-1.32	-4.80	0.68	247.5
0.75	3.97	-1.43	-6.28	0.86	243.1
0.90	10.24	-2.68	-12.44	1.68	242.7

20.5   ADDITIONAL CONCEPTS AND READING

We provided examples of a few different types of link functions (i.e., logit, probit, and ALD CDF) for binary regression in this chapter. However, many alternatives to these link functions exist. In fact, any CDF is bounded between zero and one and could serve as a link function, depending on the type of problem and characteristics of the relationship between the response and predictor variables. For example, Wang and Dey (2010) described a link function for binary regression based on the generalized extreme value (GEV) distribution. Their approach provides an alternative way to account for skewness in the relationship.

We also can generalize binary data models to accommodate repeated measurements. For example, suppose that we observe the binary response variables y_{ij}, for $i = 1,\ldots,n$ and $j = 1,\ldots,J_i$, where the number of replicates J_i is fixed and known and the data are assumed to arise from the same distribution for each replicate. Then, a reasonable model for the data is $y_{ij} \sim \text{Bern}(p_i)$, where we account for heterogeneity in p_i using an appropriate link function and covariates as before. If the replicates are conditionally independent, then we could alternatively model the sum of the binary data over the replicates as $y_i = \sum_{j=1}^{J_i} y_{ij} \sim \text{Binom}(J_i, p_i)$. We return to these types of model specifications in Chapters 23 and 24 in the presentation of occupancy models and N-mixture models.

It may not be obvious how to generalize GLMs such as logistic/probit regression to accommodate additional spatial or temporal dependence beyond that provided by the covariates. In fact, directly formulating binary data models to include dependence is challenging, but we can do it (e.g., autologistic processes; Hughes et al. 2011). However, a more typical approach is to specify hierarchical binary data models that have a latent Gaussian process. This concept is sometimes referred to as a generalized linear mixed model (GLMM) because it includes both fixed and random effects. To model the random effects with a Gaussian process, we can use the same type of temporal or spatial models that we introduced in Chapters 16 and 17. We return to spatially explicit models for binary data in Chapter 26.

Auxiliary variable models as we described in this chapter for probit regression can be unstable in some situations. Especially in cases where the sample size and number of parameters grow, recent studies have scrutinized the convergence complexity of MCMC (Qin and Hobert 2019).

21 Count Data Regression

21.1 POISSON REGRESSION

After binary data, the second most common type of ecological data are counts or non-negative integers. In wildlife ecology, counts often represent the number of observed individuals in a population (i.e., relative abundance). Counts could also arise as the number of events in a certain region of space or time. For example, the number of feeding events by a mountain lion (*Puma concolor*) in a week is a count that is bounded below by zero.

To specify a Bayesian model for counts, we first consider a probability distribution that may give rise to our observed counts. The Poisson distribution is a natural choice as a model for counts because it arises as a result of counting the number of randomly distributed events in an interval of time or region of space.[1] The Poisson distribution has a mean–variance relationship (i.e., the mean and variance are equal in the Poisson distribution) and therefore has only a single parameter.

Using the GLM framework described in Chapter 20, we can formulate a Poisson regression model for count observations y_i, for $i = 1, \ldots, n$, as $y_i \sim \text{Pois}(\lambda_i)$, where λ_i is the mean and variance of the distribution and the most commonly used link function is the natural logarithm. Thus, we link the conditional mean of the data to the covariates by $\log(\lambda_i) = \mathbf{x}_i' \boldsymbol{\beta}$. To complete the Bayesian model specification, we can specify a multivariate normal prior for the regression coefficients $\boldsymbol{\beta}$ such that $\boldsymbol{\beta} \sim \text{N}(\boldsymbol{\mu}_\beta, \boldsymbol{\Sigma}_\beta)$ (where, in our MCMC algorithm, we assume $\boldsymbol{\Sigma}_\beta \equiv \sigma_\beta^2 \mathbf{I}$).

The resulting posterior distribution for $\boldsymbol{\beta}$ is

$$[\boldsymbol{\beta}|\mathbf{y}] \propto \left(\prod_{i=1}^{n} [y_i|\boldsymbol{\beta}] \right) [\boldsymbol{\beta}], \tag{21.1}$$

which does not have an analytically tractable form that matches a known distribution. However, we can construct a MCMC algorithm to sample $\boldsymbol{\beta}$ as we did with logistic regression in the previous chapter. If we follow the previous chapter and use a random walk symmetric multivariate normal proposal for $\boldsymbol{\beta}^{(*)}$, the resulting M-H ratio is

$$mh = \frac{\left(\prod_{i=1}^{n} \left[y_i|\boldsymbol{\beta}^{(*)} \right] \right) \left[\boldsymbol{\beta}^{(*)} \right]}{\left(\prod_{i=1}^{n} \left[y_i|\boldsymbol{\beta}^{(k-1)} \right] \right) \left[\boldsymbol{\beta}^{(k-1)} \right]}. \tag{21.2}$$

---

[1] Assuming the randomness is governed by a uniform distribution over the interval of time or region of space.

In R, the MCMC algorithm called `pois.reg.mcmc.R` is as follows:

BOX 21.1   MCMC ALGORITHM FOR POISSON REGRESSION

```
pois.reg.mcmc <- function(y,X,beta.mn,beta.var,
    beta.tune,n.mcmc){


###
###   Preliminary Variables
###

n.burn=round(n.mcmc/10)
X=as.matrix(X)
y=as.vector(y)
n=length(y)
p=dim(X)[2]

beta.save=matrix(0,p,n.mcmc)
lam.save=matrix(0,n,n.mcmc)
Davg=0

###
###   Starting Values
###

beta=c(coef(glm(y~0+X,family=poisson)))
beta.sd=sqrt(beta.var)
lam=exp(X%*%beta)
Sig.beta.inv=solve(beta.var*diag(p))

beta.save[,1]=beta
lam.save[,1]=lam

###
###   MCMC Loop
###

for(k in 1:n.mcmc){

  ###
  ### Sample beta
  ###
```

(Continued)

BOX 21.1 (Continued) MCMC ALGORITHM FOR POISSON REGRESSION

```
39   beta.star=rnorm(p,beta,beta.tune)
40   lam.star=exp(X%*%beta.star)
41
42   mh1=sum(dpois(y,lam.star,log=TRUE))-0.5*t
43     (beta.star-beta.mn)%*%Sig.beta.inv%*%
44     (beta.star-beta.mn)
45   mh2=sum(dpois(y,lam,log=TRUE))-0.5*t
46     (beta-beta.mn)%*%Sig.beta.inv%*%(beta-beta.mn)
47   mh=exp(mh1-mh2)
48   if(mh > runif(1)){
49     beta=beta.star
50     lam=lam.star
51   }
52
53   ###
54   ### DIC Calculations
55   ###
56
57   if(k>n.burn){
58     Davg=Davg-2*(sum(dpois(y,lam,log=TRUE)))/
59       (n.mcmc-n.burn)
60   }
61
62   ###
63   ### Save Samples
64   ###
65
66   beta.save[,k]=beta
67   lam.save[,k]=lam
68
69 }
70
71 ###
72 ###   Calculate DIC
73 ###
74
75 postlam.mn=apply(lam.save[,-(1:n.burn)],1,mean)
76 Dhat=-2*(sum(dpois(y,postlam.mn,log=TRUE)))
```

(Continued)

BOX 21.1 (Continued)   MCMC ALGORITHM FOR POISSON REGRESSION

```
77 pD=Davg-Dhat
78 DIC=2*Davg-Dhat
79
80 ###
81 ###  Write output
82 ###
83
84 list(y=y,X=X,n.mcmc=n.mcmc,beta.save=beta.save,
85    lam.save=lam.save,DIC=DIC)
86
87 }
```

In this R code, notice that we selected starting values for β by computing the maximum likelihood point estimates $\hat{\beta}$ using the glm function. This selection allows us to find a reasonable starting place for the Markov chain so that it minimizes burn-in time when we fit the model. Recall that starting values (and tuning parameters) do not affect the posterior results as $K \to \infty$, but reasonable starting values can be helpful to reduce computing time.

We illustrate the Bayesian Poisson regression model by analyzing a data set consisting of a response variable (y_i, for $i = 1, \ldots, n$) for recorded bird species richness (Hooten and Cooch 2019) in each state in the United States of America (USA) plus the District of Columbia and a set of predictor variables corresponding to state area, mean temperature, and total annual precipitation (all standardized to have mean zero and variance one). The state area and precipitation covariates were highly negatively correlated, so we constructed a set of $L = 6$ design matrices associated with the three single covariates and then the three combinations of two covariates (including an intercept in all of them) to compare in terms of predictive ability using DIC. We used hyperparameters $\mu_\beta = 0$ and $\Sigma \equiv \sigma_\beta^2 \mathbf{I}$, where $\sigma_\beta^2 = 1000$. The following R code fits each of the $L = 6$ Bayesian Poisson regression models to produce the posterior means and DIC values listed in Table 21.1:

BOX 21.2   READ IN DATA AND FIT POISSON REGRESSION MODELS

```
1 bird.df=read.csv("birds.csv")
2 idx.outlier=(1:51)[(bird.df$spp==min(bird.df$spp)
3    | bird.df$area==max(bird.df$area))]
4 bird.df=bird.df[-idx.outlier,]
```

(Continued)

BOX 21.2 (Continued) READ IN DATA AND FIT POISSON REGRESSION MODELS

```
5  y=bird.df$spp
6  n=length(y)
7  L=6
8  X=vector("list",L)
9  X[[1]]=cbind(rep(1,n),scale(bird.df[,5]))
10 X[[2]]=cbind(rep(1,n),scale(bird.df[,6]))
11 X[[3]]=cbind(rep(1,n),scale(bird.df[,7]))
12 X[[4]]=cbind(rep(1,n),scale(bird.df[,5:6]))
13 X[[5]]=cbind(rep(1,n),scale(bird.df[,c(5,7)]))
14 X[[6]]=cbind(rep(1,n),scale(bird.df[,6:7]))
15
16 out.list=vector("list",L)
17 source("pois.reg.mcmc.R")
18 n.mcmc=20000
19 DIC.vec=rep(0,L)
20 set.seed(1)
21 for(l in 1:L){
22   out.list[[l]]=pois.reg.mcmc(y=y,X=X[[l]],
23     beta.mn=rep(0,dim(X[[l]])[2]),beta.var=1000,
24     beta.tune=.01,n.mcmc=n.mcmc)
25   DIC.vec[l]=out.list[[l]]$DIC
26 }
```

The DIC was lowest for model 4, the Bayesian Poisson regression model including both state area and average temperature as covariates. In this model, both regression coefficients were positive (area 95% CI: [0.08, 0.11]; temperature 95% CI: [0.04, 0.07]), which suggests that bird species richness increases as the state size and temperature increase.

21.2 RESOURCE SELECTION FUNCTIONS AND SPECIES DISTRIBUTION MODELS

In Chapter 17, we described spatial statistical models for two types of spatially explicit data: Continuous spatial processes and areal spatial processes. The third major type of spatial data are spatial point processes. Spatial point processes (SPPs) result in data that are observed spatial locations of events in a region of space ($\mathscr{S}$; the spatial domain, usually a study area or home range) $\mathbf{s}_i$ for $i = 1,\ldots,n$. Hooten et al. (2017) showed how to model these data in a variety of ways to learn about the generative processes that gave rise to the observed locations. Ecologists often refer to spatial

Table 21.1

The Resulting Marginal Posterior Means for β and DIC from $L = 6$ Bayesian Poisson Regression Models Fit to the Continental USA Bird Species Richness Data

Model	Coefficient				DIC
	Intercept	Area	Temperature	Precip	
1	5.76	0.10	—	—	571.2
2	5.76	—	0.07	—	691.1
3	5.76	—	—	−0.05	706.2
4	5.75	0.09	0.06	—	526.7
5	5.76	0.11	—	0.02	567.8
6	5.75	—	0.12	−0.11	536.7

Note: An intercept was included in all models and "—" indicates which covariates were excluded in each model.

point process data as "presence-only data" because they view the data as presence or absence data with missing absences rather than as observed points in space.

A rich class of SPP models has been developed to provide most types of desired inference using SPP data. Conditioning on the total number of observations n, the most common way to model a SPP is with a weighted distribution formulation (Patil and Rao 1976, 1977, 1978). The weighted distribution approach can be thought of as designing a custom probability distribution for the observed points $\mathbf{s}_i$ over the space of interest $\mathscr{S}$. This custom probability distribution needs to be proper in that it should integrate to one over the support $\mathscr{S}$ and have a non-negative PDF. Also, we can learn about the features of a landscape that correlate with the likelihood of a point arising in a certain place if we allow the custom probability distribution to depend on spatially referenced covariates. Thus, we write a simple weighted distribution SPP model as

$$[\mathbf{s}_i|\boldsymbol{\beta}] = \frac{f(\mathbf{x}'(\mathbf{s}_i)\boldsymbol{\beta})}{\int_{\mathscr{S}} f(\mathbf{x}'(\mathbf{s})\boldsymbol{\beta})d\mathbf{s}}, \tag{21.3}$$

where $f(\cdot)$ is a non-negative function over $\mathscr{S}$ and the integral in the denominator of (21.3) normalizes the density function so that it integrates to one. The vector of covariates $\mathbf{x}$ in (21.3) does not include an intercept because it would cancel in the numerator and denominator and thus the parameters $\boldsymbol{\beta} \equiv (\beta_1, \ldots, \beta_p)'$ increase or decrease the intensity of the SPP depending on the value of the covariates at a certain location.

Two ecological data types that coincide with the SPP model in (21.3) are animal telemetry data and species distribution data. In animal telemetry data, researchers track individual animals with a device that permits observation of the position of individual animals in space (Hooten et al. 2017) and in species distribution data the position of a species is recorded when an observer detects it (assuming each point represents a different individual; Hefley and Hooten 2016). Although both types of ecological data are commonly collected and the ecological questions differ among them (Johnson 1980), we can use the same model (21.3) to analyze them and provide inference.

In the case of animal telemetry data, the function $f(\mathbf{x}'(\mathbf{s}_i)\boldsymbol{\beta})$ in (21.3) is referred to as a "resource selection function" (RSF) because the coefficients $\boldsymbol{\beta}$ indicate how an individual may select resources differentially to what is available to them (i.e., out of proportion with resources in $\mathscr{S}$; Lele and Keim 2006). To estimate the RSF, we fit the Bayesian SPP model to data using the same techniques as in previous chapters where we write the posterior for $\boldsymbol{\beta}$ as

$$[\boldsymbol{\beta}|\{\mathbf{s}_i, \forall i\}] \propto \left(\prod_{i=1}^{n}[\mathbf{s}_i|\boldsymbol{\beta}]\right)[\boldsymbol{\beta}], \tag{21.4}$$

$$\propto \left(\prod_{i=1}^{n}\frac{f(\mathbf{x}'(\mathbf{s}_i)\boldsymbol{\beta})}{\int_{\mathscr{S}}f(\mathbf{x}'(\mathbf{s})\boldsymbol{\beta})d\mathbf{s}}\right)[\boldsymbol{\beta}]. \tag{21.5}$$

The denominator in (21.5) is a critical element of the posterior distribution because it involves the coefficient vector $\boldsymbol{\beta}$ and must be retained in the proportionality. Thus, to fit the model using MCMC, we need to use M-H updates for $\boldsymbol{\beta}$ because the posterior in (21.5) is not usually analytically tractable. In calculating the numerator and denominator of the M-H ratio, we must approximate the integral $\int_{\mathscr{S}}f(\mathbf{x}'(\mathbf{s})\boldsymbol{\beta})d\mathbf{s}$, or find a way to avoid calculating it directly.

When we specify the RSF as an exponential function $\exp(\mathbf{x}'(\mathbf{s}_i)\boldsymbol{\beta})$, there are three main approaches for fitting the SPP model:

1. Numerical integration: With an approach called "numerical quadrature" (Chapter 2), we grid the study area $\mathscr{S}$ and compute an approximation of the integral as the sum of $f(\mathbf{x}'(\mathbf{s})\boldsymbol{\beta})$ over a large number of grid cells multiplied by the grid cell area. We must compute this large sum at each MCMC iteration.

2. Logistic regression: We treat the observed positions of the individual as a data set consisting of ones and then take a large and uniformly distributed background sample of points from $\mathscr{S}$ that we will label as zero. We then fit a logistic regression model to the entire data set comprised of the observed ones and background zeros (both with known covariate values). This method is probably the most commonly used approach (Baddeley et al. 2010; Warton and Shepherd 2010; Fithian and Hastie 2013).

3. Poisson regression: We grid the region $\mathscr{S}$ as in the numerical quadrature approach previously described, but let the data set consist of counts of observed telemetry data in each grid cell (most of which will be zero). We then fit a Poisson regression model to the data using the covariates in each grid cell as the predictor variables in the model (Aarts et al. 2012).

(a)   (b)

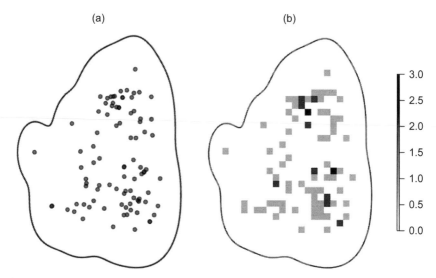

Figure 21.1   (a) Observed telemetry data (s_i, $i = 1, \ldots, n$) for an individual mountain lion in Colorado, USA (outline indicates $\mathscr{S}$) and (b) gridded cell counts (y_i for $i = 1, \ldots, m$ grid cells) of telemetry data in $\mathscr{S}$.

To fit the SPP model and make inference on the coefficients $\boldsymbol{\beta}$, we can use either the Bayesian logistic regression model in the previous chapter or we can use the Bayesian Poisson regression model presented in this chapter following the relevant procedure previously listed in either case. In either the logistic regression or Poisson regression approach, we use an intercept to account for the size of the background sample or number of grid cells used in the approximation of the likelihood. As the number of grid cells approaches infinity (and hence the grid cell size approaches zero) the resulting inference will converge to the correct inference under the exact SPP model.[2] In what follows, we demonstrate the Poisson regression approach to fitting the SPP model using telemetry data from an individual mountain lion in Colorado, USA (Figure 21.1a). The spatially explicit topographic covariates potentially associated with mountain lion space use (i.e., elevation, slope, and exposure) are shown in Figure 21.2. Based on hyperparameters $\boldsymbol{\mu}_\beta = \mathbf{0}$ and $\boldsymbol{\Sigma}_\beta \equiv \sigma_\beta^2 \mathbf{I}$ where $\sigma_\beta^2 = 100$, we fit the SPP model using the MCMC algorithm (`pois.reg.mcmc.R`) that we presented in the previous section for Poisson regression to the mountain lion data using the following R code:

---

[2] However, the sensitivity should be checked relative to the grid cell size to ensure the resolution is fine enough to approximate the likelihood well (Northrup et al. 2013).

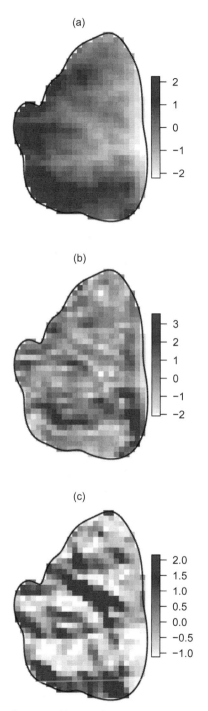

Figure 21.2 Standardized topographic covariates potentially associated with mountain lion movement: (a) elevation, (b) slope, (c) exposure (a function of aspect).

BOX 21.3   FIT SPP MODEL USING POISSON REGRESSION

```
 1  load("ml_pois.RData")
 2  y=c(ml.df$y[!is.na(ml.df$y)])
 3  X=model.matrix(y~elev+slope+exposure,
 4    family=poisson(link="log"),data=ml.df)
 5  source("pois.reg.mcmc.R")   # Code Box 21.1
 6  n.mcmc=40000
 7  mcmc.out=pois.reg.mcmc(y=y,X=X,beta.mn=rep
 8    (0,dim(X)[2]),beta.var=100,beta.tune=.1,n.mcmc)
 9
10
11  layout(matrix(1:4,2,2,byrow=TRUE))
12  hist(mcmc.out$beta.save[1,],prob=TRUE,breaks=40,
13    main="a",xlab=bquote(beta[0]))
14  curve(dnorm(x,0,10),lwd=2,add=TRUE)
15  hist(mcmc.out$beta.save[2,],prob=TRUE,breaks=40,
16    main="b",xlab=bquote(beta[1]))
17  curve(dnorm(x,0,10),lwd=2,add=TRUE)
18  hist(mcmc.out$beta.save[3,],prob=TRUE,breaks=40,
19    main="c",xlab=bquote(beta[2]))
20  curve(dnorm(x,0,10),lwd=2,add=TRUE)
21  hist(mcmc.out$beta.save[4,],prob=TRUE,breaks=40,
22    main="d",xlab=bquote(beta[3]))
23  curve(dnorm(x,0,10),lwd=2,add=TRUE)
```

After creating the gridded mountain lion cell counts (Figure 21.1b), our data set included $m = 654$ response observations and associated covariates for the intercept and three predictor variables. The previous R code results in the marginal posterior histograms shown in Figure 21.3. The posterior results in Figure 21.3 indicate that the individual mountain lion selects for lower elevation (Figure 21.3b) and steeper areas (Figure 21.3c), although we do not have enough evidence to infer that it selects more exposed areas (because the distribution overlaps zero substantially in Figure 21.3d).

This same procedure also works to fit species distribution models to species observation data that we described at the beginning of this section (e.g., Warton and Shepherd 2010; Aarts et al. 2012). In fact, recent research has shown that the point process models have strong connections to other commonly used approaches for making inference about species distributions (e.g., MaxEnt; Renner and Warton 2013).

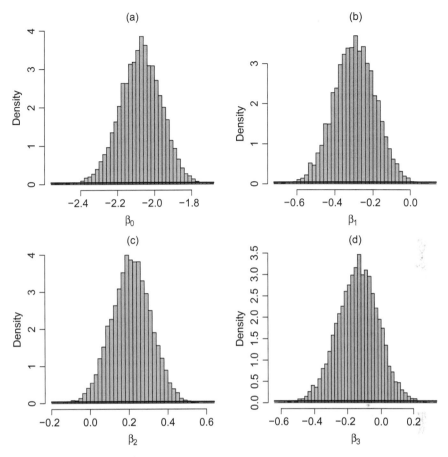

Figure 21.3 Marginal posterior histograms for β based on fitting the SPP model to the mountain lion data using Poisson regression: (a) intercept, (b) elevation, (c) slope, and (d) exposure (a function of aspect). Priors are shown as a black line.

21.3 STEP SELECTION FUNCTIONS

In terms of other types of animal telemetry data, it has become common for studies to collect high temporal resolution satellite telemetry data that depict the actual trajectory of animals as they move. Although the development of many models provide inference using these data (e.g., Hooten et al. 2017; Johnson et al. 2008) described a more general spatio-temporal point process approach to model the selection and movement of animals simultaneously. Johnson et al. (2013b) and Brost et al. (2015) developed extensions to this work as well as closely connected concepts known as step selection function (SSF; Fortin et al. 2005) and continuous-time discrete-space movement models (Hooten et al. 2010; Hanks et al. 2015a).

A weighted distribution formulation for a spatio-temporal SPP model is

$$[\mathbf{s}_i|\boldsymbol{\beta}] = \frac{f(\mathbf{x}'(\mathbf{s}_i)\boldsymbol{\beta})g(\mathbf{s}_i|\mathbf{s}_{i-1})}{\int_{\mathscr{S}} f(\mathbf{x}'(\mathbf{s})\boldsymbol{\beta})g(\mathbf{s}|\mathbf{s}_{i-1})d\mathbf{s}}, \tag{21.6}$$

where $g(\cdot)$ is an availability function conditional on the last known position $(\mathbf{s}_{i-1})$ of the individual and determines where the individual can move by the next time. We can simplify the convolution in the denominator of (21.6) in various ways depending on the choice of availability function $g(\cdot)$ and the availability function also may depend on a set of parameters that control the dynamics of movement (e.g., the scale of a movement kernel centered on the previous position) and the time between observed positions $\mathbf{s}_{i-1}$ and $\mathbf{s}_i$. Often, however, we calculate the availability function $g(\cdot)$ *a priori* (or empirically) as a function of typical step-lengths and turning angles for the individual animal; perhaps estimated as a disk centered on $\mathbf{s}_{i-1}$ with radius equal to a maximum realistic distance moved in the given telemetry data during the elapsed time between $\mathbf{s}_{i-1}$ and $\mathbf{s}_i$.

We can fit the spatio-temporal point process model given telemetry data using the data model (21.6) and prior for $\boldsymbol{\beta}$ as well as any parameters in the availability function assumed to be unknown. However, we cannot use the Bayesian logistic or Poisson GLM directly as we did in the previous example to estimate the RSF. Instead, we must either use numerical integration to calculate the n integrals separately, or use a procedure like the one described to fit the static SPP model where we augment the data structure and use a modified version of logistic or Poisson regression. However, because our availability function changes over time, we need to make n augmentations of the data and then use a technique called conditional regression.

The development of conditional regression was for case–control studies where we "match" a case with one or more controls in a regression equivalent of a paired t-test. It is possible to implement the conditional regression approach to fitting the spatio-temporal point process model (21.6) with a modified version of either logistic or Poisson regression. In either case, we begin with unconditional regression and allow the intercept $\beta_{0,i}$ to vary with each telemetry observation, for $i = 1, \ldots, n$. In what follows, we outline the approximation procedure based on Poisson regression and a simplified version of the availability function $g(\cdot)$ that we can specify before the model is fit. For example, we assume that the availability function is a uniform disk around $\mathbf{s}_{i-1}$

$$g(\mathbf{s}|\mathbf{s}_{i-1}) = \frac{1_{\{||\mathbf{s}-\mathbf{s}_{i-1}||<r_i\}}}{\pi r_i^2}, \tag{21.7}$$

where $1/(\pi r_i^2)$ is the height of the disk defined by $|\mathbf{s}_i - \mathbf{s}_{i-1}| < r_i$ with radius r_i if $g(\mathbf{s}|\mathbf{s}_{i-1})$ is a proper PDF. Then, for each observed point $\mathbf{s}_{i-1}$ ($i = 2, \ldots, n$), we grid the circular region defined by $||\mathbf{c}_{ij} - \mathbf{s}_{i-1}|| < r_i$ where $\mathbf{c}_{ij}$ are the grid cell centers for $j = 1, \ldots, J_i$ (where J_i should be very large; Northrup et al. 2013). For each grid cell, we let $y_{ij} = 0$, except for the grid cell that contains s_{i-1}, in which case we let $y_{ij} = 1$.

After building the data structure that consists of y_{ij} for $i = 2, \ldots, n$ and $j = 1, \ldots, J_i$, using the procedure described, we specify a Poisson GLM model as $y_{ij} \sim \text{Pois}(\lambda_{ij})$ with link function

$$\log(\lambda_{ij}) = \beta_{0,i} + \mathbf{x}'_{ij}\boldsymbol{\beta}, \qquad (21.8)$$

where the covariate vectors $\mathbf{x}_{ij}$ do not have a first element as one and the coefficients $\boldsymbol{\beta}$ do not contain an intercept because $\beta_{0,i}$ accommodates the intercept. For priors, we use $\boldsymbol{\beta} \sim \text{N}(\boldsymbol{\mu}_\beta, \boldsymbol{\Sigma}_\beta)$ and $\beta_{0,i} \sim \text{N}(\mu_0, \sigma_0^2)$ for $i = 1, \ldots, n$.

The Bayesian Poisson regression model with link function (21.8) is a GLM with an intercept that varies by group. In our case, the groups are the individual "steps" in the data (for $i = 2, \ldots, n$) and their associated availability regions. Note that this model contains as many parameters as original observations plus regression coefficients ($n + p - 1$ total parameters) and we can fit it using the same MCMC algorithm that we developed in this chapter because we can write the link function jointly as

$$\log(\boldsymbol{\lambda}) = \mathbf{W}\boldsymbol{\beta}_0 + \mathbf{X}\boldsymbol{\beta}, \qquad (21.9)$$

where "log" corresponds to an element-wise natural logarithmic function, $\boldsymbol{\lambda}$ is a $(\sum_{i=2}^n J_i) \times 1$ vector containing all λ_{ij}, and $\mathbf{W}$ is a $(\sum_{i=2}^n J_i) \times (n-1)$ matrix of indicator variables with ones indicating the step. Then $\boldsymbol{\beta}_0 \equiv (\beta_{0,2}, \ldots, \beta_{0,n})'$ is a vector containing the step-specific intercepts and $\mathbf{X}$ is a $(\sum_{i=2}^n J_i) \times p$ design matrix where p is number of covariates in the model. We can rewrite the joint link function in (21.9) as

$$\log(\boldsymbol{\lambda}) = \tilde{\mathbf{X}}\tilde{\boldsymbol{\beta}}, \qquad (21.10)$$

where $\tilde{\mathbf{X}} \equiv (\mathbf{W}, \mathbf{X})$ is a combined $(\sum_{i=2}^n J_i) \times (n-1+p)$ design matrix that first contains the set of intercept columns $\mathbf{W}$ and then the covariate columns $\mathbf{X}$. The associated combined coefficient vector is $\tilde{\boldsymbol{\beta}} \equiv (\boldsymbol{\beta}_0', \boldsymbol{\beta}')'$. Thus, (21.10) has the required form of a Poisson GLM with log link function and we can use the same MCMC algorithm that we developed earlier in this chapter to fit the model.

The Poisson GLM that we specified to approximate the spatio-temporal point process model requires a very large number of parameters that increases with the size of the data set (i.e., the number of steps, $n - 1$) and the associated MCMC algorithm may require substantial tuning to yield well-mixed Markov chains. Thus, a common alternative is to condition on a sufficient statistic that reduces the number of parameters by allowing us to cancel the step-specific intercepts in the likelihood.

To derive the conditional likelihood, recall that the likelihood component associated with step i is

$$[\mathbf{y}_i|\beta_0,\boldsymbol{\beta}] = \prod_{j=1}^{J_i} \frac{\lambda_{ij}^{y_{ij}} e^{-\lambda_{ij}}}{y_{ij}!}, \tag{21.11}$$

$$= \prod_{j=1}^{J_i} \frac{\left(e^{\beta_{0,i}+\mathbf{x}_{ij}'\boldsymbol{\beta}}\right)^{y_{ij}} e^{-e^{\beta_{0,i}+\mathbf{x}_{ij}'\boldsymbol{\beta}}}}{y_{ij}!}, \tag{21.12}$$

$$= \frac{\left(e^{\beta_{0,i}\sum_{j=1}^{J_i} y_{ij}+\sum_{j=1}^{J_i} \mathbf{x}_{ij}'\boldsymbol{\beta}y_{ij}}\right) e^{-\sum_{j=1}^{J_i} e^{\beta_{0,i}+\mathbf{x}_{ij}'\boldsymbol{\beta}}}}{\prod_{j=1}^{J_i} y_{ij}!}. \tag{21.13}$$

The expression in (21.13) reveals two sufficient statistics at the step level: $\sum_{j=1}^{J_i} y_{ij}$ and $\sum_{j=1}^{J_i} \mathbf{x}_{ij}y_{ij}$. We can condition on the first sufficient statistic that involves the response data y_{ij} only because $\sum_{j=1}^{J_i} y_{ij} = 1$. Thus, the likelihood component for step i that conditions on $\sum_{j=1}^{J_i} y_{ij} = 1$ is

$$\left[\mathbf{y}_i \middle| \beta_0, \boldsymbol{\beta}, \sum_{j=1}^{J_i} y_{ij} = 1\right] = \frac{[\mathbf{y}_i|\beta_0,\boldsymbol{\beta}]}{\sum_{\tilde{\mathbf{y}}_i'\mathbf{1}=1}[\tilde{\mathbf{y}}_i|\beta_0,\boldsymbol{\beta}]}, \tag{21.14}$$

where the sum in the denominator of (21.14) includes all values of $\tilde{\mathbf{y}}$ such that $\tilde{\mathbf{y}}_i'\mathbf{1} = 1$, which implies that we cycle through $j = 1,\dots,J_i$ setting $\tilde{y}_{ij} = 1$ and the others equal to zero. Substituting (21.13) into (21.14) and using the constraints $\sum_{j=1}^{J_i} y_{ij} = 1$ and $\sum_{j=1}^{J_i} \tilde{y}_{ij} = 1$ results in

$$\left[\mathbf{y}_i \middle| \beta_0, \boldsymbol{\beta}, \sum_{j=1}^{J_i} y_{ij} = 1\right] = \frac{e^{\sum_{j=1}^{J_i} \mathbf{x}_{ij}'\boldsymbol{\beta}y_{ij}}}{\sum_{\tilde{\mathbf{y}}_i'\mathbf{1}=1} e^{\sum_{j=1}^{J_i} \mathbf{x}_{ij}'\boldsymbol{\beta}\tilde{y}_{ij}}}, \tag{21.15}$$

$$= \frac{e^{\mathbf{x}_{i1}'\boldsymbol{\beta}}}{\sum_{j=1}^{J_i} e^{\mathbf{x}_{ij}'\boldsymbol{\beta}}}, \tag{21.16}$$

because the $y_{ij}!$, $e^{\beta_{0,i}}$, and $e^{-\sum_{j=1}^{J_i} e^{\beta_{0,i}+\mathbf{x}_{ij}'\boldsymbol{\beta}}}$ terms from (21.13) cancel in the numerator and denominator of (21.14). In (21.16), we assume that the data are organized such that $y_{i1} = 1$ (and thus $y_{ij} = 0$ for $j = 2,\dots,J_i$, and $\mathbf{x}_{i1}$ are the covariates associated with telemetry observation i), which results in a discretized version of a weighted distribution form for the point process model (21.3).[3] Thus, our complete conditional likelihood for all steps is

$$\left[\mathbf{y} \middle| \boldsymbol{\beta}, \left\{\sum_{j=1}^{J_i} y_{ij} = 1, \forall i\right\}\right] = \prod_{i=2}^{n} \frac{e^{\mathbf{x}_{i1}'\boldsymbol{\beta}}}{\sum_{j=1}^{J_i} e^{\mathbf{x}_{ij}'\boldsymbol{\beta}}}, \tag{21.17}$$

---

[3] We arrive at the denominator in (21.16) because, as we evaluate a case where $\tilde{\mathbf{y}}_i'\mathbf{1} = 1$ and substitute those $\tilde{y}_{ij}$ into the sum $\sum_{j=1}^{J_i} \mathbf{x}_{ij}'\boldsymbol{\beta}\tilde{y}_{ij}$, it will equal $\mathbf{x}_{ij}'\boldsymbol{\beta}$ when the jth element in $\tilde{\mathbf{y}}_i$ is equal to one. There are only J_i unique cases where $\tilde{\mathbf{y}}_i'\mathbf{1} = 1$, thus we can sum $e^{\mathbf{x}_{ij}'\boldsymbol{\beta}}$ over them.

which does not involve any intercepts $\beta_{0,i}$, and thus we can use it directly in a MCMC algorithm in place of the Poisson PMF when we sample $\boldsymbol{\beta}$. Using the same multivariate normal random walk proposal for $\boldsymbol{\beta}$ that we used throughout this chapter results in the M-H ratio

$$mh = \frac{[\mathbf{y}|\boldsymbol{\beta}^{(*)}, \{\sum_{j=1}^{J_i} y_{ij} = 1, \forall i\}][\boldsymbol{\beta}^{(*)}]}{[\mathbf{y}|\boldsymbol{\beta}^{(k-1)}, \{\sum_{j=1}^{J_i} y_{ij} = 1, \forall i\}][\boldsymbol{\beta}^{(k-1)}]}. \tag{21.18}$$

We constructed a MCMC algorithm called `cpr.mcmc.R` in R based on the conditional Poisson likelihood (21.17) and the multivariate normal prior for $\boldsymbol{\beta}$ that we used throughout this chapter:

BOX 21.4   MCMC ALGORITHM FOR CONDITIONAL POISSON REGRESSION

```
cpr.mcmc <- function(X,step.idx,beta.mn,beta.var,
    beta.tune,n.mcmc){

###
###   Subroutines and Libraries
###

library(survival)

ldcpr<-function(loglam,lam,first.idx,step.idx){
  sum.1=sum(loglam[first.idx])
  tmp.df=data.frame(lam,step.idx)
  sum.2=sum(log(summarise(group_by(tmp.df,step.
    idx),sum(lam))[[2]]))
  sum.1-sum.2
}

###
###   Preliminary Variables
###

X=as.matrix(X)
n=dim(X)[1]
p=dim(X)[2]

beta.save=matrix(0,p,n.mcmc)
```

(*Continued*)

BOX 21.4 (Continued)   MCMC ALGORITHM FOR CONDITIONAL POISSON REGRESSION

```
27 n.vec=rle(step.idx)$lengths
28 n.1=length(n.vec)
29 first.idx=c(1,cumsum(n.vec)[-((n.1-1):n.1)]+1)
30
31 ###
32 ###   Starting Values
33 ###
34
35 beta=coef(clogit(y~X+strata(step.idx),
36   method="exact"))
37
38 beta.sd=sqrt(beta.var)
39 loglam=X%*%beta
40 lam=exp(loglam)
41 loglik=ldcpr(loglam,lam,first.idx,step.idx)
42 beta.save[,1]=beta
43
44 ###
45 ###   MCMC Loop
46 ###
47
48 for(k in 1:n.mcmc){
49
50   ###
51   ### Sample beta
52   ###
53
54   beta.star=rnorm(p,beta,beta.tune)
55   loglam.star=X%*%beta.star
56   lam.star=exp(loglam.star)
57   loglik.star=ldcpr(loglam.star,lam.star,
58     first.idx,step.idx)
59   mh1=loglik.star+sum(dnorm(beta.star,beta.mn,
60     beta.sd,log=TRUE))
61   mh2=loglik+sum(dnorm(beta,beta.mn,beta.sd,
62     log=TRUE))
63   mh=exp(mh1-mh2)
64
65   if(mh > runif(1)){
66     beta=beta.star
```

(Continued)

BOX 21.4 (Continued)   MCMC ALGORITHM FOR CONDITIONAL POISSON REGRESSION

```
67      lam=lam.star
68      loglam=loglam.star
69      loglik=loglik.star
70    }
71
72    ###
73    ### Save Samples
74    ###
75
76    beta.save[,k]=beta
77
78  }
79
80  ###
81  ###   Write output
82  ###
83
84  list(n.mcmc=n.mcmc,beta.save=beta.save)
85
86  }
```

Notice that we created a custom version of the logarithm of the conditional Poisson likelihood function ldcpr to use in the M-H ratio in our MCMC algorithm to update β. This function requires index vectors that indicate the steps and positions of the use versus availability locations for each step. Also, notice that we update the quantities loglam and loglik (lines 68–69) so that we can avoid calculating them again on the subsequent MCMC iteration to speed up the algorithm.

To demonstrate the Bayesian conditional Poisson regression model for estimating a step selection function (which in turn approximates the spatio-temporal point process model), we analyzed a set of telemetry data recorded for an individual mountain lion in Colorado, USA, approximately every 3 hours over the period of approximately 1 month (Hooten and Johnson 2017). We used the topographic variables elevation (standardized), slope (standardized), and solar exposure (a scaled and rotated version of aspect where 1 indicates southwest orientations and -1 indicates northeast orientations) as covariates (Figure 21.4).

We specified the radius r_i of the circular movement area centered on the previous telemetry location as $r_i = \delta(t_i - t_{i-1})$ to represent the availability regions, where δ is specified as 95th quantile of the empirical distribution of the step lengths d_i divided by the time differences $t_i - t_{i-1}$. We used the following R code to prepare the data for the conditional Poisson regression analysis:

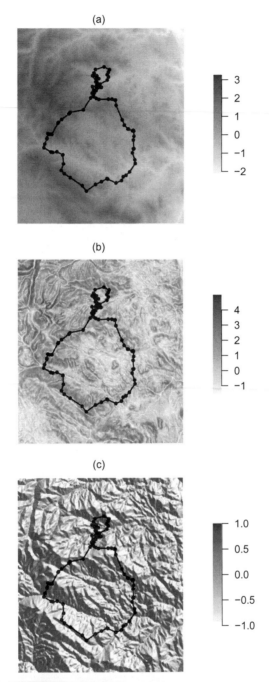

Figure 21.4 Spatial covariates: (a) elevation (standardized), (b) slope (standardized), and (c) solar exposure. Mountain lion telemetry data shown as points connected by black lines.

BOX 21.5 PREPARE DATA FOR CONDITIONAL POISSON REGRESSION

```
 1 d.vec=sqrt(apply(apply(ml.df[,1:2],2,diff)^2,1,sum))
 2 t.d.vec=diff(ml.df[,3])
 3 d.sc=d.vec/t.d.vec
 4 delta=quantile(d.sc,.95)
 5
 6 library(plotrix)
 7 plot(ELE.rast,col=gray.colors.rev(100,.3),
 8     axes=FALSE)
 9 lines(ml.df[,1:2],type="o",cex=.5,lwd=1,col=1)
10 n=dim(ml.df)[1]
11 y=NULL
12 x.1=NULL
13 x.2=NULL
14 x.3=NULL
15 r.vec=NULL
16 n.0=NULL
17 n.vec=NULL
18 for(i in 2:n){
19   if(i%%50==0) cat(i," ")
20   r.vec=c(r.vec,delta.star*t.d.vec[i-1])
21   draw.circle(ml.df[i-1,1],ml.df[i-1,2],
22     r.vec[i-1], border=rgb(0,0,0,.6))
23   x.1=c(x.1,unlist(extract(x=ELE.rast,
24     y=ml.df[i,1:2],df=TRUE)[2]))
25   x.2=c(x.2,unlist(extract(x=SLO.rast,
26     y=ml.df[i,1:2],df=TRUE)[2]))
27   x.3=c(x.3,unlist(extract(x=ASP.rast,
28     y=ml.df[i,1:2],df=TRUE)[2]))
29   y=c(y,1)
30   tmp.1.extract=extract(x=ELE.rast,y=ml.df[i,1:2],
31     buffer=r.vec[i-1],df=TRUE)[,2]
32   tmp.2.extract=extract(x=SLO.rast,y=ml.df[i,1:2],
33     buffer=r.vec[i-1],df=TRUE)[,2]
34   tmp.3.extract=extract(x=ASP.rast,y=ml.df[i,1:2],
35     buffer=r.vec[i-1],df=TRUE)[,2]
36   n.0=c(n.0,length(tmp.1.extract))
37   n.vec=c(n.vec,n.0[i-1]+1)
38   x.1=c(x.1,unlist(tmp.1.extract))
39   x.2=c(x.2,unlist(tmp.2.extract))
40   x.3=c(x.3,unlist(tmp.3.extract))
41   y=c(y,rep(0,n.0[i-1]))
42 }
```

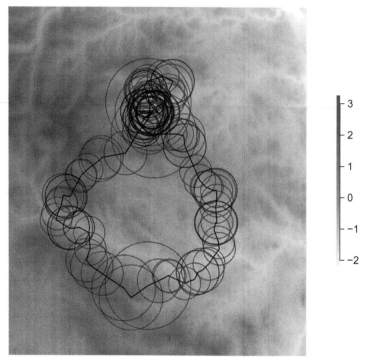

Figure 21.5 Mountain lion telemetry data (black points connected by lines) and availability regions for each step shown as circles. Map of elevation (x.1) shown in the background.

This R code creates the covariate vectors x.1, x.2, and x.3 and also creates Figure 21.5, which illustrates the availability regions for each step along the mountain lion track.

Using the data structures we created, we fit the Bayesian conditional Poisson model to the mountain lion data with the following R code:

BOX 21.6   FIT CONDITIONAL POISSON REGRESSION MODEL

```
1 X.cpr=cbind(x.1,x.2,x.3)
2 p=dim(X.cpr)[2]
3 idx=rep(1:(n-1),n.vec)
```

(Continued)

BOX 21.6 (Continued)   FIT CONDITIONAL POISSON REGRESSION MODEL

```
 4  source("cpr.mcmc.R")
 5  n.mcmc=20000
 6  set.seed(1)
 7  mcmc.out=cpr.mcmc(X=X.cpr,step.idx=idx,
 8    beta.mn=rep(0,p),beta.var=10000,
 9    beta.tune=0.05,n.mcmc=n.mcmc)
10
11  layout(matrix(1:3,1,3,byrow=TRUE))
12  hist(mcmc.out$beta.save[1,],prob=TRUE,col=8,
13    breaks=40,main="a",xlab=bquote(beta[1]))
14  curve(dnorm(x,0,10),lwd=2,add=TRUE)
15  hist(mcmc.out$beta.save[2,],prob=TRUE,col=8,
16    breaks=40,main="b",xlab=bquote(beta[2]))
17  curve(dnorm(x,0,10),lwd=2,add=TRUE)
18  hist(mcmc.out$beta.save[3,],prob=TRUE,col=8,
19    breaks=40,main="c",xlab=bquote(beta[3]))
20  curve(dnorm(x,0,10),lwd=2,add=TRUE)
21
22  apply(mcmc.out$beta.save,1,mean)
23  apply(mcmc.out$beta.save,1,quantile,
24    c(0.025,0.975))
```

This R code results in the marginal posterior distributions shown in Figure 21.6. Based on our analysis, Figure 21.6c indicates that solar exposure appears to be negatively correlated with mountain lion space use (conditioned on limited availability due to movement; $E(\beta_3|\mathbf{y}) = -0.51$ with 95% credible interval $(-0.81, -0.24)$, but we do not have enough evidence to conclude the other covariates (i.e., elevation and slope) relate to space use of this mountain lion because the posterior distributions for the associated selection coefficients overlap zero.

Conditional Poisson regression and its counterpart, conditional logistic regression, are helpful for fitting SSFs when the analyst is willing to assume availability regions can be characterized *a priori* or empirically using the data (as in our example). They provide an approximation to the spatio-temporal point process model that may be more computationally tractable in some cases. Conditional regression approaches are similar to the other GLMs that we present in this chapter, but they require extra analytical work to derive the correct likelihood.

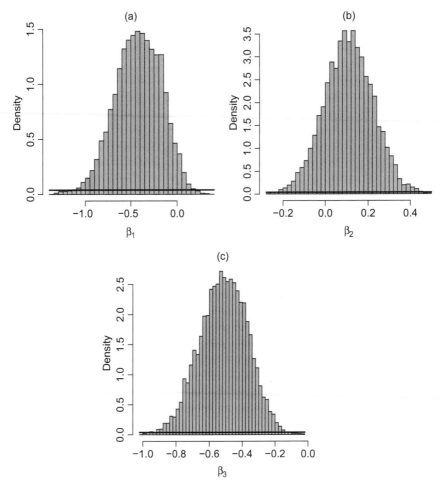

Figure 21.6 Marginal posterior distributions resulting from conditional Poisson regression to estimate the SSF for the mountain lion telemetry data.

21.4 POISSON TIME SERIES MODELS

Now that we have introduced GLMs for count data, we return to the dynamic population model that we presented in Chapter 16. Recall that, for a population measured over time, y_t, for $t = 1, \ldots, T$, we first log transformed the data and then fit a Gaussian autoregressive model with an intercept that controls the population growth rate. We arrived at that model formulation because we assumed multiplicative error. However, admittedly, the Gaussian model in Chapter 16 did not match the natural discrete support of the count data y_t.

Using the same basic concept as the GLMs that we presented earlier in this chapter, we can formulate a dynamic population model for a time series of counts. Assuming the count data are conditionally Poisson, we specify the data model as

$$y_t \sim \text{Pois}((1+r)y_{t-1}), \tag{21.19}$$

where the parameter r is the per capita Malthusian growth rate that we discussed in Chapter 16. The dynamic population model in (21.19) requires that $y_{t-1} > 0$ for $t = 2, \ldots, T$, but has only one unknown parameter r which must be larger than -1 and is also probably less than some reasonable upper bound u (based on maximum reproductive capacity). Thus, we can use a prior for r with support $(-1, u)$. For example, we could use a uniform prior on $(0, u)$. Combining the Poisson data model (21.19) and uniform prior on r results in the posterior

$$[r|\mathbf{y}] \propto \left(\prod_{t=2}^{T} [y_t|y_{t-1}, r] \right) [r], \tag{21.20}$$

which we can sample from using MCMC with M-H updates. We used a normal random walk proposal with rejection sampling for r such that $r^{(*)} \sim \text{N}(r^{(k-1)}, \sigma_{\text{tune}}^2)$. When $-1 < r^{(*)} < u$, we calculate the M-H ratio as

$$mh = \frac{\prod_{t=2}^{T} [y_t|y_{t-1}, r^{(*)}]}{\prod_{t=2}^{T} [y_t|y_{t-1}, r^{(k-1)}]}, \tag{21.21}$$

and proceed with the M-H update procedure of setting $r^{(k)} = r^{(*)}$ with probability $\min(mh, 1)$, and let $r^{(k)} = r^{(k-1)}$ otherwise.

We wrote the MCMC algorithm called pois.mc.mcmc.R in R to fit the dynamic Poisson population model described to count time series data:

BOX 21.7   MCMC ALGORITHM FOR DYNAMIC POISSON POPULATION MODEL

```
1 pois.mg.mcmc <- function(y,u,r.tune,n.mcmc){
2
3 ###
4 ###   Preliminary Variables
5 ###
6
7 n.burn=round(n.mcmc/10)
8 y=as.vector(y)
9 T=length(y)
10
11 r.save=rep(0,n.mcmc)
12 Davg=0
```

(Continued)

BOX 21.7 (Continued)   MCMC ALGORITHM FOR DYNAMIC POISSON POPULATION MODEL

```
13 ###
14 ###   Starting Values
15 ###
16
17 r=0
18
19 ###
20 ###   MCMC Loop
21 ###
22
23 for(k in 1:n.mcmc){
24
25   ###
26   ### Sample r
27   ###
28
29   r.star=rnorm(1,r,r.tune)
30   if((r.star > -1) & (r.star < u)){
31    mh1=sum(dpois(y[-1],(1+r.star)*y[-T],
32     log=TRUE))
33    mh2=sum(dpois(y[-1],(1+r)*y[-T],log=TRUE))
34    mhratio=exp(mh1-mh2)
35
36    if(mhratio > runif(1)){
37      r=r.star
38    }
39   }
40
41   ###
42   ### DIC Calculations
43   ###
44
45   Davg=Davg-2*(sum(dpois(y[-1],(1+r)*y[-T],
46     log=TRUE)))/(n.mcmc-n.burn)
47
48   ###
49   ### Save Samples
50   ###
```

(Continued)

BOX 21.7 (Continued)   MCMC ALGORITHM FOR DYNAMIC
POISSON POPULATION MODEL

```
51   r.save[k]=r
52
53 }
54
55 ###
56 ###   Calculate DIC
57 ###
58
59 r.mean=mean(r.save[-(1:n.burn)])
60 Dhat=-2*(sum(dpois(y[-1],(1+r.mean)*y[-T],
61    log=TRUE)))
62 pD=Davg-Dhat
63 DIC=2*Davg-Dhat
64
65 ###
66 ###   Write output
67 ###
68
69 list(y=y,n.mcmc=n.mcmc,r.save=r.save,DIC=DIC)
70
71 }
```

To demonstrate this MCMC algorithm, we fit the dynamic Poisson population model to the (untransformed) scaup data that we originally analyzed in Chapter 16. We used hyperparameter $u = 2$, tuning parameter $\sigma^2_{\text{tune}} = 0.05^2$, and $K = 20000$ MCMC iterations in the following R code to fit the model to the set of scaup population counts from 1980 to 2009 in the prairie pothole region (stratum 46) of the North American Waterfowl Breeding Population and Habitat Survey:

BOX 21.8   FIT DYNAMIC POISSON POPULATION MODEL

```
1 y=c(13,14,40,84,113,53,71,42,80,57,45,20,47,24,
2    49,199,173,185,133,204,308,368,185,234,618,
3    177,353,193,206,152)
4 years=1980:2009
```

(*Continued*)

BOX 21.8 (Continued)   FIT DYNAMIC POISSON POPULATION MODEL

```
5  source("pois.mg.mcmc.R")
6  n.mcmc=20000
7  set.seed(1)
8  mcmc.out=pois.mg.mcmc(y=y,u=2,r.tune=.05,
9      n.mcmc=n.mcmc)
10
11 quantile(mcmc.out$r.save,c(0.025,0.975))
12 mean(mcmc.out$r.save)
```

This R code calculates the posterior mean for r as $E(r|\mathbf{y}) \approx 0.033$ and the 95% credible interval as $(0.003, 0.063)$, which indicates a growing scaup population in stratum 46 because the credible interval does not overlap zero (if the credible interval for r included zero, it would indicate a stable population).

We can obtain a sample from the PPD using the same technique that we described in Chapter 16. The PPD for an observation $\tilde{y}_t$ in our dynamic Poisson population model is

$$[\tilde{y}_t|\mathbf{y}] = \int [\tilde{y}_t|\mathbf{y}, r][r|\mathbf{y}]dr, \qquad (21.22)$$

and we can obtain MCMC realizations for "new data" $\tilde{y}_t$ by sampling from the appropriate full-conditional at each time step $t = 1, \dots, T, \dots$ for which we want predictions. Starting with $t = 1$, recall from Chapter 16 that the full-conditional distribution for $\tilde{y}_1$ is

$$[\tilde{y}_1|\cdot] \propto [y_2|\tilde{y}_1, r], \qquad (21.23)$$

which is proportional to a Poisson PMF with mean $(1 + r)\tilde{y}_1$ and evaluated at y_2. This expression does not result in a Poisson full-conditional distribution for $\tilde{y}_1$. Thus, we must use M-H updates to sample $\tilde{y}_1^{(k)}$ in a MCMC algorithm (or a secondary algorithm that uses the output from a main model fitting MCMC algorithm). As a proposal we use a Poisson distribution with mean equal to the previous $\tilde{y}_1^{(k-1)}$ such that $\tilde{y}_1^{(*)} \sim \text{Pois}(\tilde{y}_1^{(k-1)})$. This distribution will always propose values for $\tilde{y}_1$ that have the correct support (i.e., non-negative integers) and requires no tuning. Based on the full-conditional and proposal distributions for $\tilde{y}_1$, we use the following M-H ratio in a MCMC algorithm

$$mh = \frac{[y_2|\tilde{y}_1^{(*)}, r^{(k)}][\tilde{y}_1^{(k-1)}|\tilde{y}_1^{(*)}]}{[y_2|\tilde{y}_1^{(k-1)}, r^{(k)}][\tilde{y}_1^{(*)}|\tilde{y}_1^{(k-1)}]}, \qquad (21.24)$$

where y indicates observed data and $\tilde{y}$ indicates "new data," either proposed or accepted in the Markov chain.

The procedure for obtaining posterior predictive realizations for $\tilde{y}_t$ for $1 < t < T$ is similar except that we have an extra component in the full-conditional distribution for $\tilde{y}_t$

$$[\tilde{y}_t|\cdot] \propto [y_{t+1}|\tilde{y}_t, r][\tilde{y}_t|y_{t-1}, r]. \tag{21.25}$$

Again, we use M-H updates to sample $\tilde{y}_t^{(k)}$ in a MCMC algorithm because the full-conditional distribution for $\tilde{y}_t$ is not analytically tractable. For these updates, we also use a Poisson proposal such that $\tilde{y}_t^{(*)} \sim \text{Pois}(\tilde{y}_t^{(k-1)})$ and the resulting M-H ratio is

$$mh = \frac{\left[y_{t+1}|\tilde{y}_t^{(*)}, r^{(k)}\right]\left[\tilde{y}_t^{(*)}|y_{t-1}, r^{(k)}\right]\left[\tilde{y}_t^{(k-1)}|\tilde{y}_t^{(*)}\right]}{\left[y_{t+1}|\tilde{y}_t^{(k-1)}, r^{(k)}\right]\left[\tilde{y}_t^{(k-1)}|y_{t-1}, r^{(k)}\right]\left[\tilde{y}_t^{(*)}|\tilde{y}_t^{(k-1)}\right]}, \tag{21.26}$$

for $t = 2, \ldots, T-1$.

For the last time in the data set, we also obtain predictive realizations using the predictive full-conditional distribution, which for y_T is

$$[\tilde{y}_T|\cdot] \propto [\tilde{y}_T|y_{T-1}, r]. \tag{21.27}$$

Notice that the predictive full-conditional distribution for $\tilde{y}_T$ is proportional to a Poisson with mean $(1 + r)y_{T-1}$ and evaluated at $\tilde{y}_T$, which is tractable. Thus, to obtain posterior predictive realizations for $\tilde{y}_T$, we sample $\tilde{y}_T^{(k)} \sim \text{Pois}((1+r)y_{T-1})$ directly at each iteration of a MCMC algorithm for $k = 1, \ldots, K$.

Forecasting with the dynamic Poisson population model is also straightforward. To obtain posterior forecast realizations, we use composition sampling to simulate from the predictive full-conditional distribution sequentially, which depends on the most recent predictive realization. For example, we sample the one-step ahead realization $\tilde{y}_{T+1}^{(k)} \sim \text{Pois}((1+r)y_T)$, where y_T is the last observation in the data set, but we sample the two-step ahead predictive realization by $\tilde{y}_{T+2}^{(k)} \sim \text{Pois}((1+r)\tilde{y}_{T+1}^{(k)})$ and the three-step ahead predictive realization similarly, and so on.

We wrote a secondary algorithm in R that uses the output from this model fitting algorithm to sample from the posterior predictive and forecast distributions for three time steps ahead:[4]

BOX 21.9   OBTAIN POSTERIOR PREDICTIONS

```
1 y.pred=y
2 y.pred.save=matrix(0,T+3,n.mcmc)
3
4 for(k in 1:n.mcmc){
```

(*Continued*)

---

[4] Recall that obtaining predictions and forecasts does not affect the fit of the model to data and we can do it in a separate algorithm or jointly with the main model fitting MCMC algorithm.

BOX 21.9 (Continued)   OBTAIN POSTERIOR PREDICTIONS

```
 5  ###
 6  ### Sample y.pred[1]
 7  ###
 8
 9  r=mcmc.out$r.save[k]
10
11  y1.star=rpois(1,y.pred[1])
12  mh1=dpois(y[2],(1+r)*y1.star,log=TRUE)+
13    dpois(y.pred[1],y1.star,log=TRUE)
14  mh2=dpois(y[2],(1+r)*y.pred[1],log=TRUE)+
15    dpois(y1.star,y.pred[1],log=TRUE)
16  mh=exp(mh1-mh2)
17  if(mh > runif(1)){
18    y.pred[1]=y1.star
19  }
20
21  ###
22  ### Sample y.pred[t]
23  ###
24
25  for(t in 2:(T-1)){
26    yt.star=rpois(1,y.pred[t])
27    mh1=dpois(y[t+1],(1+r)*yt.star,log=TRUE)+
28      dpois(yt.star,(1+r)*y[t-1],log=TRUE)+
29      dpois(y.pred[t],yt.star,log=TRUE)
30    mh2=dpois(y[t+1],(1+r)*y.pred[t],log=TRUE)+
31      dpois(y.pred[t],(1+r)*y[t-1],log=TRUE)
32      +dpois(yt.star,y.pred[t],log=TRUE)
33    mh=exp(mh1-mh2)
34    if(mh > runif(1)){
35      y.pred[t]=yt.star
36    }
37  }
38
39  ###
40  ### Sample y.pred[T]
41  ###
42
43  y.pred[T]=rpois(1,(1+r)*y[T-1])
```

(Continued)

BOX 21.9 (Continued)   OBTAIN POSTERIOR PREDICTIONS

```
44  ###
45  ### Sample y.pred[T+1], y.pred[T+2],y.pred[T+3]
46  ###
47
48  y.pred[T+1]=rpois(1,(1+r)*y[T])
49  y.pred[T+2]=rpois(1,(1+r)*y.pred[T+1])
50  y.pred[T+3]=rpois(1,(1+r)*y.pred[T+2])
51
52  ###
53  ### SOave Predictions
54  ###
55
56  y.pred.save[,k]=y.pred
57
58 }
59
60 y.m=apply(y.pred.save,1,mean)
61 y.u=apply(y.pred.save,1,quantile,.975)
62 y.l=apply(y.pred.save,1,quantile,.025)
63
64 years.pred=c(years,max(years)+1:3)
65
66 matplot(years.pred,cbind(y.l,y.u),type="n",
67     ylab="y",xlab="year")
68 polygon(c(years.pred,rev(years.pred)),
69     c(y.u,rev(y.l)),col=rgb(0,0,0,.2),border=NA)
70 lines(years.pred,y.m,lwd=1.5)
71 points(years,y,lwd=1.5)
72 abline(v=2009,lty=2)
```

which results in the plot of predictions shown in Figure 21.7. Although the posterior mean appears to capture the general trend of increasing scaup abundance during the time period and the forecast uncertainty increases as we predict farther into the future, the overall uncertainty envelope seems to be too narrow (Figure 21.7). That is, our PPD appears to be too precise to adequately capture the spread in the scaup abundance data. The inappropriately narrow posterior predictive credible intervals indicate that our model may be underdispersed relative to the data-generating mechanisms in nature.

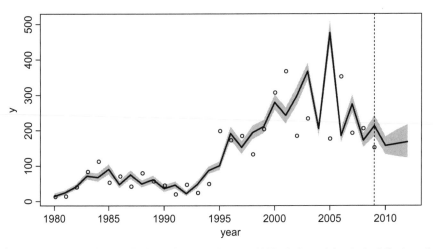

Figure 21.7 Posterior predictions, forecasts beyond 2009 (indicated by dashed line), and data (points) for observed scaup abundance from 1980 to 2012. The 95% posterior predictive credible interval is show in gray and the posterior mean is shown as a solid line.

21.5 MODEL CHECKING

The potential issue pertaining to the narrow credible intervals for our predictive distribution in the previous section provides some indication that our model may be underdispersed relative to the true data generating process. This issue may be a goodness-of-fit problem due to the inadequacy of our model assumptions. A variety of approaches exist to formally check the goodness-of-fit of Bayesian statistical models (Hobbs and Hooten 2015; Conn et al. 2018). In what follows, we describe a simple model checking procedure, using Bayesian p-values, that we can use to assess the ability of the dynamic Poisson model to properly account for dispersion in the data.

Recall the conventional definition of a p-value as the "probability of observing a more extreme statistic." Thus, in the Bayesian setting, we can think of a p-value as $P(f(\tilde{\mathbf{y}}) \geq f(\mathbf{y})|\mathbf{y})$, the probability of a statistic f being greater for predicted data $\tilde{\mathbf{y}}$ than the observed data $\mathbf{y}$, given our understanding of the model using the observed data. This is called a "posterior predictive p-value" and we can find it using

$$P(f(\tilde{\mathbf{y}}) \geq f(\mathbf{y})|\mathbf{y}) = \int \int 1_{\{f(\tilde{\mathbf{y}}) \geq f(\mathbf{y})\}} [\tilde{\mathbf{y}}|\boldsymbol{\theta}, \mathbf{y}][\boldsymbol{\theta}|\mathbf{y}] d\tilde{\mathbf{y}} d\boldsymbol{\theta}, \qquad (21.28)$$

where $\boldsymbol{\theta}$ is a generic vector of model parameters. We can approximate this posterior predictive probability using MCMC and our posterior predictive realizations by

$$P(f(\tilde{\mathbf{y}}) \geq f(\mathbf{y})|\mathbf{y}) = \frac{\sum_{k=1}^{K} 1_{\{f(\tilde{\mathbf{y}}^{(k)}) \geq f(\mathbf{y})\}}}{K}, \qquad (21.29)$$

where $\tilde{\mathbf{y}}^{(k)}$ are our posterior predictive realizations and the function $f(\cdot)$ is a statistic (i.e., function of data) that tells us something about the aspect of a model we seek to check.

For example, in our scaup population model from the previous section, we might choose to investigate dispersion using the mean squared error (MSE) statistic

$$f(\mathbf{y}) = \frac{\sum_{t=2}^{T} \left(y_t - E(y_t|y_{t-1}, r) \right)^2}{T-1}. \tag{21.30}$$

To obtain the posterior predictive p-value, we calculate the statistic based on the observed data $f(\mathbf{y})$ and the predictive realizations $f(\tilde{\mathbf{y}}^{(k)})$ and then compute the p-value (21.29). If the p-value is very small (close to 0) or large (close to 1), we would conclude lack of fit for the model because it would imply that the predicted data and the observed data are very different in terms of the statistic.

We calculated the posterior predictive p-value based on MSE to check the fit of the dynamic Poisson population model to the scaup abundance data using posterior predictive realizations from the previous section using the following R code:

BOX 21.10 CALCULATE POSTERIOR PREDICTIVE p-VALUE

```
1 > ind.TF=rep(0,n.mcmc)
2 > for(k in 1:n.mcmc){
3 +    mse.pred=mean((y.pred.save[2:T,k]-
4        (1+mcmc.out$r.save[k])*y[-T])^2)
5 +    mse.y=mean((y[-1]-(1+mcmc.out$r.save[k])*
6        y[-T])^2)
7 +    ind.TF[k]=mse.pred>mse.y
8 + }
9 > p.val=mean(ind.TF)
10 > p.val
11 [1] 0
```

The posterior predictive p-value for the Poisson model fit to the scaup data was approximately zero, which implies that the MSE is nearly always larger for the data than what the model predicts. Thus, we can conclude the model that we fit in the previous two sections is underdispersed for the scaup abundance data. To remedy the underdispersion issue, we could specify a different data model with the same mean structure, but one that allows the conditional variance to be larger than the mean (e.g., negative binomial instead of Poisson).

21.6   NEGATIVE BINOMIAL REGRESSION

As we alluded to in the previous section, we can generalize the Poisson data model to allow for overdispersion. There are several ways to introduce additional dispersion into statistical models for count data, but the simplest is to use a data model that allows the variance to be larger than the mean. For example, the negative binomial is a distribution that allows for more dispersion than the Poisson distribution at the same mean.

Recall that we often consider the negative binomial distribution as a good model for count data that arise as the number of trials until a certain number of failures occur in a sequence of binary experiments. However, we can reparameterize the negative binomial distribution in terms of its mean μ and an overdispersion parameter (often called N for convention). Using this parameterization, we model the count data y_i as

$$y_i \sim \text{NB}(\mu_i, N) \equiv \frac{\Gamma(y_i + N)}{\Gamma(N) y_i!} \left(\frac{N}{N + \mu_i} \right)^N \left(1 - \frac{N}{N + \mu_i} \right)^{y_i}, \qquad (21.31)$$

for $i = 1, \ldots, n$, $\mu_i > 0$, and $N > 0$. As $N \to \infty$, the negative binomial distribution converges to the Poisson with mean and variance μ_i. However, as N approaches zero, the variance of the negative binomial random variable increases.[5]

We can link the mean parameter μ_i to a function of other parameters, data, or covariates. For example, if we let $\log(\mu_i) = \mathbf{x}_i' \boldsymbol{\beta}$, a regression model results where instead of a Poisson distribution as we presented in the previous sections, we use a negative binomial. The posterior distribution for the negative binomial regression model is

$$[\boldsymbol{\beta}, \log(N) | \mathbf{y}] = \left(\prod_{i=1}^{n} [y_i | \boldsymbol{\beta}, N] \right) [\boldsymbol{\beta}] [\log(N)], \qquad (21.32)$$

if we specify a prior for $\log(N)$ (rather than N directly). To fit the negative binomial regression model, we need two full-conditional distributions, one for $\boldsymbol{\beta}$ and one for $\log(N)$. The full-conditional distribution for $\boldsymbol{\beta}$ is

$$[\boldsymbol{\beta} | \cdot] \propto \left(\prod_{i=1}^{n} [y_i | \boldsymbol{\beta}, N] \right) [\boldsymbol{\beta}], \qquad (21.33)$$

which is not analytically tractable. Thus, we use a M-H update to sample $\boldsymbol{\beta}$ based on a multivariate normal prior and the M-H ratio

$$mh = \frac{\left(\prod_{i=1}^{n} \left[y_i | \boldsymbol{\beta}^{(*)}, N^{(k-1)} \right] \right) \left[\boldsymbol{\beta}^{(*)} \right]}{\left(\prod_{i=1}^{n} \left[y_i | \boldsymbol{\beta}^{(k-1)}, N^{(k-1)} \right] \right) \left[\boldsymbol{\beta}^{(k-1)} \right]}, \qquad (21.34)$$

---

[5] Another way to consider the negative binomial distribution is as a Poisson with intensity that includes a multiplicative random effect that is gamma distributed. In this context, the random effect provides the overdispersion.

where we also use a multivariate normal random walk proposal for $\boldsymbol{\beta}^{(*)} \sim$ $N(\boldsymbol{\beta}^{(k-1)}, \sigma_{\text{tune}}^2 \mathbf{I})$.

The full-conditional distribution for $\log(N)$ is

$$[\log(N)|\cdot] \propto \left(\prod_{i=1}^{n} [y_i|\boldsymbol{\beta}, N] \right) [\log(N)], \tag{21.35}$$

which leads to the M-H ratio associated with the update for $\log(N)$:

$$mh = \frac{\left(\prod_{i=1}^{n} \left[y_i|\boldsymbol{\beta}^{(k)}, N^{(*)} \right] \right) \left[\log(N)^{(*)} \right]}{\left(\prod_{i=1}^{n} \left[y_i|\boldsymbol{\beta}^{(k)}, N^{(k-1)} \right] \right) \left[\log(N)^{(k-1)} \right]}, \tag{21.36}$$

if we use the normal prior $\log(N) \sim N(\mu_N, \sigma_N^2)$ and a normal random walk proposal $\log(N)^{(*)} \sim N(\log(N)^{(k-1)}, \sigma_{N,\text{tune}}^2)$.

We also can use the negative binomial data model instead of the Poisson for the dynamic population model that we described in the previous sections. To do so, we specify the mean function in

$$y_t \sim \text{NB}(\mu_t, N) \tag{21.37}$$

as $\mu_t = (1+r)y_{t-1}$, where r is the per capita growth rate as before and has prior $r \sim \text{Unif}(-1, u)$. The posterior distribution for this dynamic population model is

$$[r, \log(N)|\mathbf{y}] = \left(\prod_{t=2}^{T} [y_t|y_{t-1}, r, N] \right) [r] [\log(N)], \tag{21.38}$$

which leads to the full-conditional distributions

$$[\log(N)|\cdot] \propto \left(\prod_{t=1}^{T} [y_t|y_{t-1}, r, N] \right) [\log(N)], \tag{21.39}$$

and

$$[r|\cdot] \propto \left(\prod_{t=1}^{T} [y_t|y_{t-1}, r, N] \right) [r]. \tag{21.40}$$

The associated M-H ratios are

$$mh_{\log(N)} = \frac{\left(\prod_{t=1}^{T} \left[y_t|y_{t-1}, r^{(k)}, N^{(*)} \right] \right) \left[\log(N)^{(*)} \right]}{\left(\prod_{t=1}^{T} \left[y_t|y_{t-1}, r^{(k)}, N^{(k-1)} \right] \right) \left[\log(N)^{(k-1)} \right]}, \tag{21.41}$$

and

$$mh_r = \frac{\left(\prod_{t=1}^{T} \left[y_t|y_{t-1}, r^{(*)}, N^{(k-1)} \right] \right) \left[r^{(*)} \right]}{\left(\prod_{t=1}^{T} \left[y_t|y_{t-1}, r^{(k-1)}, N^{(k-1)} \right] \right) \left[r^{(k-1)} \right]}. \tag{21.42}$$

We constructed a MCMC algorithm in R called nb.mg.mcmc.R based on the M-H ratios for $\log(N)$ and r:

BOX 21.11   MCMC ALGORITHM FOR DYNAMIC NEGATIVE BINOMIAL POPULATION MODEL

```
nb.mg.mcmc <- function(y,u,r.tune,logN.tune,
    n.mcmc){

###
###    Preliminary Variables
###

n.burn=round(n.mcmc/10)
y=as.vector(y)
T=length(y)

mse.y=rep(0,n.mcmc)
mse.ypred=rep(0,n.mcmc)
msediffsave=rep(0,n.mcmc)
r.save=rep(0,n.mcmc)
N.save=rep(0,n.mcmc)
y.pred.save=matrix(0,T+3,n.mcmc)
msesave=rep(0,n.mcmc)
Davg=0

y.pred=y

###
###    Starting Values and Priors
###

r=0
logN=log(10)

logN.mn=log(10)
logN.sd=log(10)

###
###    MCMC Loop
###
```

(Continued)

BOX 21.11 (Continued)   MCMC ALGORITHM FOR DYNAMIC NEGATIVE BINOMIAL POPULATION MODEL

```
36 for(k in 1:n.mcmc){
37
38   ###
39   ### Sample r
40   ###
41
42   r.star=rnorm(1,r,r.tune)
43   if((r.star > -1) & (r.star < u)){
44     mh1=sum(dnbinom(y[-1],mu=(1+r.star)*y[-T],
45       size=exp(logN),log=TRUE))
46     mh2=sum(dnbinom(y[-1],mu=(1+r)*y[-T],
47       size=exp(logN),log=TRUE))
48     mhratio=exp(mh1-mh2)
49
50     if(mhratio > runif(1)){
51       r=r.star
52     }
53   }
54
55   ###
56   ### Sample logN
57   ###
58
59   logN.star=rnorm(1,logN,logN.tune)
60
61   mh1=sum(dnbinom(y[-1],mu=(1+r)*y[-T],
62     size=exp(logN.star),log=TRUE))+dnorm(logN.
63     star,logN.mn,logN.sd,log=TRUE)
64   mh2=sum(dnbinom(y[-1],mu=(1+r)*y[-T],
65     size=exp(logN),log=TRUE))+dnorm(logN,
66       logN.mn,logN.sd,log=TRUE)
67   mhratio=exp(mh1-mh2)
68
69   if(mhratio > runif(1)){
70     logN=logN.star
71   }
```

(Continued)

BOX 21.11 (Continued)   MCMC ALGORITHM FOR DYNAMIC NEGATIVE BINOMIAL POPULATION MODEL

```
72   ###
73   ### DIC Calculations
74   ###
75
76   if(k > n.burn){
77     Davg=Davg-2*(sum(dnbinom(y[-1],mu=(1+r)*
78         y[-T],size=exp(logN),log=TRUE)))/
79         (n.mcmc-n.burn)
80   }
81
82   ###
83   ### Sample y[1]
84   ###
85
86   y1.star=rpois(1,y.pred[1])
87   mh1=dnbinom(y[2],mu=(1+r)*y1.star,size=
88     exp(logN),log=TRUE)+dpois(y.pred[1],y1.star,
89     log=TRUE)
90   mh2=dnbinom(y[2],mu=(1+r)*y.pred[1],
91     size=exp(logN),log=TRUE)+dpois(y1.star,
92     y.pred[1],log=TRUE)
93   mh=exp(mh1-mh2)
94   if(mh > runif(1)){
95     y.pred[1]=y1.star
96   }
97
98   ###
99   ### Sample y[t]
100  ###
101
102  for(t in 2:(T-1)){
103    yt.star=rpois(1,y.pred[t])
104    mh1=dnbinom(y[t+1],mu=(1+r)*yt.star,
105      size=exp(logN),log=TRUE)+dnbinom(yt.star,
106          mu=(1+r)*y[t-1],size=exp(logN),
107           log=TRUE)+dpois(y.pred[t],
108        yt.star,log=TRUE)
```

(Continued)

BOX 21.11 (Continued)   MCMC ALGORITHM FOR DYNAMIC
NEGATIVE BINOMIAL POPULATION MODEL

```
109    mh2=dnbinom(y[t+1],mu=(1+r)*y.
110       pred[t],size=exp(logN),log=TRUE)+dnbinom
111       (y.pred[t],mu=(1+r)*y[t-1],
112         size=exp(logN),log=TRUE)+dpois(yt.star,
113         y.pred[t],log=TRUE)
114    mh=exp(mh1-mh2)
115    if(mh > runif(1)){
116       y.pred[t]=yt.star
117    }
118  }
119
120  ###
121  ### Sample y[T], y[T+1], y[T+2], y[T+3]
122  ###
123
124  y.pred[T]=rnbinom(1,mu=(1+r)*y[T-1],
125     size=exp(logN))
126  y.pred[T+1]=rnbinom(1,mu=(1+r)*y[T],
127     size=exp(logN))
128  y.pred[T+2]=rnbinom(1,mu=(1+r)*y.pred[T+1],
129     size=exp(logN))
130  y.pred[T+3]=rnbinom(1,mu=(1+r)*y.pred[T+2],
131     size=exp(logN))
132
133  ###
134  ### Calculate MSE
135  ###
136
137  mse.y[k]=mean((y[-1]-(1+r)*y[-T])^2)
138  mse.ypred[k]=mean((y.pred[2:T]-(1+r)*y[-T])^2)
139  msediffsave[k]=mse.ypred[k]-mse.y[k]
140  msesave[k]=mean((y.pred[1:T]-y)^2)
141
142  ###
143  ### Save Samples
144  ###
```

(Continued)

**BOX 21.11 (Continued)   MCMC ALGORITHM FOR DYNAMIC
NEGATIVE BINOMIAL POPULATION MODEL**

```
145  y.pred.save[,k]=y.pred
146  r.save[k]=r
147  N.save[k]=exp(logN)
148
149  }
150
151  ###
152  ###   Calculate DIC
153  ###
154
155  r.mean=mean(r.save[-(1:n.burn)])
156  N.mean=mean(N.save[-(1:n.burn)])
157  Dhat=-2*(sum(dnbinom(y[-1],mu=(1+r.mean)*y[-T],
158       size=N.mean,log=TRUE)))
159  pD=Davg-Dhat
160  DIC=2*Davg-Dhat
161
162  ###
163  ### Calculate P-value based on MSE
164  ###
165
166  p.value=mean((mse.ypred>mse.y)[n.burn:n.mcmc])
167
168  ###
169  ###   Write output
170  ###
171
172  list(y=y,n.mcmc=n.mcmc,r.save=r.save,p.value=
173       p.value,N.save=N.save,y.pred.save=y.
174       pred.save)
175
176  }
```

Notice that we included the posterior predictive sampling in this MCMC algorithm as well as the calculation of the posterior predictive p-value, following the same procedure described in the previous section. Also, note that we exponentiated $\log(N)^{(k)}$ to get a sample for $N^{(k)}$ when saving the results of the MCMC algorithm.

We fit the dynamic negative binomial population model to the observed scaup abundance data using the previous MCMC algorithm and hyperparameters $u = 2$, $\mu_N = \log(10)$, and $\sigma_N^2 = \log(10)^2$, using the following R code:

BOX 21.12   FIT DYNAMIC NEGATIVE BINOMIAL POPULATION MODEL

```
 1 source("nb.mg.mcmc.R")
 2 n.mcmc=50000
 3 set.seed(1)
 4 mcmc.out=nb.mg.mcmc(y=y,u=2,r.tune=0.05,
 5     logN.tune=.1,n.mcmc=n.mcmc)
 6 mcmc.out$p.value
 7
 8 years.pred=c(years,max(years)+1:3)
 9 y.m=apply(mcmc.out$y.pred.save,1,mean)
10 y.u=apply(mcmc.out$y.pred.save,1,quantile,.975)
11 y.l=apply(mcmc.out$y.pred.save,1,quantile,.025)
12
13 matplot(years.pred,cbind(y.l,y.u),type="n",
14     ylab="y",xlab="year")
15 polygon(c(years.pred,rev(years.pred)),
16     c(y.u,rev(y.l)),col=rgb(0,0,0,.2),border=NA)
17 lines(years.pred,y.m,lwd=1.5)
18 points(years,y,lwd=1.5)
19 abline(v=2009,lty=2)
```

which resulted in a posterior predictive p-value of 0.10 that is still relatively small, but much larger than the resulting p-value from the Poisson version of the model (which was ≈ 0). Thus, there is not strong evidence for lack of fit with this dynamic negative binomial population model. Figure 21.8 shows the marginal posterior distributions for r and N. The 95% credible interval for r was $(0.067, 0.723)$, which agrees with the results of the Poisson model in that there is evidence that the population is growing (Figure 21.8a) but is slightly larger at both ends of the interval. The marginal posterior distribution for N in Figure 21.8b suggests that there is evidence of overdispersion beyond what the Poisson data model can provide because N is relatively small.

Figure 21.9 shows the posterior predictions and forecasts (beyond 2009) resulting from fitting the negative binomial population model to the scaup abundance data. Notice that the predictive credible intervals now cover the majority of observed data (Figure 21.9) as compared to the resulting predictive credible intervals from the Poisson version of the model (21.7), which were much too narrow. This figure provides graphical evidence supporting the posterior predictive p-value which suggests no remaining overdispersion beyond what the negative binomial distribution can handle.

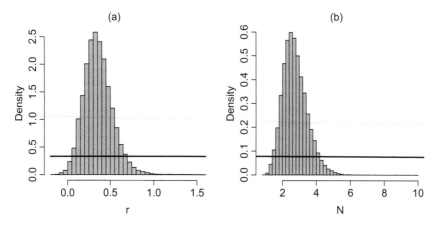

Figure 21.8 Marginal posterior histograms for r and N based on the dynamic negative binomial population model fit to the scaup abundance data. Prior PDFs shown as black lines.

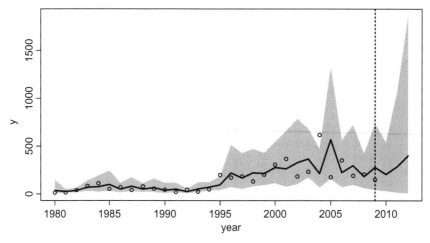

Figure 21.9 Posterior predictions, forecasts beyond 2009 (indicated by dashed line), and data (points) for observed scaup abundance from 1980 to 2012 based on the dynamic negative binomial population model. The 95% posterior predictive credible interval is show in gray and the posterior mean is shown as a solid line.

21.7 QUANTILE MODELS FOR COUNT DATA

In previous chapters, we introduced the concept of quantile regression for continuous data and binary data. Quantile regression provides a way for the analyst to study the relationship between response variables and predictor variables in other parts of the distribution besides the mean. For example, we can use median regression as a robust alternative to multiple linear regression (which is focused on the mean) by setting the quantile at $\tau = 0.5$. By using larger or smaller quantiles, we can examine the relationship between response variables and predictor variables in the tails of the distribution as well and this can provide insights about possible changes in these relationships at different quantiles.

For continuous response variables, we also showed that we could actually estimate the "quantile" (τ) by treating it as an unknown parameter in the model—a parameter that influences the skewness of the data model. In our example from Chapter 18, we found that the optimal quantile parameter for modeling white-tailed deer body mass based on age was approximately 0.25, which implied a right-skewed data model was most appropriate to model that relationship.

In the case of binary data, we showed that we can derive an appropriate link function that induces a quantile relationship by considering the auxiliary variable approach to specifying binary regression models. With this link function in hand, it is trivial to modify the Bayesian logistic regression MCMC algorithm to fit quantile regression models to binary data. We fit a sequence of binary quantile regression models to examine the relationship between *Desmodium glutinosum* and several environmental covariates and found that larger quantiles lead to better predictive models based on the DIC model selection score.

Count data are neither continuous nor binary, so we cannot use either of the previous approaches directly to perform quantile regression when the response variables are positive integers. However, Lee and Neocleous (2010) proposed an approach to Bayesian quantile regression that is quite clever and allows us to use the same algorithm that we developed for the continuous case when the counts are greater than zero. Lee and Neocleous (2010) transformed the original count data in a way that retains the ordinal structure and changes the support to the real numbers so that we can use the regular quantile regression model and MCMC algorithm with only one slight change to perform the transformation.

Specifically, for count data quantile regression, we write the original count data y_i for $i = 1, \ldots, n$ as a function of latent variables z_i and u_i as well as the quantile of interest τ, such that $y_i = e^{z_i} - u_i + \tau$ (which implies the inverse relationship $z_i = \log(y_i + u_i - \tau)$). Then, we can express the Bayesian model as

$$z_i \sim \text{ALD}(\mathbf{x}_i'\boldsymbol{\beta}, \sigma, \tau), \tag{21.43}$$

$$\boldsymbol{\beta} \sim \text{N}(\boldsymbol{\mu}_\beta, \boldsymbol{\Sigma}_\beta), \tag{21.44}$$

$$\sigma \sim \text{Unif}(0, u_\sigma), \tag{21.45}$$

where ALD is the asymmetric Laplace distribution that we introduced in Chapter 18 and the latent variables u_i arise from a known uniform distribution on (0,1). We assume the latent variables u_i are known stochastic elements in the model and sampled at every iteration of our MCMC algorithm when fitting the model, but not updated based on the data as other variables are. In fact, adding $u_i - \tau$ to the observations provides a monotonic transformation of the original data and retains the ordinal structure, thus the u_i does not affect the quantiles of the resulting distribution. The new effective response variables z_i will have continuous quantiles and an ALD can reasonably model it.

We can use the same algorithm that we developed in Chapter 18 to fit the model, except we need to recalculate the transformation $z_i^{(k)} = \log(y_i + u_i^{(k)} - \tau)$ on every MCMC iteration. We called this algorithm count.qr.mcmc.R and the R function for it is as follows:

BOX 21.13   MCMC ALGORITHM FOR COUNT DATA QUANTILE REGRESSION

```
count.qr.mcmc <- function(y,X,tau,beta.mn,
    beta.var,s.u,n.mcmc){

###
### Subroutines
###

ldasymlap <- function(y,tau,mu,sig){
  rho <- function(u){u*(tau-(u<0))}
  ld=log(tau)+log(1-tau)-log(sig)-rho((y-mu)/sig)
  ld
}

###
### Setup and Priors
###

n=dim(X)[1]
p=dim(X)[2]
n.burn=round(.1*n.mcmc)

Sig.beta.inv=solve(beta.var*diag(p))
beta.save=matrix(0,p,n.mcmc)
s2.save=rep(0,n.mcmc)

n.pred=100
y.pred.mn=rep(0,n.pred)

###
### Starting Values
###

beta=solve(t(X)%*%X)%*%t(X)%*%log(y+runif(n)-tau)
```
(*Continued*)

BOX 21.13 (Continued)   MCMC ALGORITHM FOR COUNT DATA QUANTILE REGRESSION

```
34  s.tune=.1
35  beta.tune=.1
36
37  s2=(s.u^2)/100
38  sig=sqrt(s2)
39
40  X.pred=matrix(1,n.pred,p)
41  X.pred[,2]=seq(min(X[,2]),max(X[,2]),,n.pred)
42
43  ###
44  ### MCMC Loop
45  ###
46
47  for(k in 1:n.mcmc){
48
49    ###
50    ### Sample u
51    ###
52
53    u=runif(n)
54    z=log(y+u-tau)
55
56    ###
57    ### Sample s
58    ###
59
60    s.star=rnorm(1,sig,s.tune)
61    if((s.star < s.u) & (s.star > 0)){
62      mh.1=sum(ldasymlap(z,tau,X%*%beta,s.star))
63      mh.2=sum(ldasymlap(z,tau,X%*%beta,sig))
64      mh=exp(mh.1-mh.2)
65
66      if(mh > runif(1)){
67        sig=s.star
68      }
69    }
```

(Continued)

BOX 21.13 (Continued)   MCMC ALGORITHM FOR COUNT DATA QUANTILE REGRESSION

```
70   ###
71   ### Sample beta
72   ###
73
74   beta.star=rnorm(p,beta,beta.tune)
75   mh.1=sum(ldasymlap(z,tau,X%*%beta.star,sig))
76     -0.5*t(beta.star-beta.mn)%*%Sig.beta.inv%*%
77     (beta.star-beta.mn)
78   mh.2=sum(ldasymlap(z,tau,X%*%beta,sig))-0.5*t
79     (beta-beta.mn)%*%Sig.beta.inv%*%
80     (beta-beta.mn)
81   mh=exp(mh.1-mh.2)
82
83   if(mh > runif(1)){
84     beta=beta.star
85   }
86
87   ###
88   ### Posterior Predictive Calculations
89   ###
90
91   if(k > n.burn){
92     y.pred=ceiling(tau+exp(X.pred%*%beta)-1)
93     y.pred.mn=y.pred.mn+y.pred/(n.mcmc-n.burn)
94   }
95
96   ###
97   ### Save Samples
98   ###
99
100  beta.save[,k]=beta
101  s2.save[k]=sig^2
102
103  }
104
105  ###
106  ###   Write Output
107  ###
```

(Continued)

BOX 21.13 (Continued)   MCMC ALGORITHM FOR COUNT DATA QUANTILE REGRESSION

```
108  list(beta.save=beta.save,s2.save=s2.save,y=y,
109     X=X,n.mcmc=n.mcmc,n=n,p=p,tau=tau,
110     y.pred.mn=y.pred.mn,X.pred=X.pred)
111
112  }
```

Notice that we added new lines to perform the data transformation and also computed the fitted quantiles as $\lceil e^{x'\beta} - 1 + \tau \rceil$ for the desired values of $\mathbf{x}$ (where $\lceil \cdot \rceil$ indicates the ceiling function).

We fit the count data quantile regression model to the bird species richness data that we originally analyzed using Poisson regression at the beginning of this chapter. In our analysis, we used hyperparameters $\mu_\beta = \mathbf{0}$ and $u_\sigma = 1$, and we fit the quantile regression for values of τ equal to 0.10, 0.25, 0.50, 0.75, and 0.90. We used the following R code to fit the model to the bird species richness data based on a single covariate (standardized temperature):

BOX 21.14   FIT COUNT DATA QUANTILE REGRESSION MODEL

```
1   bird.df=read.csv("birds.csv")
2   idx.outlier=(1:51)[(bird.df$spp==min(bird.df$spp)
3      | bird.df$area==max(bird.df$area))]
4   bird.df=bird.df[-idx.outlier,]
5   y=bird.df$spp
6   n=length(y)
7   X=matrix(1,n,2)
8   X[,2]=scale(bird.df$temp)
9   p=dim(X)[2]
10
11  tau.vec=c(.1,.25,.5,.75,.9)
12  n.tau=length(tau.vec)
13  out.list=vector("list",n.tau)
14  beta.mat=matrix(0,p,n.tau)
15  s.vec=rep(0,n.tau)
16  source("count.qr.mcmc.R")
17  n.mcmc=20000
18  set.seed(1)
```

(Continued)

BOX 21.14 (Continued)   FIT COUNT DATA QUANTILE
REGRESSION MODEL

```
19 for(j in 1:n.tau){
20   out.list[[j]]=count.qr.mcmc(y=y,X=X,
21     tau=tau.vec[j],beta.mn=rep(0,p),
22     beta.var=1000,s.u=1,n.mcmc=n.mcmc)
23   beta.mat[,j]=apply(out.list[[j]]$beta.save,1,
24     mean)
25   s.vec[j]=mean(sqrt(out.list[[j]]$s2.save))
26 }
27 cbind(tau.vec,t(beta.mat),s.vec)
28
29 plot(X[,2],y,xlab="temp",ylab="bird spp richness")
30 for(j in 1:n.tau){
31   lines(out.list[[j]]$X.pred[,2],out.list[[j]]$
32     y.pred.mn,lwd=1.5)
33 }
```

This R code results in the fitted quantile functions shown in Figure 21.10. Our count data quantile regression analysis for the bird species richness data indicated that the

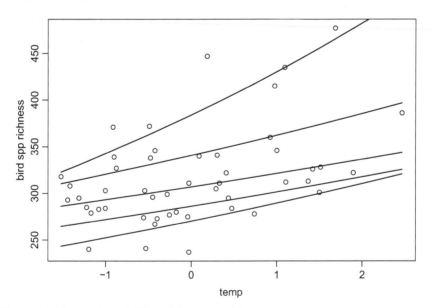

Figure 21.10 Continental USA bird species richness by state versus state temperature (points) and quantile regression lines of $E(\lceil e^{x'\beta} - 1 + \tau \rceil | \mathbf{y})$ for quantile values of τ set at 0.10, 0.25, 0.50, 0.75, and 0.90, shown as lines from bottom to top.

slope of the relationship between mean state temperature and state-level species richness is mostly similar for the lower quantiles, but seems to increase for $\tau = 0.90$ resulting in a steeper fitted quantile line (Figure 21.10). In fact the posterior mean for β_1 (the coefficient associated with the temperature covariate) was 0.11 for $\tau = 0.90$ and was 0.07, 0.05, 0.05, and 0.06 for quantiles 0.10, 0.25, 0.50, and 0.75, respectively.

We also can consider the quantile τ as an unknown skewness parameter in the model as we did in Chapter 18 and attempt to estimate it using the count data. In this case, if we let $\tau \sim \text{Unif}(0, 1)$, we can update it using M-H as we did in Chapter 18. We wrote the MCMC algorithm called count.qr.tau.mcmc.R in R, and it is as follows:

BOX 21.15   MCMC ALGORITHM FOR COUNT DATA ALD REGRESSION WITH UNKNOWN τ

```
1  count.qr.tau.mcmc <- function(y,X,tau,beta.mn,
2      beta.var,s.u,n.mcmc){
3
4  ###
5  ### Subroutines
6  ###
7
8  ldasymlap <- function(y,tau,mu,sig){
9    rho <- function(u){u*(tau-(u<0))}
10     ld=log(tau)+log(1-tau)-log(sig)-rho((y-mu)/sig)
11     ld
12  }
13
14  ###
15  ### Hyperpriors
16  ###
17
18  n=dim(X)[1]
19  p=dim(X)[2]
20  n.burn=round(.1*n.mcmc)
21
22  Sig.beta.inv=solve(beta.var*diag(p))
23  beta.save=matrix(0,p,n.mcmc)
24  s2.save=rep(0,n.mcmc)
25  tau.save=rep(0,n.mcmc)
26
27  n.pred=100
28  y.pred.mn=rep(0,n.pred)
```

(Continued)

BOX 21.15 (Continued)   MCMC ALGORITHM FOR COUNT DATA
ALD REGRESSION WITH UNKNOWN τ

```
29  ###
30  ### Starting Values
31  ###
32
33  beta=solve(t(X)%*%X)%*%t(X)%*%log(y+runif(n)-tau)
34
35  s.tune=.1
36  beta.tune=.05
37  tau.tune=.1
38
39  s2=(s.u^2)/100
40  sig=sqrt(s2)
41
42  X.pred=matrix(1,n.pred,p)
43  X.pred[,2]=seq(min(X[,2]),max(X[,2]),,n.pred)
44
45  ###
46  ### MCMC Loop
47  ###
48
49  for(k in 1:n.mcmc){
50
51    ###
52    ### Sample u
53    ###
54
55    u=runif(n)
56    z=log(y+u-tau)
57
58    ###
59    ### Sample s
60    ###
61
62    s.star=rnorm(1,sig,s.tune)
63    if((s.star < s.u) & (s.star > 0)){
64      mh.1=sum(ldasymlap(z,tau,X%*%beta,s.star))
65      mh.2=sum(ldasymlap(z,tau,X%*%beta,sig))
66      mh=exp(mh.1-mh.2)
```

(Continued)

BOX 21.15 (Continued)   MCMC ALGORITHM FOR COUNT DATA
ALD REGRESSION WITH UNKNOWN τ

```
67   if(mh > runif(1)){
68      sig=s.star
69      }
70   }
71
72   ###
73   ### Sample beta
74   ###
75
76   beta.star=rnorm(p,beta,beta.tune)
77   mh.1=sum(ldasymlap(z,tau,X%*%beta.star,sig))
78      -0.5*t(beta.star-beta.mn)%*%Sig.beta.inv%*%
79      (beta.star-beta.mn)
80   mh.2=sum(ldasymlap(z,tau,X%*%beta,sig))-0.5*t
81      (beta-beta.mn)%*%Sig.beta.inv%*%(beta-beta.mn)
82   mh=exp(mh.1-mh.2)
83
84   if(mh > runif(1)){
85      beta=beta.star
86   }
87
88   ###
89   ### Sample tau (quantile)
90   ###
91
92   tau.star=rnorm(1,tau,tau.tune)
93   if((tau.star >0) & (tau.star<1)){
94      mh.1=sum(ldasymlap(z,tau.star,X%*%beta,sig))
95      mh.2=sum(ldasymlap(z,tau,X%*%beta,sig))
96      mh=exp(mh.1-mh.2)
97
98      if(mh > runif(1)){
99         tau=tau.star
100      }
101   }
102
103   ###
104   ### Posterior Predictive Calculations
105   ###
```

(Continued)

**BOX 21.15 (Continued)   MCMC ALGORITHM FOR COUNT DATA
ALD REGRESSION WITH UNKNOWN τ**

```
106  if(k > n.burn){
107    y.pred=ceiling(tau+exp(X.pred%*%beta)-1)
108    y.pred.mn=y.pred.mn+y.pred/(n.mcmc-n.burn)
109  }
110
111  ###
112  ### Save Samples
113  ###
114
115  beta.save[,k]=beta
116  s2.save[k]=sig^2
117  tau.save[k]=tau
118
119  }
120
121  ###
122  ###  Write Output
123  ###
124
125  list(beta.save=beta.save,s2.save=s2.save,y=y,
126    X=X,n.mcmc=n.mcmc,n=n,p=p,tau.save=tau.save,
127    y.pred.mn=y.pred.mn,X.pred=X.pred,n.burn=n.
128    burn)
129
130  }
```

Using this MCMC algorithm, we fit the count data ALD regression model, treating τ as random and unknown and $K = 50000$ MCMC iterations, to the bird species richness data by state in the continental USA with the following R code:

**BOX 21.16   FIT COUNT DATA ALD REGRESSION MODEL WITH
UNKNOWN τ**

```
1  source("count.qr.tau.mcmc.R")
2  n.mcmc=50000
3  set.seed(1)
4  mcmc.out=count.qr.tau.mcmc(y=y,X=X,tau=.5,beta.mn=
5     rep(0,p),beta.var=1000,s.u=1,n.mcmc=n.mcmc)
```

(Continued)

**BOX 21.16 (Continued)   FIT COUNT DATA ALD REGRESSION
MODEL WITH UNKNOWN τ**

```
 6 keep.idx=mcmc.out$n.burn:n.mcmc
 7 layout(matrix(1:4,2,2,byrow=TRUE))
 8 hist(mcmc.out$beta.save[1,keep.idx],prob=TRUE,
 9     main="a",xlab=bquote(beta[0]))
10 curve(dnorm(x,0,sqrt(10000)),lwd=2,add=TRUE)
11 hist(mcmc.out$beta.save[2,keep.idx],prob=TRUE,
12     main="b",xlab=bquote(beta[1]))
13 curve(dnorm(x,0,sqrt(10000)),lwd=2,add=TRUE)
14 hist(sqrt(mcmc.out$s2.save[keep.idx]),prob=TRUE,
15     main="c",xlab=bquote(sigma))
16 abline(h=1/100,lwd=2)
17 hist(mcmc.out$tau.save[keep.idx],prob=TRUE,
18     xlim=c(0,1),main="d",xlab=bquote(tau))
19 abline(h=1,lwd=2)
20
21 mean(mcmc.out$tau.save[keep.idx])
22 quantile(mcmc.out$tau.save[keep.idx],
23     c(0.025,0.975))
```

The resulting marginal posterior distributions are shown in Figure 21.11. The marginal posterior distribution for τ in Figure 21.11d suggests that the quantile is likely less than 0.5 ($E(\tau|\mathbf{y}) = 0.37$ and 95% credible interval $(0.22, 0.52)$). This example illustrates the second approach to skewness regression: To fit models that best characterize the data. This inferential approach is in contrast to the first approach that we presented in this section where the goal was to examine relationships in the data from an exploratory data analysis perspective.

21.8   BINOMIAL MODELS

In this chapter, we presented a variety of models for count data that have support on the non-negative integers. However, some important ecological data sets consist of counts with support that has a finite upper bound. For example, consider a simple experiment involving larval fish where N individuals were placed in each of n tanks for 24 hours. After 24 hours, the number of surviving fish (y_i) are counted in each tank. The goal of the experiment is to learn about the survival probability (ϕ) of larval fish. If the 24 hour survival of each fish is a binary variable (1 for survival, 0 otherwise) and we assume that the survival of a fish is independent of the others, then the sum (y_i; number of surviving fish) of these binary variables is binomial with upper bound N and probability ϕ.

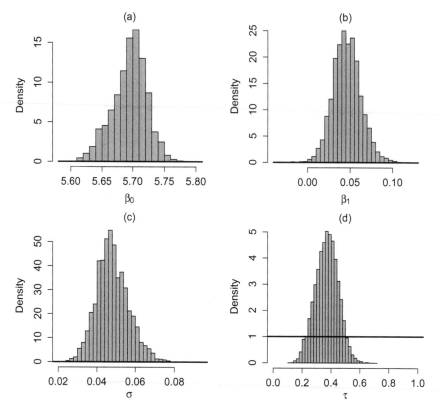

Figure 21.11   Marginal posterior histograms for (a) β_0, (b) β_1, (c) σ, and (d) τ based on fitting the count data ALD regression model to the continental USA bird species richness data with state temperature as a covariate.

We can specify a simple model for these data as

$$y_i \sim \text{Binom}(N, \phi), \tag{21.46}$$

$$\phi \sim \text{Beta}(\alpha, \beta), \tag{21.47}$$

for $i = 1, \ldots, n$ tanks. This model results in a posterior distribution that is similar to the one that we derived in Chapter 6:

$$[\phi|\mathbf{y}] \propto \prod_{i=1}^{n} [y_i|\phi][\phi], \tag{21.48}$$

$$\propto \prod_{i=1}^{n} \binom{N}{y_i} \phi^y (1-\phi)^{N-y_i} \frac{\Gamma(\alpha+\beta)}{\Gamma(\alpha)\Gamma(\beta)} \phi^{\alpha-1}(1-\phi)^{\beta-1}, \tag{21.49}$$

$$\propto \phi^{\sum_{i=1}^{n} y_i}(1-\phi)^{\sum_{i=1}^{n}(N-y_i)} \phi^{\alpha-1}(1-\phi)^{\beta-1}, \tag{21.50}$$

$$\propto \phi^{\sum_{i=1}^{n} y_i + \alpha - 1}(1-\phi)^{\sum_{i=1}^{n}(N-y_i)+\beta-1}, \tag{21.51}$$

which is Beta $(\sum_{i=1}^{n} y_i + \alpha, \sum_{i=1}^{n} (N - y_i) + \beta)$. Although this posterior is convenient, it may not be based on a realistic experiment depending on how many fish we started with, how small they are, and how much time we allot to distribute them among tanks.

A more realistic scenario is that the exact number of fish initially distributed among tanks is unknown. Suppose that the tanks are filled with λ fish "on average," but the precise number of fish per tank is unknown and labeled as N_i. Then, we could specify a modified version of the model (Royle 2004; Royle and Dorazio 2008) to estimate survival as

$$y_i \sim \text{Binom}(N_i, \phi), \tag{21.52}$$

$$N_i \sim \text{Pois}(\lambda), \tag{21.53}$$

$$\phi \sim \text{Beta}(\alpha, \beta), \tag{21.54}$$

for $i = 1, \ldots, n$ tanks. This model implies the posterior distribution

$$[\mathbf{N}, \phi | \mathbf{y}] \propto \left(\prod_{i=1}^{n} [y_i | N_i, \phi][N_i] \right) [\phi], \tag{21.55}$$

which we cannot derive analytically. To fit this model to data, we can use MCMC but need to find the full-conditional distributions for ϕ and N_i for $i = 1, \ldots, n$.

The full-conditional distribution for ϕ is similar to the posterior for the simpler model, but where the number of initial fish per tank varies: $[\phi | \cdot] \equiv \text{Beta}(\sum_{i=1}^{n} y_i + \alpha, \sum_{i=1}^{n} (N_i - y_i) + \beta)$. Thus, we can use a Gibbs update in our MCMC algorithm to sample ϕ from this full-conditional distribution.

We can find the full-conditional distribution for N_i in the same way, by retaining the components of the joint distribution in 21.55 that involves N_i:

$$[N_i | \cdot] \propto [y_i | N_i, \phi][N_i], \tag{21.56}$$

$$\propto \binom{N_i}{y_i} \phi^y (N_i - \phi)^{N_i - y_i} \frac{\lambda^{N_i} e^{-\lambda}}{N_i!}, \tag{21.57}$$

$$\propto \frac{(1 - \phi)^{N_i - y_i} \lambda^{N_i}}{(N_i - y_i)!} \lambda^{-y_i} e^{-(1-\phi)\lambda}, \tag{21.58}$$

$$\propto \frac{((1 - \phi)\lambda)^{N_i - y_i} e^{(1-\phi)\lambda}}{(N_i - y_i)!}, \tag{21.59}$$

which is a Poisson PMF, but for $N_i - y_i$ rather than N_i alone. Thus, in a MCMC algorithm, we use a Gibbs update to sample $z_i^{(k)} \sim \text{Pois}((1 - \phi^{(k)})\lambda)$ and then let $N_i^{(k)} = z_i^{(k)} + y_i$ for $i = 1, \ldots, n$.

We constructed a MCMC algorithm called `binom.beta.pois.mcmc.R` in R to sample both ϕ and N_i for $i = 1, \ldots, n$ iteratively:

BOX 21.17   MCMC ALGORITHM FOR THE BINOMIAL BETA POISSON MODEL

```r
1 binom.beta.pois.mcmc <- function(y,lambda,n.mcmc){
2
3 ####
4 ####   Setup Variables
5 ####
6
7 n=length(y)
8
9 N.save=matrix(0,n,n.mcmc)
10 N.total.save=rep(0,n.mcmc)
11 phi.save=rep(0,n.mcmc)
12
13 ####
14 ####   Priors and Starting Values
15 ####
16
17 alpha=1
18 beta=1
19
20 N=y+1
21
22 ####
23 ####   Begin MCMC Loop
24 ####
25
26 for(k in 1:n.mcmc){
27
28    ####
29    ####   Sample phi
30    ####
31
32    phi=rbeta(1,sum(y)+alpha,sum(N-y)+beta)
33
34    ####
35    ####   Sample N
36    ####
37
38    N=y+rpois(n,(1-phi)*lambda)
```

(Continued)

BOX 21.17 (Continued)   MCMC ALGORITHM FOR THE BINOMIAL BETA POISSON MODEL

```
39    ####
40    ####   Save Samples
41    ####
42
43    phi.save[k]=phi
44    N.save[,k]=N
45    N.total.save[k]=sum(N)
46
47  }
48
49  ####
50  ####   Write Output
51  ####
52
53  list(n.mcmc=n.mcmc,phi.save=phi.save,
54      N.save=N.save,N.total.save=N.total.save)
55
56  }
```

Notice that, in addition to saving the output for ϕ and N_i for $i = 1, \ldots, n$, we also saved the derived quantity $N = \sum_{i=1}^{n} N_i$ in this MCMC algorithm. This derived quantity represents the total number of fish that we started the experiment with.

To demonstrate the MCMC algorithm, we fit the fish survival model with unknown initial tank abundance to a small set of $n = 5$ counts of surviving fish. We assumed that we distributed the fish among tanks fairly evenly and that we assigned on average $\lambda = 50$ individuals to each tank. In the previous MCMC algorithm, we set the hyperparameters for the survival probability ϕ at $\alpha = \beta = 1$, implying a Unif(0,1) prior distribution. We used the following R code to fit the survival model to the data:

BOX 21.18   FIT BINOMIAL BETA POISSON MODEL

```
1  y=c(15,11,12,5,12)
2  lambda=50
3
4  source("binom.beta.pois.mcmc.R")
5  n.mcmc=20000
```

(Continued)

BOX 21.18 (Continued)   FIT BINOMIAL BETA POISSON MODEL

```
6 set.seed(1)
7 mcmc.out=binom.beta.pois.mcmc(y=y,lambda=lambda,
8     n.mcmc=n.mcmc)
9
10 layout(matrix(1:2,1,2))
11 hist(mcmc.out$phi.save[-(1:1000)],xlim=c(0,1),
12     prob=TRUE,xlab=bquote(phi),main="a")
13 hist(mcmc.out$N.total.save[-(1:1000)],
14     breaks=seq(180,330,1),prob=TRUE,xlab="N",
15     ylab="Probability",main="b")
```

This R code results in the marginal posterior histogram for ϕ and barplot for the number of total initial fish N shown in Figure 21.12. Based on the marginal posterior distribution in Figure 21.12a, we see that the survival probability had a posterior mean of $E(\phi|\mathbf{y}) = 0.22$ and 95% credible interval $(0.17, 0.29)$. Also, Figure 21.12b indicates that the total number of initial fish was likely between 219 and 280 (95% credible interval). Thus, with a minimal understanding of the initial abundance per tank and almost no *a priori* understanding of survival, using only observed surviving individuals, we can estimate a total initial population size. More generally, this type of statistical model where we observe a subset of individuals per area and we seek to estimate the total number of individuals is often referred to as an N-mixture model and we discuss it in more detail in Chapter 23.

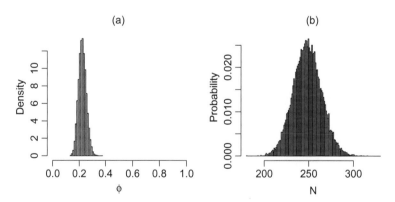

Figure 21.12   Marginal posterior histogram and barplot for (a) ϕ and (b) $N = \sum_{i=1}^{n} N_i$. For N, we use a single bar for each integer to approximate the posterior PMF because it is an integer value derived quantity.

21.9   ADDITIONAL CONCEPTS AND READING

We showed how we can use Poisson regression (and Bernoulli regression) to fit a class of point process models in this chapter. However, Gelfand and Schliep (2018) present a much broader suite of approaches to use to fit point process models. Moreover, there are several excellent resources for learning about additional statistical point process models, including books by Diggle (2013), Moller and Waagepetersen (2003), Illian et al. (2008), and Baddeley et al. (2015), that we do not cover in this book.

Our focus in this book is on Bayesian computing using MCMC methods; however, we can fit a large class of GLMs using the INLA approach (Rue et al. 2009). The procedure behind INLA relies on numerical integration and a reformulation of GLMs and GLMMs that exploits sparsity in the associated matrices to fit the models to data much more efficiently than MCMC. The resulting inference is somewhat limited to a particular class of models and focuses mainly focused on marginal inference of the parameters and/or latent variables. Implementing INLA from scratch can be somewhat complicated, but there is existing software associated with R that we can easily use to fit models to data.

A potential disadvantage with many GLMs is that conjugate priors for the parameters may not be readily apparent. For example, if we have a single count observation y and model it with a Poisson distribution as $y \sim \text{Pois}(\lambda)$, then a conjugate prior for λ is the gamma distribution. However, when $y_i \sim \text{Pois}(\lambda_i)$ for $i = 1, \ldots, n$ and $\log(\lambda_i) = \mathbf{x}_i' \boldsymbol{\beta}$ as we described in this chapter, then we need a multivariate prior for the regression coefficients $\boldsymbol{\beta}$. Bradley et al. (2018) showed how to use a multivariate log-gamma distribution as a prior for regression coefficients in this type of GLM. The log-gamma prior leads to conjugacy in the associated full-conditional distribution for $\boldsymbol{\beta}$ and this results in a MCMC algorithm that does not require tuning.

In addition to the models that we described in this chapter and the previous chapter, we also can use GLMs for other types of continuously valued data. For example, suppose we record proportions p_i for $i = 1, \ldots, n$ individuals, representing the observed data. In this case, $0 \leq p_i \leq 1$ for all $i = 1, \ldots, n$ and we want to account for the heterogeneity in the proportions using a set of measured covariates. Lepak et al. (2012) observed p_i as the proportion of walleye (*Sander vitreus*) diet comprised of rainbow trout (*Oncorhynchus mykiss*) and modeled the observations with a beta distribution such that $p_i \sim \text{Beta}(a_i, b_i)$ for $i = 1, \ldots, n$. To account for heterogeneity in the proportion of diet comprised of rainbow trout, Lepak et al. (2012) linked the mean of the data model to the measured weight of the individual as a covariate. Notice that the variance of the data model also changes with respect to the parameters a_i and b_i, thus, Lepak et al. (2012) also linked the variance of the data model to another covariate corresponding to the sex of the individual. We refer to this type of model as "beta regression" and we can specify and implement it in a variety of ways (e.g., Simas et al. 2010) to model data bounded by zero and one.

22 Zero-Inflated Models

22.1 MIXTURE MODELS FOR EXCESS ZEROS

In the previous chapter, we presented a variety of models for count data, but as rich as that class of models is, it can fail to accommodate certain situations that can arise in ecological data. In Chapter 10, we saw that mixtures of probability distributions can be useful to characterize more complicated measurement and ecological processes. In fact, the concept of mixture models sets the stage for many commonly used models in ecological data analysis (e.g., occupancy models and capture-recapture models).

For example, with ecological count data in particular, we often observe more zeros than standard GLMs like the Poisson and negative binomial regression models in Chapter 21 can accommodate. We could think of this phenomenon as a type of overdispersion (or maybe underdispersion depending on the specific situation) that we need to account for in our model specifications. However, it is important to note that just because a count data set contains many zeros, it does not imply they are "excess" zeros or the data are over- or under-dispersed (Warton 2005). In fact, the zeros could arise because the event we are studying is less prevalent or the particular area surveyed is not conducive to the response variable naturally.

Martin et al. (2005) suggest that there are two types of zeros that commonly occur in ecological data sets concerned with species distribution or abundance: True and false zeros. True zeros can occur in the following ways:

- The species does not occur at a site because of the ecological process (e.g., unsuitable habitat).
- The species does not saturate the entire suitable habitat by chance or extirpation.

By contrast, Martin et al. (2005) noted that false zeros also can occur in two ways:

- The species actually occurs at a site, but it is not present during the survey when data were collected.
- The species occurs at a site, but goes undetected during the survey (e.g., cryptic species).

This dichotomy of how zeros may arise in ecological data gives rise to two different modeling options referred to as hurdle and zero-inflated models, both of which are mixture models. Hurdle models are typically thought of as appropriate for modeling data sets with excess true zeros. In hurdle models, the response variable can arise from one of two distributions, the first is a zero with a certain probability, and the second is a distribution that does not include zero in the support.

By contrast, zero-inflated models are thought of as being used to model data sets containing excess false zeros. We can characterize zero-inflated models as allowing the data to arise from two distributions also: the first is a zero with certain probability and the second is a distribution that includes zero in the support.

Hurdle models separate the zeros into one process that occurs with a certain probability in the data set and the non-zero counts into another process that is typically modeled using a regression framework (e.g., zero-truncated Poisson regression). Thus, more often than not, in these cases, the zeros are simply omitted from the data set and then a data model that does not include zero is used to explain the variation in the remainder of positive data. This results in a simple extension of the models that we presented in the previous chapter, thus we do not cover them in more detail here.

22.2 ZERO-INFLATED POISSON MODELS

In cases where the zeros may arise from both distributions in a mixture, either from an explicit zero process or from another distribution that contains zero in the support, it is common to use a zero-inflated Poisson model (or ZI-Poisson model; Welsh et al. 1996). For a set of count data y_i for $i = 1, \ldots, n$, we can specify a ZI-Poisson model as a mixture distribution

$$y_i \sim \begin{cases} \text{Pois}(\lambda_i) & \text{, with probability } p \\ 0 & \text{, with probability } 1 - p \end{cases}. \tag{22.1}$$

This mixture distribution implies that we can write the resulting conditional PMF as

$$[y_i | \lambda_i, p] = \begin{cases} p \dfrac{\lambda_i^{y_i} e^{-\lambda_i}}{y_i!} & \text{, if } y_i > 0 \\ (1 - p) + p e^{-\lambda_i} & \text{, if } y_i = 0 \end{cases}. \tag{22.2}$$

We also can formulate the ZI-Poisson model as a mixture model using auxiliary variables like we did in the probit regression model from Chapter 20. In this case, we let the data arise from a mixture of zero or Poisson depending on a set of latent indicator variables z_i (for $i = 1, \ldots, n$) as

$$y_i \sim \begin{cases} \text{Pois}(\lambda_i) & \text{, } z_i = 1 \\ 0 & \text{, } z_i = 0 \end{cases}, \tag{22.3}$$

where $z_i \sim \text{Bern}(p)$. Notice that this version of the ZI-Poisson model implies that $z_i = 1$ if $y_i > 0$. Thus, for a portion of the data set, the auxiliary variables are known and equal to one. To complete the Bayesian ZI-Poisson model specification, we use the log link function for the Poisson mean $\log(\lambda_i) = \mathbf{x}_i' \boldsymbol{\beta}$ and specify priors for p and $\boldsymbol{\beta}$ such that $p \sim \text{Beta}(\alpha_1, \alpha_2)$ and $\boldsymbol{\beta} \sim \text{N}(\boldsymbol{\mu}_\beta, \boldsymbol{\Sigma}_\beta)$.

The posterior distribution for this ZI-Poisson regression model is

$$[\mathbf{z}, p, \boldsymbol{\beta} | \mathbf{y}] \propto \left(\prod_{i=1}^{n} \left(\frac{\lambda_i^{y_i} e^{-\lambda_i}}{y_i!} \right)^{z_i} 1_{\{y_i=0\}}^{1-z_i} [z_i | p] \right) [p][\boldsymbol{\beta}], \tag{22.4}$$

which implies that we need to sample from three sets of full-conditional distributions (for $\mathbf{z}, p$, and $\boldsymbol{\beta}$) to fit the model using MCMC. Recall that the auxiliary variable $z_i = 1$ when $y_i > 0$, thus when $y_i = 0$, the full-conditional distribution for z_i is

$$[z_i|\cdot] \propto \left(\frac{\lambda_i^{y_i}e^{-\lambda_i}}{y_i!}\right)^{z_i} 1_{\{y_i=0\}}^{1-z_i}[z_i|p], \tag{22.5}$$

$$\propto (e^{-\lambda_i})^{z_i} 1_{\{y_i=0\}}^{1-z_i} p^{z_i}(1-p)^{1-z_i}, \tag{22.6}$$

$$\propto (pe^{-\lambda_i})^{z_i}(1-p)^{1-z_i}, \tag{22.7}$$

which results in a Bernoulli distribution with probability

$$\tilde{p}_i = \frac{pe^{-\lambda_i}}{pe^{-\lambda_i}+1-p}, \tag{22.8}$$

because the expression in (22.7) is in the form of a Bernoulli PMF, but we need to normalize the probability so that the PMF sums to one. Thus, z_i has a conjugate full-conditional as we described in Chapter 10 and we can use a Gibbs update for it in the MCMC algorithm.

The full-conditional distribution for the mixture probability p involves a product over the Bernoulli distributions $[z_i|p]$ for $i = 1, \ldots, n$ and a beta prior for $[p]$, just like the posterior for the Bernoulli–beta model that we derived in Chapter 6. Thus, we can borrow the result directly from Chapter 6 for the full-conditional distribution of p:

$$[p|\cdot] \propto \left(\prod_{i=1}^{n}[z_i|p]\right)[p], \tag{22.9}$$

$$\propto p^{\sum_{i=1}^{n} z_i+\alpha_1-1}(1-p)^{\sum_{i=1}^{n}(1-z_i)+\alpha_2-1}, \tag{22.10}$$

which is a conjugate $\text{Beta}(\sum_{i=1}^{n} z_i + \alpha_1, \sum_{i=1}^{n}(1-z_i) + \alpha_2)$ distribution and we can use a Gibbs update for it in a MCMC algorithm.

We can derive the remaining full-conditional distribution for $\boldsymbol{\beta}$ similarly to how it was in Poisson regression in Chapter 21. Retaining the components of the joint distribution (22.7) that contain $\boldsymbol{\beta}$, we have

$$[\boldsymbol{\beta}|\cdot] \propto \left(\prod_{i=1}^{n}\left(\frac{\lambda_i^{y_i}e^{-\lambda_i}}{y_i!}\right)^{z_i} 1_{\{y_i=0\}}^{1-z_i}\right)[\boldsymbol{\beta}], \tag{22.11}$$

$$\propto \left(\prod_{\forall z_i=1} \lambda_i^{y_i}e^{-\lambda_i}\right)[\boldsymbol{\beta}], \tag{22.12}$$

which involves a product over the cases where $z_i = 1$. This full-conditional distribution (22.12) is non-conjugate, but we can sample from it using a M-H update. Using a random walk multivariate normal proposal distribution, we can sample $\boldsymbol{\beta}^{(*)} \sim \text{N}(\boldsymbol{\beta}^{(k-1)}, \sigma_{\text{tune}}^2\mathbf{I})$ which results in the M-H ratio

$$mh = \frac{\left(\prod_{\forall z_i=1}[y_i|\boldsymbol{\beta}^{(*)}]\right)[\boldsymbol{\beta}^{(*)}]}{\left(\prod_{\forall z_i=1}[y_i|\boldsymbol{\beta}^{(k-1)}]\right)[\boldsymbol{\beta}^{(k-1)}]},\tag{22.13}$$

where $[y_i|\boldsymbol{\beta}]$ represents the Poisson PMF (also in (22.12)).

We constructed a MCMC algorithm in R called `zip.reg.mcmc.R` to fit the ZI-Poisson regression model to zero-inflated data:

BOX 22.1   MCMC ALGORITHM TO FIT ZI-POISSON REGRESSION MODEL

```
 1 zip.reg.mcmc <- function(y,X,beta.mn,beta.var,
 2    n.mcmc){
 3
 4 ###
 5 ###  Preliminary Variables
 6 ###
 7
 8 X=as.matrix(X)
 9 y=as.vector(y)
10 n=length(y)
11 pp=dim(X)[2]
12
13 beta.save=matrix(0,pp,n.mcmc)
14 lam.save=matrix(0,n,n.mcmc)
15 z.save=matrix(0,n,n.mcmc)
16 p.save=rep(0,n.mcmc)
17
18 ###
19 ###  Starting Values
20 ###
21
22 beta.sd=sqrt(beta.var)
23 beta=beta.mn
24 z=rep(0,n)
25 z[y>0]=1
26 lam=exp(X%*%beta)
27 p=.9
28 beta.tune=.1
29
30 ###
31 ###  MCMC Loop
32 ###
```

(Continued)

BOX 22.1 (Continued)   MCMC ALGORITHM TO FIT ZI-POISSON REGRESSION MODEL

```
33 for(k in 1:n.mcmc){
34
35   ###
36   ### Sample z
37   ###
38
39   p.tmp=p*exp(-lam[y==0])/(1-p+p*exp(-lam[y==0]))
40   z[y==0]=rbinom(sum(y==0),1,p.tmp)
41
42   ###
43   ### Sample p
44   ###
45
46   p=rbeta(1,sum(z)+1,n-sum(z)+1)
47
48   ###
49   ### Sample beta
50   ###
51
52   beta.star=rnorm(pp,beta,beta.tune)
53   lam.star=exp(X%*%beta.star)
54   mh1=sum(dpois(y[z==1],lam.star[z==1],log=TRUE))
55     +sum(dnorm(beta.star,beta.mn,beta.sd,
56     log=TRUE))
57   mh2=sum(dpois(y[z==1],lam[z==1],log=TRUE))
58     +sum(dnorm(beta,beta.mn,beta.sd,log=TRUE))
59   mh=exp(mh1-mh2)
60   if(mh > runif(1)){
61     beta=beta.star
62     lam=lam.star
63   }
64
65   ###
66   ### Save Samples
67   ###
```

(Continued)

BOX 22.1 (Continued)   MCMC ALGORITHM TO FIT ZI-POISSON REGRESSION MODEL

```
68    z.save[,k]=z
69    p.save[k]=p
70    beta.save[,k]=beta
71    lam.save[,k]=lam
72
73 }
74
75 ###
76 ###   Write output
77 ###
78
79 list(y=y,X=X,n.mcmc=n.mcmc,beta.save=beta.save,
80     p.save=p.save,z.save=z.save)
81
82 }
```

We used this MCMC algorithm to fit the ZI-Poisson model to a set of 237 counts of willow tit (*Parus montanus*) in Switzerland. Royle and Dorazio (2008) analyzed these data and we used the following R code to read them in and examine their structure:

BOX 22.2   READ IN COUNT DATA

```
1 wt.c.df=read.table("wt.1.count.txt",header=TRUE)
2 y=wt.c.df$y
3 n=length(y)
4 X=cbind(rep(1,n),as.matrix(scale(wt.c.df[,-1])))
5 barplot(table(y),col=8,xlab="Count (y)",
6     ylab="Frequency")
```

Figure 22.1 shows a barplot representing the willow tit count data.

Notice the large number of zeros in this data set (approximately 170 out of 237 total observations). While being aware of our own previous warning about interpreting the marginal empirical distribution of the data directly as evidence of zero-inflation, Figure 22.1 suggests that we may want to incorporate zero-inflation in our model. In that case, if we estimate p to be close to one, that would indicate that we may not need the zero-inflation model and could just use the simpler Poisson regression model instead.

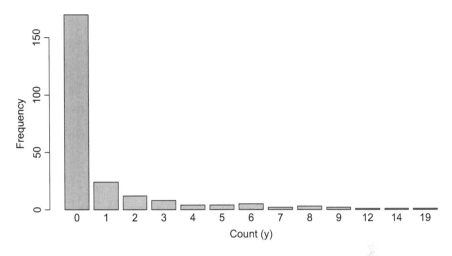

Figure 22.1 Barplot of the willow tit count data (**y**).

The willow tit count data set contains two covariates (i.e., elevation and percent forest at each location) that we standardized and used in our ZI-Poisson regression model. Thus, our model contains three regression coefficients $\boldsymbol{\beta}$, the mixture probability p, and the set of auxiliary variables $\mathbf{z}$. Using hyperparameters $\boldsymbol{\mu}_\beta \equiv \mathbf{0}$, $\boldsymbol{\Sigma}_\beta \equiv \sigma_\beta^2 \mathbf{I}$ with $\sigma_\beta^2 = 1000$, we fit the ZI-Poisson regression model to the willow tit count data in R using the previous MCMC algorithm and $K = 50000$ MCMC iterations as

---

BOX 22.3   FIT ZI-POISSON REGRESSION MODEL TO COUNT DATA

```
source("zip.reg.mcmc.R")
n.mcmc=50000
set.seed(1)
mcmc.out=zip.reg.mcmc(y=y,X=X,beta.mn=rep(0,
    dim(X)[2]),beta.var=1000,n.mcmc=n.mcmc)

layout(matrix(1:4,2,2,byrow=TRUE))
hist(mcmc.out$beta.save[1,-(1:1000)],xlab=bquote
    (beta[0]),prob=TRUE,main="a")
curve(dnorm(x,0,sqrt(1000)),lwd=2,add=TRUE)
hist(mcmc.out$beta.save[2,-(1:1000)],xlab=bquote
    (beta[1]),prob=TRUE,main="b")
curve(dnorm(x,0,sqrt(1000)),lwd=2,add=TRUE)
```

(Continued)

BOX 22.3 (Continued)   FIT ZI-POISSON REGRESSION MODEL TO COUNT DATA

```
14 hist(mcmc.out$beta.save[3,-(1:1000)],xlab=bquote
15    (beta[2]),prob=TRUE,main="c")
16 curve(dnorm(x,0,sqrt(1000)),lwd=2,add=TRUE)
17 hist(mcmc.out$p.save[-(1:1000)],xlab="p",
18    xlim=c(0,1),prob=TRUE,main="d")
19 curve(dunif(x),lwd=2,add=TRUE)
20
21 mean(mcmc.out$p.save[-(1:1000)])
22 quantile(mcmc.out$p.save[-(1:1000)],
23    c(0.025,0.975))
```

This R code results in the marginal posterior distributions for β and p shown in Figure 22.2. In particular, Figure 22.2d indicates that approximately $E(p|\mathbf{y}) = 0.49$ (95% credible interval: $[0.40, 0.60]$), which implies that approximately half of the observations are due to zero-inflation when using a ZI-Poisson regression model to describe heterogeneity in the willow tit counts. Figure 22.2b,c indicates that willow tit counts increase with both forest and elevation, which matches our understanding of this montane species. It is possible that some of the excess zeros in the willow tit data may be due to specific detectability (i.e., false zeros) issues, like with most avian species and we present methods to address that issue in the following chapter.

22.3   ZERO-INFLATED NEGATIVE BINOMIAL MODELS

Just as we can extend the Poisson regression model to accommodate additional sources of overdispersion in count data, we can do the same with the ZI-Poisson model. Although a variety of options exist for modeling overdispersion, an obvious modification to try is to exchange the Poisson component of the zero-inflation mixture model for a negative binomial model, parameterized as we described in Chapter 22 in terms of a mean and overdispersion parameter N (also sometimes referred to as a "size" parameter).

A modification of the ZI-Poisson model with auxiliary variables where we substitute in the negative binomial distribution results in the ZI-negative binomial model:

$$y_i \sim \begin{cases} \text{NB}(\mu_i, N) & , z_i = 1 \\ 0 & , z_i = 0 \end{cases}, \tag{22.14}$$

where, as before, $z_i \sim \text{Bern}(p)$ for $i = 1, \ldots, n$, $p \sim \text{Beta}(\alpha_1, \alpha_2)$, $\log(\mu_i) = \mathbf{x}_i'\boldsymbol{\beta}$, and $\boldsymbol{\beta} \sim \text{N}(\boldsymbol{\mu}_\beta, \boldsymbol{\Sigma}_\beta)$. We specify the same prior for the natural logarithm of the overdispersion parameter that we used in the negative binomial regression in Chapter 21 (i.e., $\log(N) \sim \text{N}(\mu_N, \sigma_N^2)$), but many other options exist (e.g., gamma, truncated normal, etc.). The posterior distribution is also similar, but we replace the

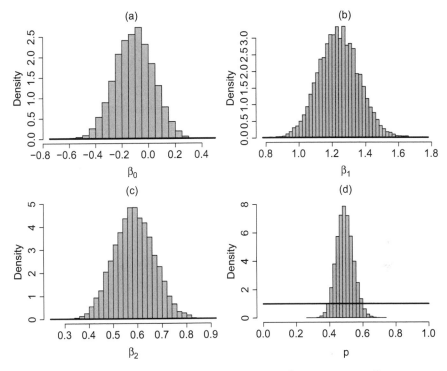

Figure 22.2 Marginal posteriors for (a) β_0 (intercept), (b) β_1 (elevation), (c) β_2 (forest), and (d) the mixture probability p that result from fitting the ZI-Poisson regression model to the willow tit count data. Prior PDFs are shown as black horizontal lines for comparison.

Poisson PMF with the negative binomial PMF and we include the prior for $\log(N)$ such that

$$[\mathbf{z}, p, \boldsymbol{\beta}, \log(N)|\mathbf{y}] \propto \left(\prod_{i=1}^{n} \left(\frac{\Gamma(y_i + N)}{\Gamma(N) y_i!} \left(\frac{N}{N + \mu_i} \right)^N \left(1 - \frac{N}{N + \mu_i} \right)^{y_i} \right)^{z_i} \times \right.$$

$$\left. 1_{\{y_i=0\}}^{1-z_i} [z_i|p] \right) [p][\boldsymbol{\beta}][\log(N)]. \tag{22.15}$$

To construct a MCMC algorithm, we derive the full-conditional distributions in the same way as we did for the ZI-Poisson regression model in the previous section. Thus, in cases where $y_i = 0$, we sample from the full-conditional distribution for the auxiliary variables (recall that we set $z_i = 1$ if $y_i > 0$, as before):

$$[z_i|\cdot] \propto \left(\frac{N}{N + \mu_i} \right)^{N z_i} p^{z_i} (1-p)^{1-z_i}, \tag{22.16}$$

$$\propto \left(p \left(\frac{N}{N + \mu_i} \right)^N \right)^{z_i} (1-p)^{1-z_i}, \tag{22.17}$$

which results in a Bernoulli distribution with probability

$$\tilde{p}_i = \frac{p\left(\frac{N}{N+\mu_i}\right)^N}{p\left(\frac{N}{N+\mu_i}\right)^N + 1 - p}. \tag{22.18}$$

Thus, we can sample the auxiliary variables z_i (when $y_i = 0$) using a Gibbs update in our MCMC algorithm as shown in Chapter 10 (as well as the previous chapter).

The full-conditional distribution for the mixture probability p is a beta distribution exactly as we derived in the previous section: $[p|\cdot] = \text{Beta}(\sum_{i=1}^n z_i + \alpha_1, \sum_{i=1}^n (1 - z_i) + \alpha_2)$. Similarly, the full-conditional distribution for the regression coefficients $\boldsymbol{\beta}$ associated with our ZI-negative binomial model takes the same form as the previous section, but with the negative binomial distribution substituted instead of the Poisson:

$$[\boldsymbol{\beta}|\cdot] \propto \left(\prod_{i=1}^n \left(\frac{\Gamma(y_i + N)}{\Gamma(N) y_i!} \left(\frac{N}{N+\mu_i} \right)^N \left(1 - \frac{N}{N+\mu_i} \right)^{y_i} \right)^{z_i} \right) [\boldsymbol{\beta}], \tag{22.19}$$

$$\propto \left(\prod_{\forall z_i = 1} \frac{\Gamma(y_i + N)}{\Gamma(N) y_i!} \left(\frac{N}{N+\mu_i} \right)^N \left(1 - \frac{N}{N+\mu_i} \right)^{y_i} \right) [\boldsymbol{\beta}]. \tag{22.20}$$

Thus, we can use a random walk multivariate normal proposal as before and update $\boldsymbol{\beta}$ using M-H like in the ZI-Poisson MCMC algorithm (adjusting the M-H ratio accordingly from the previous section).

Finally, the full-conditional distribution for the natural logarithm of the overdispersion parameter $\log(N)$ resembles the one that we derived for the same parameter in Chapter 21 for negative binomial regression, except that we only use the components of the likelihood where $z_i = 1$ so that

$$[\log(N)|\cdot] \propto \left(\prod_{i=1}^n \left(\frac{\Gamma(y_i + N)}{\Gamma(N) y_i!} \left(\frac{N}{N+\mu_i} \right)^N \left(1 - \frac{N}{N+\mu_i} \right)^{y_i} \right)^{z_i} \right) [\log(N)], \tag{22.21}$$

$$\propto \left(\prod_{\forall z_i = 1} \frac{\Gamma(y_i + N)}{\Gamma(N) y_i!} \left(\frac{N}{N+\mu_i} \right)^N \left(1 - \frac{N}{N+\mu_i} \right)^{y_i} \right) [\log(N)]. \tag{22.22}$$

Thus, we can use M-H to sample $\log(N)$ based on another random walk normal proposal which results in the M-H ratio

$$mh = \frac{\left(\prod_{z_i=1} [y_i | \boldsymbol{\beta}^{(k)}, \log(N)^{(*)}] \right) [\log(N)^{(*)}]}{\left(\prod_{z_i=1} [y_i | \boldsymbol{\beta}^{(k)}, \log(N)^{(k-1)}] \right) [\log(N)^{(k-1)}]}. \tag{22.23}$$

We constructed a MCMC algorithm to fit the ZI-negative binomial regression model to count data based on these full-conditional distributions. In R, the algorithm called `zinb.reg.mcmc.R` resembles the one we used in the previous section:

BOX 22.4   MCMC ALGORITHM FOR ZI-NEGATIVE BINOMIAL REGRESSION

```
zinb.reg.mcmc <- function(y,X,beta.mn,beta.var,
    n.mcmc){

###
###   Preliminary Variables
###

X=as.matrix(X)
y=as.vector(y)
n=length(y)
pp=dim(X)[2]

beta.save=matrix(0,pp,n.mcmc)
mu.save=matrix(0,n,n.mcmc)
z.save=matrix(0,n,n.mcmc)
p.save=rep(0,n.mcmc)
N.save=rep(0,n.mcmc)

###
###   Starting Values and Priors
###

logN.mn=log(1)
logN.sd=log(3)

N=4
beta.sd=sqrt(beta.var)
beta=beta.mn
z=rep(0,n)
z[y>0]=1
mu=exp(X%*%beta)
p=.9
beta.tune=.1
N.tune=.5

###
###   MCMC Loop
###
```

(Continued)

BOX 22.4 (Continued) MCMC ALGORITHM FOR ZI-NEGATIVE BINOMIAL REGRESSION

```
39  for(k in 1:n.mcmc){
40
41    ###
42    ### Sample z
43    ###
44
45    theta=N/(N+mu[y==0])
46    p.tmp=p*(theta^N)/(1-p+p*(theta^N))
47    z[y==0]=rbinom(sum(y==0),1,p.tmp)
48
49    ###
50    ### Sample p
51    ###
52
53    p=rbeta(1,sum(z)+1,n-sum(z)+1)
54
55    ###
56    ### Sample beta
57    ###
58
59    beta.star=rnorm(pp,beta,beta.tune)
60    mu.star=exp(X%*%beta.star)
61
62    mh1=sum(dnbinom(y[z==1],mu=mu.star[z==1],
63      size=N,log=TRUE))+sum(dnorm(beta.star,
64      beta.mn,beta.sd,log=TRUE))
65    mh2=sum(dnbinom(y[z==1],mu=mu[z==1],size=N,
66      log=TRUE))+sum(dnorm(beta,beta.mn,beta.sd,
67      log=TRUE))
68    mh=exp(mh1-mh2)
69    if(mh > runif(1)){
70      beta=beta.star
71      mu=mu.star
72    }
73
74    ###
75    ### Sample N
76    ###
```

(Continued)

**BOX 22.4 (Continued)   MCMC ALGORITHM FOR ZI-NEGATIVE
BINOMIAL REGRESSION**

```
77   N.star=exp(rnorm(1,log(N),N.tune))
78
79   mh1=sum(dnbinom(y[z==1],mu=mu[z==1],
80     size=N.star,log=TRUE))+dnorm(log(N.star),
81     logN.mn,logN.sd,log=TRUE)
82   mh2=sum(dnbinom(y[z==1],mu=mu[z==1],size=N,
83     log=TRUE))+dnorm(log(N),logN.mn,logN.sd,
84     log=TRUE)
85   mh=exp(mh1-mh2)
86   if(mh > runif(1)){
87     N=N.star
88   }
89
90   ###
91   ### Save Samples
92   ###
93
94   z.save[,k]=z
95   p.save[k]=p
96   N.save[k]=N
97   beta.save[,k]=beta
98   mu.save[,k]=mu
99
100 }
101
102 ###
103 ### Write output
104 ###
105
106 list(y=y,X=X,n.mcmc=n.mcmc,beta.save=beta.save,
107     mu.save=mu.save,p.save=p.save,z.save=z.save,
108     N.save=N.save)
109
110 }
```

We fit the Bayesian ZI-negative binomial regression model to the same willow tit data that we analyzed in the previous section (Figure 22.1) using the same hyperparameters that we did in the previous and $\mu_N = \log(1)$ and $\sigma_N^2 = \log(3)^2$. We used the following R code to fit the model:

BOX 22.5   FIT ZI-NEGATIVE BINOMIAL REGRESSION MODEL TO COUNT DATA

```
source("zinb.reg.mcmc.R")
set.seed(1)
mcmc.nb.out=zinb.reg.mcmc(y=y,X=X,beta.mn=rep(0,
    dim(X)[2]),beta.var=1000,n.mcmc=n.mcmc)

layout(matrix(1:4,2,2,byrow=TRUE))
hist(mcmc.nb.out$beta.save[1,-(1:1000)],
    xlab=bquote(beta[0]),prob=TRUE,main="a")
curve(dnorm(x,0,sqrt(1000)),lwd=2,add=TRUE)
hist(mcmc.nb.out$beta.save[2,-(1:1000)],
    xlab=bquote(beta[1]),prob=TRUE,main="b")
curve(dnorm(x,0,sqrt(1000)),lwd=2,add=TRUE)
hist(mcmc.nb.out$beta.save[3,-(1:1000)],
    xlab=bquote(beta[2]),prob=TRUE,main="c")
curve(dnorm(x,0,sqrt(1000)),lwd=2,add=TRUE)
hist(mcmc.nb.out$p.save[-(1:1000)],xlab="p",
    xlim=c(0,1),prob=TRUE,main="d")
curve(dunif(x),lwd=2,add=TRUE)

mean(mcmc.nb.out$N.save[-(1:1000)])
quantile(mcmc.nb.out$N.save[-(1:1000)],
    c(0.025,0.975))
```

This R code results in the marginal posterior distributions for β and p shown in Figure 22.3. The most obvious difference between the marginal posterior distributions resulting from the ZI-Poisson (Figure 22.2) and the ZI-negative binomial (Figure 22.3) is that we estimate the mixture probability p as closer to one in the ZI-negative binomial model. This difference occurs because we estimated the overdispersion parameter N to be relatively small ($E(N|\mathbf{y}) = 1.08$ and 95% credible interval: [0.42,2.54]) implying that we do not need as much zero-inflation in the ZI-negative binomial model because the negative binomial already accommodates a larger portion of the zeros. The coefficients (β) estimated in the ZI-negative binomial still indicate positive relationships with willow tit counts, but the coefficient associated with the effect of forest (β_2; Figure 22.3c) is slightly reduced relative to the ZI-Poisson model fit.

From a model checking perspective (Conn et al. 2018), we can compare the posterior predictive p-values for the ZI-Poisson versus ZI-negative binomial models fit to the willow tit data using MSE to gauge if there is any discrepancy in the ability of the model to accommodate overdispersion in the data beyond the excess zeros. To compute MSE for both models, we use the same equation (21.30) introduced in

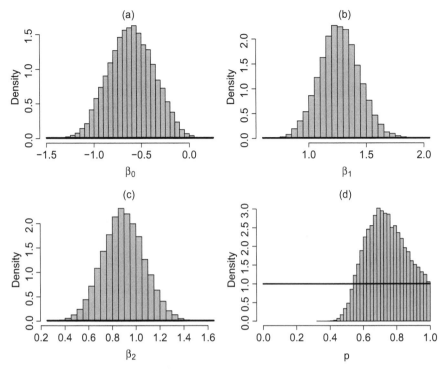

Figure 22.3 Marginal posteriors for (a) β_0 (intercept), (b) β_1 (elevation), (c) β_2 (forest), and (d) the mixture probability p resulting from fitting the ZI-negative binomial regression model to the willow tit count data. Prior PDFs are shown as black horizontal lines for comparison.

Chapter 21, where the conditional mean of the data is $p\lambda_i$ for the ZI-Poisson and $p\mu_i$ for the ZI-negative binomial model after marginalizing over z_i. The following R code uses composition sampling to obtain posterior predictive realizations resulting from each model and we calculate the posterior predictive p-value as the posterior expectation of the indicator characterizing if the MSE for the predictions is larger than that of the data:

BOX 22.6   CALCULATE POSTERIOR PREDICTIVE P-VALUES

```
p.ind=rep(0,n.mcmc)
nb.ind=rep(0,n.mcmc)
set.seed(1)
for(k in 1:n.mcmc){
```

(Continued)

BOX 22.6 (Continued) CALCULATE POSTERIOR PREDICTIVE P-VALUES

```
 5  y.pred.p=rep(0,n)
 6  z.1.p.idx=(mcmc.out$z.save[,k]==1)
 7  lam.p=exp(mcmc.out$X%*%mcmc.out$beta.save[,k])
 8  y.pred.p[z.1.p.idx]=rpois(sum(z.1.p.idx),
 9     lam.p[z.1.p.idx])
10  mse.p=mean((y.pred.p-mcmc.out$p.save[k]*
11     lam.p)^2)
12  p.ind[k]=(mse.p>mean((y-mcmc.out$p.save[k]*
13     lam.p)^2))
14
15  y.pred.nb=rep(0,n)
16  z.1.nb.idx=(mcmc.nb.out$z.save[,k]==1)
17  lam.nb=exp(mcmc.nb.out$X%*%mcmc.nb.out$beta.
18     save[,k])
19  y.pred.nb[z.1.nb.idx]=rnbinom(sum(z.1.nb.idx),
20     mu=lam.nb[z.1.nb.idx],size=mcmc.nb.out$N.
21     save[k])
22  mse.nb=mean((y.pred.nb-mcmc.nb.out$p.save[k]*
23     lam.nb)^2)
24  nb.ind[k]=(mse.nb>mean((y-mcmc.nb.out$p.save[k]*
25     lam.nb)^2))
26  }
27
28  pval.p=mean(p.ind[-(1:1000)])
29  pval.nb=mean(nb.ind[-(1:1000)])
30  pval.p
31  pval.nb
```

The resulting posterior predictive p-values were 0.03 for the ZI-Poisson model versus 0.73 for the ZI-negative binomial model. Thus, we have evidence that the ZI-Poisson model does not properly accommodate the dispersion in the data because the posterior predictive p-value was close to zero indicating the ZI-Poisson predictions were underdispersed relative to the data. However, there is no evidence for lack of fit based on dispersion using the ZI-negative binomial regression model fit to the willow tit count data and thus we can trust the resulting inference (Figure 22.3).

22.4 ADDITIONAL CONCEPTS AND READING

A variety of ecological studies have used zero-inflated models (Wenger and Freeman 2008). For example, Arab et al. (2008b) developed a ZI-Poisson model to infer the effect of sampling gear on catch of fishes in the Missouri River. In addition, studies have extended the basic ZI-Poisson model to accommodate explicit spatial structure (e.g., Lyashevska et al. 2016). Similarly, studies have used several approaches to develop spatio-temporal ZI models and have applied them to harbor seal (*Phoca vitulina*) counts in Alaska (Ver Hoef and Jansen 2007), Atlantic cod (*Gadus morhua*) in the Gulf of Maine (Wang et al. 2015), infectious disease vectors in Kenya (Amek et al. 2011), and Lyme disease in Illinois (Arab 2015).

In addition to discrete response variables, we also can extend ZI models to accommodate a mixture of zeros and continuous data (e.g., Irvine et al. 2016). For example, Li et al. (2011) developed a ZI-log normal model for skewed responses. Furthermore, it is possible to allow for inflation of other values in response variables as well. For example, Ospina and Ferrari (2012) developed an approach for zero and one inflated beta regression. Their approach extends the beta regression idea that we presented at the end of the previous chapter to accommodate extra zeros and ones in the data than could be characterized by the beta distribution.

In our ZI model examples in this chapter, we focused on checking the assumptions for the ZI-Poisson and ZI-negative binomial models to see which is most appropriate for statistical inference. Ridout et al. (2001) developed an alternative method for comparing the models based on a score test. Furthermore, in cases where neither type of ZI model appears to have invalid assumptions, we may seek to compare them in terms of predictive ability. The same concepts for model scoring and regularization that we discussed in Chapters 13 and 14 apply to ZI models, but care should be taken to derive a proper scoring function based on the integrated likelihood or PPD.

23 Occupancy Models

23.1 SIMPLE OCCUPANCY MODELS

In the previous chapter, we saw that mixture models allow us to accommodate more complicated features of ecological data sets (e.g., excess zeros) than would be possible with standard GLMs. Based on the increasing rate of use in the literature, zero-inflated models have become popular in ecology. Although we highlighted a few mechanistic reasons to select among zero-inflated models, we only focused on zero-inflated models for count data in Chapter 22. However, a large class of mixture models commonly used for ecological inference rely on binary data. For example, occupancy models and capture-recapture models accommodate zero-inflation in binary data through very specific mechanisms related to the observational process. In particular, they are models developed for making ecological inference in the presence of false zeros (i.e., false negatives).

From the Bayesian perspective, binary ZI models are most intuitively specified using a hierarchical framework where the data we observe arise from a distribution involving an underlying ecological process and that process depends on a set of parameters (as our regression models in the previous chapters did). In particular, both occupancy and capture-recapture models contain a set of parameters that affect the observational process directly. Hence, this chapter focuses on measurement error models because, with occupancy and capture-recapture models, we already know how to model the latent ecological process (usually via binary regression), but we need to account for the observational process in a way that specifically acknowledges how the data were collected.

Occupancy models were developed to learn about species distributions (presence or absence) when detectability of the species is imperfect (Hoeting et al. 2000; Tyre et al. 2003; MacKenzie et al. 2017). In particular, most occupancy studies are designed such that there is zero probability of a false positive (in principle; but see Ruiz-Gutierrez et al. 2016 and Chapter 25 of this book), but that there may be an unavoidable non-zero false negative probability in the observation of a particular species at a set of sites. With a single visit to each site, it is difficult to identify and separate both occupancy probability and detection probability (but see Lele et al. 2012). Thus, most occupancy studies aim to collect replicate data at each site, perhaps returning to survey site i a number (J_i) times for $i = 1, \ldots, n$ sites. In the classical occupancy model, we assume repeated measurements have the same underlying occupancy process. Hence, the replicates allow us to tease apart detectability from occupancy.

We present the occupancy model as a hierarchically specified zero-inflated Bernoulli model where the data y_{ij} are detections represented by zeros and ones for sites $i = 1, \ldots, n$ and replicates $j = 1, \ldots, J_i$, and we denote the underlying occupancy process as z_i, a binary process, where $z_i = 1$ implies the species is present at site i (and $z_i = 0$ implies absence). In a binary regression setting, we have already introduced the process model in Chapter 20: $z_i \sim \text{Bern}(\psi_i)$, where $\text{logit}(\psi_i) = \mathbf{x}_i'\boldsymbol{\beta}$. In this case, we have used the logit link for the occupancy probability ψ_i, but we can use other link functions as described in Chapter 20.

The data model commonly used in occupancy settings with imperfect detection conditionally on the occupancy process z_i as a mixture

$$y_{ij} \sim \begin{cases} \text{Bern}(p) & , z_i = 1 \\ 0 & , z_i = 0 \end{cases}, \tag{23.1}$$

for $i = 1,\ldots,n$ and $j = 1,\ldots,J_i$, and assuming that the detection probability (p) is homogeneous among sites (for now). Notice how similar the occupancy data model in (23.1) is to the zero-inflated data models that we introduced in the previous chapter. Also, assuming that the replicates are conditionally independent, we can rewrite the occupancy data model as

$$y_i \sim \begin{cases} \text{Binom}(J_i,p) & , z_i = 1 \\ 0 & , z_i = 0 \end{cases}, \tag{23.2}$$

where $y_i = \sum_{j=1}^{J_i} y_{ij}$ is the sum of detections at site i. Thus, in (23.2) J_i represents a known number of Bernoulli "trials" which are replicates in ideal occupancy data. Therefore, the occupancy model is a ZI-binomial regression model that is similar to, but different than, the ZI-Poisson and ZI-negative binomial models in the following important ways:

1. The occupancy model is specified to mimic a specific observational mechanism that leads to measurement error.
2. The zero-inflated models in the Chapter 22 account for variation in the parameter that occurs in the mixture component when $z_i = 1$ whereas the simplest useful occupancy model in (23.2) accounts for variation in the underlying process z_i.

Of course, we can specify the occupancy model such that the variation arises in p instead of ψ, but it would tell us mainly about variation in the detection process which we often assume to be similar among sites. However, the usual ecological questions of interest concern variation in the occupancy process itself.

To complete the Bayesian occupancy model, we specify priors for the detection probability p and occupancy regression coefficients $\boldsymbol{\beta}$ such that

$$p \sim \text{Beta}(\alpha_p, \beta_p), \tag{23.3}$$
$$\boldsymbol{\beta} \sim \text{N}(\boldsymbol{\mu}_\beta, \boldsymbol{\Sigma}_\beta), \tag{23.4}$$

which results in the posterior distribution

$$[\mathbf{z},p,\boldsymbol{\beta}|\mathbf{y}] \propto \left(\prod_{i=1}^{n} [y_i|p]^{z_i} 1_{\{y_i=0\}}^{1-z_i} [z_i|\boldsymbol{\beta}] \right) [p][\boldsymbol{\beta}]. \tag{23.5}$$

To construct a MCMC algorithm to fit the simple occupancy model to data, we need to derive the full-conditional distributions as we have for all of the models in previous chapters. Beginning with the occupancy process, as with the zero-inflated models in Chapter 22, we set $z_i = 1$ when $y_i > 0$. Thus, we only sample z_i when $y_i = 0$ because the only time when the occupancy process is unknown occurs when we never detected the species at a site. Therefore, the full-conditional distribution for z_i when $y_i = 0$ is

$$[z_i|\cdot] \propto [y_i|p]^{z_i} 1_{\{y_i=0\}}^{1-z_i} \psi_i^{z_i} (1 - \psi_i)^{1-z_i}, \tag{23.6}$$

$$\propto ((1-p)^{J_i})^{z_i} \psi_i^{z_i} (1 - \psi_i)^{1-z_i}, \tag{23.7}$$

$$\propto (\psi_i(1-p)^{J_i})^{z_i} (1 - \psi_i)^{1-z_i}, \tag{23.8}$$

where $\text{logit}(\psi_i) = \mathbf{x}_i'\boldsymbol{\beta}$. The full-conditional distribution in (23.8) implies $[z_i|\cdot] = \text{Bern}(\tilde{\psi}_i)$ where

$$\tilde{\psi}_i = \frac{\psi_i(1-p)^{J_i}}{\psi_i(1-p)^{J_i} + 1 - \psi_i}. \tag{23.9}$$

Thus, we can sample z_i for all cases where $y_i = 0$ using Gibbs updates in a MCMC algorithm.

The full-conditional distribution for p is also conjugate because

$$[p|\cdot] \propto \left(\prod_{i=1}^{n} [y_i|p]^{z_i}\right) [p], \tag{23.10}$$

$$\propto \left(\prod_{\forall z_i=1} [y_i|p]\right) [p], \tag{23.11}$$

$$\propto p^{\sum_{\forall z_i=1} y_i + \alpha_p - 1} (1 - p)^{\sum_{\forall z_i=1} (J_i - y_i) + \beta_p - 1}, \tag{23.12}$$

which results in a beta distribution where $[p|\cdot] = \text{Beta}(\sum_{\forall z_i=1} y_i + \alpha_p, \sum_{\forall z_i=1} (J_i - y_i) + \beta_p)$ that we also can sample with a Gibbs update in a MCMC algorithm.

The full-conditional distribution for $\boldsymbol{\beta}$ resembles those in zero-inflated models in Chapter 22:

$$[\boldsymbol{\beta}|\cdot] \propto \left(\prod_{i=1}^{n} [z_i|\boldsymbol{\beta}]\right) [\boldsymbol{\beta}], \tag{23.13}$$

$$\propto \left(\prod_{i=1}^{n} \psi_i^{z_i} (1 - \psi_i)^{1-z_i}\right) [\boldsymbol{\beta}], \tag{23.14}$$

which is proportional to the same full-conditional distribution for the coefficients we derived in Chapter 20 for logistic regression. Thus, we can use M-H to sample $\boldsymbol{\beta}$ and the same random walk multivariate normal proposal distribution that we used in Chapter 20 that results in the M-H ratio

$$mh = \frac{\left(\prod_{i=1}^{n}[z_i|\boldsymbol{\beta}^{(*)}]\right)[\boldsymbol{\beta}^{(*)}]}{\left(\prod_{i=1}^{n}[z_i|\boldsymbol{\beta}^{(k-1)}]\right)[\boldsymbol{\beta}^{(k-1)}]}, \qquad (23.15)$$

where $[z_i|\boldsymbol{\beta}]$ represents the Bernoulli PMF.

We constructed a MCMC algorithm in R called `occ.simple.mcmc.R` to fit the simple occupancy model to data based on the full-conditional distributions:

BOX 23.1   MCMC ALGORITHM FOR SIMPLE OCCUPANCY MODEL

```
 1 occ.simple.mcmc <- function(y,J,X,n.mcmc){
 2
 3 ####
 4 ####   Subroutines
 5 ####
 6
 7 logit.inv <- function(logit){
 8    exp(logit)/(1+exp(logit))
 9 }
10
11 ####
12 ####   Create Variables
13 ####
14
15 n=length(y)
16 pX=dim(X)[2]
17 beta.save=matrix(0,pX,n.mcmc)
18 z.mean=rep(0,n)
19 p.save=rep(0,n.mcmc)
20 N.save=rep(0,n.mcmc)
21
22 ####
23 ####   Priors and Starting Values
24 ####
25
26 alpha.p=1
27 beta.p=1
28 beta.mn=rep(0,pX)
29 beta.sd=1.5
30 beta.var=beta.sd^2
31 z=rep(0,n)
32 z[y>0]=1
```

(*Continued*)

BOX 23.1 (Continued)   MCMC ALGORITHM FOR SIMPLE OCCUPANCY MODEL

```
33 beta=as.vector(glm(z ~ 0+X,family=binomial())
34   $coefficients)
35 beta.tune=1
36
37 ####
38 ####  Begin MCMC Loop
39 ####
40
41 for(k in 1:n.mcmc){
42
43   ####
44   ####  Sample beta
45   ####
46
47   beta.star=rnorm(pX,beta,beta.tune)
48   mh1=sum(dbinom(z,1,logit.inv(X%*%beta.star),
49     log=TRUE))+sum(dnorm(beta.star,beta.mn,
50     beta.sd,log=TRUE))
51   mh2=sum(dbinom(z,1,logit.inv(X%*%beta),
52     log=TRUE))+sum(dnorm(beta,beta.mn,
53     beta.sd,log=TRUE))
54   mh=exp(mh1-mh2)
55   if(mh > runif(1)){
56     beta=beta.star
57   }
58   psi=logit.inv(X%*%beta)
59
60   ####
61   ####  Sample p
62   ####
63
64   p=rbeta(1,sum(y[z==1])+alpha.p,sum((J-y)[z==1])
65     +beta.p)
66
67   ####
68   ####  Sample z
69   ####
```

(Continued)

BOX 23.1 (Continued)   MCMC ALGORITHM FOR SIMPLE OCCUPANCY MODEL

```
70  num.tmp=psi*(1-p)^J
71  psi.tmp=num.tmp/(num.tmp+(1-psi))
72  z[y==0]=rbinom(sum(y==0),1,psi.tmp[y==0])
73
74  ####
75  ####  Save Samples
76  ####
77
78  beta.save[,k]=beta
79  p.save[k]=p
80  z.mean=z.mean+z/n.mcmc
81  N.save[k]=sum(z)
82
83  }
84
85  ####
86  ####  Write Output
87  ####
88
89  list(beta.save=beta.save,p.save=p.save,N.save
90      =N.save,z.mean=z.mean,n.mcmc=n.mcmc)
91
92  }
```

Notice in this R code that we also calculated and saved the derived quantity $N = \sum_{i=1}^{n} z_i$, which represents the total number of occupied sites. This derived quantity provides additional inference that may be useful for determining the prevalence of a species across a region.

We fit the simple occupancy model to willow warbler (*Phylloscopus trochilus*) data collected at $n = 37$ sites with varying heights and densities of willow (*Salix* spp.) in Finnmark, Norway, over $J = 3$ replicate sampling occasions (Henden et al. 2013). We specified our simple occupancy model as described, with homogeneous occupancy probability p and heterogeneous occupancy probability ψ_i varying with four standardized covariates: area, edge density, willow height, and willow density. Using hyperparameters $\alpha_p \equiv 1$, $\beta_p \equiv 1$, $\mu_\beta = 0$, and $\Sigma_\beta = \sigma_\beta^2 I$ with $\sigma_\beta^2 = 2.25$, we fit the simple occupancy model to the willow warbler data using $K = 50000$ MCMC iterations and the R code as follows:

BOX 23.2 FIT SIMPLE OCCUPANCY MODEL TO DATA

```
full.df=read.delim("PlosOne-DataFinnmark.txt",
    header=TRUE,sep="\t")
spp.idx=full.df$Species=="Willow Warbler"
y=apply(full.df[spp.idx,1:3],1,sum) # 2015 data
n=length(y)
X=matrix(1,n,5)
X[,2]=scale(full.df[spp.idx,"Pland"])
X[,3]=scale(full.df[spp.idx,"ed"])
X[,4]=scale(full.df[spp.idx,"wheight"])
X[,5]=scale(full.df[spp.idx,"wpf"])

source("occ.simple.mcmc.R")
n.mcmc=50000
set.seed(1)
mcmc.out=occ.simple.mcmc(y=y,J=3,X=X,n.mcmc=n.mcmc)

layout(matrix(1:6,3,2,byrow=TRUE))
hist(mcmc.out$beta.save[1,-(1:1000)],breaks=30,
    col=8,xlab=bquote(beta[0]),prob=TRUE,main="a")
curve(dnorm(x,0,1.5),lwd=2,add=TRUE)
hist(mcmc.out$beta.save[2,-(1:1000)],breaks=30,
    col=8,xlab=bquote(beta[1]),prob=TRUE,main="b")
curve(dnorm(x,0,1.5),lwd=2,add=TRUE)
hist(mcmc.out$beta.save[3,-(1:1000)],breaks=30,
    col=8,xlab=bquote(beta[2]),prob=TRUE,main="c")
curve(dnorm(x,0,1.5),lwd=2,add=TRUE)
hist(mcmc.out$beta.save[4,-(1:1000)],breaks=30,
    col=8,xlab=bquote(beta[3]),prob=TRUE,main="d")
curve(dnorm(x,0,1.5),lwd=2,add=TRUE)
hist(mcmc.out$beta.save[5,-(1:1000)],breaks=30,
    col=8,xlab=bquote(beta[4]),prob=TRUE,main="e")
curve(dnorm(x,0,1.5),lwd=2,add=TRUE)
hist(mcmc.out$p.save[-(1:1000)],breaks=30,col=8,
    xlab="p",xlim=c(0,1),prob=TRUE,main="f")
curve(dunif(x),lwd=2,add=TRUE)

mean(mcmc.out$p.save[-(1:1000)])
quantile(mcmc.out$p.save[-(1:1000)],c(0.025,0.975))
```

This R code results in the marginal posterior distributions for β and p shown in Figure 23.1. These results indicate that we do not have enough data to infer that edge density or willow density seem to affect willow warbler occupancy in this study because the marginal posterior distributions overlap zero substantially. Figure 23.1f

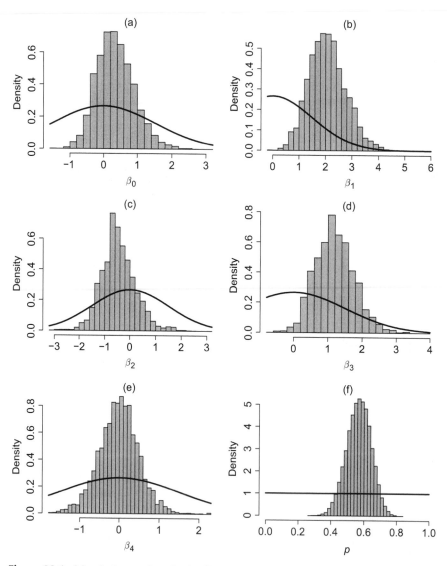

Figure 23.1 Marginal posteriors for (a) β_0 (intercept), (b) β_1 (area), (c) β_2 (edge density), (d) β_3 (willow height), (e) β_4 (willow density), and (f) the detection probability p resulting from fitting the simple occupancy model to the willow warbler data. Prior PDFs are shown as black lines for comparison.

indicates that we have a moderate probability of detecting willow warbler when they are present at a site ($E(p|\mathbf{y}) = 0.57$ and 95% credible interval $[0.42, 0.71]$). Furthermore, we can compute the power to detect the species at a given site in J sampling occasions as

$$P(y > 0|z = 1) = 1 - P(y = 0|z = 1), \tag{23.16}$$

$$= 1 - \binom{J}{0} p^0 (1 - p)^{J-0}, \tag{23.17}$$

$$= 1 - (1 - p)^J. \tag{23.18}$$

This probability depends on the detectability parameter p and the number of sampling occasions J, thus it is a derived quantity and we can make inference about it to better understand our sampling effectiveness, observer/instrument accuracy, and/or design of future occupancy monitoring efforts. In R, we can calculate the posterior mean and 95% credible interval for the power to detect in $J = 3$ occasions as

---

BOX 23.3   CALCULATE POSTERIOR POWER TO DETECT

```
1 > pow.detect=1-(1-mcmc.out$p.save[-(1:1000)])^3
2 > mean(pow.detect)
3 [1] 0.9126949
4 > quantile(pow.detect,c(0.025,0.975))
5       2.5%       97.5%
6 0.8059223 0.9751733
```

Thus, the model output indicates a 91% chance of detecting the willow warbler at a site in $J = 3$ sampling occasions if it is present (95% credible interval: $[0.81, 0.98]$).

23.2   GENERAL OCCUPANCY MODELS

The simple occupancy model that we presented in the previous section makes the fairly strong assumption that detection probability p is homogeneous over sites and occasions. The simplest model extension would allow p_i to vary by site so that certain aspects of the sites themselves or the way the data were collected at the sites could vary. Examples of this type of variation in detectability include changes in understory vegetation density that may obscure many small animals and/or different observers collecting data at the different sites. We only need to make a minor adjustment to the simple occupancy data model (23.2) to allow p_i to vary with covariates

$$y_i \sim \begin{cases} \text{Binom}(J_i, p_i) & , z_i = 1 \\ 0 & , z_i = 0 \end{cases}, \tag{23.19}$$

where $\text{logit}(p_i) = \mathbf{w}_i' \boldsymbol{\alpha}$, for a set of covariates $\mathbf{w}_i$ potentially affecting detectability at sites $i = 1, \ldots, n$ and associated coefficients $\boldsymbol{\alpha}$. The covariates used to explain variation in detectability ($\boldsymbol{\alpha}$) do not need to be the same as those used in the occupancy portion of the model ($\boldsymbol{\beta}$), but they can be. As with the occupancy coefficients $\boldsymbol{\beta}$, we can use a multivariate normal prior for the detection coefficients $\boldsymbol{\alpha}$ such that $\boldsymbol{\alpha} \sim \mathrm{N}(\boldsymbol{\mu}_\alpha, \boldsymbol{\Sigma}_\alpha)$. The remainder of the simple occupancy model presented in the previous section can be reused for this model where the latent occupancy state variables are $z_i \sim \mathrm{Bern}(\psi_i)$ for $i = 1, \ldots, n$, and $\boldsymbol{\beta} \sim \mathrm{N}(\boldsymbol{\mu}_\beta, \boldsymbol{\Sigma}_\beta)$.

The full-conditional distribution for the new parameters $\boldsymbol{\alpha}$ is similar to that we described in the previous section for $\boldsymbol{\beta}$, but involves the data model instead of the process model:

$$[\boldsymbol{\alpha}|\cdot] \propto \left(\prod_{i=1}^{n} [y_i|\boldsymbol{\alpha}]^{z_i} \right) [\boldsymbol{\alpha}], \tag{23.20}$$

$$\propto \left(\prod_{\forall z_i=1} [y_i|\boldsymbol{\alpha}] \right) [\boldsymbol{\alpha}], \tag{23.21}$$

where p_i is known implicitly when $\boldsymbol{\alpha}$ and $\mathbf{w}_i$ are known because $\mathrm{logit}(p_i) = \mathbf{w}_i'\boldsymbol{\alpha}$, and $[y_i|\boldsymbol{\alpha}]$ is a binomial PMF evaluated for observation y_i, with J_i trials and probability p_i. If we use a similar normal random walk proposal distribution for $\boldsymbol{\alpha}$, we can use the M-H ratio

$$mh = \frac{\left(\prod_{\forall z_i=1} [y_i|\boldsymbol{\alpha}^{(*)}] \right) [\boldsymbol{\alpha}^{(*)}]}{\left(\prod_{\forall z_i=1} [y_i|\boldsymbol{\alpha}^{(k-1)}] \right) [\boldsymbol{\alpha}^{(k-1)}]}. \tag{23.22}$$

The full-conditional distribution for the latent occupancy state z_i changes only slightly from the previous section, where now we have $[z_i|\cdot] = \mathrm{Bern}(\tilde{\psi}_i)$ with

$$\tilde{\psi}_i = \frac{\psi_i(1-p_i)^{J_i}}{\psi_i(1-p_i)^{J_i} + 1 - \psi_i}. \tag{23.23}$$

The full-conditional distribution for $\boldsymbol{\beta}$ is the same as the previous section because it only depends on the z_i for $i = 1, \ldots, n$ and its prior.

Using the full-conditional distributions we derived, we constructed a MCMC algorithm in R called occ.gen1.mcmc.R to fit this more general occupancy model that allows detection probability to vary over sites:

BOX 23.4   MCMC ALGORITHM TO FIT GENERAL OCCUPANCY MODEL 1

```
1 occ.gen1.mcmc <- function(y,J,W,X,n.mcmc){
2
3 ####
4 ####   Libraries and Subroutines
5 ####
```

(Continued)

**BOX 23.4 (Continued)   MCMC ALGORITHM TO FIT GENERAL
OCCUPANCY MODEL 1**

```
6  logit.inv <- function(logit){
7    exp(logit.inv)/(1+exp(logit.inv))
8  }
9
10 ####
11 ####   Create Variables
12 ####
13
14 n=length(y)
15 pX=dim(X)[2]
16 pW=dim(W)[2]
17 beta.save=matrix(0,pX,n.mcmc)
18 alpha.save=matrix(0,pW,n.mcmc)
19 z.mean=rep(0,n)
20 N.save=rep(0,n.mcmc)
21
22 ####
23 ####   Priors and Starting Values
24 ####
25
26 beta.mn=rep(0,pX)
27 alpha.mn=rep(0,pW)
28 beta.sd=1.5
29 alpha.sd=1.5
30 z=rep(0,n)
31 z[y>0]=1
32
33 beta=as.vector(glm(z ~ 0+X,family=binomial())
34    $coefficients)
35 beta.tune=2*abs(beta)
36
37 alpha=as.vector(glm(cbind(y[y>1],J[y>1]-y[y>1]) ~
38    0+W[y>1,],family=binomial())$coefficients)
39 alpha.tune=3*abs(alpha)
40
41 ####
42 ####   Begin MCMC Loop
43 ####
```

(Continued)

BOX 23.4 (Continued)   MCMC ALGORITHM TO FIT GENERAL OCCUPANCY MODEL 1

```
44 for(k in 1:n.mcmc){
45
46   ####
47   ####   Sample beta
48   ####
49
50   beta.star=rnorm(pX,beta,beta.tune)
51   mh1=sum(dbinom(z,1,logit.inv(X%*%beta.star),
52     log=TRUE))+sum(dnorm(beta.star,beta.mn,
53     beta.sd,log=TRUE))
54   mh2=sum(dbinom(z,1,logit.inv(X%*%beta),
55     log=TRUE))+sum(dnorm(beta,beta.mn,beta.sd,
56     log=TRUE))
57   mh=exp(mh1-mh2)
58   if(mh > runif(1)){
59     beta=beta.star
60   }
61   psi=logit.inv(X%*%beta)
62
63   ####
64   ####   Sample alpha
65   ####
66
67   alpha.star=rnorm(pW,alpha,alpha.tune)
68   mh1=sum(dbinom(y[z==1],J[z==1],logit.
69     inv(W[z==1,]%*%alpha.star),log=TRUE))+
70     sum(dnorm(alpha.star, alpha.mn,alpha.sd,
71     log=TRUE))
72   mh2=sum(dbinom(y[z==1],J[z==1],logit.
73     inv(W[z==1,] %*%alpha),log=TRUE))+
74     sum(dnorm(alpha,alpha.mn,alpha.sd,log=TRUE))
75   mh=exp(mh1-mh2)
76   if(mh > runif(1)){
77     alpha=alpha.star
78   }
79   p=logit.inv(W%*%alpha)
```

(Continued)

BOX 23.4 (Continued)   MCMC ALGORITHM TO FIT GENERAL OCCUPANCY MODEL 1

```
80   ####
81   ####   Sample z
82   ####
83
84   num.tmp=psi*(1-p)^J
85   psi.tmp=num.tmp/(num.tmp+(1-psi))
86   z[y==0]=rbinom(sum(y==0),1,psi.tmp[y==0])
87
88   ####
89   ####   Save Samples
90   ####
91
92   beta.save[,k]=beta
93   alpha.save[,k]=alpha
94   z.mean=z.mean+z/n.mcmc
95   N.save[k]=sum(z)
96
97 }
98
99 ####
100 ####   Write Output
101 ####
102
103 list(beta.save=beta.save,alpha.save=alpha.save,
104     N.save=N.save,z.mean=z.mean,n.mcmc=n.mcmc)
105
106 }
```

We fit the Bayesian occupancy model with heterogeneous detection probability and occupancy probability over sites to the willow warbler data (Henden et al. 2013) using standardized area and willow height as covariates in both the occupancy and detection portions of the model. In the following R code, we fit the model using hyperparameters $\mu_\alpha = 0$, $\Sigma_\alpha = \sigma_\alpha^2 I$, $\mu_\beta = 0$, $\Sigma_\beta = \sigma_\beta^2 I$, with both σ_α^2 and σ_β^2 equal to 2.25:

BOX 23.5   FIT GENERAL OCCUPANCY MODEL 1 TO DATA

```
full.df=read.delim("PlosOne-DataFinnmark.txt",
    header=TRUE,sep="\t")
spp.idx=full.df$Species=="Willow Warbler"
y=apply(full.df[spp.idx,1:3],1,sum) # 2015 data
n=length(y)
X=matrix(1,n,3)
X[,2]=scale(full.df[spp.idx,"Pland"])
X[,3]=scale(full.df[spp.idx,"wheight"])
W=matrix(1,n,3)
W[,2]=scale(full.df[spp.idx,"Pland"])
W[,3]=scale(full.df[spp.idx,"wheight"])

source("occ.gen1.mcmc.R")
n.mcmc=300000
set.seed(1)
mcmc.out=occ.gen1.mcmc(y=y,J=rep(3,n),W=W,X=X,
    n.mcmc=n.mcmc)

layout(matrix(1:2,2,1))
matplot(t(mcmc.out$alpha.save),type="l",lty=1)
matplot(t(mcmc.out$beta.save),type="l",lty=1)

layout(matrix(1:6,3,2))
hist(mcmc.out$alpha.save[1,-(1:1000)],breaks=30,
    col=8,xlab=bquote(alpha[0]),prob=TRUE,main="a")
curve(dnorm(x,0,1.5),lwd=2,add=TRUE)
hist(mcmc.out$alpha.save[2,-(1:1000)],breaks=30,
    col=8,xlab=bquote(alpha[1]),prob=TRUE,main="c")
curve(dnorm(x,0,1.5),lwd=2,add=TRUE)
hist(mcmc.out$alpha.save[3,-(1:1000)],breaks=30,
    col=8,xlab=bquote(alpha[2]),prob=TRUE,main="e")
curve(dnorm(x,0,1.5),lwd=2,add=TRUE)
hist(mcmc.out$beta.save[1,-(1:1000)],breaks=30,
    col=8,xlab=bquote(beta[0]),prob=TRUE,main="b")
curve(dnorm(x,0,1.5),lwd=2,add=TRUE)
hist(mcmc.out$beta.save[2,-(1:1000)],breaks=30,
    col=8,xlab=bquote(beta[1]),prob=TRUE,main="d")
curve(dnorm(x,0,1.5),lwd=2,add=TRUE)
```

(Continued)

**BOX 23.5 (Continued)   FIT GENERAL OCCUPANCY MODEL 1
TO DATA**

```
39  hist(mcmc.out$beta.save[3,-(1:1000)],breaks=40,
40     col=8,xlab=bquote(beta[2]),prob=TRUE,main="f")
41  curve(dnorm(x,0,1.5),lwd=2,add=TRUE)
```

This R code results in the marginal posterior graphics in Figure 23.2. The marginal posterior histograms for α and β in Figure 23.2 indicate that we have weak evidence that the covariates associated with the sites affect detection and occupancy of willow warbler. For example, we were able to reduce the uncertainty about the parameters α_1 and α_2 (coefficients for the site area and willow height effect on detection) because their marginal posterior distributions are more precise than the priors. The results are similar for the occupancy coefficients β_1 and β_2, indicating a slightly stronger effect of site area than willow height on occupancy of willow warbler.

Based on these results (Figure 23.2), we might conclude that a more appropriate model for the willow warbler data would be an occupancy model with homogeneous detection probability and prior specified directly as $p \sim \text{Beta}(\alpha_p, \beta_p)$ as described in the previous section.

We can generalize the occupancy model one step further by allowing detection probability to vary with both site and occasion (p_{ij}) as long as covariates are available for both site and occasion. Thus, the data component for this fully general occupancy model is

$$y_{ij} \sim \begin{cases} \text{Bern}(p_{ij}) & , z_i = 1 \\ 0 & , z_i = 0 \end{cases}, \tag{23.24}$$

where the detection probabilities are linked to site and sampling occasion covariates by $\text{logit}(p_{ij}) = \mathbf{w}_{ij}'\boldsymbol{\alpha}$ and the rest of the occupancy model remains the same as the former general occupancy model. The posterior distribution for this fully general Bayesian occupancy model is

$$[\mathbf{z}, \boldsymbol{\alpha}, \boldsymbol{\beta} | \mathbf{Y}] \propto \left(\prod_{i=1}^{n} \left(\prod_{j=1}^{J_i} [y_{ij}|\boldsymbol{\alpha}]^{z_i} 1_{\{y_{ij}=0\}}^{1-z_i} \right) [z_i|\boldsymbol{\beta}] \right) [\boldsymbol{\alpha}][\boldsymbol{\beta}], \tag{23.25}$$

where $[y_{ij}|\boldsymbol{\alpha}]$ is a Bernoulli PMF with probability $p_{ij} = \exp(\mathbf{w}_{ij}'\boldsymbol{\alpha})/(1 + \exp(\mathbf{w}_{ij}'\boldsymbol{\alpha}))$. Specifying the same multivariate normal priors as we used in the previous general occupancy model, only the full-conditional distributions for z_i $(i = 1, \ldots, n)$ and $\boldsymbol{\alpha}$ will be different.

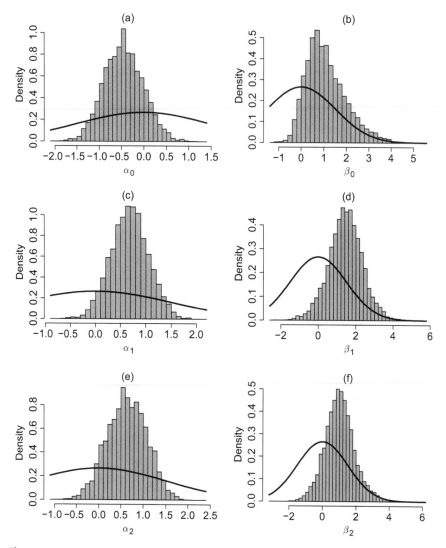

Figure 23.2 Marginal posterior histograms for the detection coefficients (a) α_0 (detection intercept), (b) β_0 (occupancy intercept) , (c) α_1 (detection area), (d) β_1 (occupancy area), (e) α_2 (detection willow height), and (f) β_2 (occupancy willow height) resulting from fitting the site-general occupancy model to the willow warbler data. Prior PDFs are shown as black lines for comparison.

For $\boldsymbol{\alpha}$, the full-conditional distribution is

$$[\boldsymbol{\alpha}|\cdot] \propto \left(\prod_{i=1}^{n} \prod_{j=1}^{J_i} [y_{ij}|\boldsymbol{\alpha}]^{z_i} \right) [\boldsymbol{\alpha}], \qquad (23.26)$$

$$\propto \left(\prod_{\forall z_i=1} \prod_{j=1}^{J_i} [y_{ij}|\boldsymbol{\alpha}] \right) [\boldsymbol{\alpha}], \tag{23.27}$$

resulting in the M-H ratio

$$mh = \frac{\left(\prod_{\forall z_i=1} \prod_{j=1}^{J_i} [y_{ij}|\boldsymbol{\alpha}^{(*)}] \right) [\boldsymbol{\alpha}^{(*)}]}{\left(\prod_{\forall z_i=1} \prod_{j=1}^{J_i} [y_{ij}|\boldsymbol{\alpha}^{(k-1)}] \right) [\boldsymbol{\alpha}^{(k-1)}]}, \tag{23.28}$$

if we use the same multivariate normal random walk proposal as before. Thus, we can use a M-H update to sample $\boldsymbol{\alpha}$ in our MCMC algorithm.

The full-conditional distribution for z_i is slightly different than the previous example because y_{ij} and p_{ij} both vary with site and sampling occasion. For the fully general occupancy model, we only sample z_i in the MCMC algorithm when $\sum_{j=1}^{J_i} y_{ij} = 0$ (otherwise we let $z_i = 1$), using the full-conditional distribution for z_i is Bernoulli with probability

$$\tilde{\psi}_i = \frac{\psi_i \prod_{j=1}^{J_i} (1-p_{ij})}{\psi_i \prod_{j=1}^{J_i} (1-p_{ij}) + 1 - \psi_i}. \tag{23.29}$$

Based on these full-conditional distributions, we constructed a MCMC algorithm called occ.gen2.mcmc.R to fit the fully general occupancy model to data in R:

BOX 23.6   MCMC ALGORITHM FOR GENERAL OCCUPANCY MODEL 2

```
1 occ.gen2.mcmc <- function(Y,J,W,X,n.mcmc){
2
3 ####
4 ####   Subroutines
5 ####
6
7 logit.inv <- function(logit){
8   exp(logit)/(1+exp(logit))
9 }
10
11 ####
12 ####   Create Variables
13 ####
14
15 n=dim(Y)[1]
16 pX=dim(X)[2]
```

(Continued)

BOX 23.6 (Continued)   MCMC ALGORITHM FOR GENERAL OCCUPANCY MODEL 2

```
17 pW=dim(W)[2]
18 beta.save=matrix(0,pX,n.mcmc)
19 alpha.save=matrix(0,pW,n.mcmc)
20 z.mean=rep(0,n)
21 N.save=rep(0,n.mcmc)
22
23 ####
24 ####   Priors and Starting Values
25 ####
26
27 y=apply(Y,1,sum,na.rm=TRUE)
28 y.vec=c(Y)
29 y0.TF=(apply(Y,1,sum)==0)
30 beta.mn=rep(0,pX)
31 alpha.mn=rep(0,pW)
32 beta.sd=1.5
33 alpha.sd=1.5
34 z=rep(0,n)
35 z[!y0.TF]=1
36
37 beta=as.vector(glm(z ~ 0+X,family=binomial())
38     $coefficients)
39 beta.tune=.5*abs(beta)
40
41 alpha=as.vector(glm(c(Y[z==1,]) ~ 0+W[rep(z==1,
42     J),],family=binomial())$coefficients)
43 alpha.tune=1*abs(alpha)
44
45 ####
46 ####   Begin MCMC Loop
47 ####
48
49 for(k in 1:n.mcmc){
50
51   ####
52   ####   Sample beta
53   ####
```

(*Continued*)

BOX 23.6 (Continued)   MCMC ALGORITHM FOR GENERAL
OCCUPANCY MODEL 2

```
54  beta.star=rnorm(pX,beta,beta.tune)
55  mh1=sum(dbinom(z,1,logit.inv(X%*%beta.star),
56    log=TRUE))+sum(dnorm(beta.star,beta.mn,
57    beta.sd,log=TRUE))
58  mh2=sum(dbinom(z,1,logit.inv(X%*%beta),log=TRUE))
59    +sum(dnorm(beta,beta.mn,beta.sd,log=TRUE))
60  mh=exp(mh1-mh2)
61  if(mh > runif(1)){
62    beta=beta.star
63  }
64  psi=logit.inv(X%*%beta)
65
66  ####
67  ####  Sample alpha
68  ####
69
70  alpha.star=rnorm(pW,alpha,alpha.tune)
71  mh1=sum(dbinom(c(Y[z==1,]),1,logit.inv(W[rep
72    (z==1,J),]%*%alpha.star),log=TRUE))+sum(dnorm
73    (alpha.star,alpha.mn,alpha.sd,log=TRUE))
74  mh2=sum(dbinom(c(Y[z==1,]),1,logit.inv
75    (W[rep(z==1,J),]%*%alpha),log=TRUE))+
76    sum(dnorm(alpha,alpha.mn,alpha.sd,log=TRUE))
77  mh=exp(mh1-mh2)
78  if(mh > runif(1)){
79    alpha=alpha.star
80  }
81  p=logit.inv(W%*%alpha)
82
83  ####
84  ####  Sample z
85  ####
86
87  num.tmp=psi*apply(1-matrix(p,,J),1,prod)
88  psi.tmp=num.tmp/(num.tmp+(1-psi))
89  z[y0.TF]=rbinom(sum(y0.TF),1,psi.tmp[y0.TF])
```

(Continued)

BOX 23.6 (Continued) MCMC ALGORITHM FOR GENERAL OCCUPANCY MODEL 2

```
90   ####
91   ####  Save Samples
92   ####
93
94   beta.save[,k]=beta
95   alpha.save[,k]=alpha
96   z.mean=z.mean+z/n.mcmc
97   N.save[k]=sum(z)
98
99 }
100
101 ####
102 ####  Write Output
103 ####
104
105 list(beta.save=beta.save,alpha.save=alpha.save,
106     N.save=N.save,z.mean=z.mean,n.mcmc=n.mcmc)
107
108 }
```

We fit the fully general occupancy model to the same set of occupancy data for willow warbler, but allowing for heterogeneity over the $J = 3$ sampling occasions. We used standardized versions of occasion index and willow height as covariates for detection probability (note that willow height does not vary over the occasions). For occupancy covariates, we used standardized site area and willow height as in the previous section. The following R code fits the fully general occupancy model to these willow warbler data using our MCMC algorithm above:

BOX 23.7 FIT GENERAL OCCUPANCY MODEL 2 TO DATA

```
1 full.df=read.delim("PlosOne-DataFinnmark.txt",
2     header=TRUE,sep="\t")
3 spp.idx=full.df$Species=="Willow Warbler"
4 Y=as.matrix(full.df[spp.idx,1:3])
5 n=dim(Y)[1]
6 J=dim(Y)[2]
7 X=matrix(1,n,3)
8 X[,2]=scale(full.df[spp.idx,"Pland"])
```

(Continued)

BOX 23.7 (Continued)   FIT GENERAL OCCUPANCY MODEL 2 TO DATA

```
 9 X[,3]=scale(full.df[spp.idx,"wheight"])
10 W=matrix(1,n*J,3)
11 W[,2]=scale(c(col(Y)))
12 W[,3]=rep(scale(full.df[spp.idx,"wheight"]),J)
13
14 source("occ.gen2.mcmc.R")
15 n.mcmc=300000
16 set.seed(1)
17 mcmc.out=occ.gen2.mcmc(Y=Y,J=J,W=W,X=X,
18     n.mcmc=n.mcmc)
19
20 layout(matrix(1:6,3,2))
21 hist(mcmc.out$alpha.save[1,-(1:1000)],breaks=30,
22     col=8,xlab=bquote(alpha[0]),prob=TRUE,main="a")
23 curve(dnorm(x,0,1.5),lwd=2,add=TRUE)
24 hist(mcmc.out$alpha.save[2,-(1:1000)],breaks=30,
25     col=8,xlab=bquote(alpha[1]),prob=TRUE,main="c")
26 curve(dnorm(x,0,1.5),lwd=2,add=TRUE)
27 hist(mcmc.out$alpha.save[3,-(1:1000)],breaks=30,
28     col=8,xlab=bquote(alpha[2]),prob=TRUE,main="e")
29 curve(dnorm(x,0,1.5),lwd=2,add=TRUE)
30 hist(mcmc.out$beta.save[1,-(1:1000)],breaks=30,
31     col=8,xlab=bquote(beta[0]),prob=TRUE,main="b")
32 curve(dnorm(x,0,1.5),lwd=2,add=TRUE)
33 hist(mcmc.out$beta.save[2,-(1:1000)],breaks=30,
34     col=8,xlab=bquote(beta[1]),prob=TRUE,main="d")
35 curve(dnorm(x,0,1.5),lwd=2,add=TRUE)
36 hist(mcmc.out$beta.save[3,-(1:1000)],breaks=30,
37     col=8,xlab=bquote(beta[2]),prob=TRUE,main="f")
38 curve(dnorm(x,0,1.5),lwd=2,add=TRUE)
39
40 mean(mcmc.out$N.save[-(1:1000)])
41 quantile(mcmc.out$N.save[-(1:1000)],
42     c(0.025,0.975))
```

The marginal posterior distributions associated with α and β are shown in Figure 23.3. The results shown in Figure 23.3 indicate that, while there only appears to be a weak effect of sampling occasion index and willow height on willow warbler detection probability, there does appear to be an effect of site area on willow warbler occupancy (as we would expect because more individuals can occupy

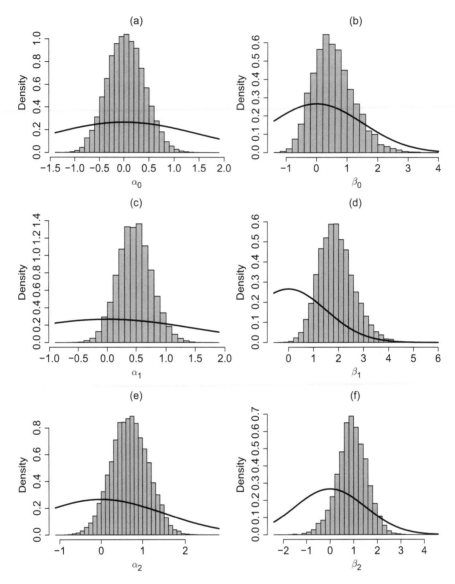

Figure 23.3 Marginal posterior histograms for the detection coefficients (a) α_0 (detection intercept), (b) β_0 (occupancy intercept), (c) α_1 (occasion index), (d) β_1 (area), (e) α_2 (detection willow height), and (f) β_2 (occupancy willow height) resulting from fitting the fully general occupancy model to the willow warbler data. Prior PDFs are shown as black lines for comparison.

larger areas). Furthermore, the posterior mean for total number of occupied sites was $E(N|\mathbf{Y}) = 20.6$ with a 95% credible interval of (18, 27). Thus, we estimated approximately 56% of the sites occupied even though willow warblers were only detected at 18 out of 37 sites (48.6%).

23.3 PROBIT OCCUPANCY MODELS

In Chapter 20, we introduced the concept of specifying hierarchical Bayesian binary data models using auxiliary variables. The basic approach was to introduce a new auxiliary variable for each observation that is real-valued and arises from a conditional normal distribution with variance one. Each auxiliary variable indicates whether the data were ones versus zeros based on whether it was larger or smaller than zero. Although originally introduced by Albert and Chib (1993), Hooten et al. (2003) were the first to apply the auxiliary variable approach to specifying hierarchical ecological models, and then Dorazio and Rodriguez (2012) and Johnson et al. (2013a) applied it in the occupancy modeling context.

The extension of the probit binary regression model (Chapter 20) to the general occupancy context is straightforward, but involves two sets of latent variables, one for the detection process (p) and another for the occupancy process (ψ). As we described in Chapter 20, we write the data model as a mixture of zeros and ones such that

$$y_{ij} = \begin{cases} 0 & , z_i = 0 \\ 0 & , z_i = 1, u_{ij} \leq 0, \\ 1 & , z_i = 1, u_{ij} > 0 \end{cases} \tag{23.30}$$

where we also specify the latent occupancy process z_i as a mixture of zeros and ones exactly as we presented it in the probit occupancy model of Chapter 20:

$$z_i = \begin{cases} 0 & , v_i \leq 0 \\ 1 & , v_i > 0 \end{cases}, \tag{23.31}$$

where $u_{ij} \sim \mathrm{N}(\mathbf{w}_{ij}'\boldsymbol{\alpha}, 1)$ and $v_i \sim \mathrm{N}(\mathbf{x}_i'\boldsymbol{\beta}, 1)$, with multivariate normal priors as we defined in the last section. Perhaps surprisingly, this auxiliary variable model formulation is equivalent to the general occupancy model from the previous section but with the link functions specified as the probit (standard normal inverse CDF; Φ^{-1}) instead of the logit such that $\Phi^{-1}(p_{ij}) = \mathbf{w}_{ij}'\boldsymbol{\alpha}$ and $\Phi^{-1}(\psi_i) = \mathbf{x}_i'\boldsymbol{\beta}$.

The general occupancy model with auxiliary variables results in the following posterior

$$[\mathbf{U}, \mathbf{v}, \mathbf{z}, \boldsymbol{\alpha}, \boldsymbol{\beta}|\mathbf{Y}] \propto \left(\prod_{i=1}^{n} \left(\prod_{j=1}^{J_i} [y_{ij}|z_i, u_{ij}][u_{ij}|\boldsymbol{\alpha}] \right) [z_i|v_i][v_i|\boldsymbol{\beta}] \right) [\boldsymbol{\alpha}][\boldsymbol{\beta}]. \tag{23.32}$$

To construct a MCMC algorithm to fit the general occupancy model with auxiliary variables we need to derive full-conditional distributions. We begin with the full-conditional distribution for the regression coefficients ($\boldsymbol{\beta}$ and $\boldsymbol{\alpha}$), then the auxiliary variables (u_{ij} and v_i, for $i = 1,\ldots,n$ and $j = 1,\ldots,J_i$), then the latent occupancy state variables z_i for all sites where $\sum_{j=1}^{J_i} y_{ij} = 0$.

The full-conditional distribution for the occupancy coefficients $\boldsymbol{\beta}$ is proportional to the components of the joint distribution (23.32) that involve $\boldsymbol{\beta}$; this results in

$$[\boldsymbol{\beta}|\cdot] \propto \left(\prod_{i=1}^{n}[v_i|\boldsymbol{\beta}]\right)[\boldsymbol{\beta}], \tag{23.33}$$

which is exactly the same form as in the Bayesian linear regression model from Chapter 11. Thus, we can borrow the result we derived there, using v_i as the response variables and setting the variance parameter equal to 1. Therefore, the full-conditional distribution for $\boldsymbol{\beta}$ is a conjugate multivariate normal such that $[\boldsymbol{\beta}|\cdot] = N(\mathbf{A}^{-1}\mathbf{b}, \mathbf{A}^{-1})$ where

$$\mathbf{b} \equiv \mathbf{X}'\mathbf{v} + \boldsymbol{\Sigma}_\beta^{-1}\boldsymbol{\mu}_\beta, \tag{23.34}$$

$$\mathbf{A} \equiv \mathbf{X}'\mathbf{X} + \boldsymbol{\Sigma}_\beta^{-1}. \tag{23.35}$$

Thus, we can sample $\boldsymbol{\beta}$ using a Gibbs update in our MCMC algorithm.

The full-conditional distribution for the detection coefficients $\boldsymbol{\alpha}$ is similar to that for $\boldsymbol{\beta}$ except that we have two products to keep track of

$$[\boldsymbol{\alpha}|\cdot] \propto \left(\prod_{i=1}^{n}\prod_{j=1}^{J_i}[u_{ij}|\boldsymbol{\alpha}]^{z_i}\right)[\boldsymbol{\alpha}], \tag{23.36}$$

$$\propto \left(\prod_{\forall z_i=1}\prod_{j=1}^{J_i}[u_{ij}|\boldsymbol{\alpha}]\right)[\boldsymbol{\alpha}], \tag{23.37}$$

$$\propto \left(\prod_{\forall z_i=1}\prod_{j=1}^{J_i}\exp\left(-\frac{1}{2}(u_{ij}-\mathbf{w}_{ij}'\boldsymbol{\alpha})^2\right)\right)\exp\left(-\frac{1}{2}(\boldsymbol{\alpha}-\boldsymbol{\mu}_\alpha)'\boldsymbol{\Sigma}_\alpha^{-1}(\boldsymbol{\alpha}-\boldsymbol{\mu}_\alpha)\right), \tag{23.38}$$

$$\propto \exp\left(-\frac{1}{2}\sum_{\forall z_i=1}\sum_{j=1}^{J_i}(u_{ij}-\mathbf{w}_{ij}'\boldsymbol{\alpha})^2\right)\exp\left(-\frac{1}{2}(\boldsymbol{\alpha}-\boldsymbol{\mu}_\alpha)'\boldsymbol{\Sigma}_\alpha^{-1}(\boldsymbol{\alpha}-\boldsymbol{\mu}_\alpha)\right), \tag{23.39}$$

$$\propto \exp\left(-\frac{1}{2}\left(-2\left(\sum_{\forall z_i=1}\sum_{j=1}^{J_i}u_{ij}\mathbf{w}_{ij}'\right)\boldsymbol{\alpha}+\boldsymbol{\alpha}'\left(\sum_{\forall z_i=1}\sum_{j=1}^{J_i}\mathbf{w}_{ij}\mathbf{w}_{ij}'\right)\boldsymbol{\alpha}-2\boldsymbol{\mu}_\alpha'\boldsymbol{\Sigma}_\alpha^{-1}\boldsymbol{\alpha}+\boldsymbol{\alpha}'\boldsymbol{\Sigma}_\alpha^{-1}\boldsymbol{\alpha}\right)\right), \tag{23.40}$$

$$\propto \exp\left(-\frac{1}{2}\left(-2\left(\sum_{\forall z_i=1}\sum_{j=1}^{J_i} u_{ij}\mathbf{w}'_{ij}+\boldsymbol{\mu}'_\alpha\boldsymbol{\Sigma}_\alpha^{-1}\right)\boldsymbol{\alpha}\right.\right.$$

$$\left.\left.+\boldsymbol{\alpha}'\left(\sum_{\forall z_i=1}\sum_{j=1}^{J_i}\mathbf{w}_{ij}\mathbf{w}'_{ij}+\boldsymbol{\Sigma}_\alpha^{-1}\right)\boldsymbol{\alpha}\right)\right), \tag{23.41}$$

$$\propto \exp\left(-\frac{1}{2}\left(-2\mathbf{b}'\boldsymbol{\alpha}+\boldsymbol{\alpha}'\mathbf{A}\boldsymbol{\alpha}\right)\right), \tag{23.42}$$

which implies that $[\boldsymbol{\alpha}|\cdot]=N(\mathbf{A}^{-1}\mathbf{b},\mathbf{A}^{-1})$ with

$$\mathbf{b}'\equiv\sum_{\forall z_i=1}\sum_{j=1}^{J_i}u_{ij}\mathbf{w}'_{ij}+\boldsymbol{\mu}'_\alpha\boldsymbol{\Sigma}_\alpha^{-1}, \tag{23.43}$$

$$\mathbf{A}\equiv\sum_{\forall z_i=1}\sum_{j=1}^{J_i}\mathbf{w}_{ij}\mathbf{w}'_{ij}+\boldsymbol{\Sigma}_\alpha^{-1}. \tag{23.44}$$

Thus, the full-conditional distribution is also conjugate and we can sample $\boldsymbol{\alpha}$ using a Gibbs update in a MCMC algorithm to fit the general occupancy model with auxiliary variables to data.

Shifting our attention now to the auxiliary variables, we can derive the full-conditional distribution for v_i, for $i=1,\ldots,n$, as

$$[v_i|\cdot]\propto[z_i|v_i][v_i|\boldsymbol{\beta}], \tag{23.45}$$

$$\propto(1_{\{v_i>0\}}1_{\{z_i=1\}}+1_{\{v_i\leq0\}}1_{\{z_i=0\}})[v_i|\boldsymbol{\beta}], \tag{23.46}$$

$$=\begin{cases}\text{TN}(\mathbf{x}'_i\boldsymbol{\beta},1)_0^\infty & ,z_i=1\\\text{TN}(\mathbf{x}'_i\boldsymbol{\beta},1)_{-\infty}^0 & ,z_i=0\end{cases}, \tag{23.47}$$

where "TN" refers to a truncated normal distribution that is truncated below by the subscript and above by the superscript. Thus, we can sample v_i for $i=1,\ldots,n$ using a Gibbs update conditioned on whether $z_i=0$ or $z_i=1$.

The full-conditional distribution for u_{ij} is similar to that of v_i except that we only sample u_{ij} for $j=1,\ldots,J_i$ when $z_i=1$ because it relates to detection probability and is only relevant when the species occupies a site and we can detect it. Thus, the full-conditional distribution for u_{ij} is

$$[u_{ij}|\cdot]\propto[y_{ij}|u_{ij}][u_{ij}|\boldsymbol{\alpha}], \tag{23.48}$$

$$\propto(1_{\{u_{ij}>0\}}1_{\{y_{ij}=1\}}+1_{\{u_{ij}\leq0\}}1_{\{y_{ij}=0\}})[u_{ij}|\boldsymbol{\alpha}], \tag{23.49}$$

$$=\begin{cases}\text{TN}(\mathbf{w}'_{ij}\boldsymbol{\alpha},1)_0^\infty & ,y_{ij}=1\\\text{TN}(\mathbf{w}'_{ij}\boldsymbol{\alpha},1)_{-\infty}^0 & ,y_{ij}=0\end{cases}. \tag{23.50}$$

Thus, we also can sample u_{ij} using a Gibbs sample conditioned on y_{ij}.

Finally, we need to derive the full-conditional distribution for the latent occupancy process z_i. As with previous latent binary variables in this book, the associated full-conditional distribution is Bernoulli. In fact, the full-conditional distribution for z_i is the same as the one we worked out for the general occupancy model in the previous chapter where $[z_i|\cdot] = \text{Bern}(\tilde{\psi}_i)$ and

$$\tilde{\psi}_i = \frac{\psi_i \prod_{j=1}^{J_i}(1-p_{ij})}{\psi_i \prod_{j=1}^{J_i}(1-p_{ij}) + 1 - \psi_i},\tag{23.51}$$

such that $\psi_i = \Phi(\mathbf{x}_i'\boldsymbol{\beta})$ and we only sample z_i in our MCMC algorithm when $\sum_{j=1}^{J_i} y_{ij} = 0$.

These full-conditional distributions indicate that we can fit a general Bayesian occupancy model using a fully Gibbs MCMC algorithm where we can sample all parameters from known distributions. The resulting MCMC algorithm is substantially more complicated than the MCMC algorithm that we used to fit the occupancy models with logit link functions in the previous sections, but the probit link induced through the auxiliary variables eliminates the need to tune the algorithm. Thus, we can apply the probit version of the general occupancy model in situations with different sets of covariates for detection and occupancy without any changes to the MCMC algorithm. We also can simplify the general occupancy model, for example, by replacing the detection auxiliary variables with a single homogeneous detection probability p as we used in the previous. This outcome still results in a fully Gibbs MCMC algorithm because the full-conditional distribution for p will be conjugate beta as we showed previously.

We constructed a MCMC algorithm called `occ.aux.mcmc.R` in R to fit the Bayesian probit occupancy model to data:

BOX 23.8   MCMC ALGORITHM FOR GENERAL OCCUPANCY MODEL WITH AUXILIARY VARIABLES

```
1 occ.aux.mcmc <- function(Y,J,W,X,n.mcmc){
2
3 ####
4 ####   Libraries and Subroutines
5 ####
6
7 rtn <- function(n,mu,sig2,low,high){
8   flow=pnorm(low,mu,sqrt(sig2))
9   fhigh=pnorm(high,mu,sqrt(sig2))
10   u=runif(n)
11   tmp=flow+u*(fhigh-flow)
```

(Continued)

BOX 23.8 (Continued)   MCMC ALGORITHM FOR GENERAL OCCUPANCY MODEL WITH AUXILIARY VARIABLES

```
12    x=qnorm(tmp,mu,sqrt(sig2))
13    x
14  }
15
16  ####
17  ####  Setup Variables
18  ####
19
20  n=dim(Y)[1]
21  p.X=dim(X)[2]
22  p.W=dim(W)[2]
23  n.burn=round(0.25*n.mcmc)
24
25  z.mean=rep(0,n)
26  beta.save=matrix(0,p.X,n.mcmc)
27  alpha.save=matrix(0,p.W,n.mcmc)
28  N.save=rep(0,n.mcmc)
29
30  ####
31  ####  Priors and Starting Values
32  ####
33
34  beta=rep(0,p.X)
35  alpha=rep(0,p.W)
36  Xbeta=X%*%beta
37  Walpha=W%*%alpha
38  psi=pnorm(Xbeta)
39  p=pnorm(Walpha)
40  Y.sum=apply(Y,1,sum)
41  z=rep(0,n)
42  z[Y.sum==0]=0
43  n0=sum(z==0)
44  n1=sum(z==1)
45  z.1.big=rep(z==1,J)
46
47  v=rep(0,n)
48  u=rep(0,n*J)
```

(Continued)

BOX 23.8 (Continued)   MCMC ALGORITHM FOR GENERAL OCCUPANCY MODEL WITH AUXILIARY VARIABLES

```
49 y=as.vector(Y)
50 ny0=sum(y==0)
51 ny1=sum(y==1)
52
53 Sig.beta=(1^2)*diag(p.X)
54 Sig.alpha=(1^2)*diag(p.W)
55 Sig.beta.inv=solve(Sig.beta)
56 Sig.alpha.inv=solve(Sig.alpha)
57
58 mu.beta=rep(0,p.X)
59 mu.alpha=rep(0,p.W)
60
61 XprimeX=t(X)%*%X
62 A.beta=XprimeX+Sig.beta.inv
63 A.beta.chol=chol(A.beta)
64 Sig.beta.inv.times.mu.beta=Sig.beta.inv%*%mu.beta
65 Sig.alpha.inv.times.mu.alpha=Sig.alpha.inv
66     %*%mu.alpha
67
68 ####
69 ####  Begin MCMC Loop
70 ####
71
72 for(k in 1:n.mcmc){
73
74   ####
75   ####  Sample v
76   ####
77
78   v[z==0]=rtn(n0,Xbeta[z==0],1,-Inf,0)
79   v[z==1]=rtn(n1,Xbeta[z==1],1,0,Inf)
80
81   ####
82   ####  Sample U
83   ####
```

(*Continued*)

BOX 23.8 (Continued) MCMC ALGORITHM FOR GENERAL OCCUPANCY MODEL WITH AUXILIARY VARIABLES

```
84  u[y==0]=rtn(ny0,Walpha[y==0],1,-Inf,0)
85  u[y==1]=rtn(ny1,Walpha[y==1],1,0,Inf)
86
87  ####
88  ####  Sample beta
89  ####
90
91  b.beta=t(X)%*%v+Sig.beta.inv.times.mu.beta
92  beta=backsolve(A.beta.chol, backsolve
93     (A.beta.chol,b.beta, transpose = TRUE) +
94     rnorm(p.X))
95  Xbeta=X%*%beta
96  psi=pnorm(Xbeta)
97
98  ####
99  ####  Sample alpha
100 ####
101
102 W.tmp=W[z.1.big,]
103 A.alpha=t(W.tmp)%*%W.tmp+Sig.alpha.inv
104 A.alpha.chol=chol(A.alpha)
105 b.alpha=t(W.tmp)%*%u[z.1.big]+Sig.alpha.
106    inv.times.mu.alpha
107 alpha=backsolve(A.alpha.chol, backsolve
108    (A.alpha.chol, b.alpha,transpose=TRUE)
109    +rnorm(p.W))
110 Walpha=W%*%alpha
111 p=pnorm(Walpha)
112
113 ####
114 ####  Sample z
115 ####
116
117 psi.numer=psi*apply(1-matrix(p,n,J),1,prod)
118 psi.tmp=psi.numer/(psi.numer+1-psi)
119 z[Y.sum==0]=rbinom(sum(Y.sum==0),1,
120    psi.tmp[Y.sum==0])
121 z[Y.sum>0]=1
```

(Continued)

BOX 23.8 (Continued)   MCMC ALGORITHM FOR GENERAL OCCUPANCY MODEL WITH AUXILIARY VARIABLES

```
122  n0=sum(z==0)
123  n1=sum(z==1)
124  z.1.big=rep(z==1,J)
125
126  ####
127  ####   Save Samples
128  ####
129
130  beta.save[,k]=beta
131  alpha.save[,k]=alpha
132  if(k > n.burn){
133    z.mean=z.mean+z/(n.mcmc-n.burn)
134  }
135  N.save[k]=sum(z)
136
137  }
138
139  ####
140  ####   Write Output
141  ####
142
143  list(alpha.save=alpha.save,beta.save=beta.save,
144    z.mean=z.mean,N.save=N.save,n.mcmc=n.mcmc)
145
146  }
```

Notice that we used our own function ($\texttt{rtn}$) to sample truncated normal variables v_i and u_{ij} as well as the "backsolve" technique to sample correlated multivariate normal random vectors $\boldsymbol{\alpha}$ and $\boldsymbol{\beta}$. Also notice that we sampled all of the u_{ij} (even where $z_i = 0$) because it was slightly easier to code and it does not affect the results, but it is not necessary to do that. We assumed that the number of replicates were the same for all sites ($J_i = J$ for $i = 1, \ldots, n$), but we could generalize this by restructuring the detection-related design matrix $\mathbf{W}$. Finally, we assumed that $\boldsymbol{\Sigma}_\alpha \equiv \mathbf{I}$ and $\boldsymbol{\Sigma}_\beta \equiv \mathbf{I}$ in our MCMC algorithm, which implies that all the regression coefficients are independent with variance equal to one a priori.

We fit the general Bayesian occupancy model to the same willow warbler data from Norway (Henden et al. 2013) that we introduced in the previous section using the R code:

BOX 23.9 FIT GENERAL OCCUPANCY MODEL WITH AUXILIARY VARIABLES TO DATA

```
1  full.df=read.delim("PlosOne-DataFinnmark.txt",
2      header=TRUE,sep="\t")
3  spp.idx=full.df$Species=="Willow Warbler"
4  Y=as.matrix(full.df[spp.idx,1:3])
5  n=dim(Y)[1]
6  J=dim(Y)[2]
7  X=matrix(1,n,3)
8  X[,2]=scale(full.df[spp.idx,"Pland"])
9  X[,3]=scale(full.df[spp.idx,"wheight"])
10 W=matrix(1,n*J,3)
11 W[,2]=scale(c(col(Y)))
12 W[,3]=rep(scale(full.df[spp.idx,"wheight"]),J)
13
14 source("occ.aux.mcmc.R")
15 n.mcmc=100000
16 mcmc.out=occ.aux.mcmc(Y=Y,J=J,W=W,X=X,n.mcmc=n.
      mcmc)
17
18 layout(matrix(1:6,3,2))
19 hist(mcmc.out$alpha.save[1,-(1:1000)],breaks=30,
20     col=8,xlab=bquote(alpha[0]),prob=TRUE,main="a")
21 curve(dnorm(x,0,1.5),lwd=2,add=TRUE)
22 hist(mcmc.out$alpha.save[2,-(1:1000)],breaks=30,
23     col=8,xlab=bquote(alpha[1]),prob=TRUE,main="c")
24 curve(dnorm(x,0,1.5),lwd=2,add=TRUE)
25 hist(mcmc.out$alpha.save[3,-(1:1000)],breaks=30,
26     col=8,xlab=bquote(alpha[2]),prob=TRUE,main="e")
27 curve(dnorm(x,0,1.5),lwd=2,add=TRUE)
28 hist(mcmc.out$beta.save[1,-(1:1000)],breaks=30,
29     col=8,xlab=bquote(beta[0]),prob=TRUE,main="b")
30 curve(dnorm(x,0,1.5),lwd=2,add=TRUE)
31 hist(mcmc.out$beta.save[2,-(1:1000)],breaks=30,
32     col=8,xlab=bquote(beta[1]),prob=TRUE,main="d")
33 curve(dnorm(x,0,1.5),lwd=2,add=TRUE)
34 hist(mcmc.out$beta.save[3,-(1:1000)],breaks=30,
35     col=8,xlab=bquote(beta[2]),prob=TRUE,main="f")
36 curve(dnorm(x,0,1.5),lwd=2,add=TRUE)
37
38 mean(mcmc.out$N.save[-(1:1000)])
39 quantile(mcmc.out$N.save[-(1:1000)],
40     c(0.025,0.975))
```

This R code produces the marginal posterior histograms for α and β shown in Figure 23.4. The inference resulting from a fit of the probit occupancy model to the willow warbler data in Figure 23.4 is essentially the same as that we obtained using the logit version of the same model (Figure 23.3). The occasion index and

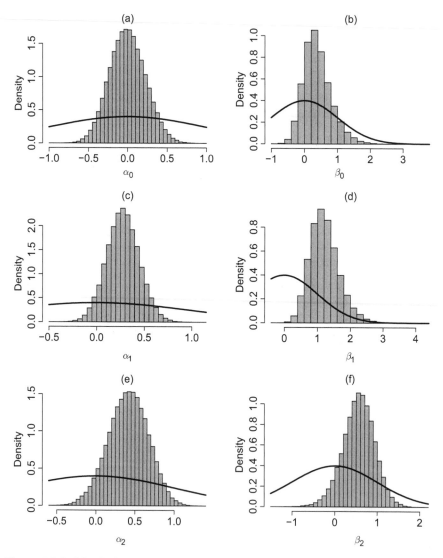

Figure 23.4 Marginal posterior histograms for the detection coefficients (a) α_0 (detection intercept), (b) β_0 (occupancy intercept), (c) α_1 (occasion index), (d) β_1 (area), (e) α_2 (detection willow height), and (f) β_2 (occupancy willow height) resulting from fitting the fully general occupancy model with auxiliary variables to the willow warbler data. Prior PDFs are shown as black lines for comparison.

willow height do not appear to strongly affect our detection probability, but the site area does appear to correlate with greater willow warbler occupancy. Notice that the actual values for the marginal posterior distributions of α and β are slightly different between the two different models, but the relative magnitude and uncertainty are similar. These differences are due to the logit versus probit link function. In this case, the choice of link function does not affect our inference. In fact, the posterior mean total occupied sites was $E(N|\mathbf{Y}) = 21.8$ with 95% credible interval $(18,27)$, which is similar to the posterior inference using the general occupancy model with logit link in the previous section which had $E(N|\mathbf{Y}) = 20.7$ and 95% credible interval $(18,27)$.

23.4 ADDITIONAL CONCEPTS AND READING

Occupancy models have been wildly popular among ecologists since the early 2000s despite the fact that they are not commonly discussed or used by many statisticians. The false negative mechanism in occupancy studies that leads to binary measurement error is somewhat unique to wildlife ecology in particular, although the models have been used in a variety of other fields at this point.[1] In fact, a variety of extensions and applications of occupancy models have been developed. To name a few: multispecies occupancy models (e.g., Dorazio et al. 2006; Kery et al. 2009; Dorazio et al. 2010; Zipkin et al. 2010; Broms et al. 2016), spatially explicit occupancy models (Johnson et al. 2013a; also Chapter 26 in this book), dynamic and multiseason occupancy models (e.g., Royle and Kéry 2007; Hines et al. 2014), and adjustments to the data component of occupancy models to account for false positives as well as false negatives (e.g., Miller et al. 2011; Ruiz-Gutierrez et al. 2016; Banner et al. 2018; also Chapter 25 in this book).

Extensions to occupancy models focusing on answering community ecology questions account for multiple species and, possibly, their interactions (e.g., Rota et al. 2016). In multispecies occupancy models, we can choose to fix the total number of species or allow it to be an unknown variable in the model to be estimated. Uncertainty about the total number of species in occupancy models creates challenges for implementation, some of which we addressed using data augmentation strategies that we describe in the next chapter. Other challenges with multispecies occupancy models involve the way in which we check the models and compare them (Broms et al. 2016).

There is a large body of literature that describes approaches for checking occupancy models. For example, MacKenzie and Bailey (2004) developed a goodness-of-fit statistic based on a parametric bootstrap procedure in a non-Bayesian setting. Wright et al. (2016) proposed an alternative approach based on a joint count test adapted from spatial statistics to assess goodness-of-fit in occupancy models. Finally, Warton et al. (2017) presented a method for computing residuals in occupancy models and showed how to graphically display them to assess goodness-of-fit.

---

[1] These modeling concepts have spread to laboratory sciences (Brost et al. 2018) and disease ecology (Minuzzi-Souza et al. 2018), among others.

24 Abundance Models

24.1 CAPTURE-RECAPTURE MODELS

Many questions in wildlife ecology involve animal population sizes and demographics over time and space. Theoretical ecologists, population biologists, and management and conservation agencies often study population dynamics to make inference about features such as population growth rates and carrying capacities. However, in most natural settings it is not feasible to count all individuals in a population. Thus, a suite of methods were developed over the past several decades to estimate population size (abundance) when we can count only a subset of the population. Conventional methods for estimating abundance relied on physically capturing, then marking the individual animal, then releasing it, and returning to recapture it at a later time (hence the phrase "capture-recapture" or "capture-mark-recapture"; Otis et al. 1978). Doing this for multiple individuals over a set of "encounter" occasions (usually trapping events) results in sequences of zeros and ones as data for a set of n observed individuals. These data resemble occupancy data except that, with capture-recapture data, we never observe all zero encounter histories because we have to first capture an individual and mark it before we can recapture it. Thus, the fundamental inferential problem in capture-recapture models is to estimate the "size" of the data set (i.e., number of rows) if it had all zero encounter histories in it representing the individuals not detected in any of the sampling occasions.

The simplest class of capture-recapture models relies on a closed-population assumption, meaning that the population being surveyed is not changing in abundance due to immigration, emigration, births, or deaths during the survey period. That is, the population size is not changing with respect to demography or movement. We normally build models for the data that include the parameter of interest. Thus, we need a model for the data where population abundance N is unknown. A heuristic example of this model is

$$\mathbf{Y} \sim [\mathbf{Y}|N,p], \tag{24.1}$$

$$N \sim [N|\psi], \tag{24.2}$$

$$\psi \sim [\psi], \tag{24.3}$$

$$p \sim [p], \tag{24.4}$$

where $\mathbf{Y}$ contains the observations y_{ij} (where y_{ij} is binary with $y_{ij} = 1$ if individual i is detected on occasion j) and p represents the capture (or detection) probability for individuals $i = 1,\ldots,n$ and $j = 1,\ldots,J$ encounter occasions. Now suppose that the model for true population size is $N \sim \text{Binom}(M, \psi)$, where M is a known upper bound and $\psi \sim \text{Unif}(0,1)$ is an inclusion probability.[1] The unconditional implied prior for N can be found by integration such that

$$[N] = \int_0^1 [N|\psi][\psi]d\psi, \tag{24.5}$$

$$= \int_0^1 \binom{M}{N} \psi^N (1-\psi)^{M-N} 1 d\psi, \tag{24.6}$$

$$= \binom{M}{N} \int_0^1 \psi^N (1-\psi)^{M-N} 1 d\psi, \tag{24.7}$$

$$= \frac{M!}{N!(M-N)!} \frac{\Gamma(N+1)\Gamma(M-N+1)}{\Gamma(M-N+N+1)}, \tag{24.8}$$

$$= \frac{\Gamma(M+1)}{\Gamma(N+1)\Gamma(M-N+1)} \frac{\Gamma(N+1)\Gamma(M-N+1)}{\Gamma(M-N+N+2)}, \tag{24.9}$$

$$= \frac{\Gamma(M+1)}{\Gamma(M+2)}, \tag{24.10}$$

$$= \frac{1}{M+1}, \tag{24.11}$$

which is a constant over the support of N $(0,1,\ldots,M)$, and thus $[N] \equiv \text{DiscUnif}(0,M)$. The fact that we can write the distribution for N conditionally or unconditionally implies that we can write the posterior in the following two ways

$$[N,p,\psi|\mathbf{Y}] \propto [\mathbf{Y}|N,p][N|\psi][\psi][p], \tag{24.12}$$

$$[N,p|\mathbf{Y}] \propto [\mathbf{Y}|N,p][N][p]. \tag{24.13}$$

Both posteriors allow us to make inference on N and p. However, another way to write the posterior is

$$[\psi,p|\mathbf{Y}] \propto [\mathbf{Y}|\psi,p][\psi][p], \tag{24.14}$$

which arises if we sum (24.12) over the possible values for N. If $y_i = \sum_{j=1}^{J_i} y_{ij}$ and $J_i = J$ for all individuals i, the posterior in (24.14) is associated with a mixture model formulated as

$$y_i \sim \begin{cases} 0 & \text{, with probability } 1-\psi \\ \text{Binom}(J,p) & \text{, with probability } \psi \end{cases}, \tag{24.15}$$

[1] "Inclusion" in the sense that there is a superpopulation of M individuals from which we select N to be included in our population of interest.

that is written in terms of the mixture probability ψ directly for $i = 1, \ldots, M$, and where $y_i = 0$ for $i = n+1, \ldots, M$ (for $N < M$). In Chapter 10, we showed how to write (24.15) as a hierarchical model that depends on latent binary variables. Thus, we can formulate the model in (24.15) as

$$y_i \sim \begin{cases} 0 & , z_i = 0 \\ \text{Binom}(J, p) & , z_i = 1 \end{cases}, \tag{24.16}$$

$$z_i \sim \text{Bern}(\psi), \tag{24.17}$$

$$\psi \sim \text{Beta}(\alpha_\psi, \beta_\psi), \tag{24.18}$$

$$p \sim \text{Beta}(\alpha_p, \beta_p), \tag{24.19}$$

for $i = 1, \ldots, M$. Thus, we augment the capture-recapture data with all zero encounter histories and fit the model above, which is a fully homogeneous occupancy model. This approach is referred to as parameter expanded data augmentation (PX-DA; Royle 2009; Royle and Dorazio 2012). To fit the model above using MCMC, we need to derive full-conditional distributions for p and ψ. The full-conditional distribution for ψ has the same Bernoulli-beta form that we described in Chapter 6: $[\psi|\cdot] = \text{Beta}(\sum_{i=1}^{M} z_i + \alpha_\psi, \sum_{i=1}^{M}(1 - z_i) + \beta_\psi)$. Similarly, the full-conditional distribution for p is also beta: $[p|\cdot] = \text{Beta}(\sum_{\forall z_i = 1} y_i + \alpha_p, \sum_{\forall z_i = 1}(J - y_i) + \beta_p)$, thus we can sample both p and ψ using Gibbs updates in a MCMC algorithm. The final full-conditional distribution for z_i for $i = n+1, \ldots, M$ (when $y_i = 0$) is $[z_i|\cdot] = \text{Bern}(\tilde{\psi}_i)$, where

$$\tilde{\psi}_i = \frac{\psi(1-p)^J}{\psi(1-p)^J + 1 - \psi}. \tag{24.20}$$

Thus, the MCMC algorithm to fit the basic capture-recapture model has all Gibbs updates for parameters and requires no tuning.

To obtain inference for N, the true population abundance, we sum the latent binary variables: $N = \sum_{i=1}^{M} z_i$. Thus, even though population abundance is not directly in the model that we fit when using PX-DA, it is a derived quantity that we can easily calculate and obtain inference for using MCMC samples. Our discussion of the unconditional prior for N before indicates that the implicit prior for N in this capture-recapture model is $[N] \equiv \text{DiscUnif}(0, M)$ when the hyperparameters are specified such that $\alpha_\psi = 1$ and $\beta_\psi = 1$. However, Link (2013) recommended using a "scale prior" for N where $[N] \propto 1/N$. The easiest way to approximate this prior within the PX-DA framework is to set the hyperparameters $\alpha_\psi \to 0$ and $\beta_\psi = 1$.

We constructed a MCMC algorithm called `cr.scale.mcmc.R` in R to fit this basic capture-recapture model to data:

BOX 24.1   MCMC ALGORITHM FOR CAPTURE-RECAPTURE
MODEL

```
 1  cr.scale.mcmc <- function(y,J,n.mcmc){
 2
 3  #####
 4  #####   Setup Variables
 5  #####
 6
 7  n=length(y)
 8
 9  psi.save=rep(0,n.mcmc)
10  p.save=rep(0,n.mcmc)
11  N.save=rep(0,n.mcmc)
12  z.mean=rep(0,n)
13
14  #####
15  #####   Priors and Starting Values
16  #####
17
18  p=mean(y[y>0]/J[y>0])
19  psi=sum(y>0)/n+p*sum(y==0)/n
20  z=rep(0,n)
21  z[y>0]=1
22  alpha=.001
23  beta=1
24
25  #####
26  #####   Begin MCMC Loop
27  #####
28
29  for(k in 1:n.mcmc){
30
31     #####
32     #####   Sample z
33     #####
34
35     psi.temp=(psi*(1-p)^J)/(psi*(1-p)^J + 1-psi)
36     z[y==0]=rbinom(sum(y==0),1,psi.temp[y==0])
37
38     #####
39     #####   Sample p
40     #####
```

(Continued)

**BOX 24.1 (Continued)   MCMC ALGORITHM FOR
CAPTURE-RECAPTURE MODEL**

```
41   alpha.tmp=sum(y[z==1])+1
42   beta.tmp=sum((J-y)[z==1])+1
43   p=rbeta(1,alpha.tmp,beta.tmp)
44
45   #####
46   #####  Sample psi
47   #####
48
49   alpha.tmp=sum(z)+alpha
50   beta.tmp=n-sum(z)+beta
51   psi=rbeta(1,alpha.tmp,beta.tmp)
52
53   #####
54   #####  Save Samples
55   #####
56
57   z.mean=z.mean+z/n.mcmc
58   p.save[k]=p
59   psi.save[k]=psi
60   N.save[k]=sum(z)
61
62 }
63
64 #####
65 #####  Write Output
66 #####
67
68 list(z.mean=z.mean,N.save=N.save,p.save=p.save,
69     psi.save=psi.save,n.mcmc=n.mcmc)
70
71 }
```

We illustrate the PX-DA approach to implementing the closed-population capture-recapture model using this MCMC algorithm and two sets of data corresponding to encounter histories for two species of salamander in a 15×15 m fenced plot in Great Smoky Mountains National Park (GSMNP; Bailey et al. 2004). The majority of salamanders in GSMNP comprise two species: the red-cheeked salamander (*Plethodon jordani*) or pygmy salamander (*Desmognathus wrighti*). However, it is thought that both species have fairly low detection probabilities ($p < 0.5$), thus researchers are unable to count all individuals of these species, even within a truly closed population.[2] Using a uniform prior with support on (0,1) for p and an approximate scale prior induced through a beta distribution with $\alpha_\psi = 0.001$ and $\beta_\psi = 1$, we fit this basic capture-recapture model to the data for each species, the first of which had $n = 93$ and $n = 132$ detected individuals, respectively. The following R code reads in the data, augments the data with all zero encounter histories up to $M = 1500$ (a reasonable upper bound) for both species, and then fits the Bayesian capture-recapture model using PX-DA:

BOX 24.2 FIT CAPTURE-RECAPTURE MODEL TO DATA

```
sal.df=read.table("sal_data.txt",header=TRUE)
y.0=apply(sal.df[sal.df$spp==0,1:4],1,sum)
    # P. jordani
y.1=apply(sal.df[sal.df$spp==1,1:4],1,sum)
    # D. wrighti


J=4
M=1500
n.0=length(y.0)
n.1=length(y.1)
y.0.aug=c(y.0,rep(0,M-n.0))
y.1.aug=c(y.1,rep(0,M-n.1))


source("cr.scale.mcmc.R")
n.mcmc=50000
set.seed(1)
mcmc.0.out=cr.scale.mcmc(y=y.0.aug,J=rep(J,M),
    n.mcmc=n.mcmc)
mcmc.1.out=cr.scale.mcmc(y=y.1.aug,J=rep(J,M),
    n.mcmc=n.mcmc)


layout(matrix(1:4,2,2,byrow=TRUE))
hist(mcmc.0.out$p.save,prob=TRUE,xlim=c(0,1),
    xlab="p",main="a")
text(.5,10,labels="P. jordani",pos=4)
```

(Continued)

---

[2] In this case, closure involves a fenced area and short sampling period so that we can assume individuals do not leave, die, or reproduce.

**BOX 24.2 (Continued)   FIT CAPTURE-RECAPTURE MODEL
TO DATA**

```
29 hist(mcmc.0.out$N.save,prob=TRUE,breaks=10000,
30     xlim=c(0,M),ylab="Probability",xlab="N",
31        main="b")
32 text(800,.15,labels="P. jordani",pos=4)
33 hist(mcmc.1.out$p.save,prob=TRUE,xlim=c(0,1),
34     xlab="p",main="c")
35 text(.5,13,labels="D. wrighti",pos=4)
36 hist(mcmc.1.out$N.save,prob=TRUE,breaks
37     =seq(0,1500,1),xlim=c(0,M),ylab="Probability",
38     xlab="N",main="d")
39 text(750,.07,labels="D. wrighti",pos=4)
40
41 mean(mcmc.0.out$N.save)
42 quantile(mcmc.0.out$N.save,c(0.025,0.975))
43 mean(mcmc.1.out$N.save)
44 quantile(mcmc.1.out$N.save,c(0.025,0.975))
45
46 mean(mcmc.0.out$N.save<mcmc.1.out$N.save)
```

This R code results in the marginal posterior histograms for the detection probability (p) and total abundance (N) for each species shown in Figure 24.1. The marginal posterior histograms in Figure 24.1 indicate that the detection probability for *P. jordani* is slightly greater than for *P. wrighti*, but there is not enough evidence to say they differ. However, because of the slight difference in detection probabilities and number of captured individuals, the estimated total abundance for *D. wrighti* ($E(N|\mathbf{y}) = 450.9$ with 95% credible interval $[307, 681]$) is nearly twice that of *P. jordani* ($E(N|\mathbf{y}) = 229.6$ with 95% credible interval $[163, 334]$). Although the credible intervals do overlap slightly at the 95% level, the probability that abundance for *D. wrighti* is larger than *P. jordani* is $E(1_{\{N_0<N_1\}}|\mathbf{y}) = 0.995$ (where N_0 corresponds to *P. jordani* and N_1 corresponds to *D. wrighti*) assuming the species detections and abundances are independent.

The PX-DA approach allows us to use a homogeneous occupancy model to estimate abundance because of the similarities of the two data types (Royle 2009) and it has been extended to several other settings (Royle and Dorazio 2012). Although PX-DA provides a convenient reparameterization that some may find more intuitive than traditional capture-recapture models, it is not necessary because we can derive the integrated likelihood for many capture-recapture models without the use of latent state variables $\mathbf{z}$. From a philosophical perspective, some have argued that it is reasonable to directly incorporate latent processes into model specifications when they

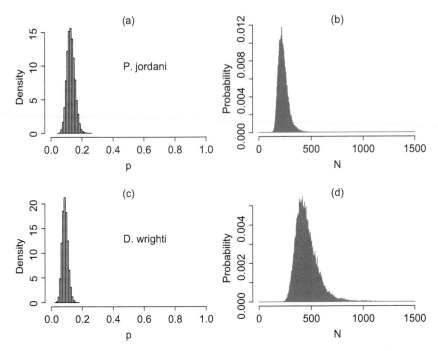

Figure 24.1 Marginal posterior histograms for the detection coefficients (a) p (*P. jordani*), (b) N (*P. jordani*), (c) p (*D. wrighti*), and (d) N (*D. wrighti*), resulting from fitting the closed-population capture-recapture model using PX-DA to both sets of salamander data.

exist and can be interpreted. However, we can also implement the model by integrating latent processes out for computational reasons if we can show the integrated version is equivalent. Usually we can recover a latent process after the fact if the model is Bayesian and we used MCMC to fit the model. However, if the model is non-Bayesian, we may not find it trivial to recover the latent process because we do not have a MCMC sample to work with.

24.2 DISTANCE SAMPLING MODELS

As we eluded to in the previous section, an entire class of models have arisen as generalizations of the homogeneous capture-recapture model (sometimes called the "dot model" or the M_0 model) that we described. We return to some of these in the Additional Concepts and Reading section at the end of this chapter, but it is important to mention a key difference between capture-recapture models and occupancy models, besides N being unknown. In occupancy models, often a focus is on variation in occupancy probability ψ, whereas in capture-recapture the interest and heterogeneity often is focused on the detection probability p. The design of these models is to characterize and account for individual-level heterogeneity in capture probability to more accurately estimate abundance (N).

Heterogeneity in capture (or detection) probability often is accompanied by auxiliary sources of data recorded during the study (for example, individual-level characteristics such as covariates pertaining to body mass or previous captures/detections). Another type of useful measurement may come in the form of spatial information. In particular, some studies leverage spatial information, assumptions about the distribution of animals, and the way individuals are detected to help estimate population abundance. Distance sampling and spatial-capture recapture models (Chapter 26) are examples of methods that rely on spatially explicit information to facilitate abundance estimation. In this section, we focus on distance sampling models (e.g., Buckland et al. 2001, 2004; Johnson et al. 2010).

As in the previous section, we develop a model for distance sampling based on the PX-DA concept. Following Royle and Dorazio (2008), suppose that we make pairs of observations while conducting a visual survey along a transect. As the observers move along the transect, they record $y_i = 1$ when they observe an individual i in the population residing within the study area. As the observers detect an individual, they also record its perpendicular distance x_i (for $x_i > 0$) from the observers on the transect. As the observers move along the transect, they accumulate detections and distances for $i = 1, \ldots, n$ observed individuals. Thus, the original data are pairs (y_i, x_i) for $i = 1, \ldots, n$.

Using PX-DA, we augment the data with zeros $y_i = 0$ for $i = n+1, \ldots, M$ for which the associated distances x_i are unknown. We can think of these augmented data points as pseudo-individuals that, together with the observed individuals, make up the superpopulation from which the model can select from to form the population of interest.

Thus, we can write a hierarchical Bayesian model for these data using a similar mixture framework as what we specified in the last two chapters

$$y_i \sim \begin{cases} 0 & , z_i = 0 \\ \text{Bern}(p_i) & , z_i = 1 \end{cases}, \qquad (24.21)$$

where $z_i \sim \text{Bern}(\psi)$ and the prior for membership probability is $\psi \sim \text{Beta}(\alpha_\psi, \beta_\psi)$. To account for heterogeneity in detectability due to distance from the transect, a common mathematical formulation specifies detection probability as decaying with distance using a half-normal function such that

$$\log(p_i) = -\frac{x_i^2}{\sigma^2}, \qquad (24.22)$$

where σ^2 acts as a range parameter such that 2σ indicates approximately 95% of the detectability distance. We specify a proper uniform prior for σ such that $\sigma \sim \text{Unif}(0, u_\sigma)$.

Critical to this model is the fact that we do not observe distances x_i for the augmented pseudo-individuals ($i = n+1, \ldots, M$), thus they need a prior. If we assume that individuals in the superpopulation are distributed uniformly at random (called "complete spatial random" in spatial statistics) throughout the study region, then the distances should be uniformly distributed such that $x_i \sim \text{Unif}(0, u_x)$ with u_x chosen *a priori* based on the study area and reasonable upper bound for which individuals.

The distance sampling model that we described differs from classical capture-recapture and occupancy models in that we do not necessarily observe replicates (i.e., there is no j index) unless we have multiple observers because of the mechanics of how we conduct the survey. Normally we need to rely on replicates (i.e., multiple survey occasions) to help separate detection from occupancy, but in distance sampling, the auxiliary information about distance and an associated decaying distance function usually provide enough structure to help us learn about abundance with only a single survey.

The posterior distribution associated with our distance sampling model is

$$[\mathbf{x}, \sigma, \mathbf{z}, \psi | \mathbf{y}] \propto \left(\prod_{i=1}^{M} [y_i | x_i, \sigma]^{z_i} 1_{\{y_i = 0\}}^{1-z_i} [z_i | \psi][x_i] \right) [\psi][\sigma], \qquad (24.23)$$

where $\mathbf{x}$ is a $(M - n) \times 1$ vector containing only the distances for individuals we did not observe.

To implement the model with MCMC, we need to derive the full-conditional distributions for the unknown parameters in our distance sampling model. Thus, beginning with the full-conditional distribution for z_i for $i = n+1, \ldots, M$ (recall that $z_i = 1$ for $i = 1, \ldots, n$ because we observed the individual directly) is

$$[z_i | \cdot] \propto [y_i | x_i, \sigma]^{z_i} 1_{\{y_i = 0\}}^{1-z_i} [z_i | \psi], \qquad (24.24)$$

$$\propto (1 - p_i)^{z_i} \psi^{z_i} (1 - \psi)^{1-z_i}, \qquad (24.25)$$

$$\propto (\psi(1 - p_i))^{z_i} (1 - \psi)^{1-z_i}, \qquad (24.26)$$

which implies that $[z_i | \cdot] = \text{Bern}(\tilde{\psi}_i)$, where

$$\tilde{\psi}_i = \frac{\psi(1 - p_i)}{\psi(1 - p_i) + 1 - \psi}, \qquad (24.27)$$

and $\log(p) = -\frac{x_i^2}{\sigma^2}$ as previously defined.

The full-conditional distribution for the membership probability ψ is exactly as we worked out for previous models $[\psi | \cdot] = \text{Beta}(\sum_{i=1}^{M} z_i + \alpha_\psi, \sum_{i=1}^{M} (1 - z_i) + \beta_\psi)$.

The parameter controlling the decay of the detection function σ is straightforward to sample in MCMC, but we need to use M-H updates for it because the full-conditional distribution is

$$[\sigma | \cdot] \propto \prod_{i=1}^{M} [y_i | x_i, \sigma] 1_{\{0 \le \sigma \le u_\sigma\}}, \qquad (24.28)$$

$$\propto \prod_{\forall z_i = 1} \text{Bern}\left(y_i \Big| e^{-\frac{x_i^2}{\sigma^2}} \right) 1_{\{0 \le \sigma \le u_\sigma\}}, \qquad (24.29)$$

where "Bern" stands for the Bernoulli PMF evaluated at y_i and the indicator $1_{\{0 \le \sigma \le u_\sigma\}}$ effectively truncates the distribution with respect to σ below by 0 and above by u_σ. If we use a normal random walk proposal distribution where $\sigma^{(k)} \sim N(\sigma^{(k-1)}, \sigma^2_{\sigma\text{tune}})$ and rejection sampling for which we skip the update if σ falls outside the interval $(0, u_\sigma)$, the M-H ratio is

$$mh = \frac{\prod_{\forall z_i=1} [y_i | x_i^{(k-1)}, \sigma^{(*)}]}{\prod_{\forall z_i=1} [y_i | x_i^{(k-1)}, \sigma^{(k-1)}]}. \tag{24.30}$$

Finally, we need to sample x_i when $y_i = 0$. In general, the full-conditional distribution for x_i is

$$[x_i|\cdot] \propto [y_i | x_i, \sigma]^{z_i} [x_i], \tag{24.31}$$

which is a uniform distribution on $(0, u_x)$ when $z_i = 0$ and is proportional to

$$[x_i|\cdot] \propto \left(1 - e^{-\frac{x_i^2}{\sigma^2}}\right), \tag{24.32}$$

when $z_i = 1$. Thus, we can sample x_i directly from the uniform distribution when $y_i = 0$ and $z_i = 0$, and when $y_i = 0$ and $z_i = 1$, we can use the M-H ratio

$$mh = \frac{\left(1 - e^{-\frac{x_i^{2(*)}}{\sigma^{2(k)}}}\right)}{\left(1 - e^{-\frac{x_i^{2(k-1)}}{\sigma^{2(k)}}}\right)}, \tag{24.33}$$

assuming we use another normal random walk proposal such that $x_i^{(*)} \sim N(x_i^{(k-1)}, \sigma^2_{x,\text{tune}})$ and reject the proposed value if it falls outside the interval $(0, u_x)$.

We constructed the following MCMC algorithm called ds.mcmc.R in R based on these full-conditional distributions:

BOX 24.3   MCMC ALGORITHM FOR DISTANCE SAMPLING MODEL

```
ds.mcmc <- function(y,x,d,u.x,u.s,n.mcmc){

####
#### Setup Variables
####

M=length(y)
```

(Continued)

BOX 24.3 (Continued)   MCMC ALGORITHM FOR DISTANCE SAMPLING MODEL

```
 8 s.save=rep(0,n.mcmc)
 9 psi.save=rep(0,n.mcmc)
10 N.save=rep(sum(y),n.mcmc)
11 D.save=rep(0,n.mcmc)
12
13 n.burn=round(.2*n.mcmc)
14
15 ####
16 #### Priors, Starting Values, and Tuning
17 ####
18
19 alpha=1
20 beta=1
21
22 z=y
23
24 n.0=sum(y==0)
25 x[y==0]=runif(n.0,0,u.x)
26 s=mean(x[y==1])/2
27 psi=(M-n.0)/M
28
29 p=exp(-(x^2)/(s^2))
30
31 s.tune=.1*u.s
32 x.tune=.1*u.x
33
34 ####
35 #### Begin MCMC Loop
36 ####
37
38 for(k in 1:n.mcmc){
39
40    ####
41    ####  Sample z
42    ####
43
44    psi.tmp=psi*(1-p)/(psi*(1-p)+1-psi)
45    z[y==0]=rbinom(n.0,1,psi.tmp[y==0])
```

(*Continued*)

**BOX 24.3 (Continued)   MCMC ALGORITHM FOR DISTANCE
SAMPLING MODEL**

```
46   ####
47   ####   Sample psi
48   ####
49
50   psi=rbeta(1,sum(z)+alpha,sum(1-z)+beta)
51
52   ####
53   ####   Sample s
54   ####
55
56   s.star=rnorm(1,s,s.tune)
57   if((s.star > 0) & (s.star < u.s)){
58     p.star=exp(-(x^2)/(s.star^2))
59     mh1=sum(dbinom(y[z==1],1,p.star[z==1],
60          log=TRUE))
61     mh2=sum(dbinom(y[z==1],1,p[z==1],log=TRUE))
62     mh=exp(mh1-mh2)
63     if(mh > runif(1)){
64       s=s.star
65       p=p.star
66     }
67   }
68
69   ####
70   ####   Sample x
71   ####
72
73   x.star=rnorm(n.0,x[y==0],x.tune)
74   tmp.idx=((x.star > 0) & (x.star < u.x))
75   p.star=p
76   p.star[y==0][tmp.idx]=exp(-(x.star[tmp.idx]^2)/
77     (s^2))
78   mh1=z*log(1-p.star)
79   mh2=z*log(1-p)
80   keep.idx=(exp(mh1-mh2)>runif(M))
81   comb.idx=(tmp.idx & keep.idx[y==0])
82   x[y==0][comb.idx]=x.star[comb.idx]
83   p[y==0][comb.idx]=p.star[y==0][comb.idx]
```

(Continued)

BOX 24.3 (Continued)   MCMC ALGORITHM FOR DISTANCE SAMPLING MODEL

```
84   ####
85   ####   Save Samples
86   ####
87
88   s.save[k]=s
89   psi.save[k]=psi
90   N.save[k]=sum(z)
91   D.save[k]=sum(z)/(d*2*u.x)
92
93   }
94
95   ####
96   #### Write Output
97   ####
98
99   list(s.save=s.save,psi.save=psi.save,
100      N.save=N.save,D.save=D.save)
101
102  }
```

Notice in this R code that we have calculated the abundance as a derived quantity $(N = \sum_{i=1}^{M} z_i)$, as in the previous section on capture-recapture models with PX-DA. Also, notice that we can calculate the density (i.e., individuals per unit area) by dividing abundance by the sampled area. Under our model specification, the sampled area is the length of the transect (d) times twice the maximum detection distance u_x (i.e., $D \equiv N/(2du_x)$). Thus, using the MCMC sample we can make inference on population abundance and density.

To illustrate this MCMC algorithm, we fit the distance sampling model in this section to a well-known data set, analyzed by Burnham et al. (1980), that includes distance sampling detections for 73 individual impala along a 60 km transect in Africa. The original data were recorded as sighting distance and angle from the transect which we converted to perpendicular distance from the transect before analyzing it. We augmented the data set with zeros up to $M = 1000$ and specified hyperparameters $u_x = 400$ and $u_s = 400$ for $K = 50000$ MCMC iterations. The following R code reads in the data, augments it with zeros, and then fits the model using the MCMC algorithm:

BOX 24.4 FIT DISTANCE SAMPLING MODEL TO DATA

```
1  x=c(71.933980, 26.047227, 58.474341, 92.349221,
2      163.830409, 84.523652, 163.830409, 157.330098,
3      22.267696, 72.105330, 86.986979, 50.795047,
4      0.000000, 73.135370, 0.000000, 128.557522,
5      163.830409, 71.845104, 30.467336, 71.073909,
6      150.960702, 68.829172, 90.000000, 64.983827,
7      165.690874, 38.008322, 378.207430, 78.146226,
8      42.127052, 0.000000, 400.000000, 175.386612,
9      30.467336, 35.069692, 86.036465, 31.686029,
10     200.000000, 271.892336, 26.047227, 76.604444,
11     41.042417, 200.000000, 86.036465, 0.000000,
12     93.969262, 55.127471, 10.458689, 84.523652,
13     0.000000, 77.645714, 0.000000, 96.418141,
14     0.000000, 64.278761, 187.938524, 0.000000,
15     160.696902, 150.453756, 63.603607, 193.185165,
16     106.066017, 114.906666, 143.394109,
17     128.557522, 245.745613, 123.127252,
18     123.127252, 153.208889, 143.394109, 34.202014,
19     96.418141, 259.807621, 8.715574)
20
21 n=length(x)
22
23 d=60000
24 u.x=400
25 u.s=400
26 M=1000
27 imp.aug.df=data.frame(y=rep(0,M),x=rep(0,M))
28 imp.aug.df[1:n,1]=1
29 imp.aug.df[1:n,2]=x
30
31 source("ds.mcmc.R")
32 n.mcmc=50000
33 set.seed(1)
34 mcmc.out=ds.mcmc(y=imp.aug.df$y,x=imp.aug.df$x,
35     d=d,u.x=u.x,u.s=u.s,n.mcmc=n.mcmc)
36
37 layout(matrix(1:2,1,2,byrow=TRUE))
38 hist(mcmc.out$s.save,prob=TRUE,xlab=bquote(sigma),
39     main="a")
```

(Continued)

**BOX 24.4 (Continued)   FIT DISTANCE SAMPLING MODEL
TO DATA**

```
40 hist(mcmc.out$N.save,prob=TRUE,breaks=seq
      (0,1000,1),xlim=c(0,M),ylab="Probability",
41     xlab="N",main="b")
42
43 mean(mcmc.out$N.save[-(1:1000)])
44 quantile(mcmc.out$N.save[-(1:1000)],
45     c(0.025,0.975))
46 mean(2*mcmc.out$s.save[-(1:1000)])
47 quantile(2*mcmc.out$s.save[-(1:1000)],
48     c(0.025,0.975))
```

This R code results in the posterior histograms for σ and N shown in Figure 24.2. Figure 24.2b indicates that there were approximately 2.5 times the number of impala in the study region than were detected ($E(N|\mathbf{y}) = 179.9$ with 95% credible interval $[139, 229]$). Recall that the effective range of detection[3] in our model is approximately 2σ, thus Figure 24.2a indicates that the effective range of detection is $E(2\sigma|\mathbf{y}) = 373.8$ meters with a 95% credible interval $(314.8, 452.3)$. Furthermore the probability of detection function over a range of distances. The following R code results in Figure 24.3 showing the 95% credible interval (gray) and posterior mean of the derived quantity $p = \exp(-x^2/\sigma^2)$ for a range of distances x (in meters):

BOX 24.5   POSTERIOR DETECTION FUNCTION

```
1 n.seq=100
2 x.seq=seq(0,u.x,,n.seq)
3 p.save=matrix(0,n.seq,n.mcmc)
4 for(k in 1:n.mcmc){
5   p.save[,k]=exp(-(x.seq^2)/(mcmc.out$s.save[k]^2))
6 }
7 p.l=apply(p.save,1,quantile,0.025)
8 p.u=apply(p.save,1,quantile,0.975)
9 p.mn=apply(p.save,1,mean)
```

(Continued)

---

[3] Recall that detectability will drop below 0.05 after approximately 2σ for a Gaussian shaped detection function.

BOX 24.5 (Continued)   POSTERIOR DETECTION FUNCTION

```
10 plot(x.seq,p.mn,type="n",main="",ylim=c(0,1),
11    xlab="x",ylab="p")
12 polygon(c(x.seq,rev(x.seq)),c(p.u,rev(p.l)),
13    border=NA,col=8)
14 lines(x.seq,p.mn,lwd=1.5)
```

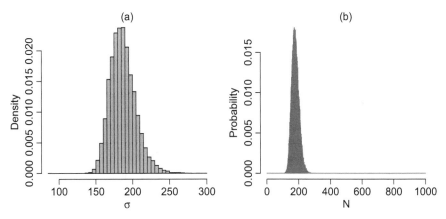

Figure 24.2   Marginal posterior histograms for (a) σ (detection range parameter) and (b) N (abundance), resulting from fitting the distance sampling model using PX-DA on the impala data.

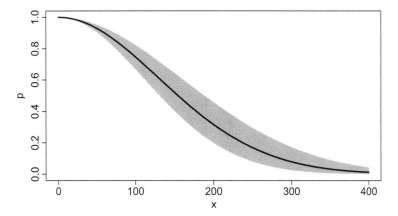

Figure 24.3   Posterior 95% credible interval (gray) and posterior mean (black line) of the detection function p based on fitting the distance sampling model using PX-DA on the impala data.

24.3   SURVIVAL MODELS

Many wildlife biologists and managers are interested in two main population vital rates: abundance and survival probability.[4] As we discussed in previous sections, we can estimate abundance using several different techniques customized for different study designs and data types, but accounting for imperfect detectability of animals is a key feature. To estimate a survival rate, we also need to account for imperfect detection. We introduced a simplified survival model for larval fish in Chapter 21 based on a binomial process over one time step and survival probability ϕ. However, we can generalize the survival process to accommodate multiple time periods and imperfect detectability of the species to estimate ϕ.

The so-called Cormack-Jolly-Seber (CJS) model is an individual-based model[5] like the capture-recapture and distance sampling models that we presented earlier in this chapter. The data portion of the CJS model is a zero-inflated Bernoulli mixture like the models that we introduced previously in this chapter. In CJS models, we are not interested in estimating population abundance, only survival. We assume a capture-recapture study design where individuals are captured (or encountered), marked, and released then recapture is attempted over a sequence of future occasions. In CJS models, we condition on the first capture, thus $y_{i1} = 1$, for individuals $i = 1,\dots,n$. The subsequent data set resembles capture-recapture data except that we label them y_{it} for $t = 2,\dots,T$ for a total number of capture occasions T. The CJS data model is

$$y_{it} \sim \begin{cases} 0 & , z_{i,t} = 0 \\ \text{Bern}(p) & , z_{i,t} = 1 \end{cases}, \tag{24.34}$$

where z_{it} represents the state of the individual (one for alive, zero for dead). The latent survival process then is a binary autoregressive process where

$$z_{it} \sim \begin{cases} 0 & , z_{i,t-1} = 0 \\ \text{Bern}(\phi) & , z_{i,t-1} = 1 \end{cases}, \tag{24.35}$$

for $t = 2,\dots,T$ and $z_{i1} = 1$ because we know the individual was alive on first capture. The process model for the state variables z_{it} in (24.35) implies that, when the individual is dead at the previous time, it must stay dead. However, when the individual is alive at the previous time, it has probability ϕ of surviving to the next time. To complete the model statement, we specify priors for p and ϕ such that

$$p \sim \text{Beta}(\alpha_p, \beta_p), \tag{24.36}$$

$$\phi \sim \text{Beta}(\alpha_\phi, \beta_\phi). \tag{24.37}$$

---

[4] Although survival is often thought of as individual-specific, a survival probability is represented by a parameter (or set of parameters) at the population level in capture-recapture models.

[5] An individual-based model implies that we need to individually recognize and record individuals rather than aggregations such as counts of individuals.

This Bayesian CJS model statement results in the posterior distribution

$$[\mathbf{Z}, p, \phi | \mathbf{Y}] \propto \left(\prod_{i=1}^{n} \prod_{t=2}^{T} [y_{it}|p]^{z_{it}} 1_{\{y_{it}=0\}}^{1-z_{it}} [z_{it}|\phi]^{z_{i,t-1}} 1_{\{z_{it}=0\}}^{1-z_{i,t-1}} \right) [p][\phi], \qquad (24.38)$$

where $\mathbf{Z}$ and $\mathbf{Y}$ are $n \times T$ matrices of zeros and ones. Thus, we can implement the model using MCMC by deriving the full-conditional distributions for $\mathbf{Z}$, p, and ϕ. The full-conditional distribution for the detection probability p is

$$[p|\cdot] \propto \left(\prod_{\forall z_{it}=1} [y_{it}|p] \right) [p], \qquad (24.39)$$

$$\propto \left(\prod_{\forall z_{it}=1} p^{y_{it}} (1-p)^{1-y_{it}} \right) p^{\alpha_p - 1} (1-p)^{\beta_p - 1}, \qquad (24.40)$$

$$\propto p^{\sum_{\forall z_{it}=1} y_{it} + \alpha_p - 1} (1-p)^{\sum_{\forall z_{it}=1} (1-y_{it}) + \beta_p - 1}, \qquad (24.41)$$

which implies that the full-conditional distribution is $[p|\cdot] = \text{Beta}(\sum_{z_{it}=1} y_{it} + \alpha_p, \sum_{z_{it}=1} (1 - y_{it}) + \beta_p)$. As with the other capture-recapture and occupancy models that we have presented, we can use a Gibbs update to sample p from this full-conditional distribution in a MCMC algorithm.

Similarly, we can find the full-conditional distribution for ϕ in the same way

$$[\phi|\cdot] \propto \left(\prod_{i=1}^{n} \prod_{t=2}^{T} [z_{it}|\phi]^{z_{i,t-1}} \right) [\phi], \qquad (24.42)$$

$$\propto \left(\prod_{\forall z_{i,t-1}=1} [z_{it}|\phi] \right) [\phi], \qquad (24.43)$$

$$\propto \phi^{\sum_{\forall z_{i,t-1}=1} z_{it} + \alpha_\phi - 1} (1-\phi)^{\sum_{\forall z_{i,t-1}=1} (1-z_{it}) + \beta_\phi - 1}, \qquad (24.44)$$

which implies that $[\phi|\cdot] = \text{Beta}(\sum_{\forall z_{i,t-1}=1} z_{it} + \alpha_\phi, \sum_{\forall z_{i,t-1}=1} (1 - z_{it}) + \beta_\phi)$ and we can sample ϕ with a Gibbs update in a MCMC algorithm. In this full-conditional distribution for ϕ, the sums are over all i and $t-1$ such that $z_{i,t-1} = 1$.

Finally, the full-conditional distribution for the latent state of the individual requires special attention. In general,

$$[z_{it}|\cdot] \propto [y_{it}|p]^{z_{it}} 1_{\{y_{it}=0\}}^{1-z_{it}} [z_{i,t+1}|\phi]^{z_{it}} 1_{\{z_{i,t+1}=0\}}^{1-z_{it}} [z_{it}|\phi]^{z_{i,t-1}} 1_{\{z_{it}=0\}}^{1-z_{i,t-1}}, \qquad (24.45)$$

$$\propto (p^{y_{it}}(1-p)^{1-y_{it}})^{z_{it}} (\phi^{z_{i,t+1}}(1-\phi)^{1-z_{i,t+1}})^{z_{it}} (\phi^{z_{it}}(1-\phi)^{1-z_{it}})^{z_{i,t-1}}, \qquad (24.46)$$

$$\propto (p^{y_{it}}(1-p)^{1-y_{it}} \phi^{z_{i,t+1}}(1-\phi)^{1-z_{i,t+1}} \phi^{z_{i,t-1}})^{z_{it}} ((1-\phi)^{z_{i,t-1}})^{1-z_{it}}, \qquad (24.47)$$

but we also have several special conditions to adhere to because of the constraints on individual-level survival processes. In particular, if we observe the individual

i alive on a particular sampling occasion τ, then all state variables z_{it} at that time and before must be equal to one (i.e., if $z_{i\tau} = 1$, then $z_{it} = 1$, $\forall t \leq \tau$). Similarly, if the state variable of the individual i indicates that it is dead at the previous sampling occasion, then it must remain dead thereafter (i.e., if $z_{i,t-1} = 0$, then $z_{it} = 0$). Finally, if the state variable of the individual i indicates the individual is alive at a future sampling occasion, then it must be alive at previous sampling occasions (i.e., if $z_{i,t+1} = 1$, then $z_{it} = 1$). Thus, the conditions under which we actually obtain a sample for z_{it} in a MCMC algorithm are:

1. $\sum_{\tau=t}^{T} y_{i\tau} = 0$
2. $z_{i,t-1} = 1$
3. $z_{i,t+1} = 0$

With the initial state assumed known ($z_{i1} = 1$ for $i = 1, \ldots, n$), the general full-conditional distribution in (24.47) simplifies to

$$[z_{it}|\cdot] \propto ((1-p)(1-\phi)\phi)^{z_{it}}(1-\phi)^{1-z_{it}}, \tag{24.48}$$

for $t = 2, \ldots, T-1$, which is a conjugate Bernoulli distribution with $[z_{it}|\cdot] = \text{Bern}(\tilde{\phi}_{it})$, where

$$\tilde{\phi}_{it} = \frac{(1-p)(1-\phi)\phi}{(1-p)(1-\phi)\phi + 1 - \phi}, \tag{24.49}$$

$$= \frac{(1-p)\phi}{(1-p)\phi + 1}. \tag{24.50}$$

The full-conditional distribution for the final state in the latent survival process z_{iT} is similar:

$$[z_{iT}|\cdot] \propto ((1-p)\phi)^{z_{it}}(1-\phi)^{1-z_{it}}, \tag{24.51}$$

which is also a conjugate Bernoulli distribution with $[z_{iT}|\cdot] = \text{Bern}(\tilde{\phi}_{iT})$, where

$$\tilde{\phi}_{iT} = \frac{(1-p)\phi}{(1-p)\phi + 1 - \phi}. \tag{24.52}$$

Thus, we can sample all of the required latent state variables with Gibbs updates in a MCMC algorithm. In addition to the detection and survival probabilities, the full-conditional distributions for z_{it} imply that our MCMC algorithm is fully Gibbs and will not require any tuning.

We constructed a MCMC algorithm called CJS.mcmc.R in R based on this conjugate full-conditional distributions. This MCMC algorithm contains a few extra steps beyond those that we developed for occupancy and basic capture-recapture models because of the very specific set of conditions that we must follow to sample the latent survival state (z_{it}):

BOX 24.6   MCMC ALGORITHM FOR CJS MODEL

```
1  cjs.mcmc <- function(Y,n.mcmc){
2
3  ####
4  ####   Setup Variables
5  ####
6
7  n=dim(Y)[1]
8  T=dim(Y)[2]
9
10 p.save=rep(0,n.mcmc)
11 phi.save=rep(0,n.mcmc)
12
13 ####
14 ####   Priors and Starting Values
15 ####
16
17 alpha.p=1
18 beta.p=1
19
20 alpha.phi=1
21 beta.phi=1
22
23 p=0.4
24 phi=0.65
25
26 Z=matrix(0,n,T)
27 Z[,1]=1
28 Z[Y==1]=1
29 for(i in 1:n){
30   if(sum(Y[i,])>0){
31     idx.tmp=max((1:T)[Y[i,]==1])
32     Z[i,1:idx.tmp]=1
33   }
34 }
35
36 ####
37 ####   Begin MCMC Loop
38 ####
```

(Continued)

BOX 24.6 (Continued)   MCMC ALGORITHM FOR CJS MODEL

```
39  for(k in 1:n.mcmc){
40
41    ####
42    ####  Sample z
43    ####
44
45    for(i in 1:n){
46      for(t in 2:(T-1)){
47        if(sum(Y[i,t:T])==0){
48          if(Z[i,t-1]==1 & Z[i,t+1]==0){
49            psi.tmp=phi*(1-p)
50            psi.tmp=psi.tmp/(psi.tmp+1)
51            Z[i,t]=rbinom(1,1,psi.tmp)
52          }
53          if(Z[i,t-1]==0){
54            Z[i,t]=0
55          }
56          if(Z[i,t+1]==1){
57            Z[i,t]=1
58          }
59        }
60      }
61      if(Z[i,T-1]==1 & Y[i,T]==0){
62        psi.tmp=phi*(1-p)
63        psi.tmp=psi.tmp/(psi.tmp+1-phi)
64        Z[i,T]=rbinom(1,1,psi.tmp)
65      }
66      if(Z[i,T-1]==0 & Y[i,T]==0){
67        Z[i,T]=0
68      }
69      if(Y[i,T]==1){
70        Z[i,T]=1
71      }
72    }
73
74    ####
75    ####  Sample p
76    ####
```

(Continued)

BOX 24.6 (Continued)   MCMC ALGORITHM FOR CJS MODEL

```
77   p=rbeta(1,sum(Y[,-1][Z[,-1]==1])+alpha.p,
78     sum(1-Y[,-1][Z[,-1]==1])+beta.p)
79
80   ####
81   ####  Sample phi
82   ####
83
84   phi=rbeta(1,sum(Z[,-1][Z[,-T]==1])+alpha.phi,
85     sum(1-Z[,-1][Z[,-T]==1])+beta.phi)
86
87   ####
88   ####  Save Samples
89   ####
90
91   p.save[k]=p
92   phi.save[k]=phi
93
94 }
95
96 ####
97 ####  Write Output
98 ####
99
100 list(p.save=p.save,phi.save=phi.save,
101     n.mcmc=n.mcmc)
102
103 }
```

To demonstrate this MCMC algorithm, we fit the Bayesian CJS model to a set of capture-recapture data on yellow warbler (*Setophaga petechia*) to estimate detection and survival probability. These data were collected as part of a North American collaborative effort to monitor avian productivity and survivorship (MAPS) and previous analyzed by Royle and Dorazio (2008). These data contain annual recapture information annually over a period of 12 years for 10 yellow warbler individuals. To estimate survival probability while accounting for imperfect detectability, we used uniform priors for p and ϕ with support $(0,1)$ and $K = 50000$ MCMC iterations in the following R code to fit the CJS model to these data:

BOX 24.7   FIT CJS MODEL TO DATA

```
1  warb=matrix(c(1, 1, 1, 1, 1, 1, 1, 1, 1, 1, 0, 0,
2              0, 1, 1, 0, 0, 0, 0, 0, 0, 0, 0, 1,
3              0, 0, 0, 0, 0, 0, 0, 0, 0, 1, 0, 0, 0,
4              0, 0, 0, 0, 0, 0, 0, 1, 0, 0, 0, 0, 0,
5              0, 0, 0, 0, 1, 0, 0, 0, 0, 0, 0, 0, 0,
6              0, 1, 0, 0, 0, 0, 0, 0, 0, 0, 0, 1, 0,
7              0, 0, 0, 0, 0, 0, 0, 0, 1, 0, 0, 0, 0,
8              0, 0, 0, 0, 0, 0, 0, 0, 0, 0, 0, 0, 0,
9              0, 0, 0, 0, 0, 0, 0, 0, 0, 0, 0, 0, 0,
10             0, 0, 0, 0, 0),10)
11
12 source("cjs.mcmc.R")
13 n.mcmc=50000
14 set.seed(1)
15 mcmc.out=cjs.mcmc(Y=warb,n.mcmc=n.mcmc)
16
17 layout(matrix(1:2,1,2))
18 hist(mcmc.out$p.save,breaks=50,col=8,xlim=c(0,1),
19     xlab="p")
20 hist(mcmc.out$phi.save,breaks=50,col=8,
21     xlim=c(0,1),xlab=bquote(phi))
22
23 apply(cbind(mcmc.out$p.save,mcmc.out$phi.save)
24     [-(1:2000),],2,mean)
25 apply(cbind(mcmc.out$p.save,mcmc.out$phi.save)
26     [-(1:2000),],2,quantile,c(0.025,0.975))
```

This R code results in the marginal posterior histograms for the detection (p) and survival (ϕ) probabilities shown in Figure 24.4. The results of our survival analysis indicate that, even for this relatively small number of individuals, we obtain some learning about the capture probability p (Figure 24.4a) and survival probability ϕ (Figure 24.4b). In particular, the posterior mean survival probability was $E(\phi|\mathbf{Y}) = 0.57$ with a 95% credible interval $(0.36, 0.77)$. Thus, although there is still substantial variability in our understanding of yellow warbler survival probability, these data indicate that it is most likely just above 50%. Furthermore, we can rule out very low capture probability as a potential cause of low recaptures, because the posterior mean of p is $E(p|\mathbf{Y}) = 0.65$ with a 95% credible interval $(0.35, 0.91)$.

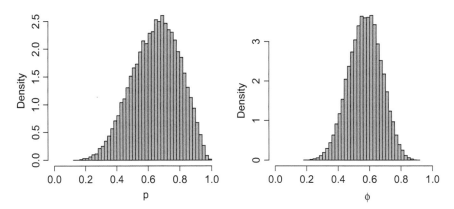

Figure 24.4 Marginal posterior histograms for (a) p (detection probability) and (b) ϕ (survival), resulting from fitting the Bayesian CJS model to the yellow warbler capture-recapture data.

24.4 N-MIXTURE MODELS

Classical approaches for estimating abundance and abundance–environment relationships have focused on individual-based methods using a conventional capture-recapture study design. However, tremendous amounts of data are available on animal populations in the form of counts of individuals, but without the ability to individually identify them on later sampling occasions. For example, consider traditional point count approaches for bird surveys. Typical point count surveys involve an observer that returns to a site a number of times during a breeding season and counts the individual members of a particular species they are interested in on each occasion. Because passerines can be difficult to individually recognize, it is not usually reliable to visually "mark" them as we might larger, more distinct animals. However, such counts represent a massive amount of data that may be useful for better understanding populations and species distributions.

So-called N-mixture models were developed to accommodate replicated count data[6] and account for imperfect detectability while inferring relationships between populations of animals and their environment (Royle 2004). Because these counts represent a subset of the total population at a site in any one sampling occasion, we denote these counts y_{ij} for $i = 1,\ldots,n$ sites and $j = 1,\ldots,J_i$ occasions. If we subsample the observed individuals randomly from the larger set of members from the population at the site (N_i), then

$$y_{ij} \sim \text{Binom}(N_i, p),\tag{24.53}$$

---

[6] These types of count data often are referred to as relative abundance data.

where p is a homogeneous per capita probability of detecting an individual member of the population N_i (i.e., a detection probability). Then, without further model structure, we would have $n + 1$ total parameters to estimate in this model. Furthermore, if the number of sampling occasions (J_i) is 1 or close to 1 in general, then it may be more challenging to separate the abundance N_i and detection probability p (but see Sólymos et al. 2012).

If we knew the abundance N_i at each site, we could fit ecological statistical models to the set of counts as described in Chapter 21 to make inference about abundance–environment relationships. However, because we do not know the abundances N_i exactly, we can treat them as part of a latent process for which we seek inference. A hierarchical model provides an ideal inferential framework when we observe an ecological process of interest with measurement error, but seek inference on a latent stochastic process in the model. Thus, the process component of the N-mixture model is

$$N_i \sim \text{Pois}(\lambda_i), \tag{24.54}$$

where heterogeneity in the abundance intensity λ_i can be linked to a set of environmental covariates as we described in Chapter 21 using the log link function $\log(\lambda_i) = \mathbf{x}_i' \boldsymbol{\beta}$. The set of regression coefficients $\boldsymbol{\beta}$ have the same interpretation as they did in Chapter 21 and allow us to reduce the number of unknown parameters in the process model component from n to p_β, where p_β is the dimension of $\boldsymbol{\beta}$.

If we use the priors

$$p \sim \text{Beta}(\alpha_p, \beta_p), \tag{24.55}$$

$$\boldsymbol{\beta} \sim \text{N}(\boldsymbol{\mu}_\beta, \boldsymbol{\Sigma}_\beta), \tag{24.56}$$

the posterior distribution for the N-mixture model is

$$[\mathbf{N}, p, \boldsymbol{\beta} | \mathbf{Y}] \propto \left(\prod_{i=1}^{n} \left(\prod_{j=1}^{J_i} [y_{ij} | N_i, p] \right) [N_i | \boldsymbol{\beta}] \right) [\boldsymbol{\beta}][p], \tag{24.57}$$

which results in the following full-conditional distributions for p, N_i for $i = 1, \dots, n$, and $\boldsymbol{\beta}$.

First, for the detection probability p, the full-conditional distribution is

$$[p|\cdot] \propto \left(\prod_{i=1}^{n} \prod_{j=1}^{J_i} [y_{ij} | N_i, p] \right) [p], \tag{24.58}$$

$$\propto p^{\sum_{i=1}^{n} \sum_{j=1}^{J_i} y_{ij}} (1 - p)^{\sum_{i=1}^{n} \sum_{j=1}^{J_i} (N_i - y_{ij})} p^{\alpha_p - 1} (1 - p)^{\beta_p - 1}, \tag{24.59}$$

$$\propto p^{\sum_{i=1}^{n} \sum_{j=1}^{J_i} y_{ij} + \alpha_p - 1} (1 - p)^{\sum_{i=1}^{n} \sum_{j=1}^{J_i} (N_i - y_{ij}) + \beta_p - 1}, \tag{24.60}$$

which results in a beta distribution $[p|\cdot] = \text{Beta}(\sum_{i=1}^{n} \sum_{j=1}^{J_i} y_{ij} + \alpha_p, \sum_{i=1}^{n} \sum_{j=1}^{J_i} (N_i - y_{ij}) + \beta_p)$. Thus, we can sample p using a Gibbs update in a MCMC algorithm.

The full-conditional distribution for N_i is analytically tractable only if $J_i = 1$, however, for more general situations it is not because

$$[N_i|\cdot] \propto \left(\prod_{j=1}^{J_i}[y_{ij}|N_i,p]\right)[N_i|\boldsymbol{\beta}], \qquad (24.61)$$

which has a product over the binomial data model. However, we can use a Poisson random walk proposal with rejection sampling where $N_i^{(*)} \sim \text{Pois}(N_i^{(k-1)} + c)$ and c is a small constant to eliminate the possibility of a zero intensity in the proposal distribution. We skip the M-H update for N_i when $N_i^{(*)} < \max(\mathbf{y}_i)$, considering only proposals for which N_i is at least as large as the largest observed relative abundance for site i.[7] This proposal distribution does not require tuning and results in the following M-H ratio

$$mh = \frac{\left(\prod_{j=1}^{J_i}[y_{ij}|N_i^{(*)},p^{(k-1)}]\right)[N_i^{(*)}|\boldsymbol{\beta}^{(k-1)}][N_i^{(k-1)}|N_i^{(*)}]}{\left(\prod_{j=1}^{J_i}[y_{ij}|N_i^{(k-1)},p^{(k-1)}]\right)[N_i^{(k-1)}|\boldsymbol{\beta}^{(k-1)}][N_i^{(*)}|N_i^{(k-1)}]}, \qquad (24.62)$$

where $[N_i^{(*)}|N_i^{(k-1)}]$ is the Poisson proposal distribution evaluated at the proposed value $N_i^{(*)}$ using intensity $N_i^{(k-1)} + c$. Thus, we can sample N_i for $i = 1,\ldots,n$ with a M-H update in a MCMC algorithm.

The final unknown elements of the Bayesian N-mixture model are the regression coefficients $\boldsymbol{\beta}$ that characterize the relationship between abundance and the site-level covariates. Because the process component of the hierarchical N-mixture model is the same as the posterior for the Poisson regression model we presented in Chapter 21, the full-conditional distribution for $\boldsymbol{\beta}$ is exactly the same as we derived in that chapter. Thus, if we use a multivariate normal random walk proposal for the coefficient vector such that we draw $\boldsymbol{\beta}^{(*)} \sim \text{N}(\boldsymbol{\beta}^{(k-1)}, \sigma_{\text{tune}}^2\mathbf{I})$, we can use the M-H ratio

$$mh = \frac{\left(\prod_{i=1}^{n}[N_i^{(k)}|\boldsymbol{\beta}^{(*)}]\right)[\boldsymbol{\beta}^{(*)}]}{\left(\prod_{i=1}^{n}[N_i^{(k)}|\boldsymbol{\beta}^{(k-1)}]\right)[\boldsymbol{\beta}^{(k-1)}]}. \qquad (24.63)$$

Assuming that $J_i = J$ for $i = 1,\ldots,n$ for simplicity, and using full-conditional distributions, we constructed a MCMC algorithm called Nmix.mcmc.R in R to fit the N-mixture model to replicate count data at a set of sites:

---

[7] We skip the update for N_i because $N_i < \max(\mathbf{y}_i)$ is not possible according to the data model and the probability of that event is zero which implies that the M-H ratio will be zero.

BOX 24.8   MCMC ALGORITHM FOR THE N-MIXTURE MODEL

```
Nmix.mcmc <- function(Y,X,n.mcmc){

###
### Setup Variables
###

n=dim(Y)[1]
J=dim(Y)[2]
pp=dim(X)[2]

p.save=rep(0,n.mcmc)
beta.save=matrix(0,pp,n.mcmc)
N.save=matrix(0,n,n.mcmc)

###
### Priors and Starting Values
###

beta.mn=rep(0,pp)
beta.sd=sqrt(1000)
alpha.p=1
beta.p=1

p=.9
N=round((apply(Y,1,max)+1)/p)
beta=solve(t(X)%*%X)%*%t(X)%*%log(N)
lam=exp(X%*%beta)
beta.tune=.05
sum.1=sum(Y)+alpha.p
const=1
y.max=apply(Y,1,max)

###
### Begin MCMC Loop
###

for(k in 1:n.mcmc){
```

(Continued)

BOX 24.8 (Continued)   MCMC ALGORITHM FOR THE N-MIXTURE MODEL

```
38   ###
39   ### Sample N
40   ###
41
42   N.star=rpois(n,N+const)
43   mh.1=apply(dbinom(Y,N.star,p,log=TRUE),1,sum)
44      +dpois(N.star,lam,log=TRUE)+dpois(N,N.star
45      +const,log=TRUE)
46   mh.2=apply(dbinom(Y,N,p,log=TRUE),1,sum)
47      +dpois(N,lam,log=TRUE)+dpois(N.star,N
48      +const,log=TRUE)
49   mh=exp(mh.1-mh.2)
50   keep.idx=((mh>runif(n)) & (N.star>=y.max))
51   N[keep.idx]=N.star[keep.idx]
52
53   ###
54   ### Sample beta
55   ###
56
57   beta.star=rnorm(pp,beta,beta.tune)
58   lam.star=exp(X%*%beta.star)
59   mh1=sum(dpois(N,lam.star,log=TRUE))+sum(dnorm
60      (beta.star,beta.mn,beta.sd,log=TRUE))
61   mh2=sum(dpois(N,lam,log=TRUE))+sum(dnorm(beta,
62      beta.mn,beta.sd,log=TRUE))
63   mh=exp(mh1-mh2)
64
65   if(mh > runif(1)){
66      beta=beta.star
67      lam=lam.star
68   }
69
70   ###
71   ### Sample p
72   ###
73
74   sum.2=sum(matrix(N,n,J)-Y)+beta.p
75   p=rbeta(1,sum.1,sum.2)
```

(Continued)

BOX 24.8 (Continued)   MCMC ALGORITHM FOR THE N-MIXTURE MODEL

```
76   ###
77   ### Save Samples
78   ###
79
80   p.save[k]=p
81   N.save[,k]=N
82   beta.save[,k]=beta
83
84 }
85
86 ###
87 ### Write Output
88 ###
89
90 list(beta.save=beta.save,N.save=N.save,
91     p.save=p.save)
92
93 }
```

To demonstrate the MCMC algorithm, we fit the model to a set of data from the Swiss amphibian monitoring program on yellow-bellied toad relative abundance (*Bombina variegata*; Roth et al. 2016). Roth et al. (2016) analyzed relative abundance data for $n = 481$ sites with $J_i = 2$ (for $i = 1, \ldots, n$) nocturnal surveys per site to understand relationships between species abundance and environmental variables. We focused on inferring relationships between abundance of yellow-bellied toads and four site-level covariates: elevation (standardized), surface water area (standardized), fish presence (binary), and vegetative cover (binary). We used hyperparameters $\mu_\beta = 0$ and $\Sigma_\beta = \sigma_\beta^2 I$ with $\sigma_\beta^2 = 1000$, and $\alpha_p = \beta_p = 1$ to fit the hierarchical N-mixture model to the yellow-bellied toad observed relative abundance data using the following R code and $K = 50000$ MCMC iterations:

BOX 24.9   FIT N-MIXTURE MODEL TO DATA

```
1 amph.df=read.csv("amphibian.csv")
2 n=dim(amph.df)[1]
3 Y=as.matrix(amph.df[,2:3])
4 X=matrix(1,n,5)
5 X[,2]=scale(amph.df$elevation)
```

(Continued)

BOX 24.9 (Continued)   FIT N-MIXTURE MODEL TO DATA

```
 6 X[,3]=scale(amph.df$surface)
 7 X[,4]=amph.df$fish
 8 X[,5]=amph.df$cover
 9
10 source("Nmix.mcmc.R")
11 n.mcmc=50000
12 set.seed(1)
13 mcmc.out=Nmix.mcmc(Y=Y,X=X,n.mcmc=n.mcmc)
14
15 apply(mcmc.out$beta.save[,-(1:10000)],1,mean)
16 apply(mcmc.out$beta.save[,-(1:10000)],1,quantile,
17     c(0.025,0.975))
18 mean(mcmc.out$p.save[-(1:10000)])
19 quantile(mcmc.out$p.save[-(1:10000)],c
20     (0.025,0.975))
21 mean(apply(mcmc.out$N.save,2,sum)[-(1:10000)])
22 quantile(apply(mcmc.out$N.save,2,sum)[-(1:10000)],
23     c(0.025,0.975))
```

We calculated posterior means and 95% credible intervals for p, β, and the total abundance of yellow-bellied toad across all sites using the derived quantity $N_{\text{total}} = \sum_{i=1}^{n} N_i$ (Table 24.1). The posterior results in Table 24.1 indicate strong relationships between all covariates that we used in the model, with elevation, presence of fish, vegetation cover negatively correlated with abundance and surface water area positively correlated with abundance. We estimated detection probability to be approximately 0.52, and the total number of yellow-bellied toads across sites to be 3865.

Table 24.1
Marginal Posterior Means and 95% Credible Intervals for β, p, and N_{total}

	Covariate						
	Intercept	Elevation	Surface	Fish	Cover	p	N_{total}
Mean	2.31	−0.44	0.13	−1.05	−0.31	0.52	3865
CI lower	2.24	−0.48	0.10	−1.23	−0.38	0.49	3719
CI upper	2.38	−0.39	0.16	−0.88	−0.23	0.54	4015

We can generalize N-mixture models to allow the detection probability also to vary among the sites associated with covariate relationships. Like occupancy models, we can use a logit link to connect detection probability to a potentially different set of environmental variables through the relationship $\text{logit}(p_i) = \mathbf{w}_i'\boldsymbol{\alpha}$. In the MCMC algorithm to fit this more general model, we would use a M-H update to sample $\boldsymbol{\alpha}$ as we described in Chapter 23 for the general occupancy model (but using the N-mixture likelihood instead of the occupancy likelihood).

24.5   ADDITIONAL CONCEPTS AND READING

The PX-DA approach is very flexible and can accommodate many different capture-recapture model specifications. Royle and Dorazio (2012) provide an overview of possible capture-recapture models that can be implemented using PX-DA. A common generalization of the simple capture-recapture model that we presented first in this chapter is the so-called M_h model which accounts for individual heterogeneity in the capture probability. A classical M_h model specification allows p_i to vary with individual i as a random effect with unknown mean and variance. Another well-known capture-recapture model that allows for individual heterogeneity is specified as a mixture model on p_i, where individuals are clustered into groups based on their capture probability (Pledger 2005).

One of the most common capture-recapture model extensions seeks to relax the closed-population assumption (Jolly 1965; Seber 1982; Pollock et al. 1990). So-called open robust capture-recapture models were developed to account for births and deaths as well as emigration and immigration (e.g., Kendall et al. 1997; Kendall and Bjorkland 2001; Schaub et al. 2004). These types of capture-recapture model extensions have also been translated to the Bayesian hierarchical setting (e.g., Gardner et al. 2018). In fact, more recent extensions to accommodate spatial information associated with recaptures have been made in a class of models called spatial capture-recapture models (Royle et al. 2013a; also see Chapter 27 in this book).

As the data sets grow and associated models become more complex, the PX-DA approach to fitting hierarchical capture-recapture models may become unstable. However, efforts are under way to improve MCMC algorithms for implementing complicated hidden Markov models such as dynamic capture-recapture models like the CJS model (Turek et al. 2016).

Like the capture-recapture model implementations based on integrated likelihoods, we can also marginalize over the N_i to obtain an N-mixture integrated likelihood that may facilitate inference (Dennis et al. 2015). In fact, when Royle (2004) introduced N-mixture models, he used an integrated likelihood in his implementation of them.

The sensitivity of assumptions associated with N-mixture models has been examined recently (Barker et al. 2018; Link et al. 2018). In particular, Link et al. 2018 noted that double-counting and heterogeneity of abundance and detection probability in N-mixture models may lead to non-identifiable components. However, others have argued that N-mixture models provide reliable inference in many settings (Kéry 2018) and that there is value in constructing models hierarchically based on mechanistic concepts of how the data were generated (Royle and Dorazio 2008). Regardless of sensitivity to assumptions, we can improve the resulting inference from N-mixture models by incorporating auxiliary data sources (e.g., such as traditional capture-recapture data or telemetry data) in the models (e.g., Ketz et al. 2018). We describe approaches for accommodating multiple data types in hierarchical models in Chapter 25.

Section V

Expert Models and Concepts

25 Integrated Population Models

25.1 DATA RECONCILIATION

As we have seen, Bayesian hierarchical models can accommodate ecological and observational mechanisms while also accounting for various sources of uncertainty. As a result, we can construct realistic ecological statistical models containing unknowns that may require more than one data set to estimate. Therefore, to obtain useful statistical inference, we often need to use multiple sources of data in a single model to help estimate all of the model parameters.

Fortunately, we can construct Bayesian models to accommodate multiple data sources (Hobbs and Hooten 2015). By connecting multiple data sources with process models that share parameters, we can often reduce the uncertainty about the shared parameters more than if we used any one data source alone.[1] Furthermore, separate data sources have their own sets of strengths and weakness and we often can leverage the strengths of each to improve inference using a procedure that Hanks et al. (2011) referred to as "data reconciliation."

For example, Hanks et al. (2011) were interested in predicting forest disease over large landscapes based on a survey conducted by Minnesota Department of Natural Resources (MDNR) at Forest Inventory and Analysis (FIA) plots. However, the MDNR surveys were not focused completely on dwarf mistletoe (*Arceuthobium pusillum*) in black spruce (*Picea mariana*) and resulted in disease incidence data that have broad extent and coverage but that are also subject to measurement error. Forest disease experts conducted a smaller, intensive survey, but due to limited resources they focused on a subset of the FIA locations surveyed by the MDNR. Thus, Hanks et al. (2011) developed a multiple data source model that reconciled the strengths of both data sets (i.e., the extent and coverage of the FIA data combined with the accuracy of the disease focused survey) to obtain more accurate large-scale predictions of forest disease.

To demonstrate how the model proposed by Hanks et al. (2011) is constructed, we specify a similar model and fit it to simulated data in what follows. Suppose y_i represents the set of binary data subject to measurement error, but with large spatial extent and complete coverage for $i = 1, \ldots, n$ total sites. Then, we denote the smaller and more accurate set of data as z_i for a subset of m of the total n sites. Because y_i is known to be a noisy version of z_i, we can model it as a mixture distribution that accommodates the possibility of both false negatives (as usual in occupancy models) and false positives:

---

[1] We also could have multiple data sources that share an underlying process directly.

$$y_i \sim \begin{cases} \text{Bern}(p_1) & , z_i = 1 \\ \text{Bern}(p_0) & , z_i = 0 \end{cases}, \tag{25.1}$$

where p_1 and p_0 represent the detection probability ($1 - p_1$ is the false negative probability) and false positive probability, respectively.

We assume the more accurate data are perfect measurements of the underlying binary process z_i, thus we specify the binary regression model $z_i \sim \text{Bern}(\psi_i)$, where $g(\psi_i) = \mathbf{x}_i'\boldsymbol{\beta}$ is a suitable link function. If we use the auxiliary variable approach that we described in Chapters 20 and 23, then we can derive conjugate full-conditional distributions for all unknown variables in the model. Thus, using the auxiliary variable formulation, the process model (based on partially observed data) is

$$z_i = \begin{cases} 0 & , v_i \leq 0 \\ 1 & , v_i > 0 \end{cases}, \tag{25.2}$$

where $v_i \sim \text{N}(\mathbf{x}_i'\boldsymbol{\beta}, 1)$. For priors, we use $p_0 \sim \text{Beta}(\alpha_0, \beta_0)$, $p_1 \sim \text{Beta}(\alpha_1, \beta_1)$, and $\boldsymbol{\beta} \sim \text{N}(\boldsymbol{\mu}_\beta, \boldsymbol{\Sigma}_\beta)$.

This Bayesian data reconciliation model results in the following posterior distribution

$$[\mathbf{z}_{\text{unobs}}, \mathbf{v}, p_0, p_1, \boldsymbol{\beta} | \mathbf{y}, \mathbf{z}_{\text{obs}}] \propto \left(\prod_{i=1}^{n} [y_i|p_1]^{z_i} [y_i|p_0]^{1-z_i} [z_i|v_i][v_i|\boldsymbol{\beta}] \right) [p_0][p_1][\boldsymbol{\beta}], \tag{25.3}$$

where $\mathbf{z}_{\text{unobs}}$ and $\mathbf{z}_{\text{obs}}$ represent the unobserved and observed ecological process of interest (e.g., disease presence or absence, species occupancy, etc.), respectively. To implement this model, we can construct a MCMC algorithm based on the full-conditional distributions for each parameter (which are all conjugate).

The full-conditional distributions for z_i in our Bayesian data reconciliation model are slightly different than for the occupancy models that we presented in Chapter 23 because we now have only a single sampling occasion for the less accurate survey, but we also have a subset of the sites measured perfectly without error. Also, unlike conventional occupancy models, we have the possibility of two sources of measurement error: false negatives and false positives. Thus, the case where $y_i = 1$ does not preclude us from sampling z_i as it did in the conventional occupancy models that we presented in Chapter 23. However, we need only sample z_i for sites where it was not observed as part of the accurate survey. Thus, the full-conditional distribution for z_i when it was not observed perfectly is

$$[z_i|\cdot] \propto [y_i|p_1]^{z_i}[y_i|p_0]^{1-z_i}[z_i|\psi_i], \tag{25.4}$$

$$\propto (p_1^{y_i}(1-p_1)^{1-y_i})^{z_i}(p_0^{y_i}(1-p_0)^{1-y_i})^{1-z_i}\psi_i^{z_i}(1-\psi_i)^{1-z_i}, \tag{25.5}$$

$$\propto (\psi_i p_1^{y_i}(1-p_1)^{1-y_i})^{z_i}((1-\psi_i)p_0^{y_i}(1-p_0)^{1-y_i})^{1-z_i}, \tag{25.6}$$

which implies that $[z_i|\cdot] = \text{Bern}(\tilde{\psi}_i)$ where

$$\tilde{\psi}_i = \frac{\psi_i p_1^{y_i}(1-p_1)^{1-y_i}}{\psi_i p_1^{y_i}(1-p_1)^{1-y_i} + (1-\psi_i)p_0^{y_i}(1-p_0)^{1-y_i}}. \tag{25.7}$$

Thus, we can sample z_i using a Gibbs update in our MCMC algorithm for the sites where it was not observed directly.

We derived the exact full-conditional distribution for v_i in Chapters 20 and 23, thus we restate it here as

$$[v_i|\cdot] = \begin{cases} \text{TN}(\mathbf{x}_i'\boldsymbol{\beta}, 1)_0^\infty & , z_i = 1 \\ \text{TN}(\mathbf{x}_i'\boldsymbol{\beta}, 1)_{-\infty}^0 & , z_i = 0 \end{cases}, \tag{25.8}$$

where $\text{TN}(\cdot, \cdot)_a^b$ indicates a truncated normal distribution truncated below at a and above at b.

Similarly, we derived the full-conditional distribution for the regression coefficient vector in Chapters 20 and 23 as $[\boldsymbol{\beta}|\cdot] = \text{N}(\mathbf{A}^{-1}\mathbf{b}, \mathbf{A}^{-1})$ where

$$\mathbf{A} = \mathbf{X}'\mathbf{X} + \boldsymbol{\Sigma}_\beta^{-1}, \tag{25.9}$$

$$\mathbf{b} = \mathbf{X}'\mathbf{v} + \boldsymbol{\Sigma}_\beta^{-1}\boldsymbol{\mu}_\beta. \tag{25.10}$$

Thus, we can also sample $\boldsymbol{\beta}$ using a Gibbs update in our MCMC algorithm.

The full-conditional distributions for the detection probability is the same as it was in the homogeneous occupancy model from Chapter 23: $[p_1|\cdot] = \text{Beta}(\sum_{\forall z_i=1} y_i + \alpha_1, \sum_{\forall z_i=1}(1-y_i) + \beta_1)$. However, the full-conditional distribution for the false positive probability p_0 is slightly different, but derived in the same way as the detection probability:

$$[p_0|\cdot] \propto \left(\prod_{i=1}^n [y_i|p_0]^{1-z_i}\right)[p_0], \tag{25.11}$$

$$\propto \left(\prod_{\forall z_i=0} [y_i|p_0]\right)[p_0], \tag{25.12}$$

$$\propto p_0^{\sum_{\forall z_i=0} y_i}(1-p_0)^{\sum_{\forall z_i=0}(1-y_i)} p_0^{\alpha_0-1}(1-p_0)^{\beta_0-1}, \tag{25.13}$$

$$\propto p_0^{\sum_{\forall z_i=0} y_i+\alpha_0-1}(1-p_0)^{\sum_{\forall z_i=0}(1-y_i)+\beta_0-1}, \tag{25.14}$$

which results in a beta full-conditional distribution where $[p_0|\cdot] = \text{Beta}(\sum_{\forall z_i=0} y_i + \alpha_0, \sum_{\forall z_i=0}(1-y_i) + \beta_0)$. Thus, we can sample each probability parameter (p_0 and p_1) using a Gibbs update in our MCMC algorithm to fit the Bayesian data reconciliation model.

To fit the Bayesian data reconciliation model, we constructed a MCMC algorithm called `bdr.mcmc.R` in R using Gibbs updates based on these full-conditional distributions:

BOX 25.1 MCMC ALGORITHM FOR DATA RECONCILIATION MODEL

```
1  bdr.mcmc <- function(y,z.obs,z.idx,z.true,
2    X,n.mcmc){
3
4  ####
5  ####   Libraries and Subroutines
6  ####
7
8  rtn <- function(n,mu,sig2,low,high){
9    flow=pnorm(low,mu,sqrt(sig2))
10   fhigh=pnorm(high,mu,sqrt(sig2))
11   u=runif(n)
12   tmp=flow+u*(fhigh-flow)
13   x=qnorm(tmp,mu,sqrt(sig2))
14   x
15 }
16
17 ####
18 ####   Setup Variables
19 ####
20
21 n=length(y)
22 p.X=dim(X)[2]
23 n.burn=round(0.25*n.mcmc)
24
25 z.mean=rep(0,n)
26 psi.mean=rep(0,n)
27 beta.save=matrix(0,p.X,n.mcmc)
28 p0.save=rep(0,n.mcmc)
29 p1.save=rep(0,n.mcmc)
30 N.save=rep(0,n.mcmc)
31 pred.acc.save=rep(0,n.mcmc)
32
33 ####
34 ####   Priors and Starting Values
35 ####
36
37 beta=rep(0,p.X)
38 p0=.25
39 p1=.75
```

(*Continued*)

**BOX 25.1 (Continued)   MCMC ALGORITHM FOR DATA
RECONCILIATION MODEL**

```
40 Xbeta=X%*%beta
41 psi=pnorm(Xbeta)
42 z=rep(0,n)
43 z[z.idx]=z.obs
44 n0=sum(z==0)
45 n1=sum(z==1)
46
47 v=rep(0,n)
48
49 ny0=sum(y==0)
50 ny1=sum(y==1)
51
52 alpha.0=1
53 beta.0=1
54 alpha.1=1
55 beta.1=1
56
57 Sig.beta=diag(p.X)
58 Sig.beta.inv=solve(Sig.beta)
59 mu.beta=rep(0,p.X)
60
61 XprimeX=t(X)%*%X
62 A.beta=XprimeX+Sig.beta.inv
63 A.beta.chol=chol(A.beta)
64 Sig.beta.inv.times.mu.beta=Sig.beta.inv%*%mu.beta
65
66 ####
67 ####  Begin MCMC Loop
68 ####
69
70 for(k in 1:n.mcmc){
71
72    ####
73    ####  Sample v
74    ####
75
76    v[z==0]=rtn(n0,Xbeta[z==0],1,-Inf,0)
77    v[z==1]=rtn(n1,Xbeta[z==1],1,0,Inf)
```

(Continued)

BOX 25.1 (Continued)   MCMC ALGORITHM FOR DATA RECONCILIATION MODEL

```
 78  ####
 79  ####  Sample beta
 80  ####
 81
 82  b.beta=t(X)%*%v+Sig.beta.inv.times.mu.beta
 83  beta=backsolve(A.beta.chol,backsolve(A.beta.
 84    chol,b.beta,transpose=TRUE)+rnorm(p.X))
 85  Xbeta=X%*%beta
 86  psi=pnorm(Xbeta)
 87
 88  ####
 89  ####  Sample p0
 90  ####
 91
 92  sum.1=sum(y[z==0])+alpha.0
 93  sum.2=sum(1-y[z==0])+beta.0
 94  p0=rbeta(1,sum.1,sum.2)
 95
 96  ####
 97  ####  Sample p1
 98  ####
 99
100  sum.1=sum(y[z==1])+alpha.1
101  sum.2=sum(1-y[z==1])+beta.1
102  p1=rbeta(1,sum.1,sum.2)
103
104  ####
105  ####  Sample z
106  ####
107
108  psi.numer=psi*(p1^y)*((1-p1)^(1-y))
109  psi.tmp=psi.numer/(psi.numer+(1-psi)*(p0^y)
110    *((1-p0)^(1-y)))
111  psi.tmp[psi.tmp>1]=1
112  z[-z.idx]=rbinom(n-length(z.idx),1,
113    psi.tmp[-z.idx])
114  n0=sum(z==0)
115  n1=sum(z==1)
```

(Continued)

BOX 25.1 (Continued)   MCMC ALGORITHM FOR DATA RECONCILIATION MODEL

```
116  ####
117  ####  Prediction Accuracy
118  ####
119
120  pred.acc.save[k]=sum((z.true-z)[-z.idx]==0)/
121    (n-length(z.idx))
122
123  ####
124  ####  Save Samples
125  ####
126
127  beta.save[,k]=beta
128  p0.save[k]=p0
129  p1.save[k]=p1
130  if(k > n.burn){
131    z.mean=z.mean+z/(n.mcmc-n.burn)
132    psi.mean=psi.mean+psi/(n.mcmc-n.burn)
133  }
134  N.save[k]=sum(z)
135
136 }
137
138 ####
139 ####  Write Output
140 ####
141
142 list(p0.save=p0.save,p1.save=p1.save,beta.save
143    =beta.save,z.mean=z.mean,psi.mean=psi.mean,
144    N.save=N.save,pred.acc.save=pred.acc.save,
145    n.mcmc=n.mcmc)
146
147 }
```

In this MCMC algorithm, we also saved a sample for the derived quantity that allows us to assess prediction accuracy (line 120) by comparing our predictions of the latent process $\mathbf{z}_{\text{unobs}}$ to true simulated values as

$$\frac{\sum_{i=1}^{n_{\text{unobs}}} 1_{\{z_{i,\text{unobs}}^{(k)} = z_{i,\text{unobs}}^{\text{true}}\}}}{n_{\text{unobs}}}, \tag{25.15}$$

where $z_{i,\text{unobs}}^{(k)}$ is the kth MCMC draw ($k = 1, \ldots, K$) for the ith unobserved occupancy, $z_{i,\text{unobs}}^{\text{true}}$ is the true simulated occupancy status, and n_{unobs} is the number of sites for which the less accurate data were observed but the accurate data were not. We can use this approach only with simulated data, or using cross-validation based on a held out subset of $\mathbf{z}_{\text{obs}}$, but it provides a rigorous quantitative way to assess the accuracy of our predictions.

We simulated true occupancy statuses ($\mathbf{z}$) and observed occupancy with measurement error ($\mathbf{y}$) for $n = 200$ sites, with $m = 50$ selected randomly to comprise $\mathbf{z}_{\text{obs}}$ and the remaining $n - m = 150$ to use for assessing predictive accuracy. We simulated heterogeneity in occupancy probability among sites by sampling n independent standard normal realizations to serve as a covariate $\mathbf{x}$. We used values for $\boldsymbol{\beta} = (-0.5, 0.5)'$, $p_0 = 0.25$, and $p_1 = 0.75$. The following R code was used to simulate the data:

BOX 25.2   SIMULATE DATA FOR DATA RECONCILIATION MODEL

```
 1 n=200
 2 m=50
 3 p1=.75
 4 p0=.25
 5
 6 set.seed(101)
 7 X=matrix(1,n,2)
 8 X[,2]=rnorm(n)
 9 beta=c(-.5,.5)
10
11 psi=pnorm(X%*%beta)
12 z=rbinom(n,1,psi)
13 y=rbinom(n,1,z*p1+(1-z)*p0)
14
15 z.idx=sort(sample(1:n,m))
16 z.obs=z[z.idx]
```

In these data, $\mathbf{z}_{\text{obs}}$ contained 33 zero values and 17 one values. By contrast, out of the total number of observed sites (n) for the simulated data with measurement error, $\mathbf{y}$ contained 115 zeros and 85 ones. We fit the Bayesian data reconciliation model to our simulated data using $K = 20000$ MCMC iterations and the following R code:

BOX 25.3 FIT DATA RECONCILIATION MODEL TO DATA

```
1  source("bdr.mcmc.R")
2  n.mcmc=20000
3  set.seed(1)
4  mcmc.out=bdr.mcmc(y=y,z.obs=z.obs,
5      z.idx=z.idx,z.true=z,X=X,n.mcmc=n.mcmc)
6
7  layout(matrix(1:4,2,2,byrow=TRUE))
8  hist(mcmc.out$beta.save[1,-(1:1000)]xlab=bquote
9      (beta[0]),prob=TRUE,main="a")
10 curve(dnorm(x,0,1),lwd=2,add=TRUE)
11 abline(v=beta[1],lty=2,lwd=2)
12 hist(mcmc.out$beta.save[2,-(1:1000)]xlab=bquote
13     (beta[1]),prob=TRUE,main="b")
14 curve(dnorm(x,0,1),lwd=2,add=TRUE)
15 abline(v=beta[2],lty=2,lwd=2)
16 hist(mcmc.out$p0.save[-(1:1000)],xlab=
17     bquote(p[0]),xlim=c(0,1),prob=TRUE,main="c")
18 curve(dbeta(x,1,1),lwd=2,add=TRUE)
19 abline(v=p0,lty=2,lwd=2)
20 hist(mcmc.out$p1.save[-(1:1000)],xlab=
21     bquote(p[1]),xlim=c(0,1),prob=TRUE,main="d")
22 curve(dbeta(x,1,1),lwd=2,add=TRUE)
23 abline(v=p1,lty=2,lwd=2)
24
25 mean(mcmc.out$N.save[-(1:1000)])
26 quantile(mcmc.out$N.save[-(1:1000)],
27     c(0.025,0.975))
28 sum(z)
```

For hyperparameters, we used $\boldsymbol{\mu}_\beta \equiv (0,0)'$, $\boldsymbol{\Sigma}_\beta \equiv \mathbf{I}$, $\alpha_0 \equiv \beta_0 \equiv \alpha_1 \equiv \beta_1 \equiv 1$.

The results of fitting the Bayesian data reconciliation model to the simulated data are shown as marginal posterior histograms in Figure 25.1. Figure 25.1 indicates that the posterior distributions capture the true parameter values used to simulate the data well. Also, using both sets of data in the same Bayesian data reconciliation model, the posterior mean total number of occupied sites was $E(N|\mathbf{y}, \mathbf{z}_{\text{obs}}) \approx 69.9$ (with posterior 95% credible interval [53, 88]). The true number of occupied sites in our simulated data was $N = 67$.

One of the key advantages to using the Bayesian data reconciliation model proposed by Hanks et al. (2011) is to borrow strength from the two observed data sets to predict the true occupancy process at the full set of sites. In fact, we can compare the predictive accuracy of the approach to predictions based on two data sets

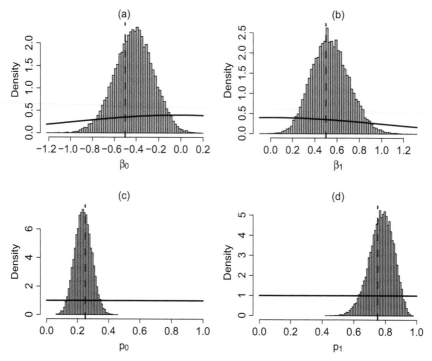

Figure 25.1 Marginal posterior histograms for (a) β_0, (b) β_1, (c) p_0, and (d) p_1 resulting from fitting the Bayesian data reconciliation model to the simulated data. Prior PDFs are shown as solid black lines and true parameter values used to simulate the data are shown as vertical dashed lines.

individually. To analyze the data without measurement error ($\mathbf{z}_{\text{obs}}$) based on the subset of covariates ($\mathbf{X}_{\text{obs}}$) associated with the observed data, we used the probit regression model introduced in Chapter 20 and obtained a sample from the posterior predictive distribution for $\mathbf{z}_{\text{unobs}}$ using $\mathbf{z}_{\text{unobs}}^{(k)} \sim \text{Bern}(\Phi(\mathbf{X}_{\text{unobs}}\boldsymbol{\beta}^{(k)}))$ for $k = 1, \ldots, K$. The updated MCMC algorithm (`probit.reg.mcmc.R`) to fit the probit regression model and compute the classification accuracy score is:

BOX 25.4   MCMC ALGORITHM FOR PROBIT REGRESSION

```
1 probit.reg.mcmc <- function(z,X,X.pred,z.true,
2     n.mcmc){
3
4 ###
5 ###   Subroutines
6 ###
```

(Continued)

BOX 25.4 (Continued)   MCMC ALGORITHM FOR PROBIT REGRESSION

```
7  rtn <- function(n,mu,sig2,low,high){
8    flow=pnorm(low,mu,sqrt(sig2))
9    fhigh=pnorm(high,mu,sqrt(sig2))
10   u=runif(n)
11   tmp=flow+u*(fhigh-flow)
12   x=qnorm(tmp,mu,sqrt(sig2))
13   x
14 }
15
16 ###
17 ###   Preliminary Variables
18 ###
19
20 X=as.matrix(X)
21 z=as.vector(z)
22 n=length(z)
23 p=dim(X)[2]
24 n.pred=dim(X.pred)[1]
25
26 beta.save=matrix(0,p,n.mcmc)
27 v.save=matrix(0,n,n.mcmc)
28 pred.acc.save=rep(0,n.mcmc)
29 N.save=rep(0,n.mcmc)
30
31 z1=(z==1)
32 z0=(z==0)
33
34 ###
35 ###   Priors and Starting Values
36 ###
37
38 beta.mn=rep(0,p)
39 beta.var=1
40
41 Sig.beta.inv=solve(beta.var*diag(p))
42 beta=beta.mn
```

(Continued)

BOX 25.4 (Continued)   MCMC ALGORITHM FOR PROBIT REGRESSION

```
43 ###
44 ###   MCMC Loop
45 ###
46
47 for(k in 1:n.mcmc){
48
49   ###
50   ### Sample v
51   ###
52
53   v=rep(0,n)
54   v[z1]=rtn(sum(z),(X%*%beta)[z1],1,0,Inf)
55   v[z0]=rtn(sum(1-z),(X%*%beta)[z0],1,-Inf,0)
56
57   ###
58   ### Sample beta
59   ###
60
61   tmp.chol=chol(t(X)%*%X + Sig.beta.inv)
62   beta=backsolve(tmp.chol,backsolve(tmp.chol,
63     t(X)%*%v + Sig.beta.inv%*%beta.mn,
64     transpose=TRUE)+rnorm(p))
65
66   ###
67   ### Sample Predictions
68   ###
69
70   z.pred=rbinom(n.pred,1,pnorm(X.pred%*%beta))
71   pred.acc.save[k]=sum((z.true-z.pred)==0)/n.pred
72
73   ###
74   ### Save Samples
75   ###
76
77   beta.save[,k]=beta
78   v.save[,k]=v
79   N.save[k]=sum(z)+sum(z.pred)
80
81 }
```

(Continued)

BOX 25.4 (Continued)   MCMC ALGORITHM FOR PROBIT REGRESSION

```
82 ###
83 ###   Write output
84 ###
85
86 list(z=z,X=X,n.mcmc=n.mcmc,v.save=v.save,
87     beta.save=beta.save,pred.acc.save
88     =pred.acc.save,N.save=N.save)
89
90 }
```

Notice that we used the same MCMC algorithm as introduced in Chapter 20 for probit regression with auxiliary variables, but we changed the response variable from y to z for consistency with this Bayesian data reconciliation example. Also notice that we added a portion of R code to the MCMC algorithm to sample from the posterior predictive distribution of $\mathbf{z}_{unobs}$ on lines 70–71 and calculate the derived quantity (pred.acc.save) that allows us to infer classification accuracy.

To analyze the data with measurement error ($\mathbf{y}_{obs}$) at the sites where we observed $\mathbf{z}_{obs}$, we used the probit occupancy model introduced in Chapter 23. This model assumes that only false negative detection errors occur in the data (not false positive errors because the false positive probability parameter is not identifiable without additional data). Because the occupancy model treats latent occupancy variables (z_i) for which the corresponding $y_i = 0$ as unknown, we automatically get a MCMC sample for $\mathbf{z}_{unobs}$ without having to sample them separately from a posterior predictive distribution.

We modified the MCMC algorithm to fit the probit occupancy model with auxiliary variables based on a simplified model with homogeneous detection probability. The updated algorithm in R is called occ.aux.p.mcmc.R:

BOX 25.5   MCMC ALGORITHM FOR PROBIT OCCUPANCY MODEL

```
1 occ.aux.p.mcmc <- function(y,J,X,z.true,z.idx,
2     n.mcmc){
3
4 ####
5 ####   Libraries and Subroutines
6 ####
```

(Continued)

BOX 25.5 (Continued) MCMC ALGORITHM FOR PROBIT OCCUPANCY MODEL

```
 7 rtn <- function(n,mu,sig2,low,high){
 8    flow=pnorm(low,mu,sqrt(sig2))
 9    fhigh=pnorm(high,mu,sqrt(sig2))
10    u=runif(n)
11    tmp=flow+u*(fhigh-flow)
12    x=qnorm(tmp,mu,sqrt(sig2))
13    x
14 }
15
16 ####
17 ####   Setup Variables
18 ####
19
20 n=length(y)
21 p.X=dim(X)[2]
22 n.burn=round(0.25*n.mcmc)
23
24 z.mean=rep(0,n)
25 beta.save=matrix(0,p.X,n.mcmc)
26 N.save=rep(0,n.mcmc)
27 p.save=rep(0,n.mcmc)
28 pred.acc.save=rep(0,n.mcmc)
29
30 ####
31 ####   Priors and Starting Values
32 ####
33
34 p=.5
35 beta=rep(0,p.X)
36 Xbeta=X%*%beta
37 psi=pnorm(Xbeta)
38 z=rep(0,n)
39 z[y==0]=0
40 n0=sum(z==0)
41 n1=sum(z==1)
```

(Continued)

BOX 25.5 (Continued)   MCMC ALGORITHM FOR PROBIT OCCUPANCY MODEL

```
42  v=rep(0,n)
43
44  ny0=sum(y==0)
45  ny1=sum(y==1)
46
47  alpha.p=1
48  beta.p=1
49
50  mu.beta=rep(0,p.X)
51  Sig.beta=diag(p.X)
52  Sig.beta.inv=solve(Sig.beta)
53
54  XprimeX=t(X)%*%X
55  A.beta=XprimeX+Sig.beta.inv
56  A.beta.chol=chol(A.beta)
57  Sig.beta.inv.times.mu.beta=Sig.beta.inv%*%mu.beta
58
59  ####
60  ####  Begin MCMC Loop
61  ####
62
63  for(k in 1:n.mcmc){
64
65     ####
66     ####  Sample v
67     ####
68
69     v[z==0]=rtn(n0,Xbeta[z==0],1,-Inf,0)
70     v[z==1]=rtn(n1,Xbeta[z==1],1,0,Inf)
71
72     ####
73     ####  Sample beta
74     ####
75
76     b.beta=t(X)%*%v+Sig.beta.inv.times.mu.beta
77     beta=backsolve(A.beta.chol, backsolve(A.beta.
78        chol,b.beta, transpose = TRUE) + rnorm(p.X))
79     Xbeta=X%*%beta
80     psi=pnorm(Xbeta)
```

(Continued)

BOX 25.5 (Continued)   MCMC ALGORITHM FOR PROBIT OCCUPANCY MODEL

```
81  ####
82  ####  Sample z
83  ####
84
85  psi.numer=psi*(1-p)^J
86  psi.tmp=psi.numer/(psi.numer+1-psi)
87  z[y==0]=rbinom(sum(y==0),1,psi.tmp[y==0])
88  z[y>0]=1
89  n0=sum(z==0)
90  n1=sum(z==1)
91
92  ####
93  ####  Sample p
94  ####
95
96  p=rbeta(1,sum(y[z==1])+alpha.p,sum(J-y[z==1])
97    +beta.p)
98
99  ####
100 ####  Calculate Pred Accuracy
101 ####
102
103 pred.acc.save[k]=sum((z.true-z[-z.idx])==0)/
104   (n-length(z.idx))
105
106 ####
107 ####  Save Samples
108 ####
109
110 beta.save[,k]=beta
111 if(k > n.burn){
112   z.mean=z.mean+z/(n.mcmc-n.burn)
113 }
114 N.save[k]=sum(z)
115 p.save[k]=p
116
117 }
```

(Continued)

BOX 25.5 (Continued)   MCMC ALGORITHM FOR PROBIT OCCUPANCY MODEL

```
118 ####
119 ####   Write Output
120 ####
121
122 list(beta.save=beta.save,z.mean=z.mean,
123    N.save=N.save,pred.acc.save=pred.acc.save,
124    n.mcmc=n.mcmc)
125
126 }
```

Notice that we compute the same classification accuracy score on line 103 of R code.

Now that we have three different approaches for predicting occupancy (z_{unobs}) at the set of sites for which we do not have perfect observations, we can compare them in terms of classification accuracy. The following R code fits each model and characterizes the posterior distribution of the classification accuracy score:

BOX 25.6   FIT THREE MODELS AND COMPARE CLASSIFICATION ACCURACY

```
 1 n.mcmc=20000
 2 source("bdr.mcmc.R")
 3 set.seed(1)
 4 mcmc.out=bdr.mcmc(y=y,z.obs=z.obs,z.idx=z.idx,
 5    z.true=z,X=X,n.mcmc=n.mcmc)
 6 mean(mcmc.out$pred.acc.save[-(1:1000)])
 7 quantile(mcmc.out$pred.acc.save[-(1:1000)],
 8    c(0.025,0.975))
 9
10 source("probit.reg.mcmc.R")
11 set.seed(1)
12 mcmc.z.out=probit.reg.mcmc(z=z.obs,X=X[z.idx,],
13    X.pred=X[-z.idx,],z.true=z[-z.idx],
14    n.mcmc=n.mcmc)
15 mean(mcmc.z.out$pred.acc.save[-(1:1000)])
16 quantile(mcmc.z.out$pred.acc.save[-(1:1000)],
17    c(0.025,0.975))
```

(Continued)

**BOX 25.6 (Continued)   FIT THREE MODELS AND COMPARE
CLASSIFICATION ACCURACY**

```
18 source("occ.aux.p.mcmc.R")
19 set.seed(1)
20 mcmc.y.out=occ.aux.p.mcmc(y=y,J=1,X=X,
21     z.true=z[-z.idx],z.idx=z.idx,
22     n.mcmc=n.mcmc)
23 mean(mcmc.y.out$pred.acc.save[-(1:1000)])
24 quantile(mcmc.y.out$pred.acc.save[-(1:1000)],
25     c(0.025,0.975))
```

Table 25.1
Posterior Mean and 95% Credible Intervals for Classification Scores Associated with Three Models for Predicting Occupancy: Bayesian Data Reconciliation (BDR), Probit Regression (PR), and Probit Occupancy Model (PO)

Model	Mean	95% CI
BDR	0.71	(0.63, 0.78)
PR	0.63	(0.52, 0.72)
PO	0.61	(0.44, 0.74)

Note: Scores indicate proportion of sites where occupancy was correctly classified.

R code results in the posterior means and 95% credible intervals for the classification score (25.15) shown in Table 25.1. Table 25.1 indicates that the Bayesian data reconciliation approach outperforms the other two methods that use only one of the data sources each. The posterior mean classification score of 0.71 implies that Bayesian data reconciliation can correctly predict occupancy for 71% of the unobserved sites, as compared with the other two approaches which only predict occupancy correctly at 63% and 61% of the sites.

25.2   FALSE POSITIVE MODELS WITH AUXILIARY DATA

As we showed in the previous section, in some cases, the coupling of multiple data sets facilitates the statistical estimation of parameters that would be difficult to estimate otherwise. However, unlike the situation with the Bayesian data reconciliation model, data sets are commonly collected that are spatially and/or temporally disjoint but still relevant for informing at least a subset of the overall parameters. For example, Ruiz-Gutierrez et al. (2016) described a situation involving occupancy data that are subject to both false negative and false positive measurement errors (also see Royle and Link 2006), resembling **y** from the previous section, but possibly with multiple replicates such that $y_i = \sum_{j=1}^{J} y_{ij}$, where y_{ij} is a binary measurement of occupancy at observed site i and survey occasion j (for $j = 1, \ldots, J$; recall that $J = 1$ in the previous section).

In the context of species identification data for monitoring birds and amphibians, Ruiz-Gutierrez et al. (2016) noted that there are often excellent auxiliary data sets for use in helping to estimate false negative and false positive probabilities. For example, suppose that we provide a separate survey to a field observer to "test" their ability to correctly measure species occupancy. In such surveys, the truth is known; for example, as a prerecorded acoustic sample of a known species played back to assess identification capability of the observer (Miller et al. 2012). The terms $1 - p_1$ and p_0 correspond to false negative and false positive detection probability respectively, given that the observer heard a sound resembling a related species. These are similar to the interpretations of p_0 and p_1 in the previous section; however, the lack of familiar sound at that site during the survey also may cause the failure to detect a species at a site. Thus, we may wish to introduce a third detection parameter p to account for imperfect detection due to other factors.

Thus, modifying the Bayesian data reconciliation model from the previous section, we have

$$y_i \sim \begin{cases} \text{Binom}(J, p_1 p) & , z_i = 1 \\ \text{Binom}(J, p_0) & , z_i = 0 \end{cases}, \tag{25.16}$$

where z_i represents the latent occupancy state as before. Unlike the Bayesian data reconciliation situation, in this case, we do not have perfect measurements for a subset of z_i to rely on, but we may have auxiliary data c_1 that correspond to the number of correctly answered identification questions on an independent observer survey with M_1 questions where the sample recording was the species of interest. For another independent set of M_0 questions where the sample recording was not the species of interest, we also may have auxiliary data c_0 corresponding to the number of times the species of interest was incorrectly given as the answer. Thus, we have two additional data models: $c_1 \sim \text{Binom}(M_1, p_1)$ and $c_0 \sim \text{Binom}(M_0, p_0)$, that are conditioned on parameters in the main occupancy model (25.16).

We specify the process model using the probit link function based on auxiliary variables as in the previous section:

$$z_i = \begin{cases} 0 & , v_i \leq 0 \\ 1 & , v_i > 0 \end{cases}, \tag{25.17}$$

where $v_i \sim \text{N}(\mathbf{x}_i'\boldsymbol{\beta}, 1)$. For priors, we use $p_0 \sim \text{Beta}(\alpha_0, \beta_0)$, $p_1 \sim \text{Beta}(\alpha_1, \beta_1)$, $p \sim \text{Beta}(\alpha_p, \beta_p)$, and $\boldsymbol{\beta} \sim \text{N}(\boldsymbol{\mu}_\beta, \boldsymbol{\Sigma}_\beta)$. Therefore, the posterior distribution for this false positive occupancy model is

$$[\mathbf{z}, \mathbf{v}, p_0, p_1, p, \boldsymbol{\beta} | \mathbf{y}, c_1, c_0] \propto \left(\prod_{i=1}^{n} [y_i | p, p_1]^{z_i} [y_i | p_0]^{1-z_i} [z_i | v_i] [v_i | \boldsymbol{\beta}] \right) \times$$

$$[c_0 | p_0][c_1 | p_1][p_0][p_1][p][\boldsymbol{\beta}], \tag{25.18}$$

where the entire latent set of occupancy states $\mathbf{z}$ is unknown.

To fit the model using MCMC, we need to find the full-conditional distributions for latent variables and parameters in the model. Fortunately, the full-conditional distributions for $\mathbf{v}$ and $\boldsymbol{\beta}$ are exactly as we derived in the previous Bayesian data reconciliation model. Thus, we use a Gibbs update in our MCMC algorithm to sample v_i, for $i = 1, \ldots, n$, from

$$[v_i | \cdot] = \begin{cases} \text{TN}(\mathbf{x}_i' \boldsymbol{\beta}, 1)_0^{\infty} & , z_i = 1 \\ \text{TN}(\mathbf{x}_i' \boldsymbol{\beta}, 1)_{-\infty}^{0} & , z_i = 0 \end{cases}. \tag{25.19}$$

We also use a Gibbs update to sample $\boldsymbol{\beta}$ from $[\boldsymbol{\beta} | \cdot] = \text{N}(\mathbf{A}^{-1} \mathbf{b}, \mathbf{A}^{-1})$ where

$$\mathbf{A} = \mathbf{X}' \mathbf{X} + \boldsymbol{\Sigma}_{\beta}^{-1}, \tag{25.20}$$

$$\mathbf{b} = \mathbf{X}' \mathbf{v} + \boldsymbol{\Sigma}_{\beta}^{-1} \boldsymbol{\mu}_{\beta}. \tag{25.21}$$

The full-conditional distributions for the detection probability parameters differ from those in the previous section because we now have two auxiliary sources of data to incorporate. However, we can still use a Gibbs update to sample the false positive probability p_0 because

$$[p_0 | \cdot] \propto \left(\prod_{i=1}^{n} [y_i | p_0]^{1-z_i} \right) [c_0 | p_0][p_0], \tag{25.22}$$

$$\propto \left(\prod_{\forall z_i = 0} [y_i | p_0] \right) [c_0 | p_0][p_0], \tag{25.23}$$

$$\propto p_0^{\sum_{\forall z_i = 0} y_i} (1 - p_0)^{\sum_{\forall z_i = 0} (1-y_i)} p_0^{c_0} (1 - p_0)^{M_0 - c_0} p_0^{\alpha_0 - 1} (1 - p_0)^{\beta_0 - 1}, \tag{25.24}$$

$$\propto p_0^{\sum_{\forall z_i = 0} y_i + c_0 + \alpha_0 - 1} (1 - p_0)^{\sum_{\forall z_i = 0} (1-y_i) + M_0 - c_0 + \beta_0 - 1}, \tag{25.25}$$

which results in a beta full-conditional distribution where $[p_0 | \cdot] = \text{Beta}(\sum_{\forall z_i = 0} y_i + c_0 + \alpha_0, \sum_{\forall z_i = 0} (1 - y_i) + M_0 - c_0 + \beta_0)$.

The full-conditional distribution for p_1 is

$$[p_1 | \cdot] \propto \left(\prod_{i=1}^{n} [y_i | p, p_1]^{z_i} \right) [c_1 | p_1][p_1], \tag{25.26}$$

$$\propto (p_1 p)^{\sum_{\forall z_i = 1} y_i} (1 - p_1 p)^{\sum_{\forall z_i = 1} (J - y_i)} p_1^{c_1} (1 - p_1)^{M_1 - c_1} p_1^{\alpha_1 - 1} (1 - p_1)^{\beta_1 - 1}, \tag{25.27}$$

$$\propto p_1^{\sum_{\forall z_i = 1} y_i + c_1 + \alpha_1 - 1} (1 - p_1 p)^{\sum_{\forall z_i = 1} (J - y_i)} (1 - p_1)^{M_1 - c_1 + \beta_1 - 1}, \tag{25.28}$$

which cannot be simplified further. Thus, the full-conditional distribution for p_1 is non-conjugate and we use a M-H update to sample p_1 based on a proposal where $p_1^{(*)} \sim \text{Beta}(\alpha_1, \beta_1)$ arises from the prior which results in the M-H ratio

$$mh = \frac{\left(\prod_{i=1}^{n} [y_i | p^{(k-1)}, p_1^{(*)}]^{z_i^{(k)}}\right) [c_1 | p_1^{(*)}]}{\left(\prod_{i=1}^{n} [y_i | p^{(k-1)}, p_1^{(k-1)}]^{z_i^{(k)}}\right) [c_1 | p_1^{(k-1)}]}. \tag{25.29}$$

Similarly, the full-conditional distribution for p is

$$[p|\cdot] \propto \left(\prod_{i=1}^{n} [y_i | p, p_1]^{z_i}\right) [p], \tag{25.30}$$

$$\propto (p_1 p)^{\sum_{\forall z_i = 1} y_i} (1 - p_1 p)^{\sum_{\forall z_i = 1}(J - y_i)} p^{\alpha_p - 1} (1 - p)^{\beta_p - 1}, \tag{25.31}$$

$$\propto p^{\sum_{\forall z_i = 1} y_i + \alpha_p - 1} (1 - p_1 p)^{\sum_{\forall z_i = 1}(J - y_i)} (1 - p_1)^{\beta_p - 1}, \tag{25.32}$$

which cannot be simplified further. Thus, the full-conditional distribution for p is non-conjugate and we use a M-H update to sample p based on a proposal where $p^{(*)} \sim \text{Beta}(\alpha_p, \beta_p)$ arises from the prior which results in the M-H ratio

$$mh = \frac{\prod_{i=1}^{n} [y_i | p^{(*)}, p_1^{(k)}]^{z_i^{(k)}}}{\prod_{i=1}^{n} [y_i | p^{(k-1)}, p_1^{(k)}]^{z_i^{(k)}}}. \tag{25.33}$$

Finally, the latent occupancy state z_i fits into the general framework that we described in Chapter 10 because the data model is a two component mixture. Thus, the full-conditional distribution for z_i is

$$[z_i|\cdot] \propto [y_i | p, p_1]^{z_i} [y_i | p_0]^{1 - z_i} [z_i | \psi_i], \tag{25.34}$$

$$\propto (\psi_i [y_i | p, p_1])^{z_i} ((1 - \psi_i)[y_i | p_0])^{1 - z_i}, \tag{25.35}$$

which implies that $[z_i|\cdot] = \text{Bern}(\tilde{\psi}_i)$ where

$$\tilde{\psi}_i = \frac{\psi_i [y_i | p, p_1]}{\psi_i [y_i | p, p_1] + (1 - \psi_i)[y_i | p_0]}, \tag{25.36}$$

where $[y_i | p, p_1]$ and $[y_i | p_0]$ are the binomial PMFs in the data model (25.16). Thus, we can sample z_i using a Bernoulli Gibbs update in our MCMC algorithm for all sites $i = 1, \ldots, n$.

We constructed the MCMC algorithm called occ.fp.aux.mcmc.R in R based on these full-conditional distributions:

BOX 25.7   MCMC ALGORITHM FOR FALSE POSITIVE
OCCUPANCY MODEL

```
occ.fp.aux.mcmc <- function(y,J,X,z.true,c1,M1,c0,
   M0,n.mcmc){

####
####    Libraries and Subroutines
####

rtn <- function(n,mu,sig2,low,high){
  flow=pnorm(low,mu,sqrt(sig2))
  fhigh=pnorm(high,mu,sqrt(sig2))
  u=runif(n)
  tmp=flow+u*(fhigh-flow)
  x=qnorm(tmp,mu,sqrt(sig2))
  x
}

####
####    Setup Variables
####

n=length(y)
p.X=dim(X)[2]
n.burn=round(0.25*n.mcmc)

z.mean=rep(0,n)
beta.save=matrix(0,p.X,n.mcmc)
N.save=rep(0,n.mcmc)
p.save=rep(0,n.mcmc)
p1.save=rep(0,n.mcmc)
p0.save=rep(0,n.mcmc)
pred.acc.save=rep(0,n.mcmc)

####
####    Priors and Starting Values
####

p=.5
p1=.5
```

(Continued)

**BOX 25.7 (Continued)   MCMC ALGORITHM FOR FALSE POSITIVE
OCCUPANCY MODEL**

```
39  p0=.5
40  beta=rep(0,p.X)
41  Xbeta=X%*%beta
42  psi=pnorm(Xbeta)
43  z=rep(0,n)
44  z[y==0]=0
45  n0=sum(z==0)
46  n1=sum(z==1)
47
48  v=rep(0,n)
49  ny0=sum(y==0)
50  ny1=sum(y==1)
51
52  alpha.p=2
53  beta.p=1
54  alpha.1=1
55  beta.1=1
56  alpha.0=1
57  beta.0=1
58
59  mu.beta=rep(0,p.X)
60  Sig.beta=diag(p.X)
61  Sig.beta.inv=solve(Sig.beta)
62
63  XprimeX=t(X)%*%X
64  A.beta=XprimeX+Sig.beta.inv
65  A.beta.chol=chol(A.beta)
66  Sig.beta.inv.times.mu.beta=Sig.beta.inv%*%mu.beta
67
68  ####
69  ####  Begin MCMC Loop
70  ####
71
72  for(k in 1:n.mcmc){
73
74    ####
75    ####  Sample v
76    ####
```

(*Continued*)

BOX 25.7 (Continued)   MCMC ALGORITHM FOR FALSE POSITIVE OCCUPANCY MODEL

```
77   v[z==0]=rtn(n0,Xbeta[z==0],1,-Inf,0)
78   v[z==1]=rtn(n1,Xbeta[z==1],1,0,Inf)
79
80   ####
81   ####  Sample beta
82   ####
83
84   b.beta=t(X)%*%v+Sig.beta.inv.times.mu.beta
85   beta=backsolve(A.beta.chol, backsolve(A.beta.
86     chol,b.beta, transpose = TRUE) + rnorm(p.X))
87   Xbeta=X%*%beta
88   psi=pnorm(Xbeta)
89
90   ####
91   ####  Sample z
92   ####
93
94   psi.numer=psi*dbinom(y,J,p*p1)
95   psi.tmp=psi.numer/(psi.numer+(1-psi)
96     *dbinom(y,J,p0))
97   z=rbinom(n,1,psi.tmp)
98   n0=sum(z==0)
99   n1=sum(z==1)
100
101  ####
102  ####  Sample p0
103  ####
104
105  p0=rbeta(1,sum(y[z==0])+c0+alpha.0,sum(J-y[z
       ==0])+M0-c0+beta.0)
106
107  ####
108  ####  Sample p1
109  ####
110
111  p1.star=rbeta(1,alpha.1,beta.1)
112  mh.1=sum(dbinom(y[z==1],J,p*p1.star,log=TRUE))
113       +dbinom(c1,M1,p1.star,log=TRUE)
```

(Continued)

BOX 25.7 (Continued)   MCMC ALGORITHM FOR FALSE POSITIVE OCCUPANCY MODEL

```
114   mh.2=sum(dbinom(y[z==1],J,p*p1,log=TRUE))
115      +dbinom(c1,M1,p1,log=TRUE)
116   mh=exp(mh.1-mh.2)
117   if(mh > runif(1)){
118     p1=p1.star
119   }
120
121   ####
122   ####  Sample p
123   ####
124
125   p.star=rbeta(1,alpha.p,beta.p)
126   mh.1=sum(dbinom(y[z==1],J,p.star*p1,log=TRUE))
127   mh.2=sum(dbinom(y[z==1],J,p*p1,log=TRUE))
128   mh=exp(mh.1-mh.2)
129   if(mh > runif(1)){
130     p=p.star
131   }
132
133   ####
134   ####  Calculate Pred Accuracy
135   ####
136
137   pred.acc.save[k]=sum((z.true-z)==0)/n
138
139   ####
140   ####  Save Samples
141   ####
142
143   beta.save[,k]=beta
144   if(k > n.burn){
145     z.mean=z.mean+z/(n.mcmc-n.burn)
146   }
147   N.save[k]=sum(z)
148   p.save[k]=p
149   p1.save[k]=p1
150   p0.save[k]=p0
151
152 }
```

(Continued)

BOX 25.7 (Continued) MCMC ALGORITHM FOR FALSE POSITIVE OCCUPANCY MODEL

```
153 ####
154 ####   Write Output
155 ####
156
157 list(beta.save=beta.save,p.save=p.save,
158     p1.save=p1.save,p0.save=p0.save,
159     z.mean=z.mean,N.save=N.save,pred.acc.save=pred
          .acc.save,n.mcmc=n.mcmc)
160
161 }
```

Notice that the R code for this MCMC algorithm requires no tuning even though it involves M-H updates because we used the prior as a proposal distribution for both p_1 and p.

To demonstrate the false positive occupancy model, we simulated data $\mathbf{y}$, c_1, and c_0 based on known parameter values $p_1 = 0.50$, $p = 0.75$, and $p_0 = 0.25$ for the detection probabilities and $J = 4$ replicates. As in the previous section, we simulated the covariate values for a set of $n = 200$ sites from an independent standard normal distribution and set the known value of the regression coefficients as $\boldsymbol{\beta} = (-0.50, 0.50)'$. We assumed $M_1 = M_0 = 1000$ survey questions (for a total of $M_1 + M_0 = 2000$) where the species of interest was the correct and incorrect answer, respectively. Although this number of questions may seem large, recall that the observer does not necessarily have to be at a field site to complete the survey. Also, because the questions only require the observer to note their identification of the species based on a sample recording, they can work through the survey quickly.

We used the following R code to simulate $\mathbf{z}$, $\mathbf{y}$, c_1, and c_0:

BOX 25.8 SIMULATE OCCUPANCY DATA WITH FALSE NEGATIVES AND FALSE POSITIVES

```
1 n=200
2 J=4
3 M0=1000
4 M1=1000
5 p=.75
6 p1=.5
```

(Continued)

BOX 25.8 (Continued)   SIMULATE OCCUPANCY DATA WITH FALSE NEGATIVES AND FALSE POSITIVES

```
 7  p0=.25
 8  beta=c(-.5,.5)
 9
10  set.seed(101)
11  X=matrix(1,n,2)
12  X[,2]=rnorm(n)
13  psi=pnorm(X%*%beta)
14  z=rbinom(n,1,psi)
15  y=rbinom(n,J,z*p*p1+(1-z)*p0)
16
17  c1=rbinom(1,M1,p1)
18  c0=rbinom(1,M0,p0)
```

Our simulated data set indicated true occupancy at 67 sites, but observed occupancy at 142 sites. For the independent observer survey, our simulation resulted in $c_1 = 507$ and $c_0 = 250$.

We fit the false positive occupancy model to our simulated data using the following R code, $K = 100000$ MCMC iterations, and hyperparameters $\alpha_1 = \beta_1 = \alpha_0 = \beta_0 = \beta_p = 1$, $\alpha_p = 2$, $\boldsymbol{\mu}_\beta = \mathbf{0}$, and $\boldsymbol{\Sigma}_\beta = \mathbf{I}$:

BOX 25.9   FIT FALSE POSITIVE OCCUPANCY MODEL TO DATA

```
 1  n.mcmc=100000
 2  source("occ.fp.aux.mcmc.R")
 3  set.seed(1)
 4  mcmc.out=occ.fp.aux.mcmc(y=y,J=J,X=X,z.true=z,
 5      c1=c1,M1=M1,c0=c0,M0=M0,n.mcmc=n.mcmc)
 6
 7  layout(matrix(1:6,3,2,byrow=TRUE))
 8  hist(mcmc.out$beta.save[1,-(1:1000)],xlab=bquote
 9      (beta[0]),prob=TRUE,main="a")
10  curve(dnorm(x,0,1),lwd=2,add=TRUE)
11  abline(v=beta[1],lty=2,lwd=2)
12  hist(mcmc.out$beta.save[2,-(1:1000)],xlab=bquote
13      (beta[1]),prob=TRUE,main="b")
14  curve(dnorm(x,0,1),lwd=2,add=TRUE)
```

(Continued)

**BOX 25.9 (Continued)   FIT FALSE POSITIVE OCCUPANCY MODEL
TO DATA**

```
15  abline(v=beta[2],lty=2,lwd=2)
16  hist(mcmc.out$p0.save[-(1:1000)],,xlab=
17      bquote(p[0]),xlim=c(0,1),prob=TRUE,main="c")
18  curve(dbeta(x,1,1),lwd=2,add=TRUE)
19  abline(v=p0,lty=2,lwd=2)
20  hist(mcmc.out$p1.save[-(1:1000)],xlab=
21      bquote(p[1]),xlim=c(0,1),prob=TRUE,main="d")
22  curve(dbeta(x,1,1),lwd=2,add=TRUE)
23  abline(v=p1,lty=2,lwd=2)
24  hist(mcmc.out$p.save[-(1:1000)],xlab="p",
25      xlim=c(0,1),prob=TRUE,main="e")
26  curve(dbeta(x,2,1),lwd=2,add=TRUE)
27  abline(v=p,lty=2,lwd=2)
28  hist(mcmc.out$N.save[-(1:1000)],breaks=
29      seq(0,200,1),xlab="N",ylab="Probability",
30      prob=TRUE,main="f")
31  abline(v=sum(z),lty=2,lwd=2)
32
33  mean(mcmc.out$pred.acc.save[-(1:1000)])
34  quantile(mcmc.out$pred.acc.save[-(1:1000)],
35      c(0.025,0.975))
```

This R code results in the marginal posterior histograms for model parameters and the total number of occupied sites $N = \sum_{i=1}^{n} z_i$ (a derived quantity) shown in Figure 25.2. The results of fitting the false positive occupancy model indicate that the auxiliary data from the independent observer survey are very informative for the probabilities p_1 and p_0 (Figure 25.2c and d) which allows the site occupancy data to reduce the uncertainty in p (Figure 25.2e). We also note that, while there remains substantial uncertainty in the posterior distribution for β (Figure 25.2a and b) and N (Figure 25.2f), recall that the observed number of occupied sites was 142. Thus, by accounting for multiple types of detection error, we have evidence that the true number of occupied sites is substantially less that 142. In fact, the probability that the number of occupied sites is less than 142 is $P(N < 142|\mathbf{y}, c_0, c_1) = 0.968$ with a posterior mean for N estimated as $E(N|\mathbf{y}, c_0, c_1) = 65.5$ which is close to the true number of occupied sites ($N = 67$). Of course, there is uncertainty about N after fitting the model (95% credible interval [20, 147]), but given that two of the model parameters would be completely unidentifiable under this type of study without the auxiliary data, this is understandable.

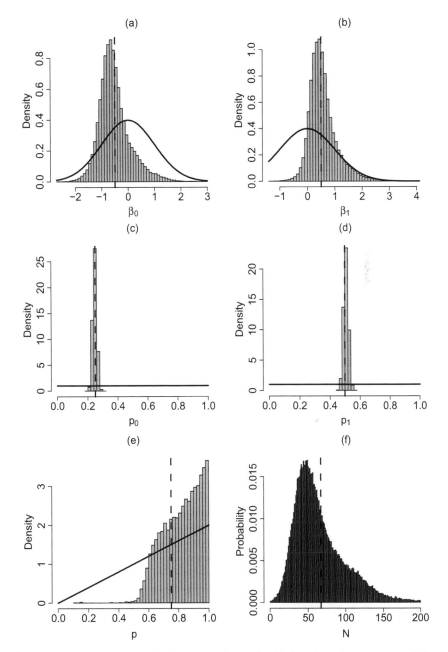

Figure 25.2 Marginal posterior histograms for (a) β_0, (b) β_1, (c) p_0, (d) p_1, (e) p, and (f) N (the total number of occupied sites) resulting from fitting the false positive occupancy model to the simulated data. Prior PDFs are shown as solid black lines and true parameter values used to simulate the data are shown as vertical dashed lines.

25.3 POPULATION VITAL RATE IPMs

In the context of wildlife population ecology, multiple data source models have become known as "integrated population models" (IPMs; Abadi et al. 2010; Schaub and Abadi 2011).[2] In wildlife analyses, IPMs often combine two or more sources of data to learn about difficult to estimate population vital rates such as survival probabilities and recruitment rates. For example, data sources that are useful for estimating population abundance (i.e., capture-recapture data) often are combined with data that are easier to collect over long time periods such as time series of count data for a set of unmarked individuals in a population (e.g., Davis et al. 2014).

Kery and Schaub (2012) described several ways to specify Bayesian IPMs for modeling population vital rates. In what follows, we describe an exemplar IPM similar to those described by Kery and Schaub (2012). We assume that two sources of data are available:

1. A time series of observed abundance data (i.e., population counts) collected without marking individuals similar to the counts that we modeled in Chapter 21.
2. Individual-based capture-recapture data from a population in the style of the survival data that we presented in Chapter 24.

We denote the observed abundance data as y_t for $t = 1, \ldots, T_y$, and assume that they are observed with the possibility of both false negative and false positive errors. Thus, we model the time series of observed abundance data as $y_t \sim \text{Pois}(N_t)$ where N_t is the true abundance. Similar to the dynamic population model we presented in Chapter 24 for counts, we assume the population size is stochastically changing over time according to mechanisms involving growth and survival such that

$$N_t \sim \text{Pois}(r\phi N_{t-1}), \tag{25.37}$$

where ϕ represents annual survival (for years $t = 1, \ldots, T_y$) and r is the expected number of offspring produced per individual per year. This dynamic process model for abundance relies on an initial abundance for $t = 0$; thus, we specify the prior $N_0 \sim \text{Gamma}(\alpha_0, \beta_0)$. We also specify a gamma prior for the population growth rate because we assume it has positive support; thus, we let $r \sim \text{Gamma}(\alpha_r, \beta_r)$. The annual survival probability ϕ is bounded between zero and one; thus, we specify a beta distribution for it as $\phi \sim \text{Beta}(\alpha_\phi, \beta_\phi)$.

As specified, we overparameterized our dynamic population model because from the data we will not be able to distinguish the growth rate r from the survival probability ϕ. However, by using a separate conditionally independent data source from the same population (or a similar one) to help estimate the survival probability, we can borrow strength from both data sources and improve estimation of model parameters as well as overall inference. Thus, for a set of n marked individuals, we have

---

[2] Not to be confused with the other IPMs: Integral projection models (Easterling et al. 2000; Ellner and Rees 2006).

resighting (or capture-recapture) data $w_{i\tau}$ for $i = 1, \ldots, n$ and $\tau = 1, \ldots, T_w$ that are subject to imperfect detection. Therefore, we can use the CJS model from Chapter 24 for the resighting data $w_{i\tau}$, which depends on a detection probability p and survival probability ϕ:

$$w_{i\tau} \sim \begin{cases} 0 & , z_{i\tau} = 0 \\ \text{Bern}(p) & , z_{i\tau} = 1 \end{cases}, \tag{25.38}$$

$$z_{i\tau} \sim \begin{cases} 0 & , z_{i\tau-1} = 0 \\ \text{Bern}(\phi) & , z_{i\tau-1} = 1 \end{cases}, \tag{25.39}$$

where the latent variable $z_{i\tau}$ indicates if individual i was alive or dead at time τ. As we described in Chapter 24, an individual that has died must stay dead, which is why the CJS model uses a dynamic zero-inflated Bernoulli model in (25.39). Also, as in Chapter 24, we assume the individuals were all alive at first detection (i.e., $z_{i1} = 1$). We can use a beta prior for the detection probability $p \sim \text{Beta}(\alpha_p, \beta_p)$.

This hierarchical population model has two data sources with separate process models that depend on a shared parameter, the survival probability ϕ. Thus, we refer to it as an IPM because it is "integrating" the multiple data sources to learn about a shared underlying process or parameter. Using two separate data sources remedies the identifiability issue that we mentioned previously in the dynamic abundance model between r and ϕ and improves inference overall. The IPM formulation also allows us to formally account for uncertainty in both data sets as well as the underlying process and parameters because we fit the IPM jointly so that feedback can properly occur between model components.

The posterior distribution for the IPM is

$$[\mathbf{N}, \mathbf{Z}, p, \phi, r | \mathbf{y}, \mathbf{W}] \propto \prod_{t=1}^{T_y} [y_t | N_t][N_t | N_{t-1}, r, \phi] \times \tag{25.40}$$

$$\prod_{i=1}^{n} \prod_{\tau=2}^{T_w} [w_{i\tau} | p]^{z_{i\tau}} 1_{\{w_{i\tau}=0\}}^{1-z_{i\tau}} [z_{i\tau} | \phi]^{z_{i\tau-1}} 1_{\{z_{i\tau}=0\}}^{1-z_{i\tau-1}} \times \tag{25.41}$$

$$[p][\phi][r][N_0], \tag{25.42}$$

where $\mathbf{N}$ is a $T_y \times 1$ vector of abundances and $\mathbf{Z}$ is a $n \times T_w$ matrix of individual alive/dead states, which we can obtain a sample from using MCMC based on the set of full-conditional distributions for processes and parameters.

To derive the full-conditional distributions, we use the same strategy as before, beginning with the initial abundance N_0

$$[N_0 | \cdot] \propto [N_1 | N_0, r, \phi][N_0], \tag{25.43}$$

$$\propto (r\phi N_0)^{N_1} e^{-r\phi N_0} N_0^{\alpha_0-1} e^{-\beta_0 N_0}, \tag{25.44}$$

$$\propto N_0^{N_1+\alpha_0-1} e^{-(r\phi+\beta_0)N_0}, \tag{25.45}$$

which takes the form of a gamma distribution where $[N_0|\cdot] = \text{Gamma}(N_1 + \alpha_0, r\phi + \beta_0)$. Thus, we can sample N_0 using a Gibbs update in our MCMC algorithm.

Unfortunately, the full-conditional distributions for the abundances at later times are not conjugate and require M-H (or alternative) updates in a MCMC algorithm. To see this, the full-conditional distributions for N_t for $t = 1, \ldots, T_y - 1$ are

$$[N_t|\cdot] \propto [y_t|N_t][N_t|N_{t-1}, r, \phi][N_{t+1}|N_t, r, \phi], \qquad (25.46)$$

which involves three separate Poisson PMFs with N_t occurring in different parts of each. Thus, we use M-H to sample N_t for $t = 1, \ldots, T_y - 1$ based on the M-H ratio

$$mh = \frac{[y_t|N_t^{(*)}][N_t^{(*)}|N_{t-1}^{(k)}, r^{(k-1)}, \phi^{(k-1)}][N_{t+1}^{(k-1)}|N_t^{(*)}, r^{(k-1)}, \phi^{(k-1)}][N_t^{(k-1)}|N_t^{(*)}]}{[y_t|N_t^{(k-1)}][N_t^{(k-1)}|N_{t-1}^{(k)}, r^{(k-1)}, \phi^{(k-1)}][N_{t+1}^{(k-1)}|N_t^{(k-1)}, r^{(k-1)}, \phi^{(k-1)}][N_t^{(*)}|N_t^{(k-1)}]},$$
$$(25.47)$$

where we use $N_{t+1}^{(k-1)}$ because we update the N_t sequentially in a "for" loop in our MCMC algorithm. For the proposal distribution, we use a conditional Poisson such that $N_t^{(*)} \sim [N_t^{(*)}|N_t^{(k-1)}] \equiv \text{Pois}(N_t^{(k-1)} + c)$ where c is fixed at a small constant value (i.e., $c = 0.1$) to allow proposals greater than zero if the previous value for $N_t^{(k-1)}$ was zero. In our example that follows, the abundance process is large enough that no samples of zero occur, but adding a small constant to the proposal (which requires no tuning) leads to a more general algorithm.[3]

Similarly, the full-conditional distribution for the final abundance in the time series N_{T_y} is

$$[N_{T_y}|\cdot] \propto [y_{T_y}|N_{T_y}][N_{T_y}|N_{T_y-1}, r, \phi], \qquad (25.48)$$

which is slightly simpler than the full-conditional distribution for N_t because it does not include the process component for the step ahead. Thus, we can update N_{T_y} using the M-H ratio

$$mh = \frac{[y_{T_y}|N_{T_y}^{(*)}][N_{T_y}^{(*)}|N_{T_y-1}^{(k)}, r^{(k-1)}, \phi^{(k-1)}][N_{T_y}^{(k-1)}|N_{T_y}^{(*)}]}{[y_{T_y}|N_{T_y}^{(k-1)}][N_{T_y}^{(k-1)}|N_{T_y-1}^{(k)}, r^{(k-1)}, \phi^{(k-1)}][N_{T_y}^{(*)}|N_{T_y}^{(k-1)}]}, \qquad (25.49)$$

in our MCMC algorithm based on the same Poisson proposal distribution: $N_{T_y}^{(*)} \sim [N_{T_y}^{(*)}|N_{T_y}^{(k-1)}] \equiv \text{Pois}(N_{T_y}^{(k-1)} + c)$.

---

[3] Note that it also is possible to use proposals for parameters with discrete support that do allow for tuning. For example, we could use a discrete uniform centered on the previous value for the parameter in the Markov chain and tunable width. Alternatively, we could use a negative binomial or Conway-Maxwell-Poisson distribution to allow for underdispersion or overdispersion in the proposal distribution. These distributions contain extra parameters that we can use to tune the proposal to yield more efficient MCMC algorithms.

The annual growth rate parameter r appears in the dynamic abundance process model and has the full-conditional distribution

$$[r|\cdot] \propto \left(\prod_{t=1}^{T_y} [N_t|N_{t-1}, r, \phi] \right) [r], \tag{25.50}$$

$$\propto \left(\prod_{t=1}^{T_y} \frac{(r\phi N_{t-1})^{N_t} e^{-r\phi N_{t-1}}}{N_t!} \right) r^{\alpha_r - 1} e^{-\beta_r r}, \tag{25.51}$$

$$\propto r^{\sum_{t=1}^{T_y} N_t + \alpha_r - 1} e^{-r(\sum_{t=1}^{T_y} \phi N_{t-1} + \beta_r)}, \tag{25.52}$$

which takes the form of a gamma distribution where $[r|\cdot] = \text{Gamma}(\sum_{t=1}^{T_y} N_t + \alpha_r, \sum_{t=1}^{T_y} \phi N_{t-1} + \beta_r)$. Thus, we can sample r using a Gibbs update in our MCMC algorithm.

The next full-conditional distributions to derive pertain to the CJS component (i.e., survival data) of the IPM. Fortunately, we already have derived the full-conditional distributions for the latent indicator variables $z_{i\tau}$ and detection probability p. However, the full-conditional distribution for ϕ now involves the additional information from the dynamic abundance process based on the abundance time series data. Thus, we write the full-conditional distribution for ϕ as

$$[\phi|\cdot] \propto \left(\prod_{t=1}^{T_y} [N_t|N_{t-1}, r, \phi] \right) \left(\prod_{i=1}^{n} \prod_{\tau=2}^{T_w} [z_{i\tau}|\phi]^{z_{i\tau}-1} \right) [\phi], \tag{25.53}$$

which is non-conjugate; thus, we use a M-H update for ϕ based on the M-H ratio

$$mh = \frac{\left(\prod_{t=1}^{T_y} [N_t^{(k)}|N_{t-1}^{(k)}, r^{(k)}, \phi^{(*)}] \right) \left(\prod_{i=1}^{n} \prod_{\tau=2}^{T_w} [z_{i\tau}^{(k)}|\phi^{(*)}]^{z_{i\tau}^{(k)}-1} \right)}{\left(\prod_{t=1}^{T_y} [N_t^{(k)}|N_{t-1}^{(k)}, r^{(k)}, \phi^{(k-1)}] \right) \left(\prod_{i=1}^{n} \prod_{\tau=2}^{T_w} [z_{i\tau}^{(k)}|\phi^{(k-1)}]^{z_{i\tau}^{(k)}-1} \right)}, \tag{25.54}$$

where we use the prior for ϕ as the proposal distribution such that $\phi^{(*)} \sim [\phi] \equiv \text{Beta}(\alpha_\phi, \beta_\phi)$. As we discussed previously, using the prior as a proposal may not be efficient, but it requires no tuning and simplifies the M-H ratio.

To sample the latent CJS state variables, we restate the full-conditional distributions for $z_{i\tau}$ (for $i = 1, \ldots, n$ and $t = 1, \ldots, T_w$) and p, which are conjugate and were derived in Chapter 24. The full-conditional distribution for $z_{i\tau}$ for $i = 1, \ldots, n$ and $\tau = 1, \ldots, T_w$ is Bernoulli (i.e., $[z_{i\tau}|\cdot] = \text{Bern}(\tilde{\phi}_{i\tau})$) where

$$\tilde{\phi}_{i\tau} = \frac{(1-p)\phi}{(1-p)\phi + 1}. \tag{25.55}$$

For the last time $\tau = T_w$, the full-conditional distribution for z_{iT_w} is also conjugate Bernoulli, but with probability

$$\tilde{\phi}_{iT_w} = \frac{(1-p)\phi}{(1-p)\phi + 1 - \phi}. \tag{25.56}$$

Thus, we can use a Gibbs update for all $z_{i\tau}$ in our MCMC algorithm, but recall that we only sample $z_{i\tau}$ when (1) all the preceding $y_{i\tau}$ equal zero, (2) when the preceding $z_{i\tau-1} = 0$, and (3) when the following $z_{i\tau+1} = 0$. For all other cases, the latent indicator variables are known (see Chapter 24 for details).

Finally, the full-conditional distribution for the detection probability p is

$$[p|\cdot] = \text{Beta}\left(\sum_{\forall z_{i\tau}=1} w_{i\tau} + \alpha_p, \sum_{\forall z_{i\tau}=1} (1 - w_{i\tau}) + \beta_p\right). \qquad (25.57)$$

Thus, we can use a Gibbs update to sample p in our MCMC algorithm.

Using these full-conditional distributions, we wrote a MCMC algorithm called ipm.mcmc.R in R to fit the Bayesian IPM to data; this algorithm is as follows:

BOX 25.10   MCMC ALGORITHM FOR IPM

```
1  ipm.mcmc <- function(y,W,n.mcmc){
2
3  ####
4  ####   Setup Variables
5  ####
6
7  Ty=length(y)
8  n=dim(W)[1]
9  Tw=dim(W)[2]
10
11 p.save=rep(0,n.mcmc)
12 phi.save=rep(0,n.mcmc)
13 r.save=rep(0,n.mcmc)
14 N.save=matrix(0,Ty+1,n.mcmc)
15
16 ####
17 ####   Priors
18 ####
19
20 alpha.p=1
21 beta.p=1
22
23 alpha.phi=1
24 beta.phi=1
25
26 alpha.r=1
27 beta.r=1
```

(Continued)

BOX 25.10 (Continued)   MCMC ALGORITHM FOR IPM

```
28  alpha.0=10
29  beta.0=.1
30
31  ####
32  ####   Starting Values
33  ####
34
35  p=0.5
36  phi=0.9
37  r=1.1
38
39  Z=matrix(0,n,Tw)
40  Z[,1]=1
41  Z[W==1]=1
42  for(i in 1:n){
43    if(sum(W[i,])>0){
44      idx.tmp=max((1:Tw)[W[i,]==1])
45      Z[i,1:idx.tmp]=1
46    }
47  }
48
49  N=c(y[1],y) # Note: add 1 when indexing N
50  const=.1
51
52  ####
53  ####   Begin MCMC Loop
54  ####
55
56  for(k in 1:n.mcmc){
57
58    ####
59    ####   Sample N_0
60    ####
61
62    N[1]=rgamma(1,N[2]+alpha.0,r*phi+beta.0)
63
64    ####
65    ####   Sample N_t
66    ####
```

(Continued)

BOX 25.10 (Continued)   MCMC ALGORITHM FOR IPM

```
67  for(t in 1:(Ty-1)){
68    Nt.star=rpois(1,N[t+1]+const)
69    mh.1=dpois(y[t],Nt.star,log=TRUE)
70        +dpois(N[t+1+1],r*phi*Nt.star,log=TRUE)
71        +dpois(Nt.star,r*phi*N[t],log=TRUE)
72        +dpois(N[t+1],Nt.star+const,log=TRUE)
73    mh.2=dpois(y[t],N[t+1],log=TRUE)
74        +dpois(N[t+1+1],r*phi*N[t+1],log=TRUE)
75        +dpois(N[t+1],r*phi*N[t],log=TRUE)
76        +dpois(Nt.star,N[t+1]
77        +const,log=TRUE)
78    mh=exp(mh.1-mh.2)
79    if(mh>runif(1)){
80      N[t+1]=Nt.star
81    }
82  }
83
84  ####
85  ####   Sample N_Ty
86  ####
87
88  NTy.star=rpois(1,N[Ty+1]+const)
89  mh.1=dpois(y[Ty],NTy.star,log=TRUE)
90    +dpois(NTy.star,r*phi*N[Ty],log=TRUE)
91    +dpois(N[Ty+1],NTy.star+const,log=TRUE)
92  mh.2=dpois(y[Ty],N[Ty+1],log=TRUE)
93    +dpois(N[Ty+1],r*phi*N[Ty],log=TRUE)
94    +dpois(NTy.star,N[Ty+1]+const,log=TRUE)
95  mh=exp(mh.1-mh.2)
96  if(mh>runif(1)){
97    N[Ty+1]=NTy.star
98  }
99
100 ####
101 ####   Sample r
102 ####
103
104 r=rgamma(1,sum(N[-1])+alpha.r,sum(phi*N[1:Ty])
105   +beta.r)
```

(Continued)

BOX 25.10 (Continued)   MCMC ALGORITHM FOR IPM

```
106   ####
107   ####   Sample z
108   ####
109
110   for(i in 1:n){
111     for(tau in 2:(Tw-1)){
112       if(sum(W[i,tau:Tw])==0){
113         if(Z[i,tau-1]==1 & Z[i,tau+1]==0){
114           psi.tmp=phi*(1-p)
115           psi.tmp=psi.tmp/(psi.tmp+1)
116           Z[i,tau]=rbinom(1,1,psi.tmp)
117         }
118         if(Z[i,tau-1]==0){
119           Z[i,tau]=0
120         }
121         if(Z[i,tau+1]==1){
122           Z[i,tau]=1
123         }
124       }
125     }
126     if(Z[i,Tw-1]==1 & W[i,Tw]==0){
127       psi.tmp=phi*(1-p)
128       psi.tmp=psi.tmp/(psi.tmp+1-phi)
129       Z[i,Tw]=rbinom(1,1,psi.tmp)
130     }
131     if(Z[i,Tw-1]==0 & W[i,Tw]==0){
132       Z[i,Tw]=0
133     }
134     if(W[i,Tw]==1){
135       Z[i,Tw]=1
136     }
137   }
138
139   ####
140   ####   Sample p
141   ####
142
143   p=rbeta(1,sum(W[,-1][Z[,-1]==1])+alpha.p,
144     sum(1-W[,-1][Z[,-1]==1])+beta.p)
```

(Continued)

BOX 25.10 (Continued)   MCMC ALGORITHM FOR IPM

```
145  ####
146  ####   Sample phi
147  ####
148
149  phi.star=rbeta(1,alpha.phi,beta.phi)
150  mh.1=sum(dpois(N[-1],r*phi.star*N[-(Ty+1)],
151      log=TRUE))+sum(dbinom(Z[,-1][Z[,-Tw]==1],
152      1,phi.star,log=TRUE))
153  mh.2=sum(dpois(N[-1],r*phi*N[-(Ty+1)],log=TRUE))
154      +sum(dbinom(Z[,-1][Z[,-Tw]==1],1,phi,
155      log=TRUE))
156  mh=exp(mh.1-mh.2)
157  if(mh>runif(1)){
158    phi=phi.star
159  }
160
161  ####
162  ####   Save Samples
163  ####
164
165  p.save[k]=p
166  phi.save[k]=phi
167  r.save[k]=r
168  N.save[,k]=N
169
170 }
171
172 ####
173 ####   Write Output
174 ####
175
176 list(p.save=p.save,phi.save=phi.save,r.save=r.
177     save,N.save=N.save,n.mcmc=n.mcmc)
178
179 }
```

Notice that we were able to reuse the R code to sample z_{it} and p from the CJS MCMC algorithm in Chapter 24 (with minor modifications of variable names).

To demonstrate the MCMC algorithm, we simulated population abundance data (y_t, for $t = 1, \ldots, 40$) and survival data (w_{it} for $i = 1, \ldots, 20$ and $t = 1, \ldots, 10$) based on the parameters $N_0 = 100$, $r = 1.1$, $\phi = 0.9$, and $p = 0.5$ using the R code:

BOX 25.11   SIMULATE ABUNDANCE AND SURVIVAL DATA

```
 1 n=20
 2 Ty=40
 3 Tw=10
 4
 5 r=1.1
 6 phi=.9
 7 p=.5
 8
 9 Z=matrix(1,n,Tw)
10 W=matrix(1,n,Tw)
11 set.seed(101)
12 for(tau in 2:Tw){
13   Z[,tau]=rbinom(n,1,phi*Z[,tau-1])
14   W[,tau]=rbinom(n,1,p*Z[,tau])
15 }
16
17 N=rep(100,Ty+1)
18 y=rep(0,Ty)
19 for(t in 1:Ty){
20   N[t+1]=rpois(1,r*phi*N[t])
21   y[t]=rpois(1,N[t+1])
22 }
```

The observed and true population abundance data are shown in Figure 25.4 and out of the $n = 20$ individuals observed in the survival data set, only 2 were observed on the last ($T_w = 10$) time period (implying we know they were alive during the entire observation period of 10 years). Most individuals in the survival data set were not observed again beyond 3 years.

We fit the IPM to these two sets of data using $K = 100000$ MCMC iterations and the following R code:

BOX 25.12   FIT IPM TO DATA

```
1 source("ipm.mcmc.R")
2 n.mcmc=100000
3 set.seed(1)
4 mcmc.out=ipm.mcmc(y=y,W=W,n.mcmc=n.mcmc)
```

(*Continued*)

BOX 25.12 (Continued)   FIT IPM TO DATA

```
 5 layout(matrix(1:4,2,2,byrow=TRUE))
 6 hist(mcmc.out$r.save[-(1:1000)],xlab="r",
 7     xlim=c(0,2),prob=TRUE,main="a")
 8 curve(dgamma(x,1,1),lwd=2,add=TRUE)
 9 abline(v=r,lty=2,lwd=2)
10 hist(mcmc.out$phi.save[-(1:1000)],xlab=
11     bquote(phi),xlim=c(0,1),prob=TRUE,main="b")
12 curve(dbeta(x,1,1),lwd=2,add=TRUE)
13 abline(v=phi,lty=2,lwd=2)
14 hist(mcmc.out$p.save[-(1:1000)],xlab="p",
15     xlim=c(0,1),prob=TRUE,main="c")
16 curve(dbeta(x,1,1),lwd=2,add=TRUE)
17 abline(v=p,lty=2,lwd=2)
18 hist(mcmc.out$N.save[1,-(1:1000)],xlab=
19     bquote(N[0]),breaks=seq(0,140,1),
20     prob=TRUE,main="d")
21 curve(dgamma(x,10,.1),lwd=2,add=TRUE)
22 abline(v=N[1],lty=2,lwd=2)
```

This R code results in the marginal posterior histograms shown in Figure 25.3. The marginal posterior histograms indicate that the IPM is able to recover the model parameters even though the annual growth rate r and survival probability ϕ would not be identifiable if we only used the observed abundance data and model. Furthermore, because we modeled the initial abundance N_0, we reduced the uncertainty about it and were able to utilize the entire time series of data.

Another benefit of using the IPM is that we can obtain valid estimates of the latent abundance process N_t. The following R code computes the posterior 95% credible interval for population abundance N_t:

BOX 25.13   POSTERIOR POPULATION ABUNDANCE

```
1 N.l=apply(mcmc.out$N.save[,-(1:1000)],1,quantile,
2     0.025)
3 N.u=apply(mcmc.out$N.save[,-(1:1000)],1,quantile,
4     0.975)
5
6 plot(1:Ty,y,xlim=c(0,Ty),ylim=range(c(N.u,N.l)),
7     type="p",lwd=2,xlab="t",ylab="Abundance")
```

(Continued)

BOX 25.13 (Continued)   POSTERIOR POPULATION ABUNDANCE

```
8  polygon(c(0:Ty,Ty:0),c(N.u,rev(N.l)),col=rgb(0,0,
9      0,.2),border=NA)
10 lines(0:Ty,N,type="o",cex=.25,lwd=2,lty=1)
11 points(1:Ty,y,lwd=2)
12 legend("bottomright",lty=c(1,NA),pch=c(NA,1),
13      lwd=2,legend=c("N","y"),bty="n")
```

Figure 25.4 shows the posterior 95% credible interval for abundance with the true simulated abundance overlaid for comparison and observed abundance data shown as points. From Figure 25.4 it is evident that the posterior distribution of N_t captures

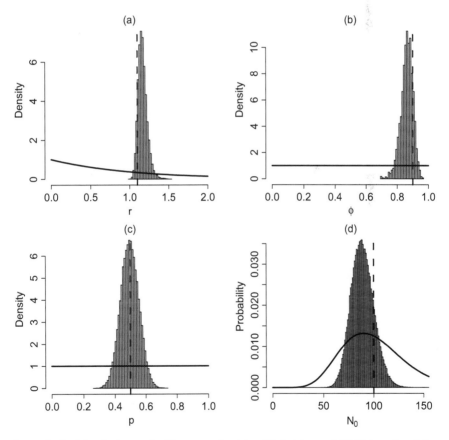

Figure 25.3   Marginal posterior histograms for (a) r, (b) ϕ, (c) p, and (d) N_0 based on fitting the IPM to the abundance time series and survival data. Priors PDFs shown as solid black lines and true parameter values used to simulate data shown as vertical dashed lines.

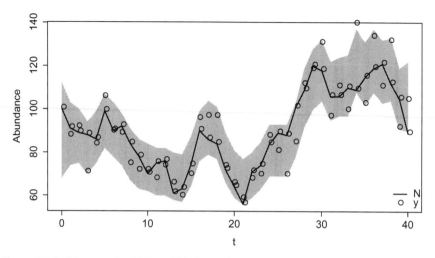

Figure 25.4 The posterior 95% credible interval (gray shaded region) for abundance (N_t, for $t = 0, \ldots, 40$) with true abundance overlaid (solid line) and observed abundance data shown as points.

the true simulated abundance well despite the measurement error present in the data y_t.

As we have shown throughout this chapter, multiple data sets can be useful when fitting ecological models. In fact, more data are usually helpful to reduce uncertainty, but different types of data can help us separate the information about different parameters, especially in cases where they appear in model specifications that would not normally be identifiable using only a single data source.

25.4   ADDITIONAL CONCEPTS AND READING

The concept of building models that contain multiple data sources is not new in statistics; in fact, it is often referred to as data fusion (Reiter 2012). However, in early work pertaining to IPMs for wildlife ecology, Besbeas et al. (2002) proposed an approach similar to the one that we present in this chapter for combining capture-recapture data and count data, which they refer to as "census" data. However, Besbeas et al. (2002) used a dynamic Gaussian linear model (see Chapter 16) for the observed abundances in two age categories over time and then integrated over this process using a Kalman filter (Kalman 1960) to obtain the likelihood which they maximized to fit the model to data. From a Bayesian perspective, Brooks et al. (2004) revisited the model proposed by Besbeas et al. (2002), but instead used dynamic binomial and Poisson specifications for the abundance process (and also relied only on "population index" data rather than the abundances directly). Brooks et al. (2004) implemented the Bayesian IPM using MCMC through WinBUGS (Spiegelhalter et al. 2003).

More recently, Hobbs et al. (2015) presented a sophisticated IPM for modeling brucellosis (*Brucella abortus*) in a wild bison (*Bison bison*) population. In their IPM,

Hobbs et al. (2015) combined multiple data sources including observed abundance data, age and sex classification data, and serology data into a single dynamic population model to understand the influence of various management actions on transmission of the disease.

In a similar management context, Arnold et al. (2018) argued for the use of IPMs in waterfowl management in North America, in part, because multiple sources of excellent long-term data exist for several waterfowl species. Arnold et al. (2018) demonstrated the value of IPMs by applying a simple model to reconcile survey and nest monitoring data from canvasbacks (*Aythya valisineria*) to obtain more accurate estimates of population vital rates.

In general, the Bayesian approach naturally accommodates multiple data sources in an IPM framework and we showed how including them can help to provide better inference. Lele and Allen (2006) described an approach involving multiple sources of data, but with at least one data set "elicited" based on expert knowledge. In their case, Lele and Allen (2006) used these data in the same spirit as we commonly use priors to capture expert knowledge about the system.

26 Spatial Occupancy Models

Occupancy models are prevalent in wildlife ecology when there are false negatives associated with observing species at a set of sites. The collection of occupancy data often is done at a set of sites in a study area, but conventional occupancy models (e.g., MacKenzie et al. 2002; Tyre et al. 2003) do not account for potential spatial dependence among sites beyond that explained by site-level covariates. As we discussed in Chapter 17, it is important to account for spatial dependence in statistical models when it exists because it can affect the inference pertaining to parameters and improve predictions.

Hoeting et al. (2000) developed a spatially explicit occupancy model that was able to account for potential spatial structure in the occupancy probability. Johnson et al. (2013a) developed a spatial occupancy model using the auxiliary variable approach that we presented in Chapters 20 and 23. Although there are several approaches to account for spatial dependence in binary data (e.g., Gumpertz et al. 1997; Huffer and Wu 1998), Johnson et al. (2013a) used the same concepts that we introduced previously by specifying spatially correlated latent random effects.

To construct a spatial occupancy model using spatially correlated latent random effects, we begin with the data model and assume the same type of data collection process as in Chapter 23. For a set of sites located at $\mathbf{s}_i$ for $i = 1,\ldots,n$, we observe binary data y_{ij} for a set of temporal replicates $j = 1,\ldots,J$. For simplicity, we assume that the replicates are independent and identically distributed for each site. Thus, we can model $y_i = \sum_{j=1}^{J} y_{ij}$ using a zero-inflated Binomial model

$$y_i \sim \begin{cases} \text{Binom}(J,p) & , z_i = 1 \\ 0 & , z_i = 0 \end{cases}, \tag{26.1}$$

where, $z_i \sim \text{Bern}(\psi_i)$ for $i = 1,\ldots,n$. We model the occupancy probability ψ_i as a function of spatially explicit covariates $\mathbf{x}_i$ and random effects η_i

$$g(\psi_i) = \mathbf{x}_i'\boldsymbol{\beta} + \eta_i. \tag{26.2}$$

In matrix notation, we can write the occupancy probability model as

$$g(\boldsymbol{\psi}) = \mathbf{X}\boldsymbol{\beta} + \boldsymbol{\eta}, \tag{26.3}$$

where $\boldsymbol{\eta} \sim \text{N}(\mathbf{0},\boldsymbol{\Sigma})$ with spatial covariance matrix $\boldsymbol{\Sigma}$ and the analyst can specify a suitable link function $g(\cdot)$ as described in Chapter 20. We can parameterize the spatial covariance matrix $\boldsymbol{\Sigma}$ using either a geostatistical model or an areal model as we presented in Chapter 17.[1]

---

[1] Recall that the occupancy data have a natural areal spatial support because they are observed at a set of "sites." However, in some cases, the size of the sites may be infinitesimal relative to the size of the study area and a continuous spatial process may provide a sufficiently flexible model to account for the spatial dependence.

We parameterize the spatial covariance matrix using a CAR structure such that $\Sigma \equiv \sigma^2 (\text{diag}(\mathbf{W1}) - \rho \mathbf{W})^{-1}$ where the spatial autocorrelation parameter is set to approach one (i.e., $\rho \to 1$). As we discussed in Chapter 17, when $\rho = 1$ the areal spatial model is referred to as an intrinsic CAR (ICAR) model. By setting ρ to be close to one (e.g., $\rho = 0.99$), our spatial occupancy model is approximately an ICAR model but is proper and the matrix inverse in Σ exists. Hooten et al. (2017) mentioned that ICAR models (or near-ICAR models as we have specified in this chapter) are popular because most real data situations involve processes with positive autocorrelation and only very large values of ρ induce visibly strong dependence in η. Thus, there is value from a parsimony perspective in fixing the autocorrelation parameter (ρ) to be a large positive value (i.e., near one). If we do not need the spatial structure to characterize variability in the data, then we will estimate the spatial variance parameter σ^2 to be small (i.e., near zero). In such cases, we can remove the spatial random vector η from the model because it will be approximately zero everywhere.

We can specify the proximity matrix $\mathbf{W}$ in many ways (Ver Hoef et al. 2018a,b) depending on the specific structure of the sites, their relative size, and how they were selected. In what follows, we specify the proximity matrix with zeros as diagonal elements, ones for all pairs of sites whose centroids are within a prespecified distance (d_{thresh}) of each other, and zero elsewhere. Thus, the elements of $\mathbf{W}$ are

$$w_{il} = \begin{cases} 0 & , i = l \\ 1 & , i \neq l, d_{il} \leq d_{\text{thresh}} \\ 0 & , i \neq l, d_{il} > d_{\text{thresh}} \end{cases} \tag{26.4}$$

for $i = 1, \ldots, n$ and $l = 1, \ldots, n$. For a regular grid of sites, d_{thresh} is usually set to induce a rook's or queen's neighborhood depending on the required amount of smoothness in the spatial process.

To implement a Bayesian version of the spatial occupancy model, we must select a link function $g(\cdot)$ as well as priors for p, $\boldsymbol{\beta}$, and σ^2. Then we can derive the full-conditional distributions for the occupancy process z_i for $i = 1, \ldots, n$, the spatial process η, and the model parameters, and then use a MCMC algorithm to fit the model. However, the usual logit link function and set of priors will not result in conjugate full-conditional distributions and we would need M-H updates for most terms in our MCMC algorithm (which will require substantial tuning because we would have to tune the M-H updates for all ψ_i in addition to the other parameters). Alternatively, we can use a modified version of the auxiliary variable formulation that we presented in Chapters 20 and 23 to induce a form of conjugacy that allows us to use Gibbs updates for all processes and parameters in the spatial occupancy model. Hooten et al. (2003) and Johnson et al. (2013a) applied this approach to model species distributions over large spatial domains.

To reformulate the spatial occupancy model using auxiliary variables, we write the process model for z_i as

$$z_i = \begin{cases} 0 & , v_i \leq 0 \\ 1 & , v_i > 0 \end{cases}, \tag{26.5}$$

where v_i for $i = 1, \ldots, n$ are a set of continuous-valued auxiliary variables arising from a normal distribution such that

$$v_i \sim N(\mathbf{x}_i'\boldsymbol{\beta} + \eta_i, 1). \tag{26.6}$$

Then the vector $\boldsymbol{\eta} \equiv (\eta_1, \ldots, \eta_n)'$ is modeled using the near-ICAR structure. For priors, we specify

$$p \sim \text{Beta}(\alpha_p, \beta_p), \tag{26.7}$$

$$\boldsymbol{\beta} \sim N(\boldsymbol{\mu}_\beta, \boldsymbol{\Sigma}_\beta), \tag{26.8}$$

$$\sigma^2 \sim \text{IG}(q, r). \tag{26.9}$$

The spatial occupancy model with auxiliary variables results in the following posterior distribution

$$[\mathbf{z}, \boldsymbol{\eta}, p, \boldsymbol{\beta}, \sigma^2 | \mathbf{y}] \propto \left(\prod_{i=1}^{n} [y_i|p]^{z_i} 1_{\{y_i=0\}}^{1-z_i} [z_i|v_i][v_i|\boldsymbol{\beta}, \eta_i] \right) [\boldsymbol{\eta}|\sigma^2][p][\boldsymbol{\beta}][\sigma^2]. \tag{26.10}$$

To implement this spatial occupancy model using MCMC, we can derive each of the full-conditional distributions similar to how we described them in Chapter 23. Beginning with the detection probability parameter p, the full-conditional distribution is

$$[p|\cdot] \propto \left(\prod_{i=1}^{n} [y_i|p]^{z_i} \right) [p], \tag{26.11}$$

$$\propto \left(\prod_{\forall z_i=1} [y_i|p] \right) [p], \tag{26.12}$$

$$\propto p^{\sum_{\forall z_i=1} y_i} (1-p)^{\sum_{\forall z_i=1} (J-y_i)} p^{\alpha_p-1} (1-p)^{\beta_p-1}, \tag{26.13}$$

$$\propto p^{\sum_{\forall z_i=1} y_i + \alpha_p - 1} (1-p)^{\sum_{\forall z_i=1} (J-y_i) + \beta_p - 1}, \tag{26.14}$$

which results in a beta distribution such that $[p|\cdot] = \text{Beta}(\sum_{\forall z_i=1} y_i + \alpha_p, \sum_{\forall z_i=1} (J - y_i) + \beta_p)$ as we have seen in previous occupancy models.

To derive the full-conditional distribution for z_i, we use the relationship $\psi_i = \Phi(\mathbf{x}_i'\boldsymbol{\beta} + \eta_i)$ where Φ is the standard normal CDF. Also, recall that $z_i = 1$ for all sites where we detected the species (i.e., if $y_i > 0$). Thus, we only sample the latent occupancy process z_i when $y_i = 0$. In Chapter 23, we showed that the full-conditional distribution for the latent occupancy process is Bernoulli such that $[z_i|\cdot] = \text{Bern}(\tilde{\psi}_i)$ where

$$\tilde{\psi}_i = \frac{\psi_i(1-p)^J}{\psi_i(1-p)^J + 1 - \psi_i}. \tag{26.15}$$

Thus, we can use a Gibbs update to sample z_i for sites where $y_i = 0$ in our MCMC algorithm.

The full-conditional distribution for $\boldsymbol{\beta}$ is similar to the one that we derived in Chapters 20 and 23, but now involves the spatial random vector $\boldsymbol{\eta}$. However, it still results in a conjugate multivariate normal distribution because the joint conditional distribution for the auxiliary variables is $\mathbf{v} \sim N(\mathbf{X}\boldsymbol{\beta} + \boldsymbol{\eta}, \mathbf{I})$ due to the conditional independence in our model specification. Thus, the full-conditional distribution for $\boldsymbol{\beta}$ in our spatial occupancy model is derived as

$$[\boldsymbol{\beta}|\cdot] \propto [\mathbf{v}|\boldsymbol{\beta}, \boldsymbol{\eta}][\boldsymbol{\beta}], \tag{26.16}$$

$$\propto \exp\left(-\frac{1}{2}(\mathbf{v} - \mathbf{X}\boldsymbol{\beta} - \boldsymbol{\eta})'(\mathbf{v} - \mathbf{X}\boldsymbol{\beta} - \boldsymbol{\eta})\right)$$

$$\exp\left(-\frac{1}{2}(\boldsymbol{\beta} - \boldsymbol{\mu}_\beta)'\boldsymbol{\Sigma}_\beta^{-1}(\boldsymbol{\beta} - \boldsymbol{\mu}_\beta)\right), \tag{26.17}$$

$$\propto \exp\left(-\frac{1}{2}(-2((\mathbf{v} - \boldsymbol{\eta})'\mathbf{X} + \boldsymbol{\mu}_\beta'\boldsymbol{\Sigma}_\beta^{-1})\boldsymbol{\beta}) + \boldsymbol{\beta}'(\mathbf{X}'\mathbf{X} + \boldsymbol{\Sigma}_\beta^{-1})\boldsymbol{\beta})\right), \tag{26.18}$$

$$\propto \exp\left(-\frac{1}{2}(-2\mathbf{b}'\boldsymbol{\beta} + \boldsymbol{\beta}'\mathbf{A}\boldsymbol{\beta})\right), \tag{26.19}$$

where

$$\mathbf{b}' \equiv (\mathbf{v} - \boldsymbol{\eta})'\mathbf{X} + \boldsymbol{\mu}_\beta'\boldsymbol{\Sigma}_\beta^{-1}, \tag{26.20}$$

$$\mathbf{A} \equiv \mathbf{X}'\mathbf{X} + \boldsymbol{\Sigma}_\beta^{-1}, \tag{26.21}$$

implies that $[\boldsymbol{\beta}|\cdot] = N(\mathbf{A}^{-1}\mathbf{b}, \mathbf{A}^{-1})$ using the same multivariate normal properties that we presented in Chapter 11. Thus, the full-conditional distribution for the probit regression coefficients is conjugate and we can sample $\boldsymbol{\beta}$ with a Gibbs update in our MCMC algorithm.

In terms of the auxiliary variables, we can derive the full-conditional distribution for v_i for $i = 1, \ldots, n$ as

$$[v_i|\cdot] \propto [z_i|v_i][v_i|\boldsymbol{\beta}, \boldsymbol{\eta}], \tag{26.22}$$

$$\propto (1_{\{v_i > 0\}} 1_{\{z_i = 1\}} + 1_{\{v_i \leq 0\}} 1_{\{z_i = 0\}})[v_i|\boldsymbol{\beta}, \boldsymbol{\eta}], \tag{26.23}$$

$$= \begin{cases} TN(\mathbf{x}_i'\boldsymbol{\beta} + \eta_i, 1)_0^\infty & , z_i = 1 \\ TN(\mathbf{x}_i'\boldsymbol{\beta} + \eta_i, 1)_{-\infty}^0 & , z_i = 0 \end{cases}, \tag{26.24}$$

where "TN" refers to a truncated normal distribution truncated below by the subscript and above by the superscript. Thus, we can sample v_i for $i = 1, \ldots, n$ using a Gibbs update conditioned on whether $z_i = 0$ or $z_i = 1$.

The full-conditional distribution for the spatial random vector $\boldsymbol{\eta}$ is similar to that for $\boldsymbol{\beta}$ because they both appear in the process model additively. Thus,

$$[\boldsymbol{\eta}|\cdot] \propto [\mathbf{v}|\boldsymbol{\beta},\boldsymbol{\eta}][\boldsymbol{\eta}|\sigma^2], \tag{26.25}$$

$$\propto \exp\left(-\frac{1}{2}(\mathbf{v}-\mathbf{X}\boldsymbol{\beta}-\boldsymbol{\eta})'(\mathbf{v}-\mathbf{X}\boldsymbol{\beta}-\boldsymbol{\eta})\right)\exp\left(-\frac{1}{2}(\boldsymbol{\eta}'\boldsymbol{\Sigma}^{-1}\boldsymbol{\eta})\right), \tag{26.26}$$

$$\propto \exp\left(-\frac{1}{2}(-2(\mathbf{v}-\mathbf{X}\boldsymbol{\beta})'\boldsymbol{\eta}+\boldsymbol{\eta}'(\mathbf{I}+\boldsymbol{\Sigma}^{-1})\boldsymbol{\eta})\right), \tag{26.27}$$

$$\propto \exp\left(-\frac{1}{2}(-2\mathbf{b}'\boldsymbol{\eta}+\boldsymbol{\eta}'\mathbf{A}\boldsymbol{\eta})\right), \tag{26.28}$$

where

$$\mathbf{b}' \equiv (\mathbf{v}-\mathbf{X}\boldsymbol{\beta})', \tag{26.29}$$

$$\mathbf{A} \equiv \mathbf{I}+\boldsymbol{\Sigma}_\beta^{-1}, \tag{26.30}$$

implies that $[\boldsymbol{\eta}|\cdot] = \mathrm{N}(\mathbf{A}^{-1}\mathbf{b},\mathbf{A}^{-1})$ using the same multivariate normal properties that we presented in Chapter 11. Thus, the full-conditional distribution for the spatial random vector is conjugate and we can sample $\boldsymbol{\eta}$ with a multivariate normal Gibbs update in our MCMC algorithm.

The final parameter in our model is the spatial variance parameter σ^2. Because we specified the auxiliary variables to be conditional normal, the full-conditional distribution for σ^2 resembles that of the variance component in the linear regression model from Chapter 11

$$[\sigma^2|\cdot] \propto [\boldsymbol{\eta}|\sigma^2][\sigma^2], \tag{26.31}$$

$$\propto |\boldsymbol{\Sigma}|^{-\frac{1}{2}}\exp\left(-\frac{1}{2}\boldsymbol{\eta}'\boldsymbol{\Sigma}^{-1}\boldsymbol{\eta}\right)(\sigma^2)^{-(q+1)}\exp\left(-\frac{1}{r\sigma^2}\right), \tag{26.32}$$

$$\propto (\sigma^2)^{-\frac{n}{2}}\exp\left(-\frac{1}{2\sigma^2}\boldsymbol{\eta}'(\mathrm{diag}(\mathbf{W1})-\rho\mathbf{W})\boldsymbol{\eta}\right)(\sigma^2)^{-(q+1)}\exp\left(-\frac{1}{r\sigma^2}\right), \tag{26.33}$$

$$\propto (\sigma^2)^{-(\frac{n}{2}+q+1)}\exp\left(-\frac{1}{\sigma^2}\left(\frac{\boldsymbol{\eta}'(\mathrm{diag}(\mathbf{W1})-\rho\mathbf{W})\boldsymbol{\eta}}{2}+\frac{1}{r}\right)\right), \tag{26.34}$$

where $\rho = 0.99$ is fixed *a priori*. Thus, letting $\tilde{q} = \frac{n}{2}+q$ and $\tilde{r} = (\frac{\boldsymbol{\eta}'(\mathrm{diag}(\mathbf{W1})-\rho\mathbf{W})\boldsymbol{\eta}}{2}+\frac{1}{r})^{-1}$, the full-conditional distribution is $[\sigma^2|\cdot] = \mathrm{IG}(\tilde{q},\tilde{r})$, which leads to another Gibbs update in our MCMC algorithm.

These full-conditional distributions imply that we can construct a fully Gibbs MCMC algorithm to fit the spatial occupancy model to data. Therefore, we wrote the MCMC algorithm called occ.sp.aux.mcmc.R in R using all Gibbs updates:

BOX 26.1   MCMC ALGORITHM FOR SPATIAL OCCUPANCY MODEL

```
1 occ.sp.aux.mcmc <- function(y,J,X,W,n.mcmc){
2
3 ####
4 ####   Subroutines
5 ####
6
7 rtn <- function(n,mu,sig2,low,high){
8    flow=pnorm(low,mu,sqrt(sig2))
9    fhigh=pnorm(high,mu,sqrt(sig2))
10   u=runif(n)
11   tmp=flow+u*(fhigh-flow)
12   x=qnorm(tmp,mu,sqrt(sig2))
13   x
14 }
15
16 ####
17 ####   Setup Variables
18 ####
19
20 n=length(y)
21 p.X=dim(X)[2]
22 n.burn=round(0.25*n.mcmc)
23
24 z.mean=rep(0,n)
25 eta.mean=rep(0,n)
26 psi.mean=rep(0,n)
27 beta.save=matrix(0,p.X,n.mcmc)
28 p.save=rep(0,n.mcmc)
29 s2.save=rep(0,n.mcmc)
30 N.save=rep(0,n.mcmc)
31
32 ####
33 ####   Priors and Starting Values
34 ####
35
36 beta=c(-.5,.5)
37 p=.25
38 Xbeta=X%*%beta
```

(Continued)

**BOX 26.1 (Continued)   MCMC ALGORITHM FOR SPATIAL
OCCUPANCY MODEL**

```
39 psi=pnorm(Xbeta)
40 z=rep(0,n)
41 z[y==0]=0
42 n0=sum(z==0)
43 n1=sum(z==1)
44
45 v=rep(0,n)
46
47 ny0=sum(y==0)
48 ny1=sum(y==1)
49
50 alpha.p=1
51 beta.p=1
52
53 r=1000
54 q=0.001
55
56 Sig.beta=diag(p.X)
57 Sig.beta.inv=solve(Sig.beta)
58 mu.beta=rep(0,p.X)
59
60 XprimeX=t(X)%*%X
61 A.beta=XprimeX+Sig.beta.inv
62 A.beta.chol=chol(A.beta)
63 Sig.beta.inv.times.mu.beta=Sig.beta.inv%*%mu.beta
64
65 rho=0.99
66 R.inv=diag(apply(W,1,sum))-rho*W
67 s2=5
68 Sigma.inv=R.inv/s2
69 eta=rep(0,n)
70
71 ####
72 #### Begin MCMC Loop
73 ####
74
75 for(k in 1:n.mcmc){
```

(Continued)

BOX 26.1 (Continued)   MCMC ALGORITHM FOR SPATIAL OCCUPANCY MODEL

```
76   ####
77   ####   Sample v
78   ####
79
80   v[z==0]=rtn(n0,(Xbeta+eta)[z==0],1,-Inf,0)
81   v[z==1]=rtn(n1,(Xbeta+eta)[z==1],1,0,Inf)
82
83   ####
84   ####   Sample beta
85   ####
86
87   b.beta=t(X)%*%(v-eta)+Sig.beta.inv.times.mu.beta
88   beta=backsolve(A.beta.chol,backsolve(A.beta.
89       chol,b.beta,transpose=TRUE)+rnorm(p.X))
90   Xbeta=X%*%beta
91   psi=pnorm(Xbeta+eta)
92
93   ####
94   ####   Sample p
95   ####
96
97   sum.1=sum(y[z==1])+alpha.p
98   sum.2=sum(J-y[z==1])+beta.p
99   p=rbeta(1,sum.1,sum.2)
100
101  ####
102  ####   Sample z
103  ####
104
105  psi.numer=psi*(1-p)^J
106  psi.tmp=psi.numer/(psi.numer+1-psi)
107  psi.tmp[psi.tmp>1]=1
108  z[y==0]=rbinom(sum(y==0),1,psi.tmp[y==0])
109  z[y>0]=1
110  n0=sum(z==0)
111  n1=sum(z==1)
```

(Continued)

**BOX 26.1 (Continued)   MCMC ALGORITHM FOR SPATIAL
OCCUPANCY MODEL**

```
112  ####
113  ####   Sample eta
114  ####
115
116  b.eta=v-Xbeta
117  A.eta.chol=chol(diag(n)+Sigma.inv)
118  eta=backsolve(A.eta.chol,backsolve(A.eta.chol,
119    b.eta,transpose=TRUE)+rnorm(n))
120  psi=pnorm(Xbeta+eta)
121
122  ####
123  ####   Sample s2
124  ####
125
126  q.tmp=n/2+q
127  r.tmp=1/(t(eta)%*%R.inv%*%eta/2+1/r)
128  s2=1/rgamma(1,q.tmp,,r.tmp)
129  Sigma.inv=R.inv/s2
130
131  ####
132  ####   Save Samples
133  ####
134
135  beta.save[,k]=beta
136  p.save[k]=p
137  s2.save[k]=s2
138  if(k > n.burn){
139    z.mean=z.mean+z/(n.mcmc-n.burn)
140    eta.mean=eta.mean+eta/(n.mcmc-n.burn)
141    psi.mean=psi.mean+psi/(n.mcmc-n.burn)
142  }
143  N.save[k]=sum(z)
144
145 }
146
147 ####
148 ####   Write Output
149 ####
```

(Continued)

BOX 26.1 (Continued)   MCMC ALGORITHM FOR SPATIAL OCCUPANCY MODEL

```
150 list(p.save=p.save,beta.save=beta.save,
151    s2.save=s2.save,z.mean=z.mean,eta.mean=
152    eta.mean,psi.mean=psi.mean,N.save=N.save,
153    n.mcmc=n.mcmc,r=r,q=q)
154
155 }
```

Notice that we used the same "backsolve" technique that we introduced in Chapter 14 to sample the multivariate normal random vectors β and η in our MCMC algorithm because it is a computationally efficient approach using base R functions.

To demonstrate our MCMC algorithm, we simulated spatially explicit occupancy data using the following R code based on a regular grid of $n = 400$ sites in the unit square:

BOX 26.2   SIMULATE SPATIAL OCCUPANCY DATA

```
1  library(mvtnorm)
2
3  n.1=20
4  n.2=20
5  n=n.1*n.2
6  J=3
7  p=.6
8  s.1.vals=seq(0,1,,n.1)
9  s.2.vals=seq(0,1,,n.2)
10 S=expand.grid(s.1.vals,s.2.vals)
11 D=as.matrix(dist(S))
12 d.thresh=.05*max(D)
13 W=ifelse(D>d.thresh,0,1)
14 diag(W)=0
15 rho=.99
16 s2=3
17 Sigma=s2*solve(diag(apply(W,1,sum))-rho*W)
18 set.seed(2)
19 eta=c(rmvnorm(1,rep(0,n),Sigma,method="chol"))
```

(Continued)

BOX 26.2 (Continued)   SIMULATE SPATIAL OCCUPANCY DATA

```
20 X=matrix(1,n,2)
21 X[,2]=rnorm(n)
22 beta=c(-.5,.5)
23
24 psi=pnorm(X%*%beta+eta)
25 z=rbinom(n,1,psi)
26 y=rbinom(n,J,z*p)
27
28 sq.cex=2.95
29 layout(matrix(1:4,2,2,byrow=TRUE))
30 plot(S,type="n",xlab="",ylab="",asp=TRUE,main="a")
31 points(S,pch=15,cex=sq.cex,col=rgb(0,0,0,(X[,2]
32     -min(X[,2]))/diff(range(X[,2]))))
33 plot(S,type="n",xlab="",ylab="",asp=TRUE,main="b")
34 points(S,pch=15,cex=sq.cex,col=rgb(0,0,0,(eta
35     -min(eta))/diff(range(eta))))
36 plot(S,type="n",xlab="",ylab="",asp=TRUE,main="c")
37 points(S,pch=15,cex=sq.cex,col=rgb(0,0,0,psi))
38 plot(S,type="n",xlab="",ylab="",asp=TRUE,main="d")
39 points(S,pch=15,cex=sq.cex,col=rgb(0,0,0,y/J))
```

We based our simulated data on $J = 3$ sampling occasions, $p = 0.6$, d_{thresh} set to 5% of the maximum distance among sites, $\sigma^2 = 3$, and $\beta = (-0.5, 0.5)'$. This R code results in the graphics depicting the covariate, spatial random vector, occupancy probability, and observed occupancy data shown in Figure 26.1. We simulated the covariate (Figure 26.1a) by sampling $n = 400$ independent realizations from a standard normal distribution.

Based on the simulated data, we fit the spatial occupancy model using hyperparameters $r = 1000$, $q = 0.001$, $\mu_\beta = 0$, $\Sigma_\beta = I$, $\alpha_p = 1$, and $\beta_p = 1$ and the following R code:

BOX 26.3   FIT SPATIAL OCCUPANCY MODEL TO DATA

```
1 source("occ.sp.aux.mcmc.R")
2 n.mcmc=40000
3 set.seed(1)
4 mcmc.out=occ.sp.aux.mcmc(y=y,J=J,X=X,W=W,
5     n.mcmc=n.mcmc)
```

(Continued)

BOX 26.3 (Continued)   FIT SPATIAL OCCUPANCY MODEL TO DATA

```
 6  dIG <- function(x,r,q){x^(-(q+1))*exp(-1/r/x)/
 7      (r^q)/gamma(q)}
 8  layout(matrix(1:4,2,2,byrow=TRUE))
 9  hist(mcmc.out$beta.save[1,-(1:1000)],xlab
10      =bquote(beta[0]),prob=TRUE,main="a")
11  curve(dnorm(x,0,1.5),lwd=2,add=TRUE)
12  abline(v=beta[1],lty=2,lwd=2)
13  hist(mcmc.out$beta.save[2,-(1:1000)],xlab
14      =bquote(beta[1]),prob=TRUE,main="b")
15  curve(dnorm(x,0,1.5),lwd=2,add=TRUE)
16  abline(v=beta[2],lty=2,lwd=2)
17  hist(mcmc.out$p.save[-(1:1000)],xlab="p",
18      xlim=c(0,1),prob=TRUE,main="c")
19  curve(dbeta(x,1,1),lwd=2,add=TRUE)
20  abline(v=p,lty=2,lwd=2)
21  hist(mcmc.out$s2.save[-(1:1000)],xlab=bquote
22      (sigma^2),prob=TRUE,main="d")
23  curve(dIG(x,mcmc.out$r,mcmc.out$q),n=1000,from=0,
24      to=14,lwd=2,add=TRUE)
25  abline(v=s2,lty=2,lwd=2)
26
27  mean(mcmc.out$N.save[-(1:1000)])
28  quantile(mcmc.out$N.save[-(1:1000)],
29      c(0.025,0.975))
```

This R code results in the marginal posterior histograms for model parameters shown in Figure 26.2. Notably, Figure 26.2 indicates that all of the marginal posterior distributions capture the true parameter values used to simulate the data. In this particular case, we estimate the coefficient associated with the relationship of the covariate and occupancy probability (β_1) to be positive and it covers the true value (Figure 26.2b). The marginal posterior histogram for detection probability p captures the true value of $p = 0.6$ used to simulate the data well (Figure 26.2c) and while the marginal posterior histogram for the spatial variance parameter σ^2 indicates substantial variability, it also captures the true parameter value used to simulate the data well (Figure 26.2d).

The observed number of occupied sites in the simulated data was $\sum_{i=1}^{n} 1_{\{y_i > 0\}} = 176$ out of a total of $n = 400$ sites. Using the spatial occupancy model, we estimated the derived quantity $N = \sum_{i=1}^{n} z_i$, which is the total number of sites actually occupied.

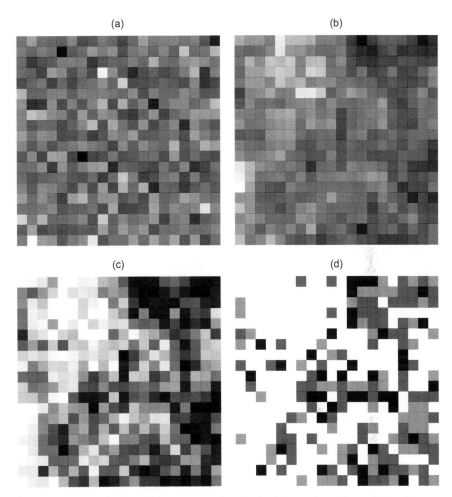

Figure 26.1 Simulated spatial occupancy data for the (a) covariate, (b) spatial random process η, (c) occupancy probabilities ψ, and (d) data **y**. Darker shades in each subfigure indicate larger values for those sites and lighter shades represent smaller values. White and black sites in (c) indicate probabilities close to zero and one, and in (d) indicate observations equal to zero and three detections, respectively.

In this case, the posterior mean number of occupied sites was $E(N|\mathbf{y}) = 187.4$ with 95% credible interval $(181, 196)$. For comparison, the true total number of occupied sites (including those undetected) was $N = 182$, which is captured by the 95% credible interval in this particular example.

We also can make inference on the spatial process η using the spatial occupancy model. The following R code compares the posterior mean of η, based on the spatial occupancy model fit to the simulated data, with the true η:

BOX 26.4   PLOT LATENT SPATIAL PROCESS η

```
1  eta.mn=mcmc.out$eta.mean
2  sq.cex=2.1
3  layout(matrix(1:2,1,2))
4  plot(S,type="n",xlab="",ylab="",asp=TRUE,main="a")
5  points(S,pch=15,cex=sq.cex,col=rgb(0,0,0,(eta
6      -min(eta))/diff(range(eta))))
7  plot(S,type="n",xlab="",ylab="",asp=TRUE,main="b")
8  points(S,pch=15,cex=sq.cex,col=rgb(0,0,0,(eta.mn
9      -min(eta.mn))/diff(range(eta.mn))))
10 points(S[y>0,],cex=.1,col=8)
```

This R code results in Figure 26.3. The posterior mean of η (Figure 26.3b) clearly characterizes the same pattern evident in the simulated η (Figure 26.3a). The pattern in Figure 26.3b for $E(\eta|\mathbf{y})$ appears smoother because it is a posterior mean, whereas

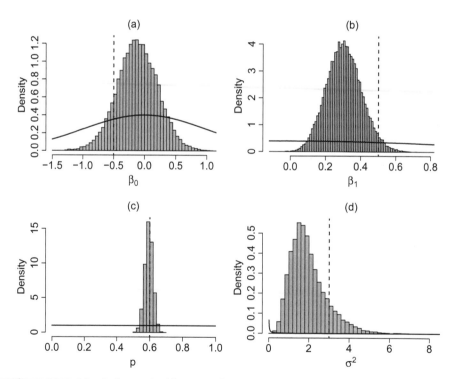

Figure 26.2 Marginal posterior histograms from the spatial occupancy model for parameters (a) β_0, (b) β_1, (c) p, and (d) σ^2. Priors shown as solid black lines and true parameters values used to simulate the data shown as vertical dashed lines.

(a)   (b)

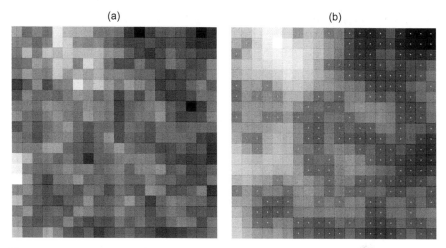

Figure 26.3 Comparison of (a) true η and (b) posterior mean $E(\eta|\mathbf{y})$ for the spatial process in the occupancy model. Sites at which the species was detected are indicated with small gray points in (b).

the simulated version of η is a realization. The spatial process η often is interpreted in two different ways: a missing covariate or as a way to account for spatial dependence due to a mechanism in the occupancy process. In either case, the model is able to recover the pattern in η and account for it while estimating the other model parameters.

26.1 ADDITIONAL CONCEPTS AND READING

We focused on binary data with measurement error in this chapter; however, for an excellent review of spatially explicit modeling approaches for binary data, see De Oliveira (2018). We can apply many of the approaches described by De Oliveira (2018) in the occupancy model setting to account for imperfect detection.

Several other studies have proposed and used occupancy models with spatial components. For example, while Broms et al. (2014) used an approach very similar to what we presented in this chapter, Webb et al. (2014) explored basis function (Hefley et al. 2017a) and generalized additive modeling (Wood 2004) approaches to account for spatial dependence among sites. Rota et al. (2016) extended these concepts in a spatially explicit multispecies occupancy model to account for both interspecific and spatial dependence among sites explicitly.

Although not relying on spatially explicit occupancy models directly, other studies have investigated the utility of occupancy models from a spatial sampling perspective (e.g., Noon et al. 2012; Linden et al. 2017; Steenweg et al. 2018). In particular, Steenweg et al. (2018) examined the effect of spatial sampling scale (i.e., the size of the areal units surveyed) on inference resulting from occupancy models.

The hierarchical framework facilitates the combination of spatial and temporal occupancy modeling (MacKenzie et al. 2003) concepts to account for ecological processes such as metapopulation dynamics. For example, Chandler et al. (2015) developed a spatio-temporal occupancy model that accounted for dispersal rates. Similarly, Broms et al. (2016) extended a spatio-temporal model for binary data (Hooten and Wikle 2010) to accommodate false negative occupancy data and applied it to study the spread of common myna (*Acridotheres tristis*) in South Africa. Bled et al. (2011) described a spatio-temporal occupancy model for Eurasian collared-dove *Streptopelia decaocto* based on a binary version of relative abundance data arising from the North American breeding bird survey (Robbins et al. 1986). Similar spatio-temporal models also have been developed based on relative abundance data in the presence of false negative measurement errors (e.g., Hooten et al. 2007; Williams et al. 2017). We discuss spatio-temporal in much more detail in Chapter 28.

27 Spatial Capture-Recapture Models

Like all ecological processes, the distribution and abundance of organisms is explicitly spatial in nature. In particular, the movement of animals is a critical component of their life history. Despite that fact, most conventional approaches to estimate animal population sizes and densities make restrictive assumptions about movement (e.g., closure to emigration and immigration) or average over processes like movement that result in smaller-scale heterogeneity. As a result, fundamental patterns in nature go undetected in many ecological studies involving wildlife populations. Although many individual-based studies in wildlife biology involve explicitly spatial information (e.g., location of capture/detection within a study area), it is often unused in conventional statistical analyses.

Originally, the development of spatial capture-recapture (SCR) models was to make use of spatially explicit information that often accompanies capture-recapture data (Royle et al. 2013a). For example, consider the situation where an array of detectors (e.g., camera traps) are arranged in a spatial pattern where each trap location $\mathbf{x}_l$ for $l = 1, \ldots, L$ is known as part of the study design. Then, for a set of N individuals in a population exposed to the detector array, each may have an activity center $\mathbf{s}_i$ for $i = 1, \ldots, N$ that is unknown and observed directly. However, as the individuals perform their daily activities, we may detect them at one or more detector locations over time. In the situation where we can detect individuals multiple times per occasion j for $j = 1, \ldots, J$, Royle and Dorazio (2008) envisioned modeling these counts as

$$\tilde{y}_{ilj} \sim \text{Pois}(\lambda g_{il}), \tag{27.1}$$

where λ is a baseline encounter intensity and g_{il} is a detection function that decays with distance for individual i and trap l. Thus, based on the Poisson model for unlimited possible detections per period j, the conditional probability that an individual goes undetected at trap location l and occasion j is $P(\tilde{y}_{ilj} = 0) = e^{-\lambda g_{il}}$, which implies that the probability of detecting the individual at least once is

$$P(\tilde{y}_{ilj} > 0) = 1 - P(\tilde{y}_{ilj} = 0), \tag{27.2}$$

$$= 1 - e^{-\lambda g_{il}}. \tag{27.3}$$

If we can detect the individual only once during occasion j, then the response variable is binary $y_{ilj} \sim \text{Bern}(p_{il})$. For a link function, Royle and Dorazio (2008) used the relationship implied by the Poisson version of the model where

$$p_{il} = 1 - e^{-\lambda g_{il}}. \tag{27.4}$$

The implied link function is the complimentary log-log link, where, instead of the more traditional logit, we have

$$\text{cloglog}(p) = \log(-\log(1-p)), \tag{27.5}$$

which, for our SCR data model, implies that

$$\text{cloglog}(p_{il}) = \log(-\log(1-p_{il})), \tag{27.6}$$

$$= \log(-\log(e^{-\lambda g_{il}})), \tag{27.7}$$

$$= \log(\lambda g_{il}), \tag{27.8}$$

$$= \log(\lambda) + \log(g_{il}). \tag{27.9}$$

As with distance sampling, there are a number of options for a detection function g_{il}. In principle, the detection function should be large when the individual's activity center $\mathbf{s}_i$ is close to a detector location $\mathbf{x}_l$ and should decrease toward zero as the individual activity center and detector location become more distant. For example, a Gaussian shaped SCR detection function is

$$g_{il}(\mathbf{s}_i, \mathbf{x}_l) = \exp\left(-\frac{(\mathbf{s}_i - \mathbf{x}_l)'(\mathbf{s}_i - \mathbf{x}_l)}{\sigma^2}\right), \tag{27.10}$$

where the numerator in (27.10) is the squared Euclidean distance between the activity center and trap location and the denominator is a parameter σ^2 that controls the rate of decay in the detection function (large σ^2 implies the activity center may be far from the detector and the individual can still be detected). Note that the parameter σ^2 in this model is not a variance parameter as it is when the data model is a Gaussian probability distribution (not a Gaussian detection function). The form of the detection function in (27.10) combined with the cloglog link function results in $\text{cloglog}(p_{il}) = \alpha + \beta d_{il}^2$, where $\alpha = \log(\lambda)$, $\beta = -1/\sigma^2$, and $d_{il}^2 = (\mathbf{s}_i - \mathbf{x}_l)'(\mathbf{s}_i - \mathbf{x}_l)$. Thus, in the situation where the abundance N and activity centers $\mathbf{s}_i$ are known for $i = 1, \ldots, N$, the SCR model is binary regression (Chapter 20) with a cloglog link function instead of logit or probit.

Furthermore, if we assume the y_{ilj} are conditionally independent for individual i and detector l and we let $y_{il} = \sum_{j=1}^{J} y_{ilj}$, then $y_{il} \sim \text{Binom}(J, p_{il})$, again, assuming abundance N is known. However, we can apply the same PX-DA strategy that we introduced in Chapter 24 by augmenting the data set with $M - n$ (where n is the number of detected individuals and M is selected to be larger than the true population size N a priori) all zero detection histories to account for abundance being unknown (Royle and Young 2008). The resulting data structure is a $M \times J$ matrix of detection counts (for the first n rows) and all zero detection histories (for rows $n+1$ through M).

To account for the zero-augmented data source, we can use a ZI binomial mixture model as we did in Chapter 24, where

$$y_{il} \sim \begin{cases} 0 & , z_i = 0 \\ \text{Binom}(J, p_{il}) & , z_i = 1 \end{cases}, \tag{27.11}$$

for $i = 1, \ldots, M$ and $l = 1, \ldots, L$, and where $\text{cloglog}(p_{il}) = \alpha + \beta (\mathbf{s}_i - \mathbf{x}_l)' (\mathbf{s}_i - \mathbf{x}_l)$. When the individual activity centers $\mathbf{s}_i$ for $i = 1, \ldots, M$ are unknown, we must specify a process model for them. As a random set of points, the activity centers comprise a point process; thus, we specify a complete spatial random point process model for $\mathbf{s}_i$. This model assumes that $\mathbf{s}_i$ arise conditionally and independently from a uniform distribution on a spatial support $\mathscr{S}$. Thus, to complete the hierarchical SCR model, we specify distributions for the remaining parameters as

$$\mathbf{s}_i \sim \text{Unif}(\mathscr{S}), \tag{27.12}$$

$$z_i \sim \text{Bern}(\psi), \tag{27.13}$$

$$\alpha \sim \text{N}(\mu_\alpha, \sigma_\alpha^2), \tag{27.14}$$

$$\beta \sim \text{N}(\mu_\beta, \sigma_\beta^2), \tag{27.15}$$

$$\psi \sim \text{Beta}(\gamma_1, \gamma_2), \tag{27.16}$$

where we inform the prior for β to have mass predominantly in negative values of the support because $\beta = -1/\sigma^2$ and $\sigma^2 > 0$. Alternatively, we could use a uniform distribution with compact negative support for β or a different model formulation that facilitates the use of other priors.

The resulting posterior distribution for this SCR model is

$$[\mathbf{S}, \mathbf{z}, \alpha, \beta, \psi | \mathbf{Y}] \propto \left(\prod_{i=1}^{M} \left(\prod_{l=1}^{L} [y_{il} | \alpha, \beta, \mathbf{s}_i]^{z_i} 1_{\{y_{il}=0\}}^{1-z_i} \right) [z_i | \psi] \right) [\alpha][\beta][\psi], \tag{27.17}$$

where $\mathbf{S}$ is a $M \times 2$ matrix (if the process is modeled in two-dimensional space) containing the activity centers, $\mathbf{z}$ is a $M \times 1$ vector containing the population membership indicator variables (as we described in Chapter 24), and $\mathbf{Y}$ is a $M \times J$ matrix of detection counts.

To fit the hierarchical SCR model to data using MCMC, we need to derive the full-conditional distributions for model parameters (α, β, and ψ) and latent state variables ($\mathbf{S}$ and $\mathbf{z}$). The full-conditional distributions for the membership probability ψ is similar to that we derived in the previous mixture models, where

$$[\psi | \cdot] \propto \left(\prod_{i=1}^{M} [z_i | \psi] \right) [\psi], \tag{27.18}$$

$$\propto \left(\prod_{i=1}^{M} \psi^{z_i} (1 - \psi)^{1-z_i} \right) \psi^{\gamma_1 - 1} (1 - \psi)^{\gamma_2 - 1}, \tag{27.19}$$

$$\propto \prod_{i=1}^{M} \psi^{\left(\sum_{i=1}^{M} z_i\right) + \gamma_1 - 1} (1 - \psi)^{\left(\sum_{i=1}^{M} (1 - z_i)\right) + \gamma_2 - 1}, \tag{27.20}$$

which results in a beta full-conditional distribution such that

$$[\psi | \cdot] = \text{Beta}\left(\left(\sum_{i=1}^{M} z_i \right) + \gamma_1, \left(\sum_{i=1}^{M} (1 - z_i) \right) + \gamma_2 \right). \tag{27.21}$$

Thus, we can sample ψ using a Gibbs update in a MCMC algorithm to fit the SCR model, as we have in the previous chapters.

The full-conditional distribution for z_i is also similar to the previous chapters that covered mixture models. Also, like in the occupancy and capture-recapture models presented in Chapters 23 and 24, we only sample the latent membership indicator z_i when there have been no detections of the individual at any of the traps and occasions. Thus, for individual i such that $\sum_{l=1}^{L} y_{il} = 0$, we have

$$[z_i|\cdot] \propto \left(\prod_{l=1}^{L} [y_{il}|\alpha,\beta,\mathbf{s}_i]^{z_i} 1_{\{y_{il}=0\}}^{1-z_i} \right) [z_i|\psi], \tag{27.22}$$

$$\propto \prod_{l=1}^{L} ((1-p_{il})^J)^{z_i} \psi^{z_i} (1-\psi)^{1-z_i}, \tag{27.23}$$

$$\propto \prod_{l=1}^{L} (\psi(1-p_{il})^J)^{z_i} (1-\psi)^{1-z_i}. \tag{27.24}$$

which results in a Bernoulli full-conditional distribution for z_i such that $[z_i|\cdot] = \text{Bern}(\tilde{\psi}_i)$, where

$$\tilde{\psi}_i = \frac{\psi \prod_{l=1}^{L} (1-p_{il})^J}{\psi \prod_{l=1}^{L} (1-p_{il})^J + 1 - \psi}. \tag{27.25}$$

Therefore, we can sample z_i using a Gibbs update in a MCMC algorithm, as we have for previous mixture models formulated using latent indicator variables. In this particular case, because we have used PX-DA to assist in the estimation of abundance, we can make inference using the derived quantity $N = \sum_{i=1}^{M} z_i$ as we discussed in Chapter 24.

The full-conditional distributions for the parameters in the detection function, α and β, will both be non-conjugate based on our model specification and therefore we must use M-H updates to sample them. For α, the full-conditional distribution is

$$[\alpha|\cdot] \propto \left(\prod_{i=1}^{M} \left(\prod_{l=1}^{L} [y_{il}|\alpha,\beta,\mathbf{s}_i]^{z_i} \right) \right) [\alpha], \tag{27.26}$$

$$\propto \left(\prod_{\forall z_i=1} \left(\prod_{l=1}^{L} [y_{il}|\alpha,\beta,\mathbf{s}_i] \right) \right) [\alpha]. \tag{27.27}$$

We sample from this full-conditional distribution using a symmetric normal random walk proposal $\alpha^{(*)} \sim \text{N}(\alpha^{(k-1)}, \sigma^2_{\alpha,\text{tune}})$ and then use the M-H ratio

$$mh = \frac{\left(\prod_{\forall z_i=1} \left(\prod_{l=1}^{L} [y_{il}|\alpha^{(*)},\beta^{(k-1)},\mathbf{s}_i^{(k-1)}] \right) \right) [\alpha^{(*)}]}{\left(\prod_{\forall z_i=1} \left(\prod_{l=1}^{L} [y_{il}|\alpha^{(k-1)},\beta^{(k-1)},\mathbf{s}_i^{(k-1)}] \right) \right) [\alpha^{(k-1)}]}, \tag{27.28}$$

where we tune the proposal to result in a well-mixed Markov chain using the tuning parameter $\sigma^2_{\alpha,\text{tune}}$.

Similarly, for β, the full-conditional distribution is

$$[\beta|\cdot] \propto \left(\prod_{i=1}^{M} \left(\prod_{l=1}^{L} [y_{il}|\alpha,\beta,\mathbf{s}_i]^{z_i} \right) \right) [\beta], \tag{27.29}$$

$$\propto \left(\prod_{\forall z_i=1} \left(\prod_{l=1}^{L} [y_{il}|\alpha,\beta,\mathbf{s}_i] \right) \right) [\beta]. \tag{27.30}$$

We can also sample from this full-conditional distribution using a symmetric normal random walk proposal $\beta^{(*)} \sim \mathrm{N}(\beta^{(k-1)}, \sigma^2_{\beta,\text{tune}})$ and then use the M-H ratio

$$mh = \frac{\left(\prod_{\forall z_i=1} \left(\prod_{l=1}^{L} [y_{il}|\alpha^{(k)},\beta^{(*)},\mathbf{s}_i^{(k-1)}] \right) \right) [\beta^{(*)}]}{\left(\prod_{\forall z_i=1} \left(\prod_{l=1}^{L} [y_{il}|\alpha^{(k)},\beta^{(k-1)},\mathbf{s}_i^{(k-1)}] \right) \right) [\beta^{(k-1)}]}, \tag{27.31}$$

where we tune the proposal to result in a well-mixed Markov chain using the tuning parameter $\sigma^2_{\beta,\text{tune}}$.

We also consider the individual activity centers as latent state variables in the hierarchical SCR model (like the population membership indicator variables z_i). Thus, we need to sample from their full-conditional distribution

$$[\mathbf{s}_i|\cdot] \propto \left(\prod_{l=1}^{L} [y_{il}|\alpha,\beta,\mathbf{s}_i]^{z_i} \right) [\mathbf{s}_i], \tag{27.32}$$

which simplifies to a uniform distribution on the spatial support $\mathscr{S}$ when $z_i = 0$ because $[\mathbf{s}_i|\cdot] \propto [\mathbf{s}_i]$. Thus, we sample $\mathbf{s}_i^{(k)} \sim \mathrm{Unif}(\mathscr{S})$ when $z_i^{(k)} = 0$ in our MCMC algorithm.

In the case when $z_i = 1$, the full-conditional distribution for the activity center becomes

$$[\mathbf{s}_i|\cdot] \propto \left(\prod_{l=1}^{L} [y_{il}|\alpha,\beta,\mathbf{s}_i]^{z_i} \right) [\mathbf{s}_i], \tag{27.33}$$

$$\propto \prod_{l=1}^{L} [y_{il}|\alpha,\beta,\mathbf{s}_i], \tag{27.34}$$

for $\mathbf{s}_i \in \mathscr{S}$. In the cases where $z_i = 1$, we use M-H updates to sample $\mathbf{s}_i$ in our MCMC algorithm with the M-H ratio

$$mh = \frac{\prod_{l=1}^{L} [y_{il}|\alpha^{(k)},\beta^{(k)},\mathbf{s}_i^{(*)}]}{\prod_{l=1}^{L} [y_{il}|\alpha^{(k)},\beta^{(k)},\mathbf{s}_i^{(k-1)}]}, \tag{27.35}$$

where we use a multivariate normal random walk proposal with rejection sampling. That is, we sample $\mathbf{s}_i^{(*)} \sim \mathrm{N}(\mathbf{s}_i^{(k-1)}, \sigma^2_{s,\text{tune}}\mathbf{I})$ so that it cancels in the M-H ratio when

$\mathbf{s}_i^{(*)} \in \mathscr{S}$; if $\mathbf{s}_i^{(*)} \notin \mathscr{S}$, we skip the update for $\mathbf{s}_i$ on that iteration and retain the previous sample $\mathbf{s}_i^{(k)} = \mathbf{s}_i^{(k-1)}$ because if the proposal falls outside the support, the M-H ratio will be zero.

Examining the full-conditional distribution for $\mathbf{s}_i$ in more detail, we can see that when $z_i = 1$, we could have $\sum_{l=1}^{L} y_{il} = 0$ or $\sum_{l=1}^{L} y_{il} > 0$. In the situation where $z_i = 1$ and $\sum_{l=1}^{L} y_{il} = 0$, the likelihood component for individual i and detector l is $[y_{il}|\mathbf{s}_i, \alpha, \beta] = (1 - p_{il})^J$, thus, the full-conditional distribution is

$$[\mathbf{s}_i|\cdot] \propto \prod_{l=1}^{L}(1 - \mathrm{cloglog}^{-1}(\alpha + \beta(\mathbf{s}_i - \mathbf{x}_l)'(\mathbf{s}_i - \mathbf{x}_l))), \qquad (27.36)$$

where $\mathrm{cloglog}^{-1}$ is the inverse cloglog link function. Thus, MCMC realizations for $\mathbf{s}_i$ likely are far away from the detectors because small distances between $\mathbf{s}_i$ and $\mathbf{x}_l$ imply large values for $1 - p_{il}$. This is intuitive because we would not expect the individual to be close to the detector array in cases where we never detected it.

By contrast, when we detected the individual ($\sum_{l=1}^{L} y_{il} > 0$), the full-conditional distribution for $\mathbf{s}_i$ is

$$[\mathbf{s}_i|\cdot] \propto \prod_{l=1}^{L}\mathrm{Binom}(y_{il}|J, p_{il}), \qquad (27.37)$$

which implies that MCMC realizations for $\mathbf{s}_i$ will be near detector $\mathbf{x}_l$ when y_{il} is large. When we detect the individual at multiple locations, we would expect the posterior distribution for $\mathbf{s}_i$ to be an average of detector locations weighted by the number of detections at each of the detectors.

To implement the MCMC algorithm for fitting the hierarchical SCR model, we iteratively sample from these full-conditional distributions. However, in practice, we must choose $\mathscr{S}$ to result in compact support for $\mathbf{s}_i$ so that it results in a proper spatial point process model for $\mathbf{s}_i$, $i = 1, \dots, M$. Also, as the support $\mathscr{S}$ increases in size with respect to the detector array, so should our number of augmented encounter histories $M - n$, which will also increase our estimated abundance N. This is reasonable because we expect the population size to increase as the area of desired inference increases. However, the size of $\mathscr{S}$ will not affect the population density $N/\mathrm{area}(\mathscr{S})$; thus, many SCR studies report population densities rather than abundances.

We constructed a MCMC algorithm called scr.mcmc.R in R based on these full-conditional distributions:

BOX 27.1   MCMC ALGORITHM FOR SCR MODEL

```
scr.mcmc <- function(Y,J,X,n.mcmc){

#### 
#### Libraries and Subroutines
####
```

(Continued)

BOX 27.1 (Continued)   MCMC ALGORITHM FOR SCR MODEL

```
6   cloglog.inv <- function(x){
7     1-exp(-exp(x))
8   }
9
10  crossdist <- function(S1,S2) {
11      c1 <- complex(real=S1[,1], imaginary=S1[,2])
12      c2 <- complex(real=S2[,1], imaginary=S2[,2])
13      dist <- outer(c1, c2, function(z1, z2)
14        Mod(z1-z2))
15      dist
16  }
17
18  ####
19  ####   Setup Variables
20  ####
21
22  M=dim(Y)[1]
23  L=dim(Y)[2]
24  idx.0=(apply(Y,1,sum)==0)
25  n=sum(!idx.0)
26  n.burn=round(.1*n.mcmc)
27
28  alpha.save=rep(0,n.mcmc)
29  beta.save=rep(0,n.mcmc)
30  psi.save=rep(0,n.mcmc)
31  N.save=rep(0,n.mcmc)
32  D.save=rep(0,n.mcmc)
33  z.mean=rep(0,M)
34  S.mean=matrix(0,M,2)
35  S1.save=matrix(0,n.mcmc,2)
36  S2.save=matrix(0,n.mcmc,2)
37
38  y.lims=c(-1,1)
39  x.lims=c(-1,1)
40
41  A=(x.lims[2]-x.lims[1])*(y.lims[2]-y.lims[1])
42
43  ####
44  ####   Priors and Starting Values
45  ####
```

(Continued)

BOX 27.1 (Continued)   MCMC ALGORITHM FOR SCR MODEL

```
46 mu.alpha=-2
47 s2.alpha=1.5^2
48
49 mu.beta=-40
50 s2.beta=10
51
52 z=as.numeric(!idx.0)
53 S=matrix(0,M,2)
54 S[-(1:n),1]=runif(M-n,x.lims[1],x.lims[2])
55 S[-(1:n),2]=runif(M-n,y.lims[1],y.lims[2])
56
57 for(i in 1:n){
58   tmp.idx=(Y[!idx.0,][i,]>0)
59   S[i,]=apply(X[tmp.idx,],2,mean)
60 }
61
62 psi=.1
63 g.1=.001  #  scale prior
64 g.2=1
65 alpha=-2
66 beta=-40
67 D2=crossdist(S,X)^2
68 P=cloglog.inv(alpha+beta*D2)
69 N=sum(z)
70
71 alpha.tune=.1
72 beta.tune=2
73 s.tune=.1
74
75 ####
76 ####  Begin MCMC Loop
77 ####
78
79 for(k in 1:n.mcmc){
80
81   ####
82   ####  Sample psi
83   ####
```

(Continued)

BOX 27.1 (Continued)   MCMC ALGORITHM FOR SCR MODEL

```
84   psi=rbeta(1,sum(z)+g.1,sum(1-z)+g.2)

85
86   ####
87   ####   Sample alpha
88   ####

89
90   alpha.star=rnorm(1,alpha,alpha.tune)
91   P.star=cloglog.inv(alpha.star+beta*D2)
92   mh1=sum(dbinom(Y[z==1,],J,P.star[z==1,],
93      log=TRUE))+dnorm(alpha.star,mu.alpha,
94      sqrt(s2.alpha),log=TRUE)
95   mh2=sum(dbinom(Y[z==1,],J,P[z==1,],log=TRUE))
96      +dnorm(alpha,mu.alpha,sqrt(s2.alpha),
97      log=TRUE)
98   mh=exp(mh1-mh2)
99   if(mh > runif(1)){
100     alpha=alpha.star
101     P=P.star
102  }

103
104  ####
105  ####   Sample beta
106  ####

107
108  beta.star=rnorm(1,beta,beta.tune)
109  P.star=cloglog.inv(alpha+beta.star*D2)
110  mh1=sum(dbinom(Y[z==1,],J,P.star[z==1,],
111     log=TRUE))+dnorm(beta.star,mu.beta,
112     sqrt(s2.beta),log=TRUE)
113  mh2=sum(dbinom(Y[z==1,],J,P[z==1,],log=TRUE))
114     +dnorm(beta,mu.beta,sqrt(s2.beta),log=TRUE)
115  mh=exp(mh1-mh2)
116  if(mh > runif(1)){
117     beta=beta.star
118     P=P.star
119  }
```

(Continued)

BOX 27.1 (Continued)   MCMC ALGORITHM FOR SCR MODEL

```
120  ####
121  ####  Sample z
122  ####
123
124  tmp.prod=psi*apply((1-P)^J,1,prod)
125  psi.tmp=tmp.prod/(tmp.prod+1-psi)
126  z[idx.0]=rbinom(sum(idx.0),1,psi.tmp[idx.0])
127  N=sum(z)
128
129  ####
130  ####  Sample s
131  ####
132
133  for(i in 1:M){
134    s.star=rnorm(2,S[i,],s.tune)
135    if((s.star[1] >  x.lims[1]) & (s.star[1]
136                   < x.lims[2]) & (s.star[2]
137                   > y.lims[1]) & (s.star[2]
138                   < y.lims[2])){
139      d.star=apply((s.star-t(X))^2,2,sum)
140      p.star=cloglog.inv(alpha+beta*d.star)
141      mh1=z[i]*sum(dbinom(Y[i,],J,p.star,
142        log=TRUE))
143      mh2=z[i]*sum(dbinom(Y[i,],J,P[i,],log=TRUE))
144      mh=exp(mh1-mh2)
145      if(mh > runif(1)){
146        S[i,]=s.star
147        P[i,]=p.star
148      }
149    }
150  }
151
152  D2=crossdist(S,X)^2
153
154  ####
155  ####  Save Samples
156  ####
```

(Continued)

BOX 27.1 (Continued)   MCMC ALGORITHM FOR SCR MODEL

```
157  alpha.save[k]=alpha
158  beta.save[k]=beta
159  psi.save[k]=psi
160  N.save[k]=N
161  D.save[k]=N/A
162  S1.save[k,]=S[19,]
163  S2.save[k,]=S[36,]
164
165  if(k > n.burn){
166     z.mean=z.mean+z/(n.mcmc-n.burn)
167     S.mean=S.mean+S/(n.mcmc-n.burn)
168  }
169
170  }
171
172  ####
173  ####   Write Output
174  ####
175
176  list(n.mcmc=n.mcmc,alpha.save=alpha.save,
177       beta.save=beta.save,psi.save=psi.save,
178       z.mean=z.mean,S.mean=S.mean,N.save=N.save,
179       D.save=D.save,S1.save=S1.save,S2.save=S2.save)
180
181  }
```

In this R code, notice that we created the function crossdist to efficiently calculate the squared distances between all activity centers s_i for $i = 1, \ldots, M$ and the detector locations x_l for $l = 1, \ldots, L$. Also, we denote the area of the support $\mathcal{S}$ as A in the R code.

To better understand the SCR process, it is illustrative to simulate data. We used a known detector array and randomly distributed individual activity centers to simulate SCR data. The following R code simulates the distribution of $N = 100$ individual activity centers and their encounters with an array of $L = 100$ detectors over a set of $J = 5$ sampling occasions.

BOX 27.2   SIMULATE SCR DATA

```
1  N=100
2  L=100
3  J=5
4  alpha=-2
5  beta=-40
6
7  cloglog.inv <- function(x){
8    1-exp(-exp(x))
9  }
10
11 x1=seq(-.5,.5,,sqrt(L))
12 x2=seq(-.5,.5,,sqrt(L))
13 x.mat=expand.grid(x1,x2)
14
15 set.seed(10)
16 s.mat=matrix(runif(N*2,-1,1),N,2)
17 p.mat=matrix(0,N,L)
18 y.mat=matrix(0,N,L)
19 for(i in 1:N){
20   for(l in 1:L){
21     p.mat[i,l]=cloglog.inv(alpha+beta*
22        sum((s.mat[i,]-x.mat[l,])^2))
23     y.mat[i,l]=rbinom(1,J,p.mat[i,l])
24   }
25 }
26
27 idx.0=(apply(y.mat,1,sum)==0)
28 n=sum(!idx.0)
29 S.tmp=s.mat[!idx.0,]
30 S.avg=matrix(0,n,2)
31
32 for(i in 1:n){
33   tmp.idx=(y.mat[!idx.0,][i,]>0)
34   S.avg[i,]=apply(as.matrix(x.mat[!idx.0,]
35      [tmp.idx,]),2,mean)
36 }
37
38 plot(s.mat,xlim=c(-1,1),ylim=c(-1,1),asp=TRUE,
39      xlab="",ylab="",type="n")
40 symbols(0,0,squares=1,add=TRUE,inches=FALSE)
```

(Continued)

BOX 27.2 (Continued)   SIMULATE SCR DATA

```
41 symbols(0,0,squares=2,lty=2,add=TRUE,inches=FALSE)
42 text(S.tmp,labels=1:n,cex=1,pos=1)
43 points(x.mat,lwd=1.5,pch=4,col=8,cex=.75)
44 points(s.mat,pch=20,cex=.4,col=rgb(0,0,0,.8))
45 points(S.tmp,pch=20,cex=1,col=rgb(0,0,0,.8))
```

The simulated activity centers for $n = 47$ out of $N = 100$ total individuals and
detector array are shown in Figure 27.1. Notice that the set of detected individuals are closer in general to the detector array whereas the individuals (small points)
that were not detected were farther away from the detector array as we would expect
(Figure 27.1). We used the following R code to prepare the simulated data for analysis

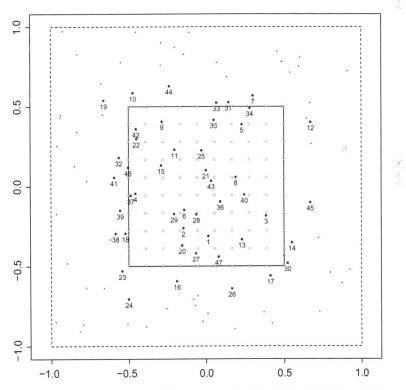

Figure 27.1   Simulated individual activity centers $\mathbf{s}_i$ for all individuals $i = 1, \ldots, N$ shown as
points, with larger points and numbers indicating the $n = 47$ observed individuals and smaller
points indicating those $N - n$ individuals not observed. The detector array (shown as gray "×"
symbols inside the smaller square region) within the study area $\mathscr{S}$ indicated by the dashed
square.

by augmenting the encounter histories for detected individuals with a set of 500 all zero encounter histories to comprise the superpopulation:

BOX 27.3   PREPARE SCR DATA FOR ANALYSIS

```
1  y.idx=(apply(y.mat,1,sum)>0)
2  Y=y.mat[y.idx,]
3  S=s.mat[y.idx,]
4  n=dim(Y)[1]
5  M=n+500
6  Y.aug=matrix(0,M,L)
7  Y.aug[1:n,]=Y
8
9  table(apply(Y>0,1,sum))
10 table(Y)
```

In this particular set of simulated data, this R code indicates that, among the $n = 47$ observed individuals, 14 were detected by only one detector, 11 by 2 detectors, 11 by 3 detectors, 6 by 4 detectors, 3 by 5 detectors, and 1 individual detected by 7 different detectors. Although several observed individuals encountered more than one detector, only 13 encountered the same detector more than once.

Using $K = 20000$ MCMC iterations, we fit the Bayesian hierarchical SCR model to the capture-recapture data we simulated using this MCMC algorithm and the following R code:

BOX 27.4   FIT SCR MODEL TO DATA

```
1  source("scr.mcmc.R")
2  n.mcmc=20000
3  set.seed(1)
4  mcmc.out=scr.mcmc(Y=Y.aug,J=J,X=x.mat,n.mcmc=n.
     mcmc)
5
6  layout(matrix(1:4,2,2))
7  hist(mcmc.out$alpha.save,prob=TRUE,xlab
8      =bquote(alpha),main="a")
9  curve(dnorm(x,-2,1.5),lwd=2,add=TRUE)
10 abline(v=alpha,col=1,lwd=2,lty=2)
11 hist(mcmc.out$beta.save,prob=TRUE,xlab=bquote
12     (beta),main="b")
```

(Continued)

BOX 27.4 (Continued)   FIT SCR MODEL TO DATA

```
13 curve(dnorm(x,-40,10),lwd=2,add=TRUE)
14 abline(v=beta,col=1,lwd=2,lty=2)
15 hist(mcmc.out$N.save,breaks=seq(60,180,1),prob=
16     TRUE,main="c",xlab="N",ylab="Probability")
17 abline(v=N,col=1,lwd=2,lty=2)
18 hist(mcmc.out$D.save,breaks=seq(10,45,1),prob=
19     TRUE,main="d",xlab="D",ylab="Probability")
20 abline(v=N/4,col=1,lwd=2,lty=2)
```

This R code results in the marginal posterior histograms for α, β, abundance N, and density D shown in Figure 27.2. Notice in Figure 27.2 that we withhold the posterior histogram for the population membership probability ψ because it is implicit in the posterior histogram for abundance N. Also, notice that the posterior histograms for all parameters capture the true values used to simulate this set of data (Figure 27.2).

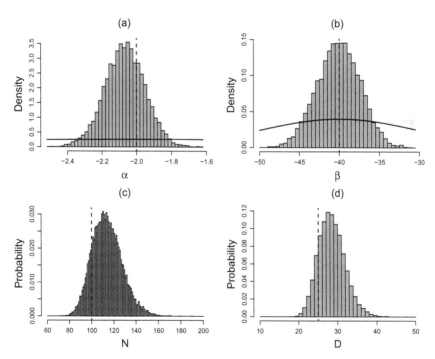

Figure 27.2 Marginal posterior histograms from the spatial capture-recapture model for (a) α, (b) β, (c) abundance N, and (d) density D. Priors for α and β are shown as black lines. True values used in simulation are shown as dashed vertical lines.

Note that the posterior distribution for D is actually a PMF because it is a derived quantity that depends on N directly and N is a discrete random variable.

To infer the effective range of detectability for individuals in our simulated population, we can calculate the posterior mean and 95% credible interval of 2σ (the distance at which the detection function drops to 95% of its maximum value). We used the following R code to compute these posterior quantities:

BOX 27.5   POSTERIOR EFFECTIVE RANGE

```
> mean(2*sqrt(-1/mcmc.out$beta.save))
[1] 0.3163047
> quantile(2*sqrt(-1/mcmc.out$beta.save),
    c(0.025,0.975))
     2.5%       97.5%
0.2969420 0.3382131
```

Thus, the maximum distance from an individual activity center at which we would expect detections of an individual to occur is approximately $E(2\sigma|\mathbf{Y}) \approx 0.32$ spatial units in our simulated example, which is approximately 11% of the maximum distance in our study area $\mathscr{S}$ (with 95% credible interval $(0.30, 0.34)$). Similarly, we can infer the baseline detection rate λ when the individual activity center is co-located with the detector location using the R code:

BOX 27.6   POSTERIOR BASELINE DETECTION RATE

```
> mean(exp(mcmc.out$alpha.save))
[1] 0.1272513
> quantile(exp(mcmc.out$alpha.save),
    c(0.025,0.975))
     2.5%       97.5%
0.09990029 0.15828149
```

These results imply that the posterior mean baseline detection probability is $E(\lambda|\mathbf{Y}) \approx 0.13$ (with 95% credible interval $(0.10, 0.16)$) in our simulated data example which suggests that our simulated species is somewhat difficult to detect in general.

Finally, in addition to inference for the population abundance, which we estimated with a 95% credible interval $(93, 143)$, we can obtain inference for the individual activity centers. For example, the R code produces Figure 27.3, which shows the posterior draws (MCMC realizations) for $\mathbf{s}_{19}$ and $\mathbf{s}_{36}$, two of the detected individuals in our simulated data.

BOX 27.7 POSTERIOR ACTIVITY CENTERS

```
1 pt.col=rgb(0,0,0,.1)
2 pt.seq=seq(1,n.mcmc,,1000)
3 plot(S,xlim=c(-1,1),ylim=c(-1,1),asp=TRUE,
4     type="n")
5 points(mcmc.out$S1.save[pt.seq,],bg=pt.col,col=0,
6     cex=1,pch=21)
7 points(mcmc.out$S2.save[pt.seq,],bg=pt.col,col=0,
8     cex=1,pch=21)
9 points(x.mat,lwd=1.5,pch=4,col=8,cex=.75)
10 points(S[c(19,36),],pch=20,lwd=4,cex=1.5,col=1)
11 symbols(0,0,squares=1,add=TRUE,inches=FALSE)
12 symbols(0,0,squares=2,lty=2,add=TRUE,inches=FALSE)
```

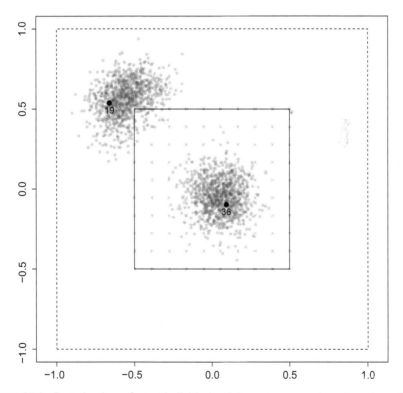

Figure 27.3 Posterior draws for two individual activity centers s_{19} and s_{36}. Locations of the true simulated activity centers are shown as black points.

The two sets of realizations indicate that we are able to learn about the activity centers for observed individuals in the study even though we only recorded a single detection for each of these individuals shown in Figure 27.3.

27.1   ADDITIONAL CONCEPTS AND READING

Spatially explicit capture-recapture models similar to the form that we presented in this chapter date back to Efford (2004), Borchers and Efford (2008), and Royle and Young (2008). Since those initial developments, a variety of studies have used SCR models, which now extend to accommodate additional ecological processes and auxiliary data sources. In fact, efforts have been made to unify SCR models with distance sampling and related methods (Borchers et al. 2015; Glennie 2018).

A natural SCR extension accounts for the uneven space use of individuals within a study area due to heterogeneity in habitat preference (e.g., Royle et al. 2013b, 2018; Proffitt et al. 2015). One approach to account for habitat preference in the SCR that we presented in this chapter is to generalize the conditional model for the activity centers. We can borrow the point process modeling approaches that we discussed in Chapter 21 to account for a higher probability of activity centers occurring in certain regions of the study area. Thus, instead of a uniform prior for s_i, we could use a weighted distribution that may depend on a set of spatially referenced covariates and associated coefficients. Furthermore, several model extensions accommodate telemetry data to help account for resource selection and animal movement in SCR and distance sampling models (e.g., Linden et al. 2018; Glennie et al. 2018).

Several alternatives to SCR models that use spatial information in a capture-recapture setting have been proposed. For example, Ivan et al. (2013) developed an approach that uses telemetry data to correct more traditional abundance estimators for movement. Royle et al. (2016) specified a model that allowed for activity centers to change over time, perhaps due to transience or dispersal. Building on that concept, Glennie (2018) developed a model-based approach to account for movement in SCR models using fast approximations to dynamic movement models embedded in the SCR framework.

28 Spatio-temporal Models

28.1 MULTIVARIATE TIME SERIES MODELS

Although every ecological process involves changes in space and time, traditional ecological statistical models often ignore explicit space and/or time dependence. More recently, attempts have been made to account for space and/or time dependence in ecological data using separable components that deal with space and time independently, but often do not incorporate explicit spatio-temporal dynamics.[1] This omission is probably due in part to the fact that many forms of historic ecological data were collected on relatively small spatial domains and limited to short-term temporal durations due to funding and field data collection constraints. Thus, for decades, ecologists have focused on models for temporal or spatial data separately, but not often jointly to capture important spatio-temporal interactions (Wikle and Hooten 2010).

By contrast, formal dynamical system models have been a strong research theme among atmospheric scientists because, for example, weather and climate systems are naturally modeled by considering spatio-temporal interactions related to fluid and gas dynamics (Cressie and Wikle 2011). Similarly, applied mathematicians and mathematical biologists have long used partial differential equations to model biological and ecological systems, but they are not always considered from a statistical perspective (e.g., Lewis et al. 2016).

Fortunately, the hierarchical Bayesian modeling framework naturally accommodates dynamic mechanisms and has become more popular recently for modeling ecological processes (Hobbs and Hooten 2015). In this chapter, we introduce spatio-temporal statistical models by extending the autoregressive time series models from Chapter 16 to a multivariate setting to illustrate a phenomenological approach. Then we demonstrate how to consider more mechanistic model formulations of spatio-temporal models using ecological diffusion as an example.

As a motivating example, we consider the case study presented by Hooten and Wikle (2007) who developed a hierarchical spatio-temporal model to better understand the growth dynamics in shortleaf pine (*Pinus echinata*) forests in Missouri and Arkansas, USA. They considered standardized tree ring width data from dendrochronologies representing a set of spatial locations and then modeled the underlying dynamical process on a lower-dimensional state-space.[2] For illustration, we consider a simplified spatio-temporal model formulation similar to that developed by Hooten and Wikle (2007) where the observations are denoted as $\mathbf{y}_t$ for $t = 1, \ldots, T$ (where $\mathbf{y}_t$

---

[1] An example of a separable spatio-temporal statistical model would include separate AR(1) specifications for a set of sites to accommodate temporal structure, but with spatially correlated errors.

[2] As we discussed previously, hierarchical time series models have been referred to historically as state-space models.

is a $n \times 1$ vector of data observed at time t). In a hierarchical modeling framework, we assume that the ecological process $\mathbf{z}_t$ (where $\mathbf{z}_t$ is $m \times 1$ dimensional) is latent and governed by dynamics that are well-described using a first-order Markov assumption similar to the AR(1) time series models in Chapter 16. We can generalize this type of model specification to accommodate more complicated dependence structures, but it serves as a good starting place.

Assuming we can model the data with a conditional multivariate normal distribution, we specify a hierarchical spatio-temporal model as

$$\mathbf{y}_t \sim \mathrm{N}(\mathbf{K}_t \mathbf{z}_t, \mathbf{\Sigma}_y), \qquad (28.1)$$

$$\mathbf{z}_t \sim \mathrm{N}(\mathbf{A}_t \mathbf{z}_{t-1}, \mathbf{\Sigma}_z), \qquad (28.2)$$

where $\mathbf{y}_t \equiv (y_{1,t}, \ldots, y_{n,t})'$, $\mathbf{z}_t \equiv (z_{1,t}, \ldots, z_{m,t})'$, $\mathbf{K}_t$ represents a $n \times m$ mapping matrix that associates elements of the data vectors $\mathbf{y}_t$ with appropriate elements of the process vectors $\mathbf{z}_t$. Typically $\mathbf{K}_t$ is a binary matrix with ones indicating places where the process was observed. The matrix $\mathbf{A}_t$ is a $m \times m$ propagator (or transition) matrix that controls the dynamics in the underlying process and could contain m^2 parameters. As we saw in the section on Animal Movement Models in Chapter 16, we can parameterize $\mathbf{A}_t$ to have a specific physical interpretation (in that case, based on the concept of rotation), which can substantially reduce the number of parameters. Alternatively, we could restrict the dynamics to be homogeneous over time ($\mathbf{A}_t = \mathbf{A}, \forall t$), which would reduce the parameter space substantially.

Three phenomenological specifications for the propagator matrix are (1) $\mathbf{A}$ fully unknown, (2) $\mathbf{A} \equiv \mathrm{diag}(\boldsymbol{\alpha}_{\mathrm{diag}})$, where $\boldsymbol{\alpha}_{\mathrm{diag}}$ is a $m \times 1$ vector along the diagonal of a $m \times m$ matrix with zeros elsewhere, and (3) $\mathbf{A} \equiv \alpha \mathbf{I}$, with scalar autocorrelation parameter α. Formulation (1) for $\mathbf{A}$, with all elements in the propagator matrix unknown, can accommodate complicated dynamics with interactions among spatial locations over time. It is the most general of the three options, but also contains m^2 unknown parameters controlling the dynamics.

As a simple example of a hierarchical Bayesian spatio-temporal model, we complete the specification based on the data and process models in (28.1)–(28.2). We let $\mathbf{\Sigma}_y \equiv \sigma_y^2 \mathbf{I}$, $\mathbf{\Sigma}_z \equiv \sigma_z^2 \mathbf{I}$, and $\boldsymbol{\alpha} \equiv \mathrm{vec}(\mathbf{A})$, where "vec" is an operator that vectorizes the matrix $\mathbf{A}$ columnwise such that $\boldsymbol{\alpha}$ is $m^2 \times 1$. In addition to the following priors for model parameters

$$\boldsymbol{\alpha} \sim \mathrm{N}(\boldsymbol{\mu}_\alpha, \mathbf{\Sigma}_\alpha), \qquad (28.3)$$

$$\sigma_y^2 \sim \mathrm{IG}(q_y, r_y), \qquad (28.4)$$

$$\sigma_z^2 \sim \mathrm{IG}(q_z, r_z), \qquad (28.5)$$

we also specify a prior for the initial state vector $\mathbf{z}_0$ such that $\mathbf{z}_0 \sim \mathrm{N}(\boldsymbol{\mu}_0, \mathbf{\Sigma}_0)$. Finally, we assume that the mapping between the data $\mathbf{y}_t$ and $\mathbf{z}_t$ is homogeneous over time which implies that $\mathbf{K}_t = \mathbf{K}, \forall t$, where $\mathbf{K}$ is usually assumed to be known as part of the study design.

This Bayesian model specification results in the following posterior distribution

$$[\mathbf{Z}, \boldsymbol{\alpha}, \sigma_y^2, \sigma_z^2 | \mathbf{Y}] \propto \left(\prod_{t=1}^{T} [\mathbf{y}_t | \mathbf{z}_t, \sigma_y^2][\mathbf{z}_t | \mathbf{z}_{t-1}, \boldsymbol{\alpha}, \sigma_z^2] \right) [\mathbf{z}_0][\boldsymbol{\alpha}][\sigma_y^2][\sigma_z^2], \quad (28.6)$$

which takes a similar form as the time series models that we described in Chapter 16, but with additional components that account for measurement error.

To construct a MCMC algorithm, we need to find the full-conditional distributions for the model parameters and latent state vectors so that we can identify appropriate strategies for the updates. Remarkably, in what follows, we show that all unknown quantities in this Bayesian spatio-temporal model have conjugate full-conditional distributions.

Beginning with the variance parameters, the full-conditional distribution for σ_y^2 is

$$[\sigma_y^2 | \cdot] \propto \left(\prod_{t=1}^{T} [\mathbf{y}_t | \mathbf{z}_t, \sigma_y^2] \right) [\sigma_y^2], \quad (28.7)$$

$$\propto \left(\prod_{t=1}^{T} |\sigma_y^2 \mathbf{I}|^{-\frac{1}{2}} \exp \left(-\frac{1}{2} (\mathbf{y}_t - \mathbf{K}\mathbf{z}_t)'(\sigma_y^2 \mathbf{I})^{-1}(\mathbf{y}_t - \mathbf{K}\mathbf{z}_t) \right) \right) (\sigma_y^2)^{-(q_y+1)} \times$$

$$\exp \left(-\frac{1}{\sigma_y^2 r_y} \right), \quad (28.8)$$

$$\propto (\sigma_y^2)^{-\left(\frac{nT}{2} + q_y + 1 \right)} \exp \left(-\frac{1}{\sigma_y^2} \left(\frac{\sum_{t=1}^{T} (\mathbf{y}_t - \mathbf{K}\mathbf{z}_t)'(\mathbf{y}_t - \mathbf{K}\mathbf{z}_t)}{2} + \frac{1}{r_y} \right) \right), \quad (28.9)$$

$$\propto (\sigma_y^2)^{-(\tilde{q}_y+1)} \exp \left(-\frac{1}{\sigma_y^2 \tilde{r}_y} \right), \quad (28.10)$$

which implies that $[\sigma_y^2 | \cdot] = \text{IG}(\tilde{q}_y, \tilde{r}_y)$ where

$$\tilde{q} \equiv \frac{nT}{2} + q_y, \quad (28.11)$$

$$\tilde{r} \equiv \left(\frac{\sum_{t=1}^{T} (\mathbf{y}_t - \mathbf{K}\mathbf{z}_t)'(\mathbf{y}_t - \mathbf{K}\mathbf{z}_t)}{2} + \frac{1}{r_y} \right)^{-1}. \quad (28.12)$$

Thus, we can sample σ_y^2 using a Gibbs update in our MCMC algorithm.

Similarly, the full-conditional distribution for σ_z^2 is

$$[\sigma_z^2 | \cdot] \propto \left(\prod_{t=1}^{T} [\mathbf{z}_t | \mathbf{z}_{t-1}, \boldsymbol{\alpha}, \sigma_z^2] \right) [\sigma_z^2], \quad (28.13)$$

$$\propto \left(\prod_{t=1}^{T} |\sigma_z^2 \mathbf{I}|^{-\frac{1}{2}} \exp \left(-\frac{1}{2} (\mathbf{z}_t - \mathbf{A}\mathbf{z}_{t-1})'(\sigma_z^2 \mathbf{I})^{-1}(\mathbf{z}_t - \mathbf{A}\mathbf{z}_{t-1}) \right) \right) \times$$

$$(\sigma_z^2)^{-(q_z+1)} \exp \left(-\frac{1}{\sigma_z^2 r_z} \right), \quad (28.14)$$

$$\propto (\sigma_z^2)^{-\left(\frac{mT}{2}+q_z+1\right)} \exp\left(-\frac{1}{\sigma_z^2}\left(\frac{\sum_{t=1}^{T}(\mathbf{z}_t-\mathbf{A}\mathbf{z}_{t-1})'(\mathbf{z}_t-\mathbf{A}\mathbf{z}_{t-1})}{2}+\frac{1}{r_z}\right)\right), \quad (28.15)$$

$$\propto (\sigma_z^2)^{-(\tilde{q}_z+1)} \exp\left(-\frac{1}{\sigma_y^2 \tilde{r}_z}\right), \quad (28.16)$$

which implies that $[\sigma_z^2|\cdot] = \text{IG}(\tilde{q}_z, \tilde{r}_z)$ where

$$\tilde{q} \equiv \frac{mT}{2} + q_z, \quad (28.17)$$

$$\tilde{r} \equiv \left(\frac{\sum_{t=1}^{T}(\mathbf{z}_t-\mathbf{A}\mathbf{z}_{t-1})'(\mathbf{z}_t-\mathbf{A}\mathbf{z}_{t-1})}{2}+\frac{1}{r_z}\right)^{-1}. \quad (28.18)$$

where we recall that $\mathbf{A}$ is a square matrix form of the vector of parameters controlling the dynamics ($\boldsymbol{\alpha}$). Thus, as with σ_y^2, we can sample σ_z^2 using a Gibbs update in our MCMC algorithm.

The full-conditional distribution for the parameter vector $\boldsymbol{\alpha}$ is

$$[\boldsymbol{\alpha}|\cdot] \propto \left(\prod_{t=1}^{T}[\mathbf{z}_t|\mathbf{z}_{t-1},\boldsymbol{\alpha},\sigma_z^2]\right)[\boldsymbol{\alpha}], \quad (28.19)$$

$$\propto \left(\prod_{t=1}^{T}\exp\left(-\frac{1}{2}(\mathbf{z}_t-\mathbf{A}\mathbf{z}_{t-1})'(\sigma_z^2\mathbf{I})^{-1}(\mathbf{z}_t-\mathbf{A}\mathbf{z}_{t-1})\right)\right) \times$$

$$\exp\left(-\frac{1}{2}(\boldsymbol{\alpha}-\boldsymbol{\mu}_\alpha)'\boldsymbol{\Sigma}_\alpha^{-1}(\boldsymbol{\alpha}-\boldsymbol{\mu}_\alpha)\right), \quad (28.20)$$

which does not appear to be analytically tractable because the parameters are in matrix form ($\mathbf{A}$) in the likelihood and vectorized ($\boldsymbol{\alpha}$) in the prior. However, by reparameterizing the process model as

$$\text{vec}(\mathbf{Z}_2) \sim \text{N}(\text{vec}(\mathbf{A}\mathbf{Z}_1), \mathbf{I}_{T\times T} \otimes \sigma_z^2\mathbf{I}), \quad (28.21)$$

where the covariance matrix $\mathbf{I}_{T\times T} \otimes \sigma_z^2\mathbf{I}$ is an $mT \times mT$ Kronecker product and the matrices $\mathbf{Z}_1$ and $\mathbf{Z}_2$ are both $m \times T$ dimensional and defined as

$$\mathbf{Z}_1 \equiv (\mathbf{z}_0, \mathbf{z}_1, \ldots, \mathbf{z}_{T-1}), \quad (28.22)$$

$$\mathbf{Z}_2 \equiv (\mathbf{z}_1, \mathbf{z}_2, \ldots, \mathbf{z}_T), \quad (28.23)$$

we can derive the full-conditional distribution of $\boldsymbol{\alpha}$. Using a vectorization result (Seber 2008), we can rewrite the mean term of the reparameterized process model as

$$\text{vec}(\mathbf{A}\mathbf{Z}_1) = (\mathbf{Z}_1' \otimes \mathbf{I}_{m\times m})\text{vec}(\mathbf{A}), \quad (28.24)$$

$$= (\mathbf{Z}_1' \otimes \mathbf{I}_{m\times m})\boldsymbol{\alpha}. \quad (28.25)$$

Thus, we use the vectorization result in (28.25) to rewrite the full-conditional for $\boldsymbol{\alpha}$ as

$$[\boldsymbol{\alpha}|\cdot] \propto [\mathbf{Z}_2|\mathbf{Z}_1, \boldsymbol{\alpha}, \sigma_z^2][\boldsymbol{\alpha}], \tag{28.26}$$

$$\propto \exp\left(-\frac{1}{2}(\mathrm{vec}(\mathbf{Z}_2) - (\mathbf{Z}_1' \otimes \mathbf{I}_{m \times m})\boldsymbol{\alpha})'(\mathbf{I}_{T \times T} \otimes \sigma_z^2\mathbf{I})^{-1}(\mathrm{vec}(\mathbf{Z}_2)\right.$$

$$\left.-(\mathbf{Z}_1' \otimes \mathbf{I}_{m \times m})\boldsymbol{\alpha})\right) \times \tag{28.27}$$

$$\exp\left(-\frac{1}{2}(\boldsymbol{\alpha} - \boldsymbol{\mu}_\alpha)'\boldsymbol{\Sigma}_\alpha^{-1}(\boldsymbol{\alpha} - \boldsymbol{\mu}_\alpha)\right), \tag{28.28}$$

$$\propto \exp\left(-\frac{1}{2}(-2\mathrm{vec}(\mathbf{Z}_2)'(\mathbf{I}_{T \times T} \otimes \sigma_z^2\mathbf{I})^{-1}(\mathbf{Z}_1' \otimes \mathbf{I}_{m \times m})\boldsymbol{\alpha}+\right. \tag{28.29}$$

$$\boldsymbol{\alpha}'(\mathbf{Z}_1' \otimes \mathbf{I}_{m \times m})'(\mathbf{I}_{T \times T} \otimes \sigma_z^2\mathbf{I})^{-1}(\mathbf{Z}_1' \otimes \mathbf{I}_{m \times m})\boldsymbol{\alpha} - 2\boldsymbol{\mu}_\alpha'\boldsymbol{\Sigma}_\alpha^{-1}\boldsymbol{\mu}_\alpha$$

$$\left.+\boldsymbol{\alpha}'\boldsymbol{\Sigma}_\alpha^{-1}\boldsymbol{\alpha}\right), \tag{28.30}$$

$$\propto \exp\left(-\frac{1}{2}(-2(\mathrm{vec}(\mathbf{Z}_2)'(\mathbf{I}_{T \times T} \otimes \sigma_z^2\mathbf{I})^{-1}(\mathbf{Z}_1' \otimes \mathbf{I}_{m \times m}) + \boldsymbol{\mu}_\alpha'\boldsymbol{\Sigma}_\alpha^{-1})\boldsymbol{\alpha}+ \tag{28.31}$$

$$\boldsymbol{\alpha}'((\mathbf{Z}_1' \otimes \mathbf{I}_{m \times m})'(\mathbf{I}_{T \times T} \otimes \sigma_z^2\mathbf{I})^{-1}(\mathbf{Z}_1' \otimes \mathbf{I}_{m \times m}) + \boldsymbol{\Sigma}_\alpha^{-1})\boldsymbol{\alpha}\right), \tag{28.32}$$

$$\propto \exp\left(-\frac{1}{2}(-2\mathbf{b}'\boldsymbol{\alpha} + \boldsymbol{\alpha}'\mathbf{B}\boldsymbol{\alpha})\right), \tag{28.33}$$

which implies that the full-conditional distribution for $\boldsymbol{\alpha}$ is $[\boldsymbol{\alpha}|\cdot] = \mathrm{N}(\mathbf{B}^{-1}\mathbf{b}, \mathbf{B}^{-1})$, where $\mathbf{b}$ and $\mathbf{B}$ are defined as

$$\mathbf{b} \equiv (\mathbf{Z}_1' \otimes \mathbf{I}_{m \times m})'(\mathbf{I}_{T \times T} \otimes \sigma_z^2\mathbf{I})^{-1}\mathrm{vec}(\mathbf{Z}_2) + \boldsymbol{\Sigma}_\alpha^{-1}\boldsymbol{\mu}_\alpha, \tag{28.34}$$

$$\mathbf{B} \equiv (\mathbf{Z}_1' \otimes \mathbf{I}_{m \times m})'(\mathbf{I}_{T \times T} \otimes \sigma_z^2\mathbf{I})^{-1}(\mathbf{Z}_1' \otimes \mathbf{I}_{m \times m}) + \boldsymbol{\Sigma}_\alpha^{-1}. \tag{28.35}$$

Thus, we can use a Gibbs update for $\boldsymbol{\alpha}$ in our MCMC algorithm.

The remaining elements in the posterior distribution (28.6) that we need to sample in our MCMC algorithm are the state vectors $\mathbf{z}_t$ for $t = 0, \ldots, T$. Starting with the initial state, the full-conditional distribution for $\mathbf{z}_0$ is

$$[\mathbf{z}_0|\cdot] \propto [\mathbf{z}_1|\mathbf{z}_0, \boldsymbol{\alpha}, \sigma_z^2][\mathbf{z}_0], \tag{28.36}$$

$$\propto \exp\left(-\frac{1}{2}(\mathbf{z}_1 - \mathbf{A}\mathbf{z}_0)'(\sigma_z^2\mathbf{I})^{-1}(\mathbf{z}_1 - \mathbf{A}\mathbf{z}_0)\right)$$

$$\exp\left(-\frac{1}{2}(\mathbf{z}_0 - \boldsymbol{\mu}_0)'\boldsymbol{\Sigma}_0^{-1}(\mathbf{z}_0 - \boldsymbol{\mu}_0)\right), \tag{28.37}$$

$$\propto \exp\left(-\frac{1}{2}(-2\mathbf{z}_1'(\sigma_z^2\mathbf{I})^{-1}\mathbf{A}\mathbf{z}_0 + \mathbf{z}_0'\mathbf{A}'(\sigma_z^2\mathbf{I})^{-1}\mathbf{A}\mathbf{z}_0\right.$$

$$\left. -2\boldsymbol{\mu}_0'\boldsymbol{\Sigma}_0^{-1}\mathbf{z}_0 + \mathbf{z}_0'\boldsymbol{\Sigma}_0^{-1}\mathbf{z}_0)\right), \tag{28.38}$$

$$\propto \exp\left(-\frac{1}{2}(-2(\mathbf{z}_1'(\sigma_z^2\mathbf{I})^{-1}\mathbf{A} + \boldsymbol{\mu}_0'\boldsymbol{\Sigma}_0^{-1})\mathbf{z}_0 + \mathbf{z}_0'(\mathbf{A}'(\sigma_z^2\mathbf{I})^{-1}\mathbf{A} + \boldsymbol{\Sigma}_0^{-1})\mathbf{z}_0)\right), \tag{28.39}$$

$$\propto \exp\left(-\frac{1}{2}(-2\mathbf{b}'\mathbf{z}_0 + \mathbf{z}_0'\mathbf{B}\mathbf{z}_0)\right), \tag{28.40}$$

which implies that $\mathbf{z}_0$ has the conjugate full-conditional distribution $[\mathbf{z}_0|\cdot] = \mathrm{N}(\mathbf{B}^{-1}\mathbf{b}, \mathbf{B}^{-1})$ where

$$\mathbf{b} \equiv \mathbf{A}'(\sigma_z^2\mathbf{I})^{-1}\mathbf{z}_1 + \boldsymbol{\Sigma}_0^{-1}\boldsymbol{\mu}_0, \tag{28.41}$$

$$\mathbf{B} \equiv \mathbf{A}'(\sigma_z^2\mathbf{I})^{-1}\mathbf{A} + \boldsymbol{\Sigma}_0^{-1}. \tag{28.42}$$

Thus, we can use a Gibbs update for the initial state $\mathbf{z}_0$ in our MCMC algorithm.

For $t = 1, \ldots, T-1$, the full-conditional distribution for the state vector $\mathbf{z}_t$ is

$$[\mathbf{z}_t|\cdot] \propto [\mathbf{y}_t|\mathbf{z}_t, \sigma_y^2][\mathbf{z}_{t+1}|\mathbf{z}_t, \boldsymbol{\alpha}, \sigma_z^2][\mathbf{z}_t|\mathbf{z}_{t-1}, \boldsymbol{\alpha}, \sigma_z^2], \tag{28.43}$$

$$\propto \exp\left(-\frac{1}{2}(\mathbf{y}_t - \mathbf{K}\mathbf{z}_t)'(\sigma_y^2\mathbf{I})^{-1}(\mathbf{y}_t - \mathbf{K}\mathbf{z}_t)\right) \times \tag{28.44}$$

$$\exp\left(-\frac{1}{2}(\mathbf{z}_{t+1} - \mathbf{A}\mathbf{z}_t)'(\sigma_z^2\mathbf{I})^{-1}(\mathbf{z}_{t+1} - \mathbf{A}\mathbf{z}_t)\right) \times \tag{28.45}$$

$$\exp\left(-\frac{1}{2}(\mathbf{z}_t - \mathbf{A}\mathbf{z}_{t-1})'(\sigma_z^2\mathbf{I})^{-1}(\mathbf{z}_t - \mathbf{A}\mathbf{z}_{t-1})\right), \tag{28.46}$$

$$\propto \exp\left(-\frac{1}{2}(-2\mathbf{y}_t'(\sigma_y^2\mathbf{I})^{-1}\mathbf{K}\mathbf{z}_t + \mathbf{z}_t'\mathbf{K}'(\sigma_y^2\mathbf{I})^{-1}\mathbf{K}\mathbf{z}_t - \right. \tag{28.47}$$

$$2\mathbf{z}_{t+1}'(\sigma_z^2\mathbf{I})^{-1}\mathbf{A}\mathbf{z}_t + \mathbf{z}_t'\mathbf{A}'(\sigma_z^2\mathbf{I})^{-1}\mathbf{A}\mathbf{z}_t - \tag{28.48}$$

$$\left. 2\mathbf{z}_{t-1}'\mathbf{A}'(\sigma_z^2\mathbf{I})^{-1}\mathbf{z}_t + \mathbf{z}_t'(\sigma_z^2\mathbf{I})^{-1}\mathbf{z}_t)\right), \tag{28.49}$$

$$\propto \exp\left(-\frac{1}{2}(-2(\mathbf{y}_t'(\sigma_y^2\mathbf{I})^{-1}\mathbf{K} + \mathbf{z}_{t+1}'(\sigma_z^2\mathbf{I})^{-1}\mathbf{A} + \mathbf{z}_{t-1}'\mathbf{A}'(\sigma_z^2\mathbf{I})^{-1})\mathbf{z}_t + \right. \tag{28.50}$$

$$\left. \mathbf{z}_t'(\mathbf{K}'(\sigma_y^2\mathbf{I})^{-1}\mathbf{K} + \mathbf{A}'(\sigma_z^2\mathbf{I})^{-1}\mathbf{A} + (\sigma_z^2\mathbf{I})^{-1})\mathbf{z}_t\right), \tag{28.51}$$

$$\propto \exp\left(-\frac{1}{2}(-2\mathbf{b}'\mathbf{z}_t + \mathbf{z}_t'\mathbf{B}\mathbf{z}_t)\right), \tag{28.52}$$

which implies that $\mathbf{z}_t$ has the conjugate full-conditional distribution $[\mathbf{z}_t|\cdot] = \mathrm{N}(\mathbf{B}^{-1}\mathbf{b}, \mathbf{B}^{-1})$ where

$$\mathbf{b} \equiv \mathbf{K}'(\sigma_y^2 \mathbf{I})^{-1}\mathbf{y}_t + \mathbf{A}'(\sigma_z^2 \mathbf{I})^{-1}\mathbf{z}_{t+1} + (\sigma_z^2 \mathbf{I})^{-1}\mathbf{A}\mathbf{z}_{t-1}, \qquad (28.53)$$

$$\mathbf{B} \equiv \mathbf{K}'(\sigma_y^2 \mathbf{I})^{-1}\mathbf{K} + \mathbf{A}'(\sigma_z^2 \mathbf{I})^{-1}\mathbf{A} + (\sigma_z^2 \mathbf{I})^{-1}. \qquad (28.54)$$

Thus, we can use a Gibbs update for the state $\mathbf{z}_t$ in our MCMC algorithm.

The full-conditional distribution for the final state vector $\mathbf{z}_T$ is

$$[\mathbf{z}_T|\cdot] \propto [\mathbf{y}_T|\mathbf{z}_T, \sigma_y^2][\mathbf{z}_T|\mathbf{z}_{T-1}, \boldsymbol{\alpha}, \sigma_z^2], \qquad (28.55)$$

$$\propto \exp\left(-\frac{1}{2}(\mathbf{y}_T - \mathbf{K}\mathbf{z}_T)'(\sigma_y^2 \mathbf{I})^{-1}(\mathbf{y}_T - \mathbf{K}\mathbf{z}_T)\right) \times \qquad (28.56)$$

$$\exp\left(-\frac{1}{2}(\mathbf{z}_T - \mathbf{A}\mathbf{z}_{T-1})'(\sigma_z^2 \mathbf{I})^{-1}(\mathbf{z}_T - \mathbf{A}\mathbf{z}_{T-1})\right), \qquad (28.57)$$

$$\propto \exp\left(-\frac{1}{2}(-2\mathbf{y}_T'(\sigma_y^2 \mathbf{I})^{-1}\mathbf{K}\mathbf{z}_T + \mathbf{z}_T'\mathbf{K}'(\sigma_y^2 \mathbf{I})^{-1}\mathbf{K}\mathbf{z}_T)-\right. \qquad (28.58)$$

$$\left. 2\mathbf{z}_{T-1}'\mathbf{A}'(\sigma_z^2 \mathbf{I})^{-1}\mathbf{z}_T + \mathbf{z}_T'(\sigma_z^2 \mathbf{I})^{-1}\mathbf{z}_T)\right), \qquad (28.59)$$

$$\propto \exp\left(-\frac{1}{2}(-2(\mathbf{y}_T'(\sigma_y^2 \mathbf{I})^{-1}\mathbf{K} + \mathbf{z}_{T-1}'\mathbf{A}'(\sigma_z^2 \mathbf{I})^{-1})\mathbf{z}_T+\right. \qquad (28.60)$$

$$\left. \mathbf{z}_T'(\mathbf{K}'(\sigma_y^2 \mathbf{I})^{-1}\mathbf{K} + (\sigma_z^2 \mathbf{I})^{-1})\mathbf{z}_T\right), \qquad (28.61)$$

$$\propto \exp\left(-\frac{1}{2}(-2\mathbf{b}'\mathbf{z}_T + \mathbf{z}_T'\mathbf{B}\mathbf{z}_T)\right), \qquad (28.62)$$

which implies that $\mathbf{z}_T$ has the conjugate full-conditional distribution $[\mathbf{z}_T|\cdot] = \mathrm{N}(\mathbf{B}^{-1}\mathbf{b}, \mathbf{B}^{-1})$ where

$$\mathbf{b} \equiv \mathbf{K}'(\sigma_y^2 \mathbf{I})^{-1}\mathbf{y}_T + (\sigma_z^2 \mathbf{I})^{-1}\mathbf{A}\mathbf{z}_{T-1}, \qquad (28.63)$$

$$\mathbf{B} \equiv \mathbf{K}'(\sigma_y^2 \mathbf{I})^{-1}\mathbf{K} + (\sigma_z^2 \mathbf{I})^{-1}. \qquad (28.64)$$

Thus, we can use a Gibbs update for the final state $\mathbf{z}_T$ in our MCMC algorithm.

Using this full-conditional distributions, we constructed a MCMC algorithm called st.mcmc.R in R:

BOX 28.1   MCMC ALGORITHM FOR SPATIO-TEMPORAL MODEL

```
1 st.mcmc <- function(Y,K,n.mcmc){
2
3 ###
4 ### Set up variables
5 ###
```

(*Continued*)

BOX 28.1 (Continued)　 MCMC ALGORITHM FOR
SPATIO-TEMPORAL MODEL

```
6  n.burn=round(.2*n.mcmc)
7  T=dim(Y)[1]
8  n=dim(Y)[2]
9  m=dim(K)[2]
10
11 s2y.save=rep(0,n.mcmc)
12 s2z.save=rep(0,n.mcmc)
13 alpha.save=matrix(0,m^2,n.mcmc)
14
15 Z=matrix(0,T+1,m)
16
17 ###
18 ### Priors and starting values
19 ###
20
21 r=1000
22 q=0.001
23 mu.alpha=rep(0,m^2)
24 Sig.alpha.inv=diag(m^2)/100
25
26 Z=matrix(0,T+1,m)
27 Z[-1,]=t(solve(t(K)%*%K)%*%t(K)%*%(t(Y)))
28 Z[1,]=Z[2,]
29
30 s2y=.1
31 s2z=.1
32
33 mu.0=rep(0,m)
34 Sig.0.inv=diag(m)/100
35
36 Im=diag(m)
37 Z.mean=matrix(0,T+1,m)
38
39 ###
40 ### Begin MCMC Loop
41 ###
```

(Continued)

BOX 28.1 (Continued)   MCMC ALGORITHM FOR SPATIO-TEMPORAL MODEL

```
42 for(k in 1:n.mcmc){
43
44   ###
45   ### Sample alpha
46   ###
47
48   tmp.chol=chol(t(Z[1:T,]%x%Im)%*%(Z[1:T,]%x%Im)/
49     s2z+ Sig.alpha.inv)
50   alpha=backsolve(tmp.chol,backsolve(tmp.chol,
51     (t(Z[1:T,]%x%Im)%*%c(t(Z[-1,])))/s2z
52     + Sig.alpha.inv%*%mu.alpha,transpose=TRUE)
53     +rnorm(m^2))
54   A=matrix(alpha,m,m)
55
56   ###
57   ### Sample s2y
58   ###
59
60   tmp.sum=sum(apply((Y-Z[-1,]%*%t(K))^2,
61     1,sum))
62   r.tmp=1/(tmp.sum/2+1/r)
63   q.tmp=n*T/2+q
64   s2y=1/rgamma(1,q.tmp,,r.tmp)
65
66   ###
67   ### Sample s2z
68   ###
69
70   tmp.sum=sum(apply((Z[-1,]-Z[1:T,]%*%t(A))^2,1
71     ,sum))
72   r.tmp=1/(tmp.sum/2+1/r)
73   q.tmp=m*T/2+q
74   s2z=1/rgamma(1,q.tmp,,r.tmp)
75
76   ###
77   ### Sample z_t
78   ###
```

(Continued)

BOX 28.1 (Continued)   MCMC ALGORITHM FOR
SPATIO-TEMPORAL MODEL

```
79  tmp.chol=chol(t(K)%*%K/s2y+t(A)%*%A/s2z+Im/s2z)
80  for(t in 1:(T-1)){
81    Z[t+1,]=backsolve(tmp.chol,backsolve(tmp.chol,
82      t(K)%*%Y[t,]/s2y+t(A)%*%Z[t+1+1,]/s2z
83      +A%*%Z[t-1+1,]/s2z,transpose=TRUE)+rnorm(m))
84  }
85
86  ###
87  ### Sample z_T
88  ###
89
90  tmp.chol=chol(t(K)%*%K/s2y+t(A)%*%A/s2z+Im/s2z)
91  Z[T+1,]=backsolve(tmp.chol,backsolve(tmp.chol,
92    t(K)%*%Y[T,]/s2y+A%*%Z[T-1+1,]/s2z,
93    transpose=TRUE)+rnorm(m))
94
95  ###
96  ### Sample z_0
97  ###
98
99  tmp.chol=chol(t(A)%*%A/s2z+Sig.0.inv)
100 Z[1,]=backsolve(tmp.chol,backsolve(tmp.chol,
101   t(A)%*%Z[1+1,]/s2z+Sig.0.inv%*%mu.0,
102   transpose=TRUE)+rnorm(m))
103
104 ###
105 ### Save Samples
106 ###
107
108 alpha.save[,k]=alpha
109 s2y.save[k]=s2y
110 s2z.save[k]=s2z
111 if(k > n.burn){
112   Z.mean=Z.mean+Z/(n.mcmc-n.burn)
113 }
114
115 }
```

(Continued)

**BOX 28.1 (Continued)   MCMC ALGORITHM FOR
SPATIO-TEMPORAL MODEL**

```
116 ###
117 ### Write Output
118 ###
119
120 list(alpha.save=alpha.save,s2y.save=s2y.save,
121     s2z.save=s2z.save,Z.mean=Z.mean,n.mcmc=n.mcmc,
122     n.burn=n.burn)
123
124 }
```

In this R code we used an indexing technique for the $T+1 \times m$ matrix Z where Z[1,] corresponds to $\mathbf{z}_0$ and Z[T+1,] corresponds to $\mathbf{z}_T$. Also, we used the same hyperparameter specifications for σ_y^2 and σ_z^2 and initial values for $\mathbf{z}_t$ based on a spatial average of the data $\mathbf{y}_t$ (more on this in the following example).

To demonstrate this MCMC algorithm, we simulated data from the hierarchical spatio-temporal model described in this section using a small spatio-temporal domain consisting of $m = 4$ spatial regions (or points, depending on the situation), $T = 50$, and 4 measurements for each spatial region and time point. Thus, the data vectors $\mathbf{y}_t$ are $n \times 1$ where $n = 16$ in our example. To construct the associated mapping matrix, we specified $\mathbf{K}$ such that

$$\mathbf{K} \equiv \begin{pmatrix} 1 & 0 & 0 & 0 \\ 1 & 0 & 0 & 0 \\ 1 & 0 & 0 & 0 \\ 1 & 0 & 0 & 0 \\ 0 & 1 & 0 & 0 \\ 0 & 1 & 0 & 0 \\ 0 & 1 & 0 & 0 \\ 0 & 1 & 0 & 0 \\ 0 & 0 & 1 & 0 \\ 0 & 0 & 1 & 0 \\ 0 & 0 & 1 & 0 \\ 0 & 0 & 1 & 0 \\ 0 & 0 & 0 & 1 \\ 0 & 0 & 0 & 1 \\ 0 & 0 & 0 & 1 \\ 0 & 0 & 0 & 1 \end{pmatrix}, \tag{28.65}$$

which is $n \times m$, as required. For a propagator matrix $\mathbf{A}$, we used a diagonal matrix with diagonal elements all less than one in absolute value (for stability; see Cressie and Wikle 2011) such that

$$\mathbf{A} \equiv \begin{pmatrix} -0.75 & 0 & 0 & 0 \\ 0 & 0 & 0 & 0 \\ 0 & 0 & 0.5 & 0 \\ 0 & 0 & 0 & 0.9 \end{pmatrix}. \tag{28.66}$$

The simulated data were based on values for the variance parameters of $\sigma_y^2 = 1$ and $\sigma_z^2 = 1$. We simulated data using the following R code:

BOX 28.2 SIMULATE SPATIO-TEMPORAL DATA

```
1  T=50
2  n=16
3  m=4
4  s2y=1
5  s2z=1
6  alpha.diag=c(-.75,0,.5,.9)
7  A=diag(alpha.diag)
8  alpha=c(A)
9
10 K=matrix(0,n,m)
11 for(j in 1:m){
12   K[(j*m-m+1):(j*m),j]=1
13 }
14
15 Z=matrix(0,T,m)
16 Y=matrix(0,T,n)
17
18 set.seed(101)
19 Z[1,]=rnorm(m)
20 Y[1,]=rnorm(n,K%*%Z[1,],sqrt(s2z))
21 for(t in 2:T){
22   Z[t,]=rnorm(m,A%*%Z[t-1,],sqrt(s2z))
23   Y[t,]=rnorm(n,K%*%Z[t,],sqrt(s2y))
24 }
```

We fit the hierarchical spatio-temporal model to the simulated data using hyperparameters $r = 1000$, $q = 0.001$, $\boldsymbol{\mu}_0 = \mathbf{0}$, $\boldsymbol{\Sigma}_0 \equiv \sigma_0^2 \mathbf{I}$ for $\sigma_0^2 = 100$, $\boldsymbol{\mu}_\alpha = \mathbf{0}$, and $\boldsymbol{\Sigma}_\alpha \equiv \sigma_\alpha^2 \mathbf{I}$, with $\sigma_\alpha^2 = 100$. In the following R code, we used 50000 MCMC iterations.

BOX 28.3   FIT SPATIO-TEMPORAL MODEL TO DATA

```
source("st.mcmc.R")
n.mcmc=50000
set.seed(1)
mcmc.out=st.mcmc(Y=Y,K=K,n.mcmc=n.mcmc)

layout(matrix(1:(m^2),m,m))
for(i in 1:4){
  hist(mcmc.out$alpha.save[i,],breaks=seq(-4,4,,
    200),col=8,prob=TRUE,xlab=bquote(alpha[.(i)]),
    main="",xlim=c(-1,1))
  curve(dnorm(x,0,10),lwd=2,add=TRUE)
  abline(v=alpha[i],lwd=2,lty=2)
}
for(i in 5:(m^2)){
  hist(mcmc.out$alpha.save[i,],breaks=seq(-4,4,,
    200),col=8,prob=TRUE,xlab=bquote(alpha[.(i)]),
    main="",ylab="",xlim=c(-1,1))
  curve(dnorm(x,0,10),lwd=2,add=TRUE)
  abline(v=alpha[i],lwd=2,lty=2)
}
```

This R code fits the spatio-temporal model to the simulated data and shows the resulting marginal posterior histograms for α arranged in a grid as they appear in the propagator matrix $\mathbf{A}$. Recall that the true values for α that we used to simulate the data were non-zero on three elements of the diagonal of $\mathbf{A}$, for spatial region 1 (α_1), 3 (α_{11}), and 4 (α_{16}). Thus, the marginal posterior distributions shown in Figure 28.1 indicate that the model does a reasonable job of recovering the nonzero dynamic parameters. In terms of the parameters for which the true values were zero, the marginal posterior distributions did not indicate they were different from zero. Thus, although we overparameterized our model with respect to the true dynamics in our simulation example, it was still able to recover the parameters using the data. Of course, with a knowledge of the true values, we could have used the diagonal parameterization for the propagator matrix ($\mathbf{A} \equiv \mathrm{diag}(\boldsymbol{\alpha}_{\mathrm{diag}})$) and that would lead to a better predictive model because fewer parameters would need to be estimated (i.e., we would not need to estimate the off-diagonal elements of $\mathbf{A}$).

We used the following R code to make inference on the two variance parameters in the model (σ_y^2 and σ_z^2) using the MCMC output:

BOX 28.4   POSTERIOR SUMMARIES FOR VARIANCE PARAMETERS

```
> n.burn=mcmc.out$n.burn
> mean(mcmc.out$s2y.save[-(1:n.burn)])
[1] 0.9382672
> quantile(mcmc.out$s2y.save[-(1:n.burn)],
    c(0.025,0.975))
      2.5%       97.5%
0.8365467 1.0522097
> mean(mcmc.out$s2z.save[-(1:n.burn)])
[1] 0.9950574
> quantile(mcmc.out$s2z.save[-(1:n.burn)],
    c(0.025,0.975))
      2.5%       97.5%
0.7496105 1.2914530
```

Recall that the true values for the variance terms σ_y^2 and σ_z^2 were 1. From this MCMC output, we can see that the marginal posterior means were $E(\sigma_y^2|\mathbf{Y}) = 0.94$ and $E(\sigma_y^2|\mathbf{Y}) = 1.00$, and the 95% equal tail marginal credible interval for σ_y^2 was $(0.84, 1.05)$ and for σ_z^2 was $(0.75, 1.29)$. Thus, the marginal posterior distributions for the variance parameters in our hierarchical spatio-temporal model capture the values used to simulate the data.

State-space models, such as the hierarchical spatio-temporal model that we presented in this section, are used commonly to separate noise from signal while estimating the latent process (i.e., state variables $\mathbf{z}_t$, for $t = 0, \ldots, T$). In non-Bayesian settings, computational approaches such as Kalman filtering and smoothing are useful for learning about the latent process $\mathbf{z}_t$ (i.e., the "signal") in these types of models. However, we can obtain inference for the latent process automatically when we fit this Bayesian model using MCMC. We used the following R code to create Figure 28.2 which shows the marginal time series trajectories for each of the $m = 4$ spatial locations in our simulation example.

BOX 28.5   POSTERIOR MEAN TRAJECTORIES FOR $\mathbf{Z}_T$

```
layout(matrix(1:m,m,1))
for(j in 1:m){
  matplot(1:T,Y[,(j*m-m+1):(j*m)],type="p",pch=20,
    cex=1.5,col=rgb(0,0,0,.4),xlab="t",
    ylab=bquote(z[.(j)]))
  lines(1:T,Z[,j],lty=1,lwd=2,col=rgb(0,0,0,.4))
  lines(0:T,mcmc.out$Z.mean[,j],lty=1,lwd=2,col=1)
}
```

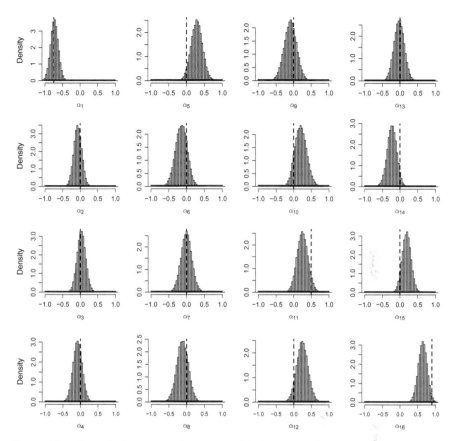

Figure 28.1 Posterior histograms for α arranged in the same order they appear in the propagator matrix **A**. Priors shown as solid black lines (bottom of each panel). The dashed vertical lines represent the true values for α used to simulate the data.

In each panel shown in Figure 28.2, the data ($\mathbf{y}_t$) for each spatial region are shown as gray points and the true underlying process trajectory is shown as the solid gray line connecting the process values at the discrete set of time points that we used to simulate the data. The marginal posterior mean trajectories for the latent process ($E(\mathbf{z}_t|\mathbf{Y})$) are shown as solid black lines in Figure 28.2. Notice that the posterior mean process recovers the true simulated process well throughout the spatio-temporal domain despite the somewhat noisy simulated data used to fit the model.

We specified a simple hierarchical spatio-temporal model in this section to illustrate how the implementation procedure works and to easily visualize the simulated data and parameters. The model that we presented also scales to much larger spatial and temporal domains. Furthermore, we are not required to have replication in the observed data. Thus, we could fit the same model if $n = m$ and we only had one replicate from each spatial region. In fact, if we were focused on a continuous spatial domain, the dimension of the latent process is technically infinite, but we could

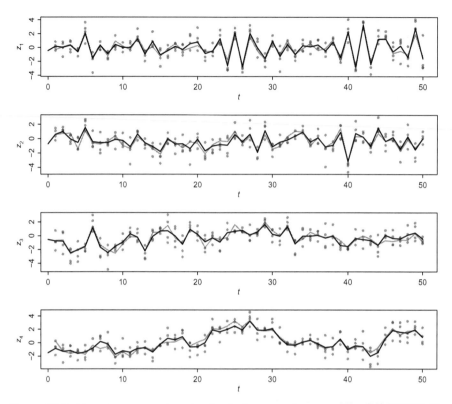

Figure 28.2 Posterior mean trajectories for the latent spatio-temporal process $E(\mathbf{z}_t|\mathbf{Y})$ for $t = 0, 1, \ldots, T$ shown for each spatial region (black lines). Simulated data $\mathbf{y}_t$ (gray points) and true latent process $\mathbf{z}_t$ (gray lines) shown for each spatial region.

still learn about $\mathbf{z}_t$ at unobserved locations if the dynamics implied by the propagator matrix were specified such that they had general meaning for the entire spatial domain and contained few enough parameters that we could estimate using the data (see the following section). If we used the previous model specification, with $\mathbf{A}$ fully parameterized, we would need strong prior information for $\boldsymbol{\alpha}$ to learn about the latent process at unmeasured locations if $n < m$.

Another common modification of the process model that we presented (i.e., $\mathbf{z}_t \sim \mathrm{N}(\mathbf{A}\mathbf{z}_{t-1}, \boldsymbol{\Sigma}_z)$) is to use a formal spatial covariance function to define $\boldsymbol{\Sigma}_z$ rather than assume a diagonal structure as we did in the previous example. Including a spatially structured process covariance allows the model to characterize structure in the process not accounted for with the dynamics alone. Spatial structure in the process covariance also may facilitate prediction of the latent process at unobserved locations in the spatial domain. However, without enough data in space, and possibly over time, it may be difficult to estimate parameters in a spatial covariance function (see Chapter 17) if the residual spatial signal is weak after accounting for the dynamics.

28.2 MECHANISTIC SPATIO-TEMPORAL MODELS

Phenomenologically specified spatio-temporal models like those that we presented in the previous section have been useful for characterizing pattern and structure in ecological and environmental data sets. In fact, we can facilitate the implementation of such models by marginalizing out dynamic structure typically expressed in the mean component of conditionally specified spatio-temporal models which forces the covariance to account for all the spatio-temporal structure in the process. Cressie and Wikle (2011) refer to marginally parameterized models as "descriptive" spatio-temporal models. However, from a scientific perspective, there is value in constructing models that directly reconcile with our understanding of the dynamics of the process we seek to learn about (Wikle and Hooten 2010). The mechanistic approach to specifying dynamic spatio-temporal models explicitly leverages the long history of contributions in applied mathematics and physics to construct statistical models that are "closer to the etiology of the phenomenon under study" (Cressie and Wikle 2011; Wikle et al. 2019).

Mechanistic spatio-temporal statistical models have been particularly valuable in fields that involve direct and well-specified physical relationships such as atmospheric science and oceanography (e.g., Wikle et al. 2001). Adaptations of these models also have become more popular in ecology (Wikle and Hooten 2010). For example, hierarchical statistical models using process specifications motivated by solutions to partial differential equations (PDEs) were originally developed to model invasive bird species distributions over large spatial regions (e.g., Wikle 2003; Hooten and Wikle 2008). More recently, these types of models have been used to characterize spreading disease in wildlife population (e.g., chronic wasting disease in white-tailed deer populations; Hefley et al. 2017c). They have also been placed in an IPM framework that utilizes multiple data sources to study the recolonization of sea otters in Glacier Bay, Alaska, USA, while accounting for imperfect detectability in aerial surveys (Williams et al. 2017).

Most mechanistic spatio-temporal models for dynamical systems are based on mathematical equations that describe ecological and environmental systems in discrete time/space (e.g., difference equations), continuous time/space (e.g., differential equations), or some combination (e.g., integro-difference equations). Such models describe how the state of a system evolves over time, space, or both. In what follows, we focus on models with dynamics in continuous time and space.

A simple univariate example of a dynamical system is Malthusian growth in continuous time, which can be expressed as the ordinary differential equation (ODE): $\frac{d\lambda(t)}{dt} = r\lambda(t)$, where $\lambda(t)$ is the expected population abundance at time t and with growth rate r. Most introductory courses and textbooks on differential equations focus on developing the mathematical tools to find analytical solutions to such equations. For the Malthusian growth model, the analytical solution to the differential equation is $\lambda(t) = \lambda(0)e^{rt}$ where $\lambda(0)$ is the population size at the initial time (i.e., $t = 0$). Mathematical equations like the Malthusian growth model are often deterministic and typically contain parameters (e.g., r) that we may wish to learn about given a set of observations of the study population. We can embed deterministic mathematical models in a statistical framework and fit to data using MCMC strategies similar to those presented throughout this book. For example, using the techniques presented in Chapter 21, this continuous-time solution to the Malthusian growth model can be fit to count data using Poisson regression by letting $y(t) \sim \text{Pois}(\lambda(t))$ and $\log(\lambda(t)) = \beta_0 + \beta_1 t$, where $\lambda(0) = e^{\beta_0}$ and $r = \beta_1$.[3] This is an alternative to the discretized form of Malthusian growth that we discussed in Chapter 21 and we show how to arrive at the discretized version in what follows.

Although ordinary differential equations (ODEs) are widely used in theoretical and applied ecology, these models are typically limited to univariate time series because they are functions of only one variable.[4] Differential equation models that involve changes in multiple dimensions (i.e., PDEs) can incorporate space and time. The PDEs most commonly used to describe ecological systems cannot be solved analytically; thus, numerical methods are required to approximate the derivatives. Finite-difference approximations are one of the most common ways to numerically solve PDEs and similar to the deterministic methods used to approximate integrals that we presented in Chapter 2 (Haberman 1987; Farlow 1993).

The first implementation we present involving finite-difference approximations is the forward difference operator

$$\frac{df(x)}{dx} \approx \frac{f(x + \Delta x) - f(x)}{\Delta x},\qquad(28.67)$$

where $f(x)$ is the univariate function we seek to differentiate and Δx is referred to as the "step size." For the Malthusian growth differential equation, using the forward difference operator involves replacing $\frac{d\lambda(t)}{dt}$ with $\frac{\lambda(t+\Delta t)-\lambda(t)}{\Delta t}$, which results in

$$\frac{\lambda(t + \Delta t) - \lambda(t)}{\Delta t} = r\lambda(t).\qquad(28.68)$$

To isolate $\lambda(t + \Delta t)$ on the left-hand side, we rearrange (28.68) such that

$$\lambda(t + \Delta t) = \lambda(t) + r\Delta t\lambda(t).\qquad(28.69)$$

Starting from the initial population size, $\lambda(0)$, we can solve the equation iteratively to obtain $\lambda(t)$ at the desired time t. Our presentation of the discrete-time Malthusian growth model in Chapter 16 is based on this use of finite-difference approximation.

The univariate Malthusian growth example illustrates that the finite-difference approximation results in an algebraic equation that is easy to understand and simple to use. The accuracy of a finite-difference approximation increases as Δt decreases. In fact, the finite-difference approximation is related to the definition of a derivative in that $\frac{df(x)}{dx} = \lim_{\Delta x \to 0} \frac{f(x+\Delta x)-f(x)}{\Delta x}$, which indicates that, as the step size decreases toward zero, the finite-difference approximation converges to the derivative.

---

[3] We can learn about $\lambda(0)$ and r as derived quantities using the MCMC samples for β_0 and β_1 as described in Chapter 11.

[4] Several ODEs can be coupled together so that they depend on each other such as in the Lotka-Volterra equation which is commonly used to model predator-prey dynamics (e.g., Cangelosi and Hooten 2009).

An important variation of the finite-difference approximation is the centered difference operator

$$\frac{df(x)}{dx} \approx \frac{f(x+\Delta x) - f(x-\Delta x)}{2\Delta x}. \tag{28.70}$$

In addition to first-order derivatives (i.e., $\frac{df(x)}{dx}$), many differential equations contain second- or higher-order derivates. Similar to the forward and centered difference operator for first-order derivatives, we can approximate second-order derivatives using the forward-difference operator

$$\frac{d^2f(x)}{dx^2} \approx \frac{f(x+2\Delta x) - 2f(x+\Delta x) + f(x)}{(\Delta x)^2} \tag{28.71}$$

or the centered difference operator

$$\frac{d^2f(x)}{dx^2} \approx \frac{f(x+\Delta x) - 2f(x) + f(x-\Delta x)}{(\Delta x)^2}. \tag{28.72}$$

Because PDEs contain multiple variables, we also need to approximate partial derivatives. We can solve some partial derivatives using the finite-difference approximations similar to those we presented previously. For example, suppose we seek to approximate $\frac{\partial f(x,y)}{\partial x}$, where $f(x,y)$ is a function of x and y. Using the centered difference operator, we can approximate $\frac{\partial f(x,y)}{\partial x}$ with $\frac{f(x+\Delta x,y) - f(x-\Delta x,y)}{2\Delta x}$.

Finite difference methods provide a simple and powerful approach to approximate derivatives. Thus, fitting statistical models involving PDEs to spatio-temporal data relies on straightforward extensions of the methods from previous chapters. In what follows, we present an example that we can use to model the spread of invasive species. For this example, we used the so-called ecological diffusion (e.g., Garlick et al. 2011; Hooten et al. 2013) PDE

$$\frac{\partial \lambda(\mathbf{s},t)}{\partial t} = \left(\frac{\partial^2}{\partial s_1^2} + \frac{\partial^2}{\partial s_2^2} \right) \delta(\mathbf{s})\lambda(\mathbf{s},t), \tag{28.73}$$

where $\lambda(\mathbf{s},t)$ is the intensity of the spreading population, $\mathbf{s} \equiv (s_1,s_2)'$ are the spatial coordinates, and t is the time. The diffusion coefficient, $\delta(\mathbf{s})$, varies over space and may depend on spatially varying covariates that control the rate of spread.[5]

To fully specify a PDE, we need to define the initial and boundary conditions. For an invasive species, the initial conditions include the location, time, and number of individuals initially released (i.e., the propagule size; Lockwood et al. 2005; Colautti et al. 2006). A common assumption in mathematical models for invasions is that the

---

[5] The "ecological" aspect ecological diffusion refers to the fact that we can derive the ecological diffusion model based on a recurrence equation involving the probability of animal movement through environmental space. See Turchin (1998), Hooten and Wikle (2010), and Hooten et al. (2017) for details on this perspective of the Fokker-Planck equation.

first individuals were introduced at a single point. We can incorporate this point of introduction into the specification of the initial conditions with

$$
\lambda(\mathbf{s}, t_0) = \begin{cases} \theta & \text{if } \mathbf{s} = \mathbf{c} \\ 0 & \text{if } \mathbf{s} \neq \mathbf{c} \end{cases}, \tag{28.74}
$$

where θ is the initial number of individuals, $\mathbf{c} \equiv (c_1, c_2)'$ are the coordinates of the location of introduction, and t_0 is the unknown time of introduction. At the boundary, we use Dirichlet (absorbing) boundary conditions and set $\lambda(\mathbf{s}, t) = 0$.

Now that we have a fully specified PDE that includes initial and boundary conditions, we can use the finite-difference method to approximate the derivatives. Similar to when we choose the step size for the forward difference operator when approximating the derivative in the Malthusian growth model, approximating the ecological diffusion PDE requires a discretization in space and time. Using the finite-difference approximation requires partitioning the spatial domain $\mathscr{S}$ into a grid with m cells that are $\Delta s_1 \times \Delta s_2$ in dimension. The temporal domain $\mathscr{T}$ is partitioned into bins of width Δt. Applying the forward difference operator to approximate $\frac{\partial \lambda(\mathbf{s}, t)}{\partial t}$ and the centered difference operator to approximate $\frac{\partial^2}{\partial s_1^2}$ and $\frac{\partial^2}{\partial s_2^2}$ results in

$$
\frac{\lambda(\mathbf{s}, t + \Delta t) - \lambda(\mathbf{s}, t)}{\Delta t} =
$$
$$
\frac{\delta(s_1 + \Delta s_1, s_2)\lambda(s_1 + \Delta s_1, s_2, t) - 2\delta(\mathbf{s})\lambda(\mathbf{s}, t) + \delta(s_1 - \Delta s_1, s_2)\lambda(s_1 - \Delta s_1, s_2, t)}{(\Delta s_1)^2} +
$$
$$
\frac{\delta(s_1, s_2 + \Delta s_2)\lambda(s_1, s_2 + \Delta s_2, t) - 2\delta(\mathbf{s})\lambda(\mathbf{s}, t) + \delta(s_1, s_2 - \Delta s_2)\lambda(s_1, s_2 - \Delta s_2, t)}{(\Delta s_2)^2}. \tag{28.75}
$$

Similar to the Malthusian growth example, we isolate $\lambda(\mathbf{s}, t + \Delta t)$ on the left-hand side so that we can solve the equation iteratively. Algebraically rearranging (28.75), we obtain

$$
\lambda(\mathbf{s}, t + \Delta t) = \Delta t \left(1 - \frac{2}{(\Delta s_1)^2 + (\Delta s_2)^2} \delta(\mathbf{s}) \right) \lambda(\mathbf{s}, t) +
$$
$$
\Delta t \frac{\delta(s_1 + \Delta s_1, s_2)}{(\Delta s_1)^2} \lambda(s_1 + \Delta s_1, s_2, t) +
$$
$$
\Delta t \frac{\delta(s_1 - \Delta s_1, s_2)}{(\Delta s_1)^2} \lambda(s_1 - \Delta s_1, s_2, t) +
$$
$$
\Delta t \frac{\delta(s_1, s_2 + \Delta s_2)}{(\Delta s_2)^2} \lambda(s_1, s_2 + \Delta s_2, t) +
$$
$$
\Delta t \frac{\delta(s_1, s_2 - \Delta s_2)}{(\Delta s_2)^2} \lambda(s_1, s_2 - \Delta s_2, t). \tag{28.76}
$$

Written as (28.76), there are five components involved in the calculation of $\lambda(\mathbf{s}, t + \Delta t)$. Notice that $\lambda(\mathbf{s}, t + \Delta t)$ depends on the values of $\lambda(\mathbf{s}, t)$ in the grid cell of interest at the previous time and the four neighbors that share a common border. Because of this structure, we can write the finite-difference approximation as a discrete time matrix equation.

The next step involves fitting a statistical model with dynamics characterized by the PDE to data and learn about unknown parameters. To demonstrate, we assume the diffusion coefficient, $\delta(\mathbf{s})$, depends on a single covariate, $x(\mathbf{s})$, that varies spatially. The diffusion rate should be positive; thus, we use a log link function: $\log(\delta(\mathbf{s})) = \beta_0 + \beta_1 x(\mathbf{s})$. We can write the data model as $y_i \sim [y_i | \lambda_i]$, where λ_i corresponds to the approximate value of $\lambda(\mathbf{s}, t)$ that is closest in space and time to the corresponding observation y_i. In this example, we consider the process λ_i as deterministic and write the data model as $y_i \sim [y_i | \boldsymbol{\beta}, \mathbf{c}, t_0, \theta]$, where $\boldsymbol{\beta} \equiv (\beta_0, \beta_1)'$ is the slope and intercept for the diffusion rate, $\mathbf{c} \equiv (c_1, c_2)'$ is the location of introduction, t_0 is the time of introduction, and θ is the propagule size. To fully specify a Bayesian model, we use the data model $y_i \sim \text{Pois}(\lambda_i)$ and priors $\boldsymbol{\beta} \sim \text{N}(\mathbf{0}, \sigma_\beta^2 \mathbf{I})$, $\theta \sim \text{Gamma}(q, r)$, $c_1 \sim \text{Unif}(l_1, u_1)$, $c_2 \sim \text{Unif}(l_2, u_2)$, and $t_0 \sim \text{Unif}(l_{t_0}, u_{t_0})$.

The posterior distribution for the ecological diffusion model is

$$[\boldsymbol{\beta}, \mathbf{c}, t_0, \theta | \mathbf{y}] \propto \prod_{i=1}^{n} [y_i | \boldsymbol{\beta}, \mathbf{c}, t_0, \theta][\boldsymbol{\beta}][\mathbf{c}][t_0][\theta], \qquad (28.77)$$

and, as with many other models that we have presented, the full-conditional distributions are not analytically tractable. Thus, we use a random walk proposal where $\boldsymbol{\beta}^{(*)} \sim \text{N}(\boldsymbol{\beta}^{(k-1)}, \sigma_{\text{tune}, \beta}^2 \mathbf{I})$ and calculate the M-H ratio as

$$mh = \frac{\left(\prod_{i=1}^{n} [y_i | \boldsymbol{\beta}^{(*)}, \mathbf{c}^{(k-1)}, t_0^{(k-1)}, \theta^{(k-1)}] \right) [\boldsymbol{\beta}^{(*)}]}{\left(\prod_{i=1}^{n} [y_i | \boldsymbol{\beta}^{(k-1)}, \mathbf{c}^{(k-1)}, t_0^{(k-1)}, \theta^{(k-1)}] \right) [\boldsymbol{\beta}^{(k-1)}]}. \qquad (28.78)$$

Similarly, we use the proposal distribution $\mathbf{c}^{(*)} \sim \text{N}(\mathbf{c}^{(k-1)}, \sigma_{\text{tune}, c}^2 \mathbf{I})$ and calculate the M-H ratio as

$$mh = \frac{\left(\prod_{i=1}^{n} [y_i | \boldsymbol{\beta}^{(k)}, \mathbf{c}^{(*)}, t_0^{(k-1)}, \theta^{(k-1)}] \right) [\mathbf{c}^{(*)}]}{\left(\prod_{i=1}^{n} [y_i | \boldsymbol{\beta}^{(k)}, \mathbf{c}^{(k-1)}, t_0^{(k-1)}, \theta^{(k-1)}] \right) [\mathbf{c}^{(k-1)}]}. \qquad (28.79)$$

The support of $\mathbf{c}$ is bounded because it must be contained within the study area $\mathscr{S}$. We can express this constraint in the uniform priors used for the elements of $\mathbf{c}$. As a consequence, it is important to remember that the M-H ratio is equal to zero when $\mathbf{c}^{(*)}$ falls outside of the support of $[\mathbf{c}]$. Next, we sample the time of introduction using the proposal $t_0^{(*)} \sim \text{N}(t_0^{(k-1)}, \sigma_{\text{tune}, t_0}^2)$ and calculate the M-H ratio as

$$mh = \frac{\left(\prod_{i=1}^{n} [y_i | \boldsymbol{\beta}^{(k)}, \mathbf{c}^{(k)}, t_0^{(*)}, \theta^{(k-1)}] \right) [t_0^{(*)}]}{\left(\prod_{i=1}^{n} [y_i | \boldsymbol{\beta}^{(k)}, \mathbf{c}^{(k)}, t_0^{(k-1)}, \theta^{(k-1)}] \right) [t_0^{(k-1)}]}. \qquad (28.80)$$

Similar to the location of introduction, the support of t_0 is bounded because the time of introduction must occur before the first observation where $y_i > 0$, which we also can express in the uniform prior used for t_0. Thus, it is important to remember that the M-H ratio is equal to zero when $t_0^{(*)}$ falls outside of the support of $[t_0]$. Finally, the proposal distribution for the propagule size is $\theta^{(*)} \sim N(\theta^{(k-1)}, \sigma_{\text{tune},\theta}^2)$ and the M-H ratio is

$$mh = \frac{\left(\prod_{i=1}^{n}[y_i|\boldsymbol{\beta}^{(k)}, \mathbf{c}^{(k)}, t_0^{(k)}, \theta^{(*)}]\right)[\theta^{(*)}]}{\left(\prod_{i=1}^{n}[y_i|\boldsymbol{\beta}^{(k)}, \mathbf{c}^{(k)}, t_0^{(k)}, \theta^{(k-1)}]\right)[\theta^{(k-1)}]}, \tag{28.81}$$

where $mh = 0$ if $\theta^{(*)}$ falls outside of the support of $[\theta]$. We constructed the MCMC algorithm called `diffusion.mcmc.R` in R based on the full-conditional distributions that we just outlined:

BOX 28.6   MCMC ALGORITHM FOR MECHANISTIC SPATIO-TEMPORAL MODEL

```
diffusion.mcmc <- function(y,x,t.sample,dt,
    beta.mn,beta.var,c.lwr,c.upr,t0.lwr,t0.upr,
    theta.q,theta.r,beta.tune,c.tune,t0.tune,
    theta.tune,beta.start,c.start,t0.start,
    theta.start,n.mcmc){

###
### Libraries and Subroutines
###

library(raster)
library(sp)
library(matrixcalc)

neighborhood <- function(raster){
  nn=matrix(,length(raster[]),4)
  for(i in 1:dim(nn)[1]){
    loc=adjacent(raster,i)[,2]
    ln=loc[which((loc+1)==i)]
    rn=loc[which((loc-1)==i)]
    bn=loc[which((loc-dim(raster)[2])==i)]
    tn=loc[which((loc+dim(raster)[2])==i)]
    nn[i,1]=if(length(ln)>0){ln}else{0}
    nn[i,2]=if(length(rn)>0){rn}else{0}
    nn[i,3]=if(length(bn)>0){bn}else{0}
```

(Continued)

BOX 28.6 (Continued)   MCMC ALGORITHM FOR MECHANISTIC SPATIO-TEMPORAL MODEL

```
26    nn[i,4]=if(length(tn)>0){tn}else{0}
27    }
28    nn
29  }
30
31  propagator <- function(NN,delta,dx,dy,dt){
32    H=matrix(0,dim(NN)[1],dim(NN)[1])
33    for(i in 1:dim(H)[1]){
34      if(length(which(NN[i,]>0))==4){
35        H[i,i]=1-2*delta[i]*(dt/dx^2 + dt/dy^2)
36        H[i,NN[i,1]]=dt/dx^2*delta[NN[i,1]]
37        H[i,NN[i,2]]=dt/dx^2*delta[NN[i,2]]
38        H[i,NN[i,3]]=dt/dy^2*delta[NN[i,3]]
39        H[i,NN[i,4]]=dt/dy^2*delta[NN[i,4]]}
40    }
41    H
42  }
43
44  calc.lam <- function(H,lambda0,t.sample){
45    T=max(t.sample)
46    lambda=lambda0
47    lambda[]=H%*%lambda0[]
48    lambda=stack(mget(rep("lambda",T)))
49    for(t in 2:T){
50      lambda[[t]][]=H%*%lambda[[t-1]][]
51    }
52    lambda[[t.sample]]
53  }
54
55  pde.solve <- function(x,beta,c.vec,t0,theta,
56    t.sample,dt,NN){
57    delta=exp(beta[1] + beta[2]*x)
58    H=propagator(NN,delta[],res(delta)[1],
59        res(delta)[2],dt)
60    lambda0=x
61    lambda0[]=0
62    lambda0[extract(lambda0,SpatialPoints
63        (t(c.vec)),cell=TRUE)[1]]=theta
```

(Continued)

BOX 28.6 (Continued)   MCMC ALGORITHM FOR MECHANISTIC SPATIO-TEMPORAL MODEL

```
64      lambda=calc.lam(H,lambda0,t.sample-round(t0))
65      lambda
66   }
67
68   ####
69   #### Setup Variables
70   ####
71
72   beta.save=matrix(,2,n.mcmc)
73   c.save=matrix(,2,n.mcmc)
74   t0.save=matrix(,1,n.mcmc)
75   theta.save=matrix(,1,n.mcmc)
76   NN=neighborhood(x)
77
78   ###
79   ### Starting Values
80   ###
81
82   beta=beta.start
83   c.vec=c.start
84   t0=t0.start
85   theta=theta.start
86
87   ###
88   ### MCMC Loop
89   ###
90
91   for(k in 1:n.mcmc){
92
93      ###
94      ### Sample beta
95      ###
96
97      beta.star=rnorm(2,beta,beta.tune)
98      lambda.star=pde.solve(x,beta.star,c.vec,t0,
99          theta,t.sample,dt,NN)
```

                                                              (*Continued*)

BOX 28.6 (Continued)   MCMC ALGORITHM FOR MECHANISTIC
SPATIO-TEMPORAL MODEL

```
100   mh1=sum(dpois(vec(y[]),vec(lambda.star[]),
101       log=TRUE)) + sum(dnorm(beta.star,beta.mn,
102       sqrt(beta.var),log=TRUE))
103   lambda=pde.solve(x,beta,c.vec,t0,theta,
104       t.sample,dt,NN)
105   mh2=sum(dpois(vec(y[]),vec(lambda[]),
106       log=TRUE)) + sum(dnorm(beta,beta.mn,sqrt
107       (beta.var),log=TRUE))
108   mh=exp(mh1-mh2)
109   if(mh > runif(1)){
110       beta=beta.star
111       lambda=lambda.star
112   }
113
114   ###
115   ### Sample c
116   ###
117
118   c.star=rnorm(2,c.vec,c.tune)
119   if((min(c.star) > c.lwr) & (max(c.star)
120       < c.upr)){
121       lambda.star=pde.solve(x,beta,c.star,t0,
122           theta,t.sample,dt,NN)
123       mh1=sum(dpois(vec(y[]),vec(lambda.star[]),
124           log=TRUE)) + sum(dunif(c.star,c.lwr,c.upr,
125           log=TRUE))
126       mh2=sum(dpois(vec(y[]),vec(lambda[]),
127           log=TRUE)) + sum(dunif(c.vec,c.lwr,c.upr,
128           log=TRUE))
129       mh=exp(mh1-mh2)
130       if(mh > runif(1)){
131           c.vec=c.star
132           lambda=lambda.star
133       }
134   }
```

(Continued)

BOX 28.6 (Continued)   MCMC ALGORITHM FOR MECHANISTIC SPATIO-TEMPORAL MODEL

```
135   ###
136   ### Sample t0
137   ###
138
139   t0.star=rnorm(1,t0,t0.tune)
140   if((t0.star > t0.lwr) & (t0.star < t0.upr)){
141     lambda.star=pde.solve(x,beta,c.vec,t0.star,
142       theta,t.sample,dt,NN)
143     mh1=sum(dpois(vec(y[]),vec(lambda.star[]),
144       log=TRUE)) + dunif(t0.star,t0.lwr,t0.upr,
145       log=TRUE)
146     mh2=sum(dpois(vec(y[]),vec(lambda[]),
147       log=TRUE)) + dunif(t0,t0.lwr,t0.upr,
148       log=TRUE)
149     mh=exp(mh1-mh2)
150     if(mh > runif(1)){
151       t0=t0.star
152       lambda=lambda.star
153     }
154   }
155
156   ###
157   ### Sample theta
158   ###
159
160   theta.star=rnorm(1,theta,theta.tune)
161   if(theta.star > 0){
162     lambda.star=pde.solve(x,beta,c.vec,t0,
163       theta.star,t.sample,dt,NN)
164     mh1=sum(dpois(vec(y[]),vec(lambda.star[]),
165       log=TRUE)) + dgamma(theta.star,theta.q,,
166       theta.r,log=TRUE)
167     mh2=sum(dpois(vec(y[]),vec(lambda[]),
168       log=TRUE)) + dgamma(theta,theta.q,,
169       theta.r,log=TRUE)
170     mh=exp(mh1-mh2)
171     if(mh > runif(1)){
172       theta=theta.star
```

(Continued)

BOX 28.6 (Continued)   MCMC ALGORITHM FOR MECHANISTIC SPATIO-TEMPORAL MODEL

```
173          lambda=lambda.star
174       }
175    }
176
177    ###
178    ### Save Samples
179    ###
180
181    beta.save[,k]=beta
182    c.save[,k]=c.vec
183    t0.save[,k]=t0
184    theta.save[,k]=theta
185
186 }
187
188 ###
189 ### Write Output
190 ###
191
192 list(beta.save=beta.save,c.save=c.save,
193    t0.save=t0.save,theta.save=theta.save)
194
195 }
```

To demonstrate the ecological diffusion model, we simulated count data based on known parameter values of $\beta_0 = -8.5$, $\beta_1 = -1.0$, $c_1 = 0.5$, $c_2 = 0.5$, $t_0 = -10$, and $\theta = 1000$. For the simulation, we defined the study area as the unit square (i.e., $\mathscr{S} \equiv (0,1) \times (0,1)$), and set $\Delta s_1 = \Delta s_2 = 0.04$ which resulted in 625 grid cells. We set $\Delta t = 1$ and observe data y_i for times $t = 1, 2, \ldots, 20$ at all spatial locations. Thus, our simulated data set includes $n = 12500$ observations. We used the following R code to simulate data y_i, which are shown in Figure 28.3. Note that this code requires the functions neighborhood, propagator, calc.lam, and solve.pde, which we specified in the main MCMC algorithm diffusion.mcmc.R.

BOX 28.7   SIMULATE MECHANISTIC SPATIO-TEMPORAL DATA

```
1 library(raster)
2 library(sp)
```

(Continued)

**BOX 28.7 (Continued)   SIMULATE MECHANISTIC
SPATIO-TEMPORAL DATA**

```
 3 library(matrixcalc)
 4 library(rasterVis)
 5 library(grDevices)
 6 source("required.functions.R") # these functions
     also appear in diffusion.mcmc.R
 7
 8 beta=c(-8.5,-1)
 9 c.vec=c(0.5,0.5)
10 t0=-10
11 theta=1000
12
13 T=20
14 dt=1
15 t.sample=seq(1,T,by=1*dt)
16 m=625
17
18 x=raster(vals=0,nrows=m^0.5,ncols=m^0.5,xmn=0,
19     xmx=1,ymn=0,ymx=1,crs=NA)
20 set.seed(1244)
21 x[]=rbinom(m,1,pnorm(-1.5+3*xyFromCell(x,1:m)
22     [,1]))
23 delta=raster(vals=0,nrows=m^0.5,ncols=m^0.5,xmn=0,
24     xmx=1, ymn=0,ymx=1,crs=NA)
25 delta[]=exp(model.matrix(~x[])%*%beta)
26 ramp.gray=colorRampPalette(c("white","gray30"))
27     (2000)
28
29 NN=neighborhood(x)
30 lambda=pde.solve(x,beta,c.vec,t0,theta,t.sample,
31     dt,NN)
32 y=raster(vals=0,nrows=m^0.5,ncols=m^0.5,xmn=0,
33     xmx=1,ymn=0,ymx=1,crs=NA)
34 y=stack(mget(rep("y",length(t.sample))))
35 set.seed(9533)
36 y[]=rpois(prod(dim(lambda)),vec(lambda[]))
37
38 levelplot(y, margin=FALSE,scales=list(draw=FALSE),
39     names.attr=paste("t =",t.sample),cuts=2000,
40     col.regions=ramp.gray)
```

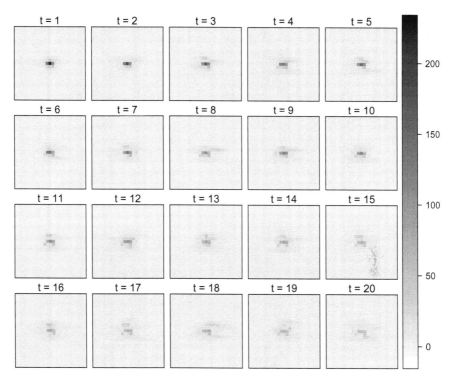

Figure 28.3 Simulated count data using ecological diffusion PDE for $t = 1, 2, \ldots, 20$.

We fit the ecological diffusion model to our simulated count data using the following R code with $K = 20000$ MCMC iterations[6] and hyperparameters $\sigma_\beta^2 = 10^6$, $l_c = 0$, $u_c = 1$, $l_{t_0} = -100$, $u_{t_0} = 0$, $q = 1$, and $r = 1000$.

BOX 28.8   FIT MECHANISTIC SPATIO-TEMPORAL MODEL TO DATA

```
1 source("diffusion.mcmc.R")
2 set.seed(3204)
3 mcmc.out=diffusion.mcmc(y=y,x=x,t.sample=t.sample,
4     dt=dt,beta.mn=0,beta.var=100,c.lwr=0,c.upr=1,
5     t0.lwr=-100,t0.upr=0,theta.q=1,theta.r=1000,
6     beta.tune=0.025,c.tune=0.025,t0.tune=1.5,
```
(Continued)

---

[6] Note that this code may require hours to run on a standard laptop computer.

BOX 28.8 (Continued)   FIT MECHANISTIC SPATIO-TEMPORAL MODEL TO DATA

```
 7      theta.tune=25,beta.start=beta,c.start=c(0.25,
 8      0.25),t0.start=-25,theta.start=theta,
 9      n.mcmc=20000)

10
11  n.burn=5000
12  layout(matrix(c(1,3,5,2,4,6),3,2))
13  hist(mcmc.out$beta.save[1,-(1:n.burn)],xlab=bquote
14      (beta[1]),main="",prob=TRUE,col=8,breaks=30,
15      las=1)
16  abline(v=beta[1],lwd=2,lty=2)
17  hist(mcmc.out$beta.save[2,-(1:n.burn)],xlab=bquote
18      (beta[2]),main="",prob=TRUE,col=8,breaks=30,
19      las=1)
20  abline(v=beta[2],lwd=2,lty=2)
21  hist(mcmc.out$c.save[1,-(1:n.burn)],xlab=bquote
22      (c[1]),main="",prob=TRUE,col=8,breaks=seq
23      (0.4,0.6,by=0.006),las=1)
24  abline(v=c.vec[1],lwd=2,lty=2)
25  hist(mcmc.out$c.save[2,-(1:n.burn)],xlab=bquote
26      (c[2]),main="",prob=TRUE,col=8,breaks
27      =seq(0.4,0.6,by=0.006),las=1)
28  abline(v=c.vec[2],lwd=2,lty=2)
29  hist(mcmc.out$t0.save[1,-(1:n.burn)],xlab=bquote
30      (t[0]),main="",prob=TRUE,col=8,breaks=seq
31      (-11,-9,by=0.05),las=1)
32  abline(v=t0,lwd=2,lty=2)
33  hist(mcmc.out$theta.save[1,-(1:n.burn)],
34      xlab=bquote(theta),main="",prob=TRUE,col=8,
35      breaks=30,las=1)
36  abline(v=theta,lwd=2,lty=2)
```

This R code results in marginal posterior histograms for model parameters shown in Figure 28.4. Based on the posterior results shown in Figure 28.4, the parameters associated with the diffusion coefficient (β) capture the true values used to simulate the data. The posterior histograms for c and t_0 indicate that there is enough information in the data to recover the initial position and time of the simulated invasion. Furthermore, the posterior histogram for θ in Figure 28.4 indicates that we also can recover the initial population size as well.

Overall, the mechanistic spatio-temporal statistical model that we presented in this section could be embedded in a hierarchical framework to accommodate

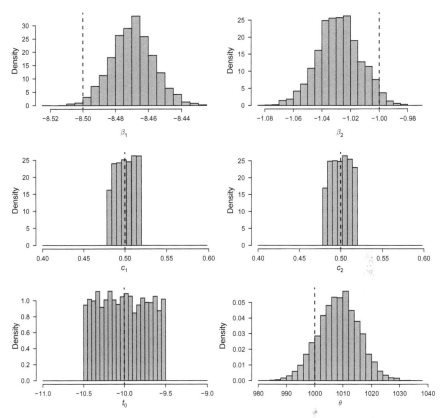

Figure 28.4 Marginal posterior histograms for all parameters. True parameters values used to simulate the data shown as dashed vertical lines.

measurement error like the phenomenological spatio-temporal model in the preceding section. For example, as long as we met the assumptions in the sampling protocol and made repeated observations of the process, we could use an N-mixture model (Chapter 24) to account for imperfect detectability in the data (Williams et al. 2017).

28.3   ADDITIONAL CONCEPTS AND READING

Spatio-temporal statistical models have been used with increasing popularity in recent years and there are several good examples in the literature of how we can apply them in different settings. For example, Hooten et al. (2007) used a discrete-time matrix model (Caswell 2001) specification for the process component of a spatio-temporal model in an N-mixture framework to model invasive species. Hu and Bradley (2018) applied spatio-temporal models using a latent multivariate log-gamma random process to model earthquake magnitudes. In an effort to forecast soil moisture, McDermott and Wikle (2016) used a spatio-temporal dynamic model based on a concept known as "analog" forecasting. Analog forecasting relates current dynamics to historical systems.

Hooten and Wikle (2010) presented an approach for modeling wildlife disease spread using discrete-time discrete-space models motivated by stochastic cellular automata dynamics.[7] Broms et al. (2016) extended this agent-based modeling framework to incorporate imperfect detectability and model the spread of invasive species.

In terms of more general references, the book by Cressie and Wikle (2011) serves as a comprehensive reference for spatio-temporal statistical models and the book by Wikle et al. (2019) provides an excellent introduction to the associated software for spatio-temporal statistics. Le and Zidek (2006) also describe spatio-temporal methods for analyzing environmental data.

In practice, there can be many complications when implementing statistical models with processes governed by PDEs. For example, we need to consider initial conditions, boundary conditions, discretization resolutions (e.g., Δt, Δs_1, Δs_2). Also, we must consider whether the spatial and temporal resolution of our data is fine enough to recover the behavior in the dynamical system and whether we have enough data to provide useful inference. Finally, over large spatial and temporal extents with fine resolutions, solving PDEs with numerical methods can be slow. This slowness creates a computational bottleneck for statistical inference because we need to solve the PDE thousands of times in a MCMC algorithm.

For computationally intensive PDEs, two approaches can be useful for making spatio-temporal statistical models more feasible to implement: homogenization and emulation. Although the details of these topics are beyond the scope of this book, we summarize them briefly in what follows.

- Homogenization: This technique involves a direct analysis of the PDE that considers multiple scales of space and time. For certain classes of PDEs, homogenization results in an approximation consistent with the original dynamics in the PDE and retains the high resolution structure in the solution while resulting in large potential savings in computation. Examples of the use of homogenization to facilitate mathematical models for wildlife disease systems are presented by Garlick et al. (2011) and Garlick et al. (2014). Hooten et al. (2013) provided an overview of the homogenization approach from a statistical perspective and Williams et al. (2017) and Hefley et al. (2017c) provided examples of its use for modeling ecological processes.

- Emulation: Statistical emulators are predictive models trained to mimic process models (potentially involving differential equations) and, as a result, are much faster to solve than the original models. For example, suppose a system of ordinary differential equations takes minutes to solve, but we can train a predictive statistical model so that it provides solutions (with some uncertainty) in seconds. We can use the statistical model in place of the original process model in a larger inferential framework. Unlike the homogenization technique,

---

[7] Hooten and Wikle (2010) referred to this as an agent-based model because the spatial regions are dynamic and connected to each other in the model, acting as agents.

most emulators are purely statistical and do not necessarily involve direct analytical work with the underlying mathematical process model. Instead, we train emulators using a computer experiment, where the process model is solved for a range of parameter settings and possibly initial conditions. Then we use the inputs (e.g., parameters, initial conditions) and outputs (process solution) from the experiment as data in a statistical model where the outputs are the response variables and some function of the inputs serve as the predictor variables. After this statistical model is fit, we can use it to predict the outputs given specific input settings. Usually, we can obtain these statistical predictions much faster than solutions of the original process model. Thus, "solving" the model thousands of times in a MCMC algorithm becomes practical. However, because the emulated solutions have uncertainty associated with them (because the statistical model is not a perfect predictor of the mathematical solution), we must account for them in the larger inferential framework. We often use Gaussian processes as emulators because of their optimal predictive properties (e.g., Higdon et al. 2008) however, other approaches based on first-order (i.e., mean) structure have also been proposed as emulators (Hooten et al. 2011).

29 Hamiltonian Monte Carlo

Bayesian computing methods for ecological and environmental models primarily consist of conventional MCMC algorithms (i.e., those involving conventional M-H and Gibbs updates), either those constructed natively in R or as part of an automated software package (e.g., JAGS or BUGS); hence, the focus on conventional MCMC algorithms in this book. However, since the development of STAN (Carpenter et al. 2016), there has been a rise in the use of Hamiltonian MC methods (HMC; Neal 2011) to fit Bayesian ecological models (e.g., Tredennick et al. 2017). Although the STAN software requires a model syntax similar, and only slightly more involved, to specify than JAGS or BUGS, some ecologists have embraced it. However, with few exceptions, most users of the STAN software are mystified by how it actually works. This may be in contrast to JAGS and BUGS because conventional MCMC algorithms are relatively easy to understand and implement (as we have shown in previous chapters).

Although the theory that gives rise to conventional MCMC algorithms is still out of reach for many users because it relies on a detailed knowledge of probability and stochastic processes, MCMC algorithms themselves are straightforward. On the other hand, the theory and algorithms increase in sophistication substantially for Hamiltonian Monte Carlo (HMC). Thus, despite recent efforts to communicate an intuition about HMC to ecologists (Monnahan et al. 2017), the theory and procedure are still poorly understood by most users of automated HMC software. In what follows, we describe a simple heuristic for HMC and demonstrate how to actually implement a version of HMC to fit a Bayesian regression model (e.g., Chapter 11). Our presentation of HMC is different than most because we seek to connect it directly to practice with a specific statistical model rather than convey the generality of where it came from and the deep connections of HMC to physics (Betancourt et al. 2017). In light of that, our presentation of HMC methods is similar to our overview of MCMC from Chapter 4, but we relegated it to this final chapter because it is more challenging to implement from scratch than conventional MCMC methods.

The essence of HMC in practice is that we use a MCMC algorithm with M-H updates based on proposals that are the endpoints of continuous-time trajectories through the parameter space. We use the discretization resolution and length of the trajectory, as well as the magnitude of the initial velocity of the trajectory, as tuning parameters in HMC. By contrast, MCMC with conventional M-H updates following the procedure that we described in Chapter 4 is based on a single value from a proposal distribution often unrelated to the target distribution of interest (i.e., the posterior distribution). For example, the normal random walk proposal we used regularly throughout this book gives rise to proposed points in the parameter space. The proposed point becomes the next value of the Markov chain if it is fairly close to the previous value in the Markov chain and if it is in a high density region of the posterior distribution.

In HMC, the proposals are more likely to arise in regions of the posterior distribution with higher density because we base the proposed value on a trajectory that responds to the gradient of the target distribution. In this way, the Hamiltonian trajectory is strikingly similar to a continuous-time animal movement model (e.g., Brillinger 2010; Hooten and Johnson 2017). Instead of geographic space, the trajectory traverses the parameter space taking cues from the shape of the posterior distribution. However, unlike most continuous-time animal movement models, each Hamiltonian trajectory is deterministic after conditioning on its starting position and initial velocity.

The path of the Hamiltonian trajectory can be expressed as a set of differential equations (i.e., the Hamiltonian equations). These Hamiltonian equations use the shape of the target distribution (i.e., the posterior or full-conditional distribution) for guidance on where to send the trajectory. Thus, the resulting trajectory will end up in a region of the parameter space that has non-negligible posterior density. The trajectory end point is a proposed value that will be accepted with greater probability than if it was proposed from a normal random walk with a large variance parameter. However, depending on the tuning parameters, the value proposed using HMC will likely be acceptable and distant from the previous value of the parameter in the Markov chain. Using conventional M-H, proposed values will either be highly acceptable and close to the previous value or they will be less acceptable and far from the previous value in the Markov chain. Thus, we can tune HMC algorithms to be much more efficient than conventional M-H algorithms for fitting the same model because they can explore the parameter space more quickly without wasting time in low density regions of the posterior.

The concept that distinguishes HMC from conventional M-H updates is the Hamiltonian trajectory. The use of the Hamiltonian trajectory to obtain M-H proposals is straightforward. In what follows, we demonstrate how to obtain a Hamiltonian trajectory associated with a distribution of interest, then we show how to use that procedure to fit a Bayesian regression model.

Suppose that we wish to obtain a Hamiltonian trajectory for θ, a $J \times 1$ parameter vector arising from a multivariate normal distribution with known mean (μ) and covariance matrix (Σ). We can express a specific type of Hamiltonian function for this distribution as

$$h(\theta, \mathbf{v}) = -\log[\theta] - \log[\mathbf{v}], \qquad (29.1)$$

where $[\theta] \equiv \mathrm{N}(\theta|\mu, \Sigma)$ is the multivariate normal PDF evaluated at θ and the negative log of it often is referred to as the potential energy function. Similarly, the negative log of $[\mathbf{v}]$ in (29.1) is referred to as the kinetic energy function and is user specified. We consider $\mathbf{v}$ as a velocity vector with the same dimension as θ; thus, the PDF $[\mathbf{v}]$ provides the initial momentum for the Hamiltonian trajectory (hence, the term "kinetic energy;" because it gets things moving). The kinetic energy function is not related inherently to the parameter vector $\boldsymbol{\theta}$; in fact, it often is specified to be independent of the potential energy function, but it is necessary to induce some motion in a trajectory for $\boldsymbol{\theta}$ that moves around in its distribution. It is common to specify $[\mathbf{v}] \equiv N(\mathbf{0}, \sigma_v^2 \mathbf{I})$, implying that the momentum in the trajectory is based on a centered multivariate normal distribution with covariance $\sigma_v^2 \mathbf{I}$.

Hamiltonian equations control the actual movement of the trajectory

$$\frac{d\boldsymbol{\theta}(t)}{dt} = \frac{\partial h(\boldsymbol{\theta}, \mathbf{v})}{\partial \mathbf{v}}, \tag{29.2}$$

$$\frac{d\mathbf{v}(t)}{dt} = -\frac{\partial h(\boldsymbol{\theta}, \mathbf{v})}{\partial \boldsymbol{\theta}}, \tag{29.3}$$

where the partial derivatives on the right-hand side represent gradient functions (from multivariate calculus). The gradient function $\frac{\partial h(\boldsymbol{\theta}, \mathbf{v})}{\partial \mathbf{v}}$ is comprised of elements $\frac{\partial h(\boldsymbol{\theta}, \mathbf{v})}{\partial v_j}$ for $j = 1, \ldots, J$. We have indexed the parameter and velocity vectors with time (t) because we can think of their trajectories as forward dynamic processes.

As with all continuous models, we must discretize the process to simulate it (or "solve" it). Thus, our resulting trajectories will have T discrete steps for which we must choose the temporal resolution Δt. Although there are several algorithms for simulating processes based on differential equation specifications (e.g., Chapter 28), the most commonly used in HMC applications is the "leap frog" algorithm. Like Euler and Runge-Kutta algorithms, the leap frog algorithm is also sequential but starts off with a half step for the velocity vector. Based on the Hamiltonian function we specified in (29.1) and Hamiltonian system in (29.2) and (29.3), the associated leap frog algorithm is

1. Choose initial velocity $\mathbf{v}(0)$
2. Update the velocity a half step in time using

$$\mathbf{v}\left(t + \frac{\Delta t}{2}\right) = \mathbf{v}(t) - \frac{\Delta t}{2} \frac{\partial - \log[\boldsymbol{\theta}(t)]}{\partial \boldsymbol{\theta}} \tag{29.4}$$

3. Update the position using

$$\boldsymbol{\theta}(t + \Delta t) = \boldsymbol{\theta}(t) + \Delta t \frac{\partial - \log\left[\mathbf{v}\left(t + \frac{\Delta t}{2}\right)\right]}{\partial \mathbf{v}} \tag{29.5}$$

4. Update the velocity again using

$$\mathbf{v}(t + \Delta t) = \mathbf{v}\left(t + \frac{\Delta t}{2}\right) - \frac{\Delta t}{2} \frac{\partial - \log[\boldsymbol{\theta}(t + \Delta t)]}{\partial \boldsymbol{\theta}} \tag{29.6}$$

5. Let $t = t + \Delta t$, go to 2, and repeat until $t = T$.

Another way of looking at this leap frog algorithm is to first update the velocity by a half step, then, inside the loop, update position and velocity by full steps, until the last iteration, at which point we update velocity by another half step. Overall, the leap frog algorithm is preferred over the standard Euler discretization procedure because of the extra half step updates.

For our example involving the multivariate normal distribution for $\boldsymbol{\theta}$, we need to find the gradient functions. Fortunately, for the multivariate normal PDF, the required gradients are analytically tractable as

$$\frac{\partial - \log[\boldsymbol{\theta}(t)]}{\partial \boldsymbol{\theta}} = \boldsymbol{\Sigma}^{-1}(\boldsymbol{\theta}(t) - \boldsymbol{\mu}), \tag{29.7}$$

$$\frac{\partial - \log[\mathbf{v}(t)]}{\partial \mathbf{v}} = (\sigma_v^2 \mathbf{I})^{-1}(\mathbf{v}(t) - \mathbf{0}), \tag{29.8}$$

$$= \frac{\mathbf{v}(t)}{\sigma_v^2}. \tag{29.9}$$

If the gradient functions were not available as previously derived results (as in the multivariate normal case), we would have to find the derivatives analytically using calculus or, if that is not possible, using numerical differentiation.

To demonstrate how to simulate the Hamiltonian trajectories when the parameters $\boldsymbol{\mu}$, $\boldsymbol{\Sigma}$, are σ_v^2 known, we used the following R code:

BOX 29.1   SIMULATE SEQUENTIAL HAMILTONIAN TRAJECTORIES

```
1  mu=rep(0,2)
2  Sig=matrix(c(1,.9,.9,1),2,2)
3  Sig.inv=solve(Sig)
4  s2v=1
5
6  n.sim=4
7  T=150
8  dt=.025
9  theta.array=array(0,c(T,2,n.sim))
10 theta.array[T,,1]=c(0.5,0.75)
11 set.seed(97)
12 for(k in 2:n.sim){
13   theta=theta.array[T,,k-1]
14   v=rnorm(2,0,sqrt(s2v))
15   theta.array[1,,k]=theta
16   for(t in 1:T){
17     v=v-(dt/2)*Sig.inv%*%(theta-mu)
18     theta=theta+dt*v/s2v
```

(Continued)

BOX 29.1 (Continued) SIMULATE SEQUENTIAL HAMILTONIAN TRAJECTORIES

```
19    v=v-(dt/2)*Sig.inv%*%(theta-mu)
20    theta.array[t,,k]=theta
21  }
22 }
23
24 library(mixtools)
25 labs=c("a","b","c")
26 layout(matrix(1:3,3,1))
27 for(k in 2:n.sim){
28    plot(0,0,type="n",xlim=c(-3.5,3.5),ylim=c
29      (-3.5,3.5),xlab=bquote(theta[1]),ylab=bquote
30      (theta[2]),asp=TRUE,main=labs[k-1])
31    ellipse(mu,Sig,.5,1000,col=8)
32    ellipse(mu,Sig,.25,1000,col=8)
33    ellipse(mu,Sig,.05,1000,col=8)
34    lines(theta.array[,,k],lwd=1.5)
35    points(theta.array[1,1,k],theta.array[1,2,k],
36      pch=20)
37    arrows(theta.array[T-2,1,k],theta.array[T-2,
38      2,k],theta.array[T,1,k],theta.array[T,2,k],
39      length=.05,lwd=1.5)
40 }
```

Thus, for a $J = 2$ position vector $\mu \equiv (0,0)'$, velocity variance $\sigma_v^2 \equiv 1$, and

$$\Sigma \equiv \begin{bmatrix} 1 & 0.9 \\ 0.9 & 1 \end{bmatrix}, \tag{29.10}$$

we simulated three trajectories starting from the initial position $\theta \equiv (0.5, 0.75)'$ and sampled the initial velocity for each from the velocity distribution (Figure 29.1). For each trajectory, we used $T = 150$ steps with a resolution of $\Delta t = 0.025$ (Figure 29.1). The trajectories in Figure 29.1 provide a visual intuition about how an HMC algorithm might work, even though they are just sequential trajectories associated with the distribution in this case. However, to visualize how an HMC algorithm works, imagine the trajectory in Figure 29.1a provides the proposal for θ (at the point of the arrow) in iteration 1 of a MCMC algorithm. Then, assuming that proposal is accepted, we begin at the previous location and simulate the trajectory forward again which provides the proposal (at the point of the arrow in Figure 29.1b) for iteration 2 of the MCMC algorithm. If that proposal is accepted, we begin there and simulate a trajectory forward again, arriving at a new location that serves as the next proposal

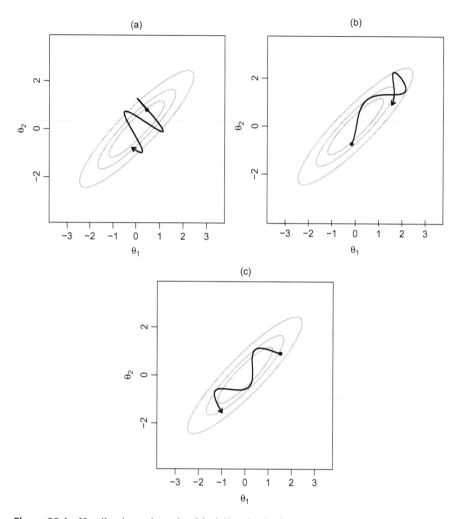

Figure 29.1 Hamiltonian trajectories (black lines beginning at point and ending at the arrow) associated with a bivariate normal distribution for $\boldsymbol{\theta}$. The trajectory for (b) begins where the trajectory for (a) ends, and the trajectory for (c) begins where the trajectory for (b) ends. Bivariate normal isopleths containing 0.50, 0.75, and 0.95 percent of the density in the distribution shown as gray ellipses.

(Figure 29.1c), etc. The stochastic aspect of each simulation occurs because the initial velocity (or momentum) $\mathbf{v}(0)$ is sampled from its distribution $[\mathbf{v}]$ which relates to the kinetic energy (i.e., direction and speed) in the trajectory.

If we retain only the ending points (at the tips of the arrows in Figure 29.1) of each sequential trajectory, then these Hamiltonian realizations will arise from our target distribution $[\boldsymbol{\theta}]$ and we can use the stochastic integration techniques (e.g., MC moment approximation) to learn about characteristics of the target distribution (Chapter 3) just as we would using a MC sample. However, if we can sample from a distribution independently, then the Hamiltonian method may not be helpful. On the other hand, when the distribution is unknown up to a normalizing constant, as we often have with Bayesian posterior and full-conditional distributions, then we cannot sample independently from the target distribution and Hamiltonian trajectories may provide proposals for high density regions of the parameter space and improve the overall efficiency of a MCMC algorithm.

An MCMC algorithm with Hamiltonian proposals is referred to as a HMC algorithm. To develop a HMC algorithm, we first let $[\boldsymbol{\theta}]$ in the Hamiltonian function represent the posterior (or full-conditional) distribution up to a multiplicative normalizing constant. Thus, we let $h(\boldsymbol{\theta}, \mathbf{v}) = -\log[\boldsymbol{\theta}|\cdot] - \log[\mathbf{v}]$, where $[\boldsymbol{\theta}|\cdot]$ is either the full-conditional distribution or a posterior (up to a normalizing constant), depending on the context, and $[\mathbf{v}]$ is the velocity distribution as specified before. The associated gradient functions and leap frog algorithm are found as before, but notice that the normalizing constant in the full-conditional distribution is irrelevant in the gradients associated with $h(\boldsymbol{\theta}, \mathbf{v})$, so we can ignore it (which implies that we do not need to know the closed-form for the full-conditional or posterior distribution, as is usually the case).

To perform a M-H update for $\boldsymbol{\theta}$ based on a Hamiltonian proposal, we use the following procedure:

1. Assign initial value for $\boldsymbol{\theta}^{(0)}$
2. Let $k = 1$
3. Sample $\mathbf{v}^{(k-1)} \sim [\mathbf{v}]$
4. Use the leap frog algorithm to simulate $\boldsymbol{\theta}^{(*)} = \boldsymbol{\theta}(T)$ and $\mathbf{v}^{(*)} = \mathbf{v}(T)$ based on T time steps (and resolution Δt), beginning from $\boldsymbol{\theta}(0) = \boldsymbol{\theta}^{(k-1)}$ and $\mathbf{v}(0) = \mathbf{v}^{(k-1)}$.
5. Calculate the M-H ratio:

$$mh = \frac{e^{-h(\boldsymbol{\theta}^{(*)}, \mathbf{v}^{(*)})}}{e^{-h(\boldsymbol{\theta}^{(k-1)}, \mathbf{v}^{(k-1)})}} \tag{29.11}$$

6. Let $\boldsymbol{\theta}^{(k)} = \boldsymbol{\theta}^{(*)}$ with probability $\min(mh, 1)$, otherwise retain the previous value $\boldsymbol{\theta}^{(k)} = \boldsymbol{\theta}^{(k-1)}$
7. Let $k = k + 1$ and go to step 3. Repeat for $k = 1, \ldots, K$, where K is large enough to approximate posterior quantities well.

In practice, we need to tune the velocity distribution (by adjusting σ_v^2) and the length (T) and resolution (Δt) of the leap frog algorithm so that the trajectory is accurate and long enough to explore the parameter space efficiently on each step. However, recall that the Hamiltonian trajectories are conditionally deterministic and will usually repeat the same path through the parameter space if allowed to be too long. Although this is not necessarily a bad thing (nor is the trajectory being too short a bad thing), it is not computationally efficient to waste time letting the trajectory retrace its steps over and over. Sophisticated approaches assist in the tuning, but a brute force approach would involve monitoring the trajectories (visually) for a trial set of iterations to assess their properties. For example, the set of trajectories shown in Figure 29.1 are good because they yield distant starting and end points and are not repetitive.

Computational methods like HMC are useful for fitting models that have high-dimensional parameter spaces, possibly with highly correlated parameters. In these situations, a MCMC algorithm with conventional M-H updates based on random walk proposals will be inefficient because they either explore too much space outside the region of high posterior density or they yield highly correlated MCMC samples because the sequential proposals are too close to the previously accepted value in the Markov chain. An example of where this discrepancy occurs is in regression models with highly correlated covariates (i.e., multicollinearity). Consider the following R code that simulates two covariates ($\mathbf{x}_1$ and $\mathbf{x}_2$) and response data ($\mathbf{y}$):

BOX 29.2   SIMULATE LINEAR REGRESSION DATA

```
set.seed(101)
n=100
x1=rnorm(n)
x2=rnorm(n,.99*x1,.1)
X=cbind(rep(1,n),x1,x2)
beta=c(.5,1,1)
sig=.1
s2=sig^2
y=rnorm(n,X%*%beta,sig)

cor(x1,x2)
```

The correlation between $\mathbf{x}_1$ and $\mathbf{x}_2$ is approximately 0.99 and this causes the associated regression coefficient estimates to be highly correlated as well. From a Bayesian perspective, the joint posterior distribution of β_1 and β_2 will have a shape similar to that shown in Figure 29.1. Even though we can derive conjugate full-conditional distributions for this model (depending on the specified priors; Chapter 11), suppose for a moment that we could not analytically derive the full-conditional distributions and had to resort to M-H updates for β in our MCMC algorithm. Because the posterior distribution is so long and thin, a random walk will usually propose values far away from the posterior distribution if the tuning parameter (i.e., proposal variance) is large. Thus, this is a case where HMC will outperform conventional M-H based on random walk proposals.

To set up this Bayesian linear regression example based on the data we simulated above, we specify the same model as described in Chapter 11:

$$\mathbf{y} \sim N(\mathbf{X}\boldsymbol{\beta}, \sigma^2\mathbf{I}), \tag{29.12}$$

$$\boldsymbol{\beta} \sim N(\boldsymbol{\mu}_\beta, \sigma_\beta^2\mathbf{I}), \tag{29.13}$$

$$\sigma^2 \sim IG(q, r), \tag{29.14}$$

where, for the data we simulated above, $\mathbf{y}$ is $n \times 1$ with $n = 100$ and $\mathbf{X}$ is $n \times p$ with $p = 3$ (i.e., one intercept and two covariates). As in the previous example, we used a standard normal distribution for $[\mathbf{v}]$ which results in the following Hamiltonian function for this model:

$$h(\boldsymbol{\beta}, \mathbf{v}) = -\log[\boldsymbol{\beta}|\cdot] - \log[\mathbf{v}], \tag{29.15}$$

$$= -\log\left([\mathbf{y}|\boldsymbol{\beta}, \sigma^2][\boldsymbol{\beta}]\right) - \log[\mathbf{v}]. \tag{29.16}$$

We can take advantage of the fact that the data model $[\mathbf{y}|\boldsymbol{\beta}, \sigma^2]$ and prior $[\boldsymbol{\beta}]$ have a multivariate normal kernel to derive the gradients associated with the Hamiltonian function. We note that we can write the potential energy portion of the Hamiltonian function as

$$-\log([\mathbf{y}|\boldsymbol{\beta}, \sigma^2][\boldsymbol{\beta}]) = -\log[\mathbf{y}|\boldsymbol{\beta}, \sigma^2] - \log[\boldsymbol{\beta}], \tag{29.17}$$

$$= \frac{1}{2}(\mathbf{y} - \mathbf{X}\boldsymbol{\beta})'\boldsymbol{\Sigma}^{-1}(\mathbf{y} - \mathbf{X}\boldsymbol{\beta}) + \frac{1}{2}(\boldsymbol{\beta} - \boldsymbol{\mu}_\beta)'\boldsymbol{\Sigma}_\beta^{-1}(\boldsymbol{\beta} - \boldsymbol{\mu}_\beta), \tag{29.18}$$

where the gradient of the first component in (29.18) is $-\mathbf{X}'\boldsymbol{\Sigma}^{-1}\mathbf{y} + \mathbf{X}'\boldsymbol{\Sigma}^{-1}\mathbf{X}\boldsymbol{\beta}$ and the gradient of the second component in (29.18) is $-\boldsymbol{\Sigma}_\beta^{-1}\boldsymbol{\mu}_\beta + \boldsymbol{\Sigma}_\beta^{-1}\boldsymbol{\beta}$ (Seber 2008). Adding these gradients together and combining like terms results in

$$\frac{\partial - \log([\mathbf{y}|\boldsymbol{\beta}, \sigma^2][\boldsymbol{\beta}])}{\partial\boldsymbol{\beta}} = (\mathbf{X}'\boldsymbol{\Sigma}^{-1}\mathbf{X} + \boldsymbol{\Sigma}_\beta^{-1})\boldsymbol{\beta} - (\mathbf{X}'\boldsymbol{\Sigma}^{-1}\mathbf{y} + \boldsymbol{\Sigma}_\beta^{-1}\boldsymbol{\mu}_\beta). \tag{29.19}$$

Also, recall that the gradient for $-\log[\mathbf{v}]$ is $\mathbf{v}/\sigma_v^2$. Thus, we have all the pieces we need to implement a leap frog algorithm to simulate trajectories from the full-conditional distribution for $\boldsymbol{\beta}$ using the equations in (29.4–29.6) in an HMC algorithm.

Because we now have (at least) three different ways to fit this Bayesian regression model (i.e., MCMC with Gibbs, M-H, and HMC updates for $\boldsymbol{\beta}$), we can construct three MCMC algorithms representing each and compare their efficiency and behavior. Critically, we note that we can use HMC to update a subset of parameters in a MCMC algorithm, leaving the other parameters to be updated using Gibbs or M-H. This outcome allows us to target certain problematic components of statistical models without forcing us to use Hamiltonian updates for all of the parameters. Thus, in each of the algorithms that follow, we use a Gibbs update for σ^2 based on the full-conditional distribution that we derived in Chapter 11:

$$[\sigma^2|\cdot] = IG\left(\frac{n}{2} + q, \left(\frac{(\mathbf{y} - \mathbf{X}\boldsymbol{\beta})'(\mathbf{y} - \mathbf{X}\boldsymbol{\beta})}{2} + \frac{1}{r}\right)^{-1}\right). \tag{29.20}$$

In what follows, we present the R code for the three different MCMC algorithms using Gibbs, M-H, and HMC updates for β. Beginning with the Gibbs sampler version of the MCMC algorithm we presented in Chapter 11, we have the R function called reg.mcmc.R:

BOX 29.3   MCMC ALGORITHM USING GIBBS UPDATES FOR LINEAR REGRESSION

```
reg.mcmc <- function(y,X,beta.mn,beta.var,n.mcmc){

###
### Setup and Priors
###

n=dim(X)[1]
p=dim(X)[2]
r=1000
q=.001
Sig.beta=beta.var*diag(p)
Sig.beta.inv=solve(Sig.beta)

beta.save=matrix(0,p,n.mcmc)
s2.save=rep(0,n.mcmc)

###
### Starting Values
###

XpX=t(X)%*%X
beta=solve(XpX)%*%t(X)%*%y

###
### MCMC Loop
###

for(k in 1:n.mcmc){

    ###
    ### Sample s2
    ###
```

(*Continued*)

BOX 29.3 (Continued) MCMC ALGORITHM USING GIBBS UPDATES FOR LINEAR REGRESSION

```
33  tmp.r=(1/r+.5*t(y-X%*%beta)%*%(y-X%*%beta))^(-1)
34  tmp.q=n/2+q
35
36  s2=1/rgamma(1,tmp.q,,tmp.r)
37
38  ###
39  ### Sample beta
40  ###
41
42  tmp.chol=chol(XpX/s2 + Sig.beta.inv)
43  beta=backsolve(tmp.chol,backsolve(tmp.chol,t(X)%
44    *%y/s2 + Sig.beta.inv%*%beta.mn,transpose=
45    TRUE)+rnorm(p))
46
47  ###
48  ### Save Samples
49  ###
50
51  beta.save[,k]=beta
52  s2.save[k]=s2
53
54 }
55
56 ###
57 ###  Write Output
58 ###
59
60 list(beta.save=beta.save,s2.save=s2.save,y=y,X=X,
61    n.mcmc=n.mcmc,n=n,r=r,q=q,p=p)
62
63 }
```

Notice that we used the backsolve technique to sample from the conjugate full-conditional distribution for β in this algorithm.

The next MCMC algorithm is called reg.mh.mcmc.R and uses M-H updates for β based on normal random walk proposals:

BOX 29.4 MCMC ALGORITHM USING M-H UPDATES FOR LINEAR REGRESSION

```
1  reg.mh.mcmc <- function(y,X,beta.mn,beta.var,beta.
      tune,n.mcmc){
2
3  ###
4  ### Setup and Priors
5  ###
6
7  n=dim(X)[1]
8  p=dim(X)[2]
9  r=1000
10 q=.001
11 Sig.beta=beta.var*diag(p)
12 Sig.beta.inv=solve(Sig.beta)
13 beta.sd=sqrt(beta.var)
14
15 beta.save=matrix(0,p,n.mcmc)
16 s2.save=rep(0,n.mcmc)
17
18 ###
19 ### Starting Values
20 ###
21
22 XpX=t(X)%*%X
23 beta=solve(XpX)%*%t(X)%*%y
24 beta.acc=0
25
26 ###
27 ### MCMC Loop
28 ###
29
30 for(k in 1:n.mcmc){
31
32    ###
33    ### Sample s2
34    ###
35
36    tmp.r=(1/r+.5*t(y-X%*%beta)%*%(y-X%*%beta))^(-1)
37    tmp.q=n/2+q
```

(*Continued*)

BOX 29.4 (Continued)   MCMC ALGORITHM USING M-H UPDATES FOR LINEAR REGRESSION

```
38  s2=1/rgamma(1,tmp.q,,tmp.r)
39
40  ###
41  ### Sample beta
42  ###
43
44  beta.star=rnorm(p,beta,beta.tune)
45  mh.1=sum(dnorm(y,X%*%beta.star,sqrt(s2),
46    log=TRUE))+sum(dnorm(beta.star,beta.mn,
47    beta.sd,log=TRUE))
48  mh.2=sum(dnorm(y,X%*%beta,sqrt(s2),log=TRUE))+
49    sum(dnorm(beta,beta.mn,beta.sd,log=TRUE))
50  mh=exp(mh.1-mh.2)
51  if(mh>runif(1)){
52    beta=beta.star
53    beta.acc=beta.acc+1
54  }
55
56  ###
57  ### Save Samples
58  ###
59
60  beta.save[,k]=beta
61  s2.save[k]=s2
62
63 }
64
65 ###
66 ### Write Output
67 ###
68
69 list(beta.save=beta.save,s2.save=s2.save,y=y,
70    X=X,n.mcmc=n.mcmc,n=n,r=r,q=q,p=p,
71    beta.acc=beta.acc)
72
73 }
```

In this algorithm, notice that we still perform block updates for the entire coefficient vector β on each iteration. Also notice that we keep track of the number of times that we accept the proposal with the variable beta.acc so that we can tune this M-H version of the algorithm to match the number of accepted proposals in the HMC version of the algorithm that follows.

Finally, the HMC version of the MCMC algorithm to fit the Bayesian regression model in R is called reg.hmc.R:

**BOX 29.5   MCMC ALGORITHM USING HMC UPDATES FOR
LINEAR REGRESSION**

```
 1 reg.hmc <- function(y,X,beta.mn,beta.var,n.mcmc){
 2
 3 ###
 4 ### Setup and Priors
 5 ###
 6
 7 n=dim(X)[1]
 8 p=dim(X)[2]
 9 T=100
10 dt=.01
11 r=1000
12 q=0.001
13 Sig.beta=beta.var*diag(p)
14 Sig.beta.inv=solve(Sig.beta)
15 beta.sd=sqrt(beta.var)
16
17 beta.save=matrix(0,p,n.mcmc)
18 s2.save=rep(0,n.mcmc)
19
20 ###
21 ### Starting Values
22 ###
23
24 XpX=t(X)%*%X
25 beta=solve(XpX)%*%t(X)%*%y
26 v.star=rnorm(p)
27
28 beta.acc=0
```

(Continued)

BOX 29.5 (Continued)   MCMC ALGORITHM USING HMC UPDATES FOR LINEAR REGRESSION

```
29 ###
30 ### MCMC Loop
31 ###
32
33 for(k in 1:n.mcmc){
34
35   ###
36   ### Sample s2
37   ###
38
39   tmp.r=(1/r+.5*t(y-X%*%beta)%*%(y-X%*%beta))^(-1)
40   tmp.q=n/2+q
41
42   s2=1/rgamma(1,tmp.q,,tmp.r)
43   sig=sqrt(s2)
44
45   ###
46   ### Sample beta
47   ###
48
49   beta.star=beta
50   v=rnorm(p)
51   v.star=v
52   for(t in 1:T){
53     v.star=v.star-(dt/2)*((XpX/s2+Sig.beta.inv)%*%
54         beta.star-(t(X)%*%y/s2+Sig.beta.inv%*%
55         beta.mn))
56     beta.star=beta.star+dt*v.star
57     v.star=v.star-(dt/2)*((XpX/s2+Sig.beta.inv)%*%
58         beta.star-(t(X)%*%y/s2+Sig.beta.inv%*%
59         beta.mn))
60   }
61
62   mh.1=sum(dnorm(y,X%*%beta.star,sig,log=TRUE))+
63       sum(dnorm(beta.star,beta.mn,beta.sd,
64       log=TRUE))+sum(dnorm(v.star,log=TRUE))
```

(Continued)

BOX 29.5 (Continued)   MCMC ALGORITHM USING HMC UPDATES FOR LINEAR REGRESSION

```
63    mh.2=sum(dnorm(y,X%*%beta,sig,log=TRUE))+
64       sum(dnorm(beta,beta.mn,beta.sd,log=TRUE))+
65       sum(dnorm(v,log=TRUE))
66    mh=exp(mh.1-mh.2)
67    if(mh>runif(1)){
68       beta=beta.star
69       beta.acc=beta.acc+1
70    }
71
72    ###
73    ### Save Samples
74    ###
75
76    beta.save[,k]=beta
77    s2.save[k]=s2
78
79 }
80
81 ###
82 ###   Write Output
83 ###
84
85 list(beta.save=beta.save,s2.save=s2.save,y=y,X=X,
86       n.mcmc=n.mcmc,n=n,r=r,q=q,p=p,T=T,dt=dt,
87       beta.acc=beta.acc)
88
89 }
```

Notice that, in this HMC version of the MCMC algorithm, we used the leap frog algorithm to simulate the trajectories for β, using the last value of the trajectory as the proposal on each MCMC iteration. We then follow that up with a M-H update based on the value of the Hamiltonian function evaluated at the proposed and previous coefficients (conditioning on the most recent value for σ^2) and velocity.

We fit the Bayesian regression model to the simulated data using all three algorithms and the following R code:

BOX 29.6   FIT LINEAR REGRESSION MODEL USING THREE APPROACHES

```
n.mcmc=2000
source("reg.hmc.R") # Code Box 29.5
set.seed(201)
hmc.out=reg.hmc(y=y,X=X,beta.mn=rep(0,3),beta.var
    =100,n.mcmc=n.mcmc)

source("reg.mh.mcmc.R") # Code Box 29.4
set.seed(201)
mh.out=reg.mh.mcmc(y=y,X=X,beta.mn=rep(0,3),beta.
    var=100,beta.tune=0.0035,n.mcmc=n.mcmc)

source("reg.mcmc.R") # Code Box 29.3
set.seed(201)
mcmc.out=reg.mcmc(y=y,X=X,beta.mn=rep(0,3),beta.
    var=100,n.mcmc=n.mcmc)

layout(matrix(1:3,3,1))
plot(hmc.out$beta.save[2,],type="l",
    ylim=c(.5,1.2),ylab=bquote(beta[1]),
    xlab="iteration",main="a")
abline(h=beta[2],col=8)
plot(mh.out$beta.save[2,],type="l",ylim=c(.5,1.2),
    ylab=bquote(beta[1]),xlab="iteration",
    main="b")
abline(h=beta[2],col=8)
plot(mcmc.out$beta.save[2,],type="l",ylim=c
    (.5,1.2),ylab=bquote(beta[1]),
    xlab="iteration",main="c")
abline(h=beta[2],col=8)

apply(hmc.out$beta.save,1,mean)
apply(mh.out$beta.save,1,mean)
apply(mcmc.out$beta.save,1,mean)
apply(hmc.out$beta.save,1,sd)
apply(mh.out$beta.save,1,sd)
apply(mcmc.out$beta.save,1,sd)

hmc.out$beta.acc
mh.out$beta.acc
```

Figure 29.2 shows trace plots for the regression coefficient associated with the first covariate (β_1; the trace plots for β_2 are similar). We used the same prior specifications for the model in each algorithm ($\boldsymbol{\mu}_\beta = \mathbf{0}$, $\sigma_\beta^2 = 100$, $q = 0.001$, and $r = 1000$). For tuning of the HMC updates, we used $T = 100$ and $\Delta t = 0.01$, and for the conventional M-H updates we used $\sigma_{\text{tune}} = 0.0035$ (where $\boldsymbol{\beta}^{(*)} \sim N(\boldsymbol{\beta}^{(k-1)}, \sigma_{\text{tune}}^2 \mathbf{I})$) to match the acceptance probability of the HMC algorithm, which was approximately 0.79. The striking comparison in Figure 29.2 is the difference between the HMC and M-H updates (Figure 29.2a versus Figure 29.2b). Clearly the conventional M-H updates yield a very highly correlated Markov chain that mixes slowly relative to those resulting from the algorithms using the HMC and Gibbs updates for $\boldsymbol{\beta}$. Normally, upon visual inspection of the M-H trace plot, we would increase the tuning parameter (σ_{tune}) to improve mixing; however, in this case, doing so quickly results in a Markov chain comprised of very few accepted proposals. Thus, we would need to substantially increase the number of MCMC iterations for the conventional M-H approach to yield the similar accuracy in posterior inference as HMC and Gibbs.[1]

In terms of posterior quantities, Table 29.1 summarizes the posterior means and standard deviations for the regression coefficients resulting from each of the three algorithms. Table 29.1 indicates that the inference arising from the MCMC algorithm based on HMC updates is similar to that resulting from Gibbs updates. However, because of the poorer mixing in the conventional M-H trace plots (Figure 29.2b), we see that, while the posterior means are close, the posterior standard deviations differ substantially between the M-H and HMC/Gibbs approaches. These results illustrate the primary advantage of HMC over conventional M-H updates: HMC proposals are usually better and that results in more efficient MCMC algorithms that take less time to converge and require fewer samples to accurately approximate posterior quantities. However, HMC algorithms require nested "for" loops because simulating the Hamiltonian trajectory requires a loop that is inside the larger MCMC loop. Programmed natively in R, the HMC will be slower than conventional M-H on a per iteration basis, but because many fewer HMC iterations are required, the overall algorithm may be faster to achieve the same accuracy. Furthermore, the loop to simulate Hamiltonian trajectory can be programmed using C++ or Rcpp, which will be much faster than a loop directly in R (Wickham 2014; Chambers 2016). This is why the STAN software interfaces with C++ to perform the Hamiltonian updates and may appear faster than JAGS or BUGS to fit the same model.

Overall, we find that the additional time deriving gradient functions and programming the HMC algorithm may be worthwhile if certain full-conditionals are particularly amenable to HMC updates and if the model needs to be fit to many data sets for separate analyses. HMC also may be worthwhile in cases where we need to fit models that have highly correlated parameters and unusually shaped posterior distributions. However, it is worth highlighting that we were able to analytically derive the associated gradient functions in our regression example. We may need to find

---

[1] In fact, two orders of magnitude more MCMC iterations (i.e., $K = 200000$) are needed in this example if we use the same M-H updates and tuning parameter to achieve similar results as the HMC and Gibbs methods!

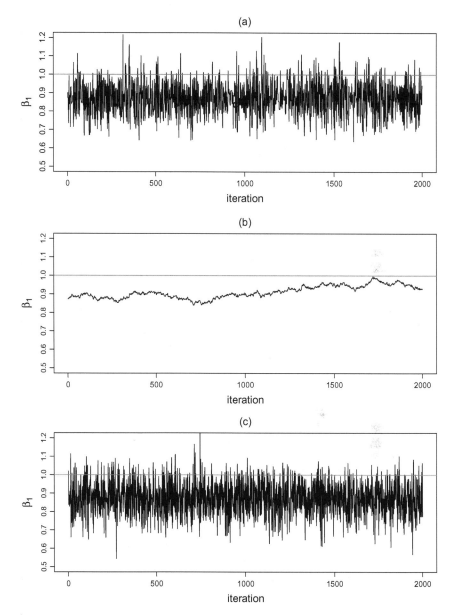

Figure 29.2 Trace plots ($K = 2000$ iterations for illustration) for the first slope coefficient β_1 resulting from fitting the Bayesian linear regression model to the simulated data using MCMC with updates for β based on (a) HMC, (b) M-H, and (c) Gibbs. The true value for the coefficient is shown as a horizontal gray line.

Table 29.1

Posterior Means and Standard Deviations for β Resulting from Fitting the Bayesian Regression Model to the Simulated Data Using MCMC Algorithms with Three Different Kinds of Updates for β: HMC, M-H, and Gibbs

Quantity	Algorithm		
	HMC	**M-H**	**Gibbs**
$E(\beta_0\|\mathbf{y})$	0.500	0.500	0.501
$E(\beta_1\|\mathbf{y})$	0.874	0.912	0.874
$E(\beta_2\|\mathbf{y})$	1.126	1.089	1.126
$\sqrt{\mathrm{Var}(\beta_0\|\mathbf{y})}$	0.009	0.008	0.009
$\sqrt{\mathrm{Var}(\beta_1\|\mathbf{y})}$	0.088	0.034	0.089
$\sqrt{\mathrm{Var}(\beta_2\|\mathbf{y})}$	0.088	0.032	0.089

the gradients associated with the Hamiltonian function for more complicated models using numerical differentiation, which will inevitably slow the computation and require additional supervised tuning. However, there is still value in understanding how HMC works because there may be convenient situations where we can use it in algorithms with other non-HMC updates. Also, for users of the STAN software, an understanding of the algorithm behind the scenes may help when troubleshooting problems with the implementation for particular models.

Finally, we note that we contrived the simulation example provided in this chapter such that the covariates would be highly collinear, resulting in a strongly correlated parameter space for β. With the large degree of collinearity in our example, we would expect to see an effect on the posterior inference for β. In particular, notice that the marginal posterior distribution for β_1 is smaller than the value used to simulate the data ($\beta_1 = 1$) and, similarly, the posterior distribution for β_2 is larger than the value used to simulate the data (β_2; Table 29.1). These outcomes happen because the two coefficients are competing to explain the same variation in the response.

Although the HMC approach allows us to fit the model with less computational issues than conventional MCMC, there are also several remedial approaches for dealing with the model structure. The traditional, and simplest, approach is to simply exclude one of the covariates from the model to avoid redundancy. A second approach is to use regularization (Hooten and Hobbs 2015) to shrink the posterior for β toward zero based on a precise prior to alleviate the effects of collinearity (i.e., inducing a Bayesian ridge regression) and yield a better predictive model. Finally, we could use principle components analysis (PCA) regression to decorrelate the covariates prior to model fitting. Using the resulting PCA scores in place of the covariates in the regression model will still retain the explanatory information from both covariates while facilitating model fitting. In this latter case, however, we lose the ability to make direct inferential statements about the original untransformed covariates.

It is important to remember that the HMC approach to fitting models is not a fundamentally new way to model ecological and environmental processes, only a way to fit the models that we already know how to specify. Thus, using it in situations with high collinearity (e.g., the example in this chapter) does not automatically remedy a pathological situation with a poorly specified statistical model.

29.1 ADDITIONAL CONCEPTS AND READING

Implementing HMC for our linear regression example in this chapter is convenient because it involved multivariate normal distributions for a random parameter vector with real support and no hard constraints. In a more general context, HMC is most useful for parameters with continuous support as in our example because the derivatives are well defined in the gradients. Thus, HMC methods are not easily amenable to models with discrete-valued parameters (but see Zhang et al. 2012; Pakman and Paninski 2013; Nishimura et al. 2017). New developments in HMC theory are arising all the time, however, and there are existing ways to handle constrained parameters spaces. For example, the variance parameter σ^2 in our regression model has positive support; thus, we would need to modify the Hamiltonian function to recognize that.

Langevin algorithms for sampling from posterior distributions are related to HMC algorithms and may also be useful. Similarly, we could use adaptive MCMC methods to more efficiently sample from posterior distributions (Givens and Hoeting 2013). The most common form of adaptive MCMC relies on conventional M-H updates, but tunes the proposal distribution to yield more optimal M-H acceptance probabilities. A simple brute force approach for adaptive MCMC would involve calculating the running acceptance probability for a certain parameter (or parameters) as a MCMC algorithm runs and allow it to increase or decrease the tuning parameter as necessary to achieve a certain acceptance rate. As long as we stop modifying the proposal at some point and discard all iterations up to that point, the resulting Markov chain still will have detailed balance (a term used in MCMC theory) and be guaranteed to converge to the ergodic stationary target distribution (Andrieu and Thoms 2008; Roberts and Rosenthal 2009).

Another, more general, approach to adaptive MCMC involves tuning the proposal covariance for block updates of parameters like those we used for coefficient vectors in regression models throughout this book. For example, if we used the proposal distribution $\boldsymbol{\beta}^{(*)} \sim N(\boldsymbol{\beta}^{(k-1)}, \boldsymbol{\Sigma}_{\text{tune}})$, then we could obtain an estimate for the covariance $\boldsymbol{\Sigma}_{\text{tune}}$ based on a sample covariance estimate using a set of samples $\boldsymbol{\beta}^{(k)}$ ($k = 1, \ldots, K_{\text{trial}}$) resulting from a trial run of the MCMC algorithm using only a diagonal proposal distribution covariance matrix (i.e., $\sigma^2_{\text{tune}}\mathbf{I}$). We could continue to refine the tuning covariance matrix using additional trial runs until we see an improvement in the trace plots or reach a desired M-H acceptance probability. Again, at some point, we need to stop adapting the proposal and discard all prior samples when making inference. These adaptive MCMC methods can be very helpful to automate the tuning for complicated MCMC algorithms where monitoring individual trace plots is infeasible. Furthermore, we can use similar adaptive MCMC strategies to tune HMC algorithms as well, which may be particularly useful in those settings because HMC algorithms often have more tuning parameters than conventional M-H. See Andrieu and Thoms (2008), Roberts and Rosenthal (2009), and Givens and Hoeting (2013) for further details and references on adaptive MCMC methods.

In our presentation of HMC in this chapter, we focused on a very simple form of HMC algorithm for a simple model where it was straightforward to tune the proposals so that they provide well-mixed Markov chains. However, for more complicated models, adaptive MCMC approaches also have been developed for use with HMC. In particular, the No-U-Turn sampler (NUTS) is based on an adaptive approach for tuning the trajectory length in HMC (Hoffman and Gelman 2014).

Tips and Tricks

automatic marginalization: Inference from a single parameter or variable from a Bayesian model is obtained from the marginal posterior distribution. This requires integrating over all other random variables in the joint posterior distribution. When using sampling-based algorithms to obtain samples from the joint posterior, the marginal posterior can be obtained automatically (Chapter 11).

auxiliary variables: Some Bayesian models can be expressed to include additional random variables that lead to more efficient algorithms (Section 20.3).

backsolve function: Obtaining samples from a multivariate normal distribution can be computationally challenging. Using the backsolve function in R is an efficient way to sample from the multivariate normal distribution (Chapter 14).

basic linear algebra subprograms: Linear algebra functions allow us to avoid using loops and are efficient for multivariate calculations. Although R comes equipped with basic linear algebra subprograms (BLAS), they may not be optimized to perform vector and matrix calculations. Upgrading to an optimized BLAS will speed up many linear algebra calculations. Several options are available such as openBLAS, and some modified installations of R come equipped with an optimized BLAS.

closed-form full-conditional: MCMC algorithms require sampling from full-conditional distributions. If a full-conditional distribution is available in closed-form, then it can be sampled from directly using a known distribution without any required tuning.

conditional likelihood: Bayesian models can contain a large number of random variables. In some situations, variables can be removed before implementing the model by deriving a likelihood function that conditions on the sufficient statistics for the random variables that need to be eliminated (Section 21.3).

data augmentation: Some Bayesian models can be implemented using algorithms that were developed for other types of models by merely augmenting the observed data in certain ways (Chapter 24).

discrete prior: Using a discrete prior for a continuous parameter allows certain steps within the algorithm to be pre-computed, which can take advantage of parallel computing resources (Section 17.1).

invariance property: Inference for a transformed parameter(s) is often desired. When using sampling-based approaches, posterior inference can be obtained by transforming the MCMC sample resulting from fitting the model (Section 3.2).

log-sum-exp: Calculating ratios can lead to numerical underflow/overflow issues. The log-sum-exp trick may help avoid such problems using a simple analytical manipulation of the ratio (Chapter 15).

multi-stage model fitting: Some hierarchical Bayesian models can be fit to data in multiple stages to decrease run-time by taking advantage of parallel computing resources (Section 19.2).

Rao-Blackwellization: Sampling from the posterior distribution can be challenging if it contains a large number of random variables. Rao-Blackwellization is a technique that can eliminate a substantial number of random variables that would otherwise need to be sampled, often leading to more efficient MCMC algorithms and well-mixed Markov chains (Section 17.4).

starting values: Many algorithms used to fit Bayesian models to data require starting values. Poor choices for starting values will influence the efficiency of the algorithm (e.g., long burn-in sequence). Generating good starting values can be difficult. The examples in this book use a wide variety of approaches to obtain reasonable starting values including: using the data directly (Chapter 10); using summary statistics of the data such as the mean or variance (Section 17.3); using point estimates obtained from fitting a simpler model (Chapter 11); and designing the algorithm so that it eliminates the need for starting values for a subset of the variables (Chapter 9).

Glossary

algorithm: a set of steps used to obtain the posterior distribution or samples from the posterior distribution of a Bayesian model

analytical solution: an equation that is exact and can be expressed using mathematical symbols

autocorrelation: dependence among random variables, often used in a spatial or temporal context

autoregression: a model often used for spatial or temporal data where the response variable is regressed on "itself" (i.e., on values of the response variable in other positions in time or space)

Bayesian model: a statistical model that treats parameters as random and is specified and implemented using conditional probability

block update: a step within a MCMC algorithm where a set of parameters (or state variables) are sampled from a multivariate full-conditional distribution simultaneously

burn-in: the initial portion of a Markov chain resulting from a MCMC algorithm that precedes convergence, and is discarded before using the chain for inference

closed-form solution: see analytical solution

convergence: when a Markov chain resulting from a MCMC algorithm reaches its stationary target distribution (i.e., the posterior distribution)

convolution: a function that is obtained by multiplying two functions of one or more variables and integrating with respect to a single variable

data augmentation: augmenting the parameter space of a model in a way that does not affect the inference, but usually facilitates the implementation; often short for parameter expanded data augmentation (PX-DA)

data fusion: see data reconciliation

data reconciliation: borrowing different strengths from multiple data sources in a single model to improve the resulting inference

derived quantity: a function of data, predictions, or model parameters whose posterior distribution can be characterized using MCMC output resulting from fitting the model

deviance information criterion, DIC: a function of data, model, and parameters used to score different models for Bayesian model selection

diffusion: a spatio-temporal process that spreads out over time, redistributing mass, energy, population size, etc., typically described by a mathematical model

fixed effect: in a Bayesian model, parameters that have a prior distribution with fixed and known hyperparameters

full-conditional: a probability distribution for a random variable in a model conditional on the data and all other random variables in the model

Gaussian: most commonly referring to a normal probability distribution for a continuous random variable, characterized by a mean and variance parameter (We use the terms Gaussian and normal interchangeably when referring to probability distributions. However, a Gaussian function could also be used to describe covariance or detectability as they decay with distance.)

generative model: a statistical model that is capable of simulating data. All models in this book are generative

Gibbs sampler: a MCMC algorithm that contains Gibbs updates for parameters (i.e., draws from analytically tractable full-conditional distributions)

hierarchical model: a joint probability model for many variables that is often specified as a sequence of simpler conditional distributions (Most commonly has three levels: data, process, and parameters (Berliner 1996).)

identifiability: in the Bayesian context, the ability to distinguish inferred parameters in a model based on the likelihood and priors

improper prior: a prior distribution that does not integrate (or sum, for discrete parameters) to one over its support

indicator function: a function that evaluates to one when a set of conditions are met, and zero otherwise (denoted as $1_{\{conditions\}}$ in this book)

integrated likelihood: a likelihood that eliminates parameters by integrating or summing over a portion of the random variables in the corresponding probability distribution that specifies the likelihood

integrated population model: a specific type of data reconciliation that involves data and models of population dynamics and demographics

joint posterior: posterior distribution of all unobserved random variables in a Bayesian model conditional on the observed data

kernel: most commonly referred to in this book as a function that is proportional to a PMF or PDF with respect to the random variable of interest

likelihood: a function that describes the shape and position of the probability or density of the response variable given the parameters

link function: a function that links the moments (e.g., expected value) of a distribution to parameters that are explicitly modeled

marginal posterior: a distribution of a single random variable in a Bayesian model that is conditional on only the data (a marginal posterior is obtained by integrating over all other random variables in the joint posterior)

Markov chain Monte Carlo, MCMC: a class of algorithms to sample from a probability distribution; most commonly used to sample from the joint posterior distribution associated with a Bayesian model

Markov process: a sequence of random variables that depend on each other only through their "neighbors"

mixing: a characterization of the autocorrelation and behavior of a Markov chain obtained resulting from a MCMC algorithm

moment: a characteristic of a random variable or data set; a quantitative description of the location or shape of a probability distribution

mixed model: a statistical model that contains both fixed and random effects

Monte Carlo: obtaining realizations of random variables by drawing them from a probability distribution

nonparametric inference: a type of statistical inference that does not rely on a specific form of probability distribution for the data

normalizing constant: a multiplicative constant in a PDF (or PMF) that allows it to integrate (or sum) to one over the support of the random variable

parametric inference: a type of statistical inference that involves a specific probability distribution as a model for the data; the functional form depends on a set of parameters that are often unknown

point process: a stochastic process where the positions (and possibly number) of the events are the random quantity of interest (in movement and species distribution models, the events are typically either the observed or true locations of the individual)

posterior distribution: probability distribution of parameters given observed data

posterior predictive distribution: probability distribution of new data given the observed data, based on a specified Bayesian model

precision: the inverse of variance (e.g., $1/\sigma^2$, or Σ^{-1} if Σ is a covariance matrix)

prior distribution: a probability distribution containing information about model parameters that does not depend on the set of data to be analyzed

probability density (or mass) function, PDF or PMF: a function expressing the stochastic nature of a continuous or discrete random variable (usually denoted as $f(y)$ or $[y]$ for random variable y)

random effect: a parameter that is modeled with a distribution that has unknown components we seek to infer

random field: a continuous stochastic process over space or time that is usually correlated

simulated data: data drawn by sampling from a generative model (simulated data are used commonly to check the assumptions of Bayesian models and check algorithms for programming errors)

spatial domain: usually the geographical area where a spatial process is observed, predicted, or conceptualized to exist

stationary distribution: in the MCMC context, a target distribution coinciding with the posterior distribution of interest

stationary process: a process with covariance structure that does not vary with location (in space or time)

state-space model: a hierarchical model that is usually temporally dynamic

support: the set of values that a random variable can assume (sometimes referred to as the sample space)

thinning: retaining a reduced set of regularly spaced values of a Markov chain resulting from a MCMC algorithm to reduce autocorrelation before making inference

trace plot: a graphic showing a MCMC sample in the order it was obtained sequentially; used to assess convergence and mixing of the Markov chain

tuning: the process of optimizing an algorithm to improve its efficiency

References

Aarts, G., J. Fieberg, and J. Matthiopoulos. 2012. Comparative interpretation of count, presence-absence, and point methods for species distribution models. *Methods in Ecology and Evolution*, 3:177–187.

Abadi, F., O. Gimenez, R. Arlettaz, and M. Schaub. 2010. An assessment of integrated population models: Bias, accuracy, and violation of the assumption of independence. *Ecology*, 88:7–14.

Aigner, D., T. Amemiya, and D. Poirier. 1976. On the estimation of production frontiers: Maximum likelihood estimation of the parameters of a discontinuous density function. *International Economic Review*, 17:377–396.

Aitkin, M. 1987. Modelling variance heterogeneity in normal regression using GLIM. *Journal of the Royal Statistical Society: Series C (Applied Statistics)*, 36:332–339.

Albert, J. and S. Chib. 1993. Bayesian analysis of binary and polychotomous response data. *Journal of the American Statistical Association*, 88:669–679.

Amek, N., N. Bayoh, M. Hamel, K. Lindblade, J. Gimnig, K. Laserson, L. Slutsker et al. 2011. Spatio-temporal modeling of sparse geostatistical malaria sporozoite rate data using a zero inflated binomial model. *Spatial and Spatio-temporal Epidemiology*, 2:283–290.

Andrieu, C., A. Doucet, and R. Holenstein. 2010. Particle Markov chain Monte Carlo methods. *Journal of the Royal Statistical Society: Series B (Statistical Methodology)*, 72:269–342.

Andrieu, C. and J. Thoms. 2008. A tutorial on adaptive MCMC. *Statistical Computing*, 18:343–373.

Arab, A. 2015. Spatial and spatio-temporal models for modeling epidemiological data with excess zeros. *International Journal of Environmental Research and Public Health*, 12:10536–10548.

Arab, A., M. Hooten, and C. Wikle. 2008a. Hierarchical spatial models. In Shekhar, S. and H. Xiong, editors, *Encyclopedia of GIS*, pp. 425–431. Springer, New York.

Arab, A., M. Wildhaber, C. Wikle, and C. Gentry. 2008b. Zero-inflated modeling of fish catch per unit area resulting from multiple gears: Application to channel catfish and shovelnose sturgeon in the Missouri River. *North American Journal of Fisheries Management*, 28:1044–1058.

Arlot, S. and A. Celisse. 2010. A survey of cross-validation procedures for model selection. *Statistics Surveys*, 4:40–79.

Arnold, T., R. Clark, D. Koons, and M. Schaub. 2018. Integrated population models facilitate ecological understanding and improved management decisions. *The Journal of Wildlife Management*, 82:266–274.

Baddeley, A., M. Berman, N. Fisher, A. Hardegen, R. Milne, D. Schuhmacher, R. Shah, and R. Turner. 2010. Spatial logistic regression and change-of-support in Poisson point processes. *Electronic Journal of Statistics*, 4:1151–1201.

Baddeley, A., E. Rubak, and R. Turner. 2015. *Spatial Point Patterns: Methodology and Applications with R*. CRC Press, Boca Raton, FL.

Bailey, L., T. Simons, and K. Pollock. 2004. Estimating detection probability parameters for plethodon salamanders using the robust capture-recapture design. *The Journal of Wildlife Management*, 68:1–13.

Banner, K. and M. Higgs. 2017. Considerations for assessing model averaging of regression coefficients. *Ecological Applications*, 27:78–93.

Banner, K., K. Irvine, T. Rodhouse, W. Wright, R. Rodriguez, and A. Litt. 2018. Improving geographically extensive acoustic survey designs for modeling species occurrence with imperfect detection and misidentification. *Ecology and Evolution*, 8:6144–6156.

Barker, R. and W. Link. 2013. Bayesian multimodel inference by RJMCMC: A Gibbs sampling approach. *The American Statistician*, 67:150–156.

Barker, R., M. Schofield, W. Link, and J. Sauer. 2018. On the reliability of N-mixture models for count data. *Biometrics*, 74:369–377.

Beaumont, M. 2010. Approximate Bayesian computation in evolution and ecology. *Annual Review in Ecology, Evolution, and Systematics*, 41:379–406.

Bedard, M. 2008. Optimal acceptance rates for Metropolis algorithms: Moving beyond 0.234. *Stochastic Processes and Their Applications*, 118:2198–2222.

Benoit, D. and D. Van Den Poel. 2012. Binary quantile regression: A Bayesian approach based on the asymmetric Laplace distribution. *Journal of Applied Econometrics*, 27:1174–1188.

Benoit, D. and D. Van Den Poel. 2017. bayesQR: A Bayesian approach to quantile regression. *Journal of Statistical Software*, 76:1–32.

Berliner, L. 1996. Hierarchical Bayesian time series models. In Hanson, K. and R. Silver, editors, *Maximum Entropy and Bayesian Methods*, pp. 15–22. Kluwer Academic Publishers, Dordrecht, the Netherlands.

Besbeas, P., S. Freeman, B. Morgan, and E. Catchpole. 2002. Integrating mark–recapture–recovery and census data to estimate animal abundance and demographic parameters. *Biometrics*, 58:540–547.

Betancourt, M. 2017. A conceptual introduction to Hamiltonian Monte Carlo. arXiv:1701.02434.

Betancourt, M., S. Byrne, S. Livingstone, and M. Girolami. 2017. The geometric foundations of Hamiltonian Monte Carlo. *Bernoulli*, 23:2257–2298.

Bickel, P., B. Li, A. Tsybakov, S. van de Geer, B. Yu, T. Valdés, C. Rivero et al. 2006. Regularization in statistics. *Test*, 15:271–344.

Bierema, A.-K. and D. Rudge. 2014. Using David Lack's observations of finch beak size to teach natural selection & the nature of science. *The American Biology Teacher*, 76(5):312–317.

Bled, F., J. Royle, and E. Cam. 2011. Hierarchical modeling of an invasive spread: The Eurasian Collared-Dove *Streptopelia decaocto* in the United States. *Ecological Applications*, 21:290–302.

Blum, M. 2010. Approximate Bayesian computation: A non-parameteric perspective. *Journal of the American Statistical Association*, 105:1178–1187.

Borchers, D. and M. Efford. 2008. Spatially explicit maximum likelihood methods for capture–recapture studies. *Biometrics*, 64:377–385.

Borchers, D., B. Stevenson, D. Kidney, L. Thomas, and T. Marques. 2015. A unifying model for capture–recapture and distance sampling surveys of wildlife populations. *Journal of the American Statistical Association*, 110:195–204.

Bose, M., J. Hodges, and S. Banerjee. 2018. Toward a diagnostic toolkit for linear models with Gaussian-process distributed random effects. *Biometrics*, 74:863–873.

Bradley, J., S. Holan, and C. Wikle. 2018. Computationally efficient multivariate spatio-temporal models for high-dimensional count-valued data. *Bayesian Analysis*, 13:253–310.

Brillinger, D. 2010. Modeling spatial trajectories. In Gelfand, A., P. Diggle, M. Fuentes, and P. Guttorp, editors, *Handbook of Spatial Statistics*. Chapman & Hall/CRC Press, Boca Raton, FL.

Broms, K., M. Hooten, R. Altwegg, and L. Conquest. 2016. Dynamic occupancy models for explicit colonization processes. *Ecology*, 97:194–204.

Broms, K., D. Johnson, R. Altwegg, and L. Conquest. 2014. Spatial occupancy models applied to atlas data show Southern Ground Hornbills strongly depend on protected areas. *Ecological Applications*, 24:363–374.

Brooks, S., R. King, and B. Morgan. 2004. A Bayesian approach to combining animal abundance and demographic data. *Animal Biodiversity and Conservation*, 27:515–529.

Brost, B., M. Hooten, E. Hanks, and R. Small. 2015. Animal movement constraints improve resource selection inference in the presence of telemetry error. *Ecology*, 96:2590–2597.

Brost, B., M. Hooten, and R. Small. 2017. Leveraging constraints and biotelemetry data to pinpoint repetitively used spatial features. *Ecology*, 98:12–20.

Brost, B., B. Mosher, and K. Davenport. 2018. A model-based solution for observational errors in laboratory studies. *Molecular Ecology Resources*, 18:580–589.

Buckland, S., D. Anderson, K. Burnham, J. Laake, and D. Borchers. 2001. *Introduction to Distance Sampling: Estimating Abundance of Biological Populations*. Oxford University Press, Oxford, UK.

Buckland, S., D. Anderson, K. Burnham, J. Laake, D. Borchers, and L. Thomas. 2004. *Advanced Distance Sampling*. Oxford University Press, Oxford, UK.

Buderman, F., M. Hooten, J. Ivan, and T. Shenk. 2016. A functional model for characterizing long distance movement behavior. *Methods in Ecology and Evolution*, 7:264–273.

Burnham, K., D. Anderson, and J. Laake. 1980. Estimation of density from line transect sampling of biological populations. *Wildlife Monographs*, 72:3–202.

Cade, B. 2015. Model averaging and muddled multimodel inferences. *Ecology*, 96:2370–2382.

Cade, B. and B. Noon. 2003. A gentle introduction to quantile regression for ecologists. *Frontiers in Ecology and the Environment*, 1:412–420.

Cangelosi, A. and M. Hooten. 2009. Models for bounded systems with continuous dynamics. *Biometrics*, 65:850–856.

Cappé, O. 2007. An overview of existing methods and recent advances in sequential Monte Carlo. *Proceedings of the IEEE*, 95:899–924.

Carlin, B. and S. Chib. 1995. Bayesian model choice via Markov chain Monte Carlo methods. *Journal of the Royal Statistical Society, Series B*, 57:473–484.

Carlin, B. and T. Louis. 2009. *Bayesian Methods for Data Analysis*, 3rd ed. CRC Press, Boca Raton, FL.

Carpenter, B., A. Gelman, M. Hoffman, D. Lee, B. Goodrich, M. Betancourt, M. Brubaker et al. 2016. Stan: A probabilistic programming language. *Journal of Statistical Software*, 20:1–37.

Casella, G. and E. George. 1992. Explaining the Gibbs sampler. *The American Statistician*, 46:167–174.

Caswell, H. 2001. *Matrix Population Models*. Wiley Online Library.

Celeux, G., F. Forbes, C. Robert, and D. Titterington. 2006. Deviance information criteria for missing data models. *Bayesian Analysis*, 1:651–673.

Chambers, S. 2016. *Extending R*. CRC Press, Boca Raton, FL.

Chandler, R., E. Muths, B. Sigafus, C. Schwalbe, C. Jarchow, and B. Hossack. 2015. Spatial occupancy models for predicting metapopulation dynamics and viability following reintroduction. *Journal of Applied Ecology*, 52:1325–1333.

Chib, S. 1995. Marginal likelihood from the Gibbs output. *Journal of the American Statistical Association*, 88:669–679.

Chib, S. and E. Greenberg. 1995. Understanding the Metropolis-Hastings algorithm. *The American Statistician*, 49:327–335.

Chopin, N. 2002. A sequential particle filter method for static models. *Biometrika*, 89:539–552.

Clark, J. 2007. *Models for Ecological Data: An Introduction*. Princeton University Press, Princeton, NJ.

Clark, J., S. Carpenter, M. Barber, S. Collins, A. Dobson, J. Foley, D. Lodge et al. 2001. Ecological forecasts: An emerging imperative. *Science*, 293:657–660.

Clyde, M., J. Ghosh, and M. Littman. 2011. Bayesian adaptive sampling for variable selection and model averaging. *Journal of Computational and Graphical Statistics*, 20:80–101.

Colautti, R., I. Grigorovich, and H. MacIssac. 2006. Propagule pressure: A null model for invasions. *Biological Invasions*, 8:1023–1037.

Conn, P., D. Johnson, P. Williams, S. Melin, and M. Hooten. 2018. A guide to Bayesian model checking for ecologists. *Ecological Monographs*, 88: 526–542.

Cook, R. and S. Weisberg. 1983. Diagnostics for heteroscedasticity in regression. *Biometrika*, 70:1–10.

Cook, S., A. Gelman, and D. Rubin. 2006. Validation of software for Bayesian models using posterior quantiles. *Journal of Computational and Graphical Statistics*, 15:675–692.

Cressie, N. 1990. The origins of Kriging. *Mathematical Geology*, 22:239–252.

Cressie, N., C. Calder, J. Clark, J. Ver Hoef, and C. Wikle. 2009. Accounting for uncertainty in ecological analysis: The strengths and limitations of hierarchical statistical modeling. *Ecological Applications*, 19:553–570.

Cressie, N. and C. Wikle. 2011. *Statistics for Spatio-temporal Data*. John Wiley & Sons, New York.

Davis, A., M. Hooten, M. Phillips, and P. Doherty. 2014. An integrated modeling approach to estimating gunnison sage-grouse population dynamics: Combining index and demographic data. *Ecology and Evolution*, 4:4247–4257.

Davis, R. A., S. H. Holan, R. Lund, and N. Ravishanker. 2016. *Handbook of Discrete-Valued Time Series*. CRC Press Boca Raton, FL.

Dawid, A. 2011. Posterior model probabilities. In Bandyopadhyay, P. S. and M. R. Forster, editors, *Philosophy of Statistics*, pages 607–630. Elsevier, Oxford, UK.

De Oliveira, V. 2018. Models for geostatistical binary data: Properties and connections. *The American Statistician*, 1–8.

de Valpine, P. and A. Hastings. 2002. Fitting population models incorporating process noise and observation error. *Ecological Monographs*, 72:57–76.

Dennis, B., J. Ponciano, S. Lele, M. Taper, and D. Staples. 2006. Estimating density dependence, process noise, and observation error. *Ecological Monographs*, 76: 323–341.

Dennis, E., B. Morgan, and M. Ridout. 2015. Computational aspects of N-mixture models. *Biometrics*, 71:237–246.

Dietze, M. 2017. *Ecological Forecasting*. Princeton University Press, Princeton, NJ.

Dietze, M., A. Fox, L. Beck-Johnson, J. Betancourt, M. Hooten, C. Jarnevich, T. Keitt et al. 2018. Iterative near-term ecological forecasting: Needs, opportunities, and challenges. *Proceedings of the National Academy of Sciences*, 115:1424–1432.

Diggle, P. 2013. *Statistical Analysis of Spatial and Spatio-temporal Point Patterns*, Third Edition. Chapman & Hall/CRC Press, Boca Raton, FL.

Dobrow, R. 2016. *Introduction to Stochastic Processes with R*. Wiley, Hoboken, NJ.

Dorazio, R., M. Kery, J. Royle, and M. Plattner. 2010. Models for inference in dynamic metacommunity systems. *Ecology*, 91:2466–2475.

Dorazio, R., B. Mukherjee, L. Zhang, M. Ghosh, H. Jelks, and F. Jordan. 2008. Modeling unobserved sources of heterogeneity in animal abundance using a Dirichlet process prior. *Biometrics*, 64:635–644.

Dorazio, R. and D. Rodriguez. 2012. A Gibbs sampler for Bayesian analysis of site-occupancy data. *Methods in Ecology and Evolution*, 3:1093–1098.

Dorazio, R., J. Royle, B. Söderström, and A. Glimskär. 2006. Estimating species richness and accumulation by modeling species occurrence and detectability. *Ecology*, 87:842–854.

Dorfman, R. 1938. A note on the δ-method for finding variance formulae. *The Biometric Bulletin*, 1:129–137.

Doucet, A., N. De Freitas, and N. Gordon. 2001. *Sequential Monte Carlo Methods in Practice*. Springer-Verlag, New York.

Easterling, M., S. Ellner, and P. Dixon. 2000. Size-specific sensitivity: Applying a new structured population model. *Ecology*, 81:694–708.

Efford, M. 2004. Density estimation in live-trapping studies. *Oikos*, 106:598–610.

Ellner, S. and M. Rees. 2006. Integral projection models for species with complex demography. *The American Naturalist*, 167:410–428.

Farlow, S. 1993. *Partial Differential Equations for Scientists and Engineers*. Courier Corporation North Chelmsford, MA.

Fieberg, J. and D. Johnson. 2015. Mmi: Multimodel inference or models with management implications? *The Journal of Wildlife Management*, 79:708–718.

Finley, A., S. Banerjee, and B. Carlin. 2015. spBayes: An R package for univariate and multivariate hierarchical point-referenced spatial models. *Journal of Statistical Software*, 63:1–28.

Fithian, W. and T. Hastie. 2013. Finite-sample equivalence in statistical models for presence-only data. *The Annals of Applied Statistics*, 7:1917–1939.

Fortin, D., H. Beyer, M. Boyce, D. Smith, T. Duchesne, and J. Mao. 2005. Wolves influence elk movements: Behavior shapes a trophic cascade in Yellowstone National Park. *Ecology*, 86:1320–1330.

Gamerman, D. and H. Lopes. 2006. *Markov Chain Monte Carlo: Stochastic Simulation for Bayesian Inference*. Chapman & Hall/CRC Press, London, UK.

Gardner, B., R. Sollmann, N. Kumar, D. Jathanna, and K. Karanth. 2018. State space and movement specification in open population spatial capture–recapture models. *Ecology and Evolution*, 8:10336–10344.

Garlick, M., J. Powell, M. Hooten, and L. McFarlane. 2011. Homogenization of large-scale movement models in ecology. *Bulletin of Mathematical Biology*, 73: 2088–2108.

Garlick, M., J. Powell, M. Hooten, and L. McFarlane. 2014. Homogenization, sex, and differential motility predict spread of chronic wasting disease in mule deer in Southern Utah. *Journal of Mathematical Biology*, 69:369–399.

Geisser, S. 1993. *Predictive Inference: An Introduction*. Chapman & Hall, New York.

Gelfand, A. and D. Dey. 1994. Bayesian model choice: Asymptotics and exact calculations. *Journal of the Royal Statistical Society, Series B*, 56:501–514.

Gelfand, A. and S. Ghosh. 1998. Model choice: A minimum posterior predictive loss approach. *Biometrika*, 85:1–11.

Gelfand, A. and E. Schliep. 2018. Bayesian inference and computing for spatial point patterns. *NSF-CBMS Regional Conference Series in Probability and Statistics*, 10: 1–117.

Gelfand, A. and A. Smith. 1990. Sampling-based approaches to calculating marginal densities. *Journal of the American Statistical Association*, 85:398–409.

Gelfand, A., A. Smith, and T. Lee. 1992. Bayesian analysis of constrained parameter and truncated data problems using Gibbs sampling. *Journal of the American Statistical Association*, 87:523–532.

Gelman, A. 2006. Prior distributions for variance parameters in hierarchical models (comment on article by Browne and Draper). *Bayesian Analysis*, 1:515–534.

Gelman, A., B. Carlin, H. Stern, D. Dunson, A. Vehtari, and D. Rubin. 2013. *Bayesian Data Analysis*, 3rd ed. Taylor & Francis Group, Oxford, UK.

Gelman, A. and J. Hill. 2006. *Data Analysis Using Regression and Multilevel/ Hierarchical Models.* Cambridge University Press, Cambridge, UK.

Gelman, A., J. Hwang, and A. Vehtari. 2014. Understanding predictive information criteria for Bayesian models. *Statistics and Computing*, 24:997–1016.

Gelman, A., X.-L. Meng, and H. Stern. 1996. Posterior predictive assessment of model fitness via realized discrepancies. *Statistica Sinica*, 6:733–760.

Geman, S. and D. Geman. 1984. Stochastic relaxation, Gibbs distributions and the Bayesian restoration of images. *IEEE Transactions of Pattern Analysis and Machine Intelligence*, 6:721–741.

George, E. and R. McCulloch. 1993. Variable selection via Gibbs sampling. *Journal of the American Statistical Association*, 88(423):881–889.

Gerber, B., W. Kendall, M. Hooten, J. Dubovsky, and R. Drewien. 2015. Optimal population prediction of sandhill crane recruitment based on climate-mediated habitat limitations. *Journal of Animal Ecology*, 84:1299–1310.

Geweke, J. 1989. Bayesian inference in econometric models using Monte Carlo integration. *Econometrica*, 57:1317–1339.

Givens, G. and J. Hoeting. 2013. *Computational Statistics*, 2nd ed. Wiley, Hoboken, NJ.

Glennie, R. 2018. Incorporating animal movement with distance sampling and spatial capture-recapture. PhD dissertation, University of St. Andrews.

Glennie, R., S. Buckland, R. Langrock, T. Gerrodette, L. Ballance, S. Chivers, M. Scott et al. 2018. Incorporating animal movement into distance sampling. *Journal of the American Statistical Association*.

Gneiting, T. and A. Raftery. 2007. Strictly proper scoring rules, prediction, and estimation. *Journal of the American Statistical Association*, 102:359–378.

Gordon, N., D. Salmond, and A. Smith. 1993. Novel approach to nonlinear/non-Gaussian Bayesian state estimation. *IEEE Proceedings F (Radar and Signal Processing)*, 140:107–113.

Green, P. 1995. Reversible jump Markov chain Monte Carlo computation and Bayesian model determination. *Biometrika*, 82:711–732.

Green, P., K. Latuszynski, M. Pereyra, and C. Robert. 2015. Bayesian computation: A summary of the current state, and samples backwards and forwards. *Statistical Computation*, 25:835–862.

Gumpertz, M., J. Graham, and J. Ristaino. 1997. Autologistic model of spatial pattern of phytophthora epidemic in bell pepper: Effects of soil variables on disease presence. *Journal of Agricultural, Biological, and Environmental Statistics*, 2:131–156.

Haberman, R. 1987. *Elementary Partial Differential Equations with Fourier Series and Boundary Value Problems*, 2nd ed. Prentice Hall, New York.

Hanks, E., M. Hooten, and M. Alldredge. 2015a. Continuous-time discrete-space models for animal movement. *Annals of Applied Statistics*, 9:145–165.

Hanks, E., M. Hooten, and F. Baker. 2011. Reconciling multiple data sources to improve accuracy of large-scale prediction of forest disease incidence. *Ecological Applications*, 24:1173–1188.

Hanks, E., E. Schliep, M. Hooten, and J. Hoeting. 2015b. Restricted spatial regression in practice: Geostatistical models, confounding, and robustness under model misspecification. *Environmetrics*, 26:243–254.

Hastings, W. 1970. Monte Carlo sampling methods using Markov chains and their application. *Biometrika*, 57:97–109.

Heaton, M., A. Datta, A. Finley, R. Furrer, R. Guhaniyogi, F. Gerber, R. Gramacy et al. 2018. Methods for analyzing large spatial data: A review and comparison. arXiv:1710.05013.

Hefley, T., K. Broms, B. Brost, F. Buderman, S. Kay, H. Scharf, J. Tipton et al. 2017a. The basis function approach to modeling autocorrelation in ecological data. *Ecology*, 98:632–646.

Hefley, T. and M. Hooten. 2016. Hierarchical species distribution models. *Current Landscape Ecology Reports*, 1:87–97.

Hefley, T., M. Hooten, J. Drake, R. Russell, and D. Walsh. 2016. When can the cause of a population decline be determined? *Ecology Letters*, 19:1353–1362.

Hefley, T., M. Hooten, E. Hanks, R. Russell, and D. Walsh. 2017b. The Bayesian group lasso for confounded spatial data. *Journal of Agricultural, Biological, and Environmental Statistics*, 22:42–59.

Hefley, T., M. Hooten, R. Russell, D. Walsh, and J. Powell. 2017c. When mechanism matters: Forecasting the spread of disease using ecological diffusion. *Ecology Letters*, 20:640–650.

Hefley, T., S. Hygnstrom, J. Gilsdorf, G. Clements, M. Clements, A. Tyre, D. Baasch et al. 2013. Effects of deer density and land use on mass of white-tailed deer. *Journal of Fish and Wildlife Management*, 4:20–32.

Held, L., B. Schrödle, and H. Rue. 2010. Posterior and cross-validatory predictive checks: A comparison of MCMC and INLA. In Kneib, T. and L. Fahrmeir, editors, *Statistical Modelling and Regression Structures*, pages 91–110. Springer, Berlin, Germany.

Henden, J., N. Yoccoz, R. Ims, and K. Langeland. 2013. How spatial variation in areal extent and configuration of labile vegetation states affect the riparian bird community in Arctic tundra. *PLoS One*, 8(5):e63312.

Hendry, A., P. Grant, R. Grant, H. Ford, M. Brewer, and J. Podos. 2006. Possible human impacts on adaptive radiation: Beak size bimodality in Darwin's finches. *Proceedings of the Royal Society of London B: Biological Sciences*, 273(1596): 1887–1894.

Hendry, A., S. Huber, L. De Leon, A. Herrel, and J. Podos. 2009. Disruptive selection in a bimodal population of Darwin's finches. *Proceedings of the Royal Society B*, 276:753–759.

Higdon, D. 1998. Auxiliary variable methods for Markov chain Monte Carlo with applications. *Journal of the American Statistical Association*, 93:585–595.

Higdon, D., J. Gattiker, B. Williams, and M. Rightley. 2008. Computer model calibration using high-dimensional output. *Journal of the American Statistical Association*, 103:570–583.

Hines, J., J. Nichols, and J. Collazo. 2014. Multiseason occupancy models for correlated replicate surveys. *Methods in Ecology and Evolution*, 5:583–591.

Hobbs, N., C. Geremia, J. Treanor, R. Wallen, P. White, M. Hooten, and J. Rhyan. 2015. State-space modeling to support adaptive management of brucellosis in the Yellowstone bison population. *Ecological Monographs*, 85:525–556.

Hobbs, N. and M. Hooten. 2015. *Bayesian Models: A Statistical Primer for Ecologists*. Princeton University Press, Princeton, NJ.

Hodges, J. and B. Reich. 2010. Adding spatially-correlated errors can mess up the fixed effect you love. *The American Statistician*, 64:325–334.

Hoerl, A. and R. Kennard. 1970. Ridge regression: Biased estimation for nonorthogonal problems. *Technometrics*, 12:55–67.

Hoeting, J., M. Leecaster, and D. Bowden. 2000. An improved model for spatially correlated binary responses. *Journal of Agricultural, Biological, and Environmental Statistics*, 5:102–114.

Hoeting, J., D. Madigan, A. Raftery, and C. Volinsky. 1999. Bayesian model averaging: A tutorial. *Statistical Science*, 14:382–401.

Hoffman, M. and A. Gelman. 2014. The No-U-turn sampler: Adaptively setting path lengths in Hamiltonian Monte Carlo. *Journal of Machine Learning Research*, 15: 1593–1623.

Hooten, M., F. Buderman, B. Brost, E. Hanks, and J. Ivan. 2016. Hierarchical animal movement models for population-level inference. *Environmetrics*, 27: 322–333.

Hooten, M. and E. Cooch. 2019. Comparing ecological models. In Brennan, L., A. Tri, and B. Marcot, editors, *Quantitative Analyses in Wildlife Science*. Johns Hopkins University Press, Baltimore, MD.

Hooten, M., M. Garlick, and J. Powell. 2013. Computationally efficient statistical differential equation modeling using homogenization. *Journal of Agricultural, Biological and Environmental Statistics*, 18:405–428.

Hooten, M. and N. Hobbs. 2015. A guide to Bayesian model selection for ecologists. *Ecological Monographs*, 85:3–28.

Hooten, M. and D. Johnson. 2017. Basis function models for animal movement. *Journal of the American Statistical Association*, 112:578–589.

Hooten, M., D. Johnson, and B. Brost. 2019. Making recursive Bayesian inference accessible. The American Statistician. arXiv: 1807.10981.

Hooten, M., D. Johnson, E. Hanks, and J. Lowry. 2010. Agent-based inference for animal movement and selection. *Journal of Agricultural, Biological and Environmental Statistics*, 15:523–538.

Hooten, M., D. Johnson, B. McClintock, and J. Morales. 2017. *Animal Movement: Statistical Models for Telemetry Data*. Chapman & Hall/CRC Press, Boca Raton, FL.

Hooten, M., D. Larsen, and C. Wikle. 2003. Predicting the spatial distribution of ground flora on large domains using a hierarchical Bayesian model. *Landscape Ecology*, 18:487–502.

Hooten, M., W. Leeds, J. Fiechter, and C. Wikle. 2011. Assessing first-order emula-
tor inference for physical parameters in nonlinear mechanistic models. *Journal of
Agricultural, Biological, and Environmental Statistics*, 16:475–494.

Hooten, M., H. Scharf, T. Hefley, A. Pearse, and M. Weegman. 2018. Animal movement models for migratory individuals and groups. *Methods in Ecology and Evolution*, 9:1692–1705.

Hooten, M. and C. Wikle. 2007. Shifts in the spatio-temporal growth dynamics of
shortleaf pine. *Environmental and Ecological Statistics*, 14:207–227.

Hooten, M. and C. Wikle. 2008. A hierarchical Bayesian non-linear spatio-temporal
model for the spread of invasive species with application to the Eurasian collared-
dove. *Environmental and Ecological Statistics*, 15:59–70.

Hooten, M. and C. Wikle. 2010. Statistical agent-based models for discrete spatio-
temporal systems. *Journal of the American Statistical Association*, 105:236–248.

Hooten, M., C. Wikle, R. Dorazio, and J. Royle. 2007. Hierarchical spatio-temporal
matrix models for characterizing invasions. *Biometrics*, 15:558–567.

Hooten, M., C. Wikle, S. Sheriff, and J. Rushin. 2009. Optimal spatio-temporal
hybrid sampling designs for ecological monitoring. *Journal of Vegetation Science*,
20:639–649.

Hu, W. and J. Bradley. 2018. A Bayesian spatial-temporal model with latent multivariate log-gamma random effects with application to earthquake magnitudes.
Stat, 7:e179.

Huffer, F. and H. Wu. 1998. Markov chain Monte Carlo for autologistic regression models with application to the distribution of plant species. *Biometrics*, 54:
509–524.

Hughes, J., M. Haran, and P. Caragea. 2011. Autologistic models for binary data on
a lattice. *Environmetrics*, 22:857–871.

Hurlbert, S. 1984. Pseudoreplication and the design of ecological field experiments.
Ecological Monographs, 54:187–211.

Illian, J., A. Penttinen, H. Stoyan, and D. Stoyan. 2008. *Statistical Analysis of Spatial
Point Patterns*. Wiley-Interscience, Chichester, UK.

Irvine, K., T. Rodhouse, and I. Keren. 2016. Extending ordinal regression with a
latent zero-augmented beta distribution. *Journal of Agricultural, Biological, and
Environmental Statistics*, 21:619–640.

Ivan, J., G. White, and T. Shenk. 2013. Using auxiliary telemetry information to
estimate animal density from capture–recapture data. *Ecology*, 94:809–816.

Jeffreys, H. 1961. *The Theory of Probability*. Oxford University Press, Oxford, UK.

Jerem, P., K. H. S. Jenni-Eiermann, D. McKeegan, D. McCafferty, and R. Nager.
2018. Eye region surface temperature reflects both energy reserves and circulating
glucocorticoids in a wild bird. *Scientific Reports*, 8:1907.

Johnson, D. 1980. The comparison of usage and availability measurements for eval-
uating resource preference. *Ecology*, 61:65–71.

Johnson, D., P. Conn, M. Hooten, J. Ray, and B. Pond. 2013a. Spatial occupancy
models for large data sets. *Ecology*, 94:801–808.

Johnson, D., M. Hooten, and C. Kuhn. 2013b. Estimating animal resource selection from telemetry data using point process models. *Journal of Animal Ecology*, 82: 1155–1164.

Johnson, D., J. Laake, and J. Ver Hoef. 2010. A model-based approach for making ecological inference from distance sampling data. *Biometrics*, 66:310–318.

Johnson, D., J. London, M. Lea, and J. Durban. 2008. Continuous-time correlated random walk model for animal telemetry data. *Ecology*, 89:1208–1215.

Johnson, D., R. Ream, R. Towell, M. Williams, and J. Guerrero. 2013c. Bayesian clustering of animal abundance trends for inference and dimension reduction. *Journal of Agricultural, Biological and Environmental Statistics*, 18:299–313.

Johnson, D. and E. Sinclair. 2017. Modeling joint abundance of multiple species using dirichlet process mixtures. *Environmetrics*, 28:e2440.

Jolly, G. 1965. Explicit estimates from capture-recapture data with both death and immigration-stochastic model. *Biometrika*, 52:225–247.

Jonsen, I., J. Flemming, and R. Myers. 2005. Robust state-space modeling of animal movement data. *Ecology*, 45:589–598.

Kalman, R. 1960. A new approach to linear filtering and prediction problems. *Transactions of the ASME-Journal of Basic Engineering*, 82:35–45.

Kass, R. and A. Raftery. 1995. Bayes factors. *Journal of the American Statistical Association*, 90:773–795.

Kendall, W. and R. Bjorkland. 2001. Using open robust design models to estimate temporary emigration from capture-recapture data. *Biometrics*, 57:1113–1122.

Kendall, W., J. Nichols, and J. Hines. 1997. Estimating temporary emigration using capture–recapture data with pollock's robust design. *Ecology*, 78:563–578.

Kéry, M. 2018. Identifiability in N-mixture models: A large-scale screening test with bird data. *Ecology*, 99:281–288.

Kéry, M. and J. Royle. 2016. *Applied Hierarchical Modeling in Ecology*. Academic Press, London, UK.

Kéry, M., J. Royle, M. Plattner, and R. Dorazio. 2009. Species richness and occupancy estimation in communities subject to temporary emigration. *Ecology*, 90: 1279–1290.

Kéry, M. and M. Schaub. 2012. *Bayesian Population Analysis Using WinBUGS: A Hierarchical Perspective*. Academic Press, New York.

Ketz, A., T. Johnson, R. Monello, J. Mack, J. George, B. Kraft, M. Wild et al. 2018. Estimating abundance of an open population with an N-mixture model using auxiliary data on animal movements. *Ecological Applications*, 28:816–825.

Knape, J. and P. de Valpine. 2012. Fitting complex population models by combining particle filters with Markov chain Monte Carlo. *Ecology*, 93:256–263.

Kneib, T. 2013. Beyond mean regression. *Statistical Modelling*, 13:275–303.

Koenker, R. and G. Basset. 1978. Regression quantiles. *Econometrica*, 46:33–50.

Koenker, R. and J. Machado. 1999. Goodness of fit and related inference processes for quantile regression. *Journal of the American Statistical Association*, 94: 1296–1310.

Laud, P. and J. Ibrahim. 1995. Predictive model selection. *Journal of the Royal Statistical Society, Series B*, 57:247–262.

Le, N. D. and J. V. Zidek. 2006. *Statistical Analysis of Environmental Space-Time Processes*. Springer Science & Business Media, New York.

Lee, D. and T. Neocleous. 2010. Bayesian quantile regression for count data with application to environmental epidemiology. *Journal of the Royal Statistical Society, Series C (Applied Statistics)*, 59:905–920.

Lele, S. and K. Allen. 2006. On using expert opinion in ecological analyses: A frequentist approach. *Environmetrics*, 17:683–704.

Lele, S. and J. Keim. 2006. Weighted distributions and estimation of resource selection probability functions. *Ecology*, 87:3021–3028.

Lele, S., M. Moreno, and E. Bayne. 2012. Dealing with detection error in site occupancy surveys: What can we do with a single survey? *Journal of Plant Ecology*, 5: 22–21.

Lepak, J., M. Hooten, and B. Johnson. 2012. The influence of external subsidies on diet, growth and Hg concentrations of freshwater sport fish: Implications for management and fish consumption advisories. *Ecotoxicology*, 21: 1878–1888.

Lewis, M., S. Petrovskii, and J. Potts. 2016. *Mathematics Behind Biological Invasions*. Springer-Verlag, Cham, Switzerland.

Li, N., D. Elashoff, W. Robbins, and L. Xun. 2011. A hierarchical zero-inflated lognormal model for skewed responses. *Statistical Methods in Medical Research*, 20: 175–189.

Linden, D., A. Fuller, J. Royle, and M. Hare. 2017. Examining the occupancy–density relationship for a low-density carnivore. *Journal of Applied Ecology*, 54:2043–2052.

Linden, D., A. Sirén, and P. Pekins. 2018. Integrating telemetry data into spatial capture–recapture modifies inferences on multi-scale resource selection. *Ecosphere*, 9:e02203.

Link, W. 2013. A cautionary note on the discrete uniform prior for the binomial N. *Ecology*, 94:2173–2179.

Link, W. and R. Barker. 2006. Model weights and the foundations of multimodel inference. *Ecology*, 87:2626–2635.

Link, W. and R. Barker. 2010. *Bayesian Inference: With Ecological Applications*. Academic Press Amsterdam, the Netherlands.

Link, W. and M. Eaton. 2012. On thinning of chains in MCMC. *Methods in Ecology and Evolution*, 3:112–115.

Link, W., M. Schofield, R. Barker, and J. Sauer. 2018. On the robustness of N-mixture models. *Ecology*, 99:1547–1551.

Liu, J. 2004. *Monte Carlo Strategies in Scientific Computing*. Springer Science & Business Media, New York.

Lockwood, J., P. Cassey, and T. Blackburn. 2005. The role of propagule pressure in explaining species invasions. *Trends in Ecology and Evolution*, 20: 223–228.

Lunn, D., J. Barrett, M. Sweeting, and S. Thompson. 2013. Fully Bayesian hierarchical modelling in two stages, with application to meta-analysis. *Journal of the Royal Statistical Society: Series C (Applied Statistics)*, 62:551–572.

Lunn, D., D. Spiegelhalter, A. Thomas, and N. Best. 2009. The BUGS project: Evolution, critique and future directions. *Statistics in Medicine*, 28:3049–3067.

Lunn, D., A. Thomas, N. Best, and D. Spiegelhalter. 2000. WinBUGS—A Bayesian modelling framework: Concepts, structure, and extensibility. *Statistics and Computing*, 10(4):325–337.

Lyashevska, O., D. Brus, and J. van der Meer. 2016. Mapping species abundance by a spatial zero-inflated Poisson model: A case study in the Wadden Sea, the Netherlands. *Ecology and Evolution*, 6:532–543.

Machado, J. and J. Santos Silva. 2005. Quantiles for counts. *Journal of the American Statistical Association*, 100:1226–1237.

MacKenzie, D. and L. Bailey. 2004. Assessing the fit of site-occupancy models. *Journal of Agricultural, Biological, and Environmental Statistics*, 9:300–318.

MacKenzie, D., J. Nichols, J. Hines, M. Knutson, and A. Franklin. 2003. Estimating site occupancy, colonization, and local extinction when a species is detected imperfectly. *Ecology*, 84:2200–2207.

MacKenzie, D., J. Nichols, G. Lachman, S. Droege, J. Royle, and C. Langtimm. 2002. Estimating site occupancy rates when detection probabilities are less than one. *Ecology*, 83:2248–2255.

MacKenzie, D., J. Nichols, J. Royle, K. Pollock, L. Bailey, and J. Hines. 2017. *Occupancy Estimation and Modeling: Inferring Patterns and Dynamics of Species Occurrence*. Elsevier, Burlington, MA.

Madigan, D. and A. Raftery. 1994. Model selection and accounting for model uncertainty in graphical models using Occam's window. *Journal of the American Statistical Association*, 89:1535–1546.

Martin, T., B. Wintle, J. Rhodes, P. Kuhnert, S. Field, S. Low-Choy, A. Tyre et al. 2005. Zero tolerance ecology: Improving ecological inference by modelling the source of zero observations. *Ecology Letters*, 8:1235–1246.

McClintock, B., D. Johnson, M. Hooten, J. Ver Hoef, and J. Morales. 2014. When to be discrete: The importance of time formulation in understanding animal movement. *Movement Ecology*, 2:21.

McDermott, P. and C. Wikle. 2016. A model-based approach for analog spatiotemporal dynamic forecasting. *Environmetrics*, 27:70–82.

Metropolis, N. 1987. The beginning of the Monte Carlo method. *Los Alamos Science*, 15:125–130.

Metropolis, N., A. Rosenbluth, M. Rosenbluth, A. Teller, and E. Teller. 1953. Equations of state calculations by fast computing machines. *Journal of Chemical Physics*, 21:1087–1092.

Miller, D., J. Nichols, B. McClintock, E. Grant, L. Bailey, and L. Weir. 2011. Improving occupancy estimation when two types of observational error occur: Nondetection and species misidentification. *Ecology*, 92:1422–1428.

Miller, D., L. Weir, B. McClintock, E. Grant, L. Bailey, and T. Simons. 2012. Experimental investigation of false positive errors in auditory species occurrence surveys. *Ecological Applications*, 22:1665–1674.

Minuzzi-Souza, T., N. Nitz, C. Cuba, L. Hagström, M. Hecht, C. Santana, M. Ribeiro et al. 2018. Surveillance of vector-borne pathogens under imperfect detection: Lessons from chagas disease risk (mis) measurement. *Scientific Reports*, 8:151.

Moller, J. and R. Waagepetersen. 2003. *Statistical Inference and Simulation for Spatial Point Processes*. Chapman & Hall/CRC Press, Boca Raton, FL.

Monnahan, C., J. Thorson, and T. Branch. 2017. Faster estimation of Bayesian models in ecology using Hamiltonian Monte Carlo. *Methods in Ecology and Evolution*, 8: 339–348.

Neal, R. 1999. Slice sampling. *Annals of Statistics*, 31:705–767.

Neal, R. 2000. Markov chain sampling methods for Dirichlet process mixture models. *Journal of Computational and Graphical Statistics*, 9:249–265.

Neal, R. MCMC using Hamiltonian dynamics. In Brooks, S., A. Gelman, G. Jones, and X. Meng, editors, *Handbook of Markov Chain Monte Carlo*. Chapman & Hall/CRC Press, Boca Raton, FL, 2011.

Newey, W. and J. Powell. 1987. Asymmetric least squares estimation and testing. *Econometrica*, 55:819–847.

Newton, M. and A. Raftery. 1994. Approximate Bayesian inference by the weighted likelihood bootstrap (with discussion). *Journal of the Royal Statistical Society, Series B*, 56:1–48.

Nishimura, A., D. Dunson, and J. Lu. 2017. Discontinuous Hamiltonian Monte Carlo for sampling discrete parameters. arXiv:1705.08510.

Noon, B., L. Bailey, T. Sisk, and K. McKelvey. 2012. Efficient species-level monitoring at the landscape scale. *Conservation Biology*, 26:432–441.

Northrup, J. and B. Gerber. 2018. A comment on priors for Bayesian occupancy models. *PloS One*, 13:e0192819.

Northrup, J., M. Hooten, C. Anderson, and G. Wittemyer. 2013. Practical guidance on characterizing availability in resource selection functions under a use-availability design. *Ecology*, 94:1456–1464.

Ospina, R. and S. Ferrari. 2012. A general class of zero-or-one inflated beta regression models. *Computational Statistics & Data Analysis*, 56:1609–1623.

Otis, D., K. Burnham, G. White, and D. Anderson. 1978. Statistical inference from capture data on closed animal populations. *Wildlife Monographs*, 62:3–135.

Pakman, A. and L. Paninski. Auxiliary-variable exact Hamiltonian Monte Carlo samplers for binary distributions. In *Advances in Neural Information Processing Systems*, pp. 2490–2498, 2013.

Parent, E. and E. Rivot. 2012. *An Introduction to Hierarchical Bayesian Modeling for Ecological Data*. CRC Press, Boca Raton, FL.

Park, J. and M. Haran. 2018. Bayesian inference in the presence of intractable normalizing functions. *Journal of the American Statistical Association*, 113:1372–1390.

Park, T. and G. Casella. 2008. The Bayesian lasso. *Journal of the American Statistical Association*, 103:681–686.

Patil, G. and C. Rao. 1976. On size-biased sampling and related form-invariant weighted distributions. *Indian Journal of Statistics*, 38:48–61.

Patil, G. and C. Rao. 1977. The weighted distributions: A survey of their applications. In Krishnaiah, P., editor, *Applications of Statistics*. North Holland Publishing Company, Amsterdam, the Netherlands.

Patil, G. and C. Rao. 1978. Weighted distributions and size-biased sampling with applications to wildlife populations and human families. *Biometrics*, 34:179–189.

Patterson, T., L. Thomas, C. Wilcox, O. Ovaskainen, and J. Matthiopoulos. 2008. State-space models of individual animal movement. *Trends in Ecology and Evolution*, 23:87–94.

Pettit, L. 1990. The conditional predictive ordinate for the normal distribution. *Journal of the Royal Statistical Society, Series B*, 52:175–184.

Pledger, S. 2005. The performance of mixture models in heterogeneous closed population capture–recapture. *Biometrics*, 61:868–873.

Pollock, K., J. Nichols, C. Brownie, and J. Hines. 1990. Statistical inference for capture-recapture experiments. *Wildlife Monographs*, 107:3–97.

Powell, L. 2007. Approximating variance of demographic parameters using the delta method: A reference for avian biologists. *The Condor*, 109:949–954.

Proffitt, K., J. Goldberg, M. Hebblewhite, R. Russell, B. Jimenez, H. Robinson, K. Pilgrim, and M. Schwartz. 2015. Integrating resource selection into spatial capture-recapture models for large carnivores. *Ecosphere*, 6:1–15.

Qin, Q. and J. Hobert. 2019. Convergence complexity analysis of Albert and Chib's algorithm for Bayesian probit regression. Annals of Statistics, In Press.

R Core Team. 2018. *R: A Language and Environment for Statistical Computing*. R Foundation for Statistical Computing, Vienna, Austria.

Raftery, A. 2016. Use and communication of probabilistic forecasts. *Statistical Analysis and Data Mining: The ASA Data Science Journal*, 9:397–410.

Raftery, A., D. Madigan, and J. Hoeting. 1997. Bayesian model averaging for linear regression models. *Journal of the American Statistical Association*, 92:179–191.

Reich, B. 2012. Spatiotemporal quantile regression for detecting distributional changes in environmental processes. *Journal of the Royal Statistical Society, Series C*, 61:535–553.

Reich, B., M. Fuentes, and D. Dunson. 2011. Bayesian spatial quantile regression. *Journal of the American Statistical Association*, 106:6–20.

Reiter, J. 2012. Bayesian finite population imputation for data fusion. *Statistica Sinica*, 22:795–811.

Renner, I. and D. Warton. 2013. Equivalence of maxent and Poisson point process models for species distribution modeling in ecology. *Biometrics*, 69:274–281.

Richardson, S. 2002. Discussion of the paper by Spiegelhalter et al. *Journal of the Royal Statistical Society, Series B*, 64:226–227.

Ridout, M., J. Hinde, and C. DeméAtrio. 2001. A score test for testing a zero-inflated Poisson regression model against zero-inflated negative binomial alternatives. *Biometrics*, 57:219–223.

Robbins, C., D. Bystrak, and P. Geissler. The breeding bird survey: Its first fifteen years, 1965–1979. Technical report, Patuxent Wildlife Research Center, Laurel, MD, 1986.

Robert, C. and G. Casella. 2011. A short history of Markov chain Monte Carlo: Subjective recollections from incomplete data. *Statistical Science*, 26:102–115.

Roberts, G. and J. Rosenthal. 2009. Examples of adaptive MCMC. *Journal of Computational and Graphical Statistics*, 18:349–367.

Roberts, G. and R. Tweedie. 1996. Exponential convergence of langevin distributions and their discrete approximations. *Bernoulli*, 2:341–363.

Ross, B., M. Hooten, J.-M. DeVink, and D. Koons. 2015. Combined effects of climate, predation, and density dependence on greater and lesser scaup population dynamics. *Ecological Applications*, 25:1606–1617.

Rota, C., M. Ferreira, R. Kays, T. Forrester, E. Kalies, W. McShea, A. Parsons, and J. Millspaugh. 2016. A multispecies occupancy model for two or more interacting species. *Methods in Ecology and Evolution*, 7:1164–1173.

Rota, C., C. Wikle, R. Kays, T. Forrester, W. McShea, A. Parsons, and J. Millspaugh. 2016. A two-species occupancy model accommodating simultaneous spatial and interspecific dependence. *Ecology*, 97:48–53.

Roth, T., C. Bühler, and V. Amrhein. 2016. Estimating effects of species interactions on populations of endangered species. *The American Naturalist*, 187:457–467.

Royle, J. 2004. N-mixture models for estimating population size from spatially replicated counts. *Biometrics*, 60:108–115.

Royle, J. 2009. Analysis of capture-recapture models with individual covariates using data augmentation. *Biometrics*, 65:267–274.

Royle, J., R. Chandler, R. Sollmann, and B. Gardner. 2013a. *Spatial Capture-Recapture*. Academic Press, Oxford, UK.

Royle, J., R. Chandler, C. Sun, and A. Fuller. 2013b. Integrating resource selection information with spatial capture–recapture. *Methods in Ecology and Evolution*, 4:520–530.

Royle, J. and R. Dorazio. 2008. *Hierarchical Modeling and Inference in Ecology: The Analysis of Data from Populations, Metapopulations and Communities*. Academic Press, Amsterdam, the Netherlands.

Royle, J. and R. Dorazio. 2012. Parameter-expanded data augmentation for Bayesian analysis of capture-recapture models. *Journal of Ornithology*, 152:521–537.

Royle, J., A. Fuller, and C. Sutherland. 2016. Spatial capture–recapture models allowing Markovian transience or dispersal. *Population Ecology*, 58:53–62.

Royle, J., A. Fuller, and C. Sutherland. 2018. Unifying population and landscape ecology with spatial capture–recapture. *Ecography*, 41:444–456.

Royle, J. and M. Kéry. 2007. A Bayesian state-space formulation of dynamic occupancy models. *Ecology*, 88:1813–1823.

Royle, J. and W. Link. 2006. Generalized site occupancy models allowing for false positive and false negative errors. *Ecology*, 87:835–841.

Royle, J. and K. Young. 2008. A hierarchical model for spatial capture-recapture data. *Ecology*, 89:2281–2289.

Rue, H., S. Martino, and N. Chopin. 2009. Approximate Bayesian inference for latent Gaussian models by using integrated nested Laplace approximations. *Journal of the Royal Statistical Society: Series B*, 71(2):319–392.

Ruiz-Gutierrez, V., M. Hooten, and E. Campbell Grant. 2016. Uncertainty in biological monitoring: A framework for data collection and analysis to account for multiple sources of sampling bias. *Methods in Ecology and Evolution*, 7: 900–909.

Sampson, P. 2010. Constructions for nonstationary spatial processes. In Gelfand, A., P. Diggle, M. Fuentes, and P. Guttorp, editors, *Handbook of Spatial Statistics*. Chapman & Hall/CRC Press, Boca Raton, FL.

Schaub, M. and F. Abadi. 2011. Integrated population models: A novel analysis framework for deeper insights into population dynamics. *Journal of Ornithology*, 152:227–237.

Schaub, M., O. Gimenez, B. Schmidt, and R. Pradel. 2004. Estimating survival and temporary emigration in the multistate capture–recapture framework. *Ecology*, 85: 2107–2113.

Seber, G. 1982. *The Estimation of Animal Abundance and Related Parameters*, 2nd ed. MacMillan, New York.

Seber, G. 2008. *A Matrix Handbook for Statisticians*. Wiley, New York.

Simas, A., W. Barreto-Souza, and A. Rocha. 2010. Improved estimators for a general class of beta regression models. *Computational Statistics & Data Analysis*, 54:348–366.

Sólymos, P., S. Lele, and E. Bayne. 2012. Conditional likelihood approach for analyzing single visit abundance survey data in the presence of zero inflation and detection error. *Environmetrics*, 23:197–205.

Spiegelhalter, D., N. Best, B. Carlin, and A. Van Der Linde. 2002. Bayesian measures of model complexity and fit. *Journal of the Royal Statistical Society: Series B*, 64:583–639.

Spiegelhalter, D. and A. Smith. 1982. Bayes factors for linear and log-linear models with vague prior information. *Journal of the Royal Statistical Society, Series B*, 44:377–387.

Spiegelhalter, D., A. Thomas, N. Best, and D. Lunn. 2003. *Winbugs User Manual*.

Steenweg, R., M. Hebblewhite, J. Whittington, P. Lukacs, and K. McKelvey. 2018. Sampling scales define occupancy and underlying occupancy–abundance relationships in animals. *Ecology*, 99:172–183.

Talts, S., M. Betancourt, D. Simpson, A. Vehtari, and A. Gelman. 2018. Validating Bayesian inference algorithms with simulation-based calibration. arXiv: 1804.06788.

Tanner, M. 1991. *Tools for Statistical Inference: Methods for the Exploration of Posterior Distributions and Likelihood Functions*. Springer, New York.

Teh, Y., M. Jordan, M. Beal, and D. Blei. 2006. Hierarchical Dirichlet processes. *Journal of the American Statistical Association*, 101:1566–1581.

Tierney, L. 1994. Markov chains for exploring posterior distributions (with discussion). *Annals of Statistics*, 22:1701–1786.

Tredennick, A., M. Hooten, and P. Adler. 2017. Do we need demographic data to forecast plant population dynamics? *Methods in Ecology and Evolution*, 8:541–551.

Turchin, P. 1998. *Quantitative Analysis of Animal Movement*. Sinauer Associates, Sunderland, MA.

Turchin, P. 2003. *Complex Population Dynamics: A Theoretical/Empirical Synthesis*. Princeton University Press, Princeton, NJ.

Turek, D., P. de Valpine, and C. Paciorek. 2016. Efficient Markov chain Monte Carlo sampling for hierarchical hidden Markov models. *Environmental and Ecological Statistics*, 23:549–564.

Tyre, A., B. Tenhumberg, S. Field, D. Niejalke, K. Parris, and H. Possingham. 2003. Improving precision and reducing bias in biological surveys: Estimating false-negative error rates. *Ecological Applications*, 13:1790–1801.

Van Dyk, D. and T. Park. 2008. Partially collapsed Gibbs samplers: Theory and methods. *Journal of the American Statistical Association*, 103:790–796.

Ver Hoef, J. 2012. Who invented the delta method? *The American Statistician*, 66: 124–127.

Ver Hoef, J. and P. Boveng. 2015. Iterating on a single model is a viable alternative to multimodel inference. *The Journal of Wildlife Management*, 79: 719–729.

Ver Hoef, J., E. Hanks, and M. Hooten. 2018b. On the relationship between conditional (CAR) and simultaneous (SAR) autoregressive models. *Spatial Statistics*, 25:68–85.

Ver Hoef, J. and J. Jansen. 2007. Space-time zero-inflated count models of harbor seals. *Environmetrics*, 18:697–712.

Ver Hoef, J., E. Peterson, M. Hooten, E. Hanks, and M.-J. Fortin. 2018a. Spatial autoregressive models for statistical inference from ecological data. *Ecological Monographs*, 88:36–59.

Waltrup, L., F. Sobotka, T. Kneib, and G. Kauermann. 2015. Expectile and quantile regression David and Goliath? *Statistical Modelling*, 15:433–456.

Wang, X., M.-H. Chen, R. Kuo, and D. Dey. 2015. Bayesian spatial-temporal modeling of ecological zero-inflated count data. *Statistica Sinica*, 25:189.

Wang, X. and D. Dey. 2010. Generalized extreme value regression for binary response data: An application to B2B electronic payments system adoption. *Annals of Applied Statistics*, 4:2000–2023.

Warton, D. 2005. Many zeros does not mean zero inflation: Comparing the goodness-of-fit of parametric models to multivariate abundance data. *Environmetrics*, 16: 275–289.

Warton, D. and L. Shepherd. 2010. Poisson point process models solve the "pseudo-absence problem" for presence-only data in ecology. *Annals of Applied Statistics*, 4:1383–1402.

Warton, D., J. Stoklosa, G. Guillera-Arroita, D. MacKenzie, and A. Welsh. 2017. Graphical diagnostics for occupancy models with imperfect detection. *Methods in Ecology and Evolution*, 8:408–419.

Watanabe, S. 2010. Asymptotic equivalence of Bayes cross validation and widely applicable information criterion in singular learning theory. *Journal of Machine Learning Research*, 11:3571–3594.

Webb, M., S. Wotherspoon, D. Stojanovic, R. Heinsohn, R. Cunningham, P. Bell, and A. Terauds. 2014. Location matters: Using spatially explicit occupancy models to predict the distribution of the highly mobile, endangered swift parrot. *Biological Conservation*, 176:99–108.

Welsh, A., R. Cunningham, C. Donnelly, and D. Lindenmayer. 1996. Modelling the abundance of rare species: Statistical models for counts with extra zeros. *Ecological Modelling*, 88:297–308.

Wenger, S. and M. Freeman. 2008. Estimating species occurrence, abundance, and detection probability using zero-inflated distributions. *Ecology*, 89:2953–2959.

Wickham, H. 2014. *Advanced R*. CRC Press, London, UK.

Wikle, C. 2003. Hierarchical Bayesian models for predicting the spread of ecological processes. *Ecology*, 84:1382–1394.

Wikle, C. 2010. Low-rank representations for spatial processes. In Gelfand, A., P. Diggle, M. Fuentes, and P. Guttorp, editors, *Handbook of Spatial Statistics*. Chapman & Hall/CRC Press, Boca Raton, FL.

Wikle, C. and M. Hooten. 2010. A general science-based framework for nonlinear spatio-temporal dynamical models. *Test*, 19:417–451.

Wikle, C., R. Milliff, D. Nychka, and L. Berliner. 2001. Spatiotemporal hierarchical Bayesian modeling: Tropical ocean surface winds. *Journal of the American Statistical Association*, 96:382–397.

Wikle, C., A. Zammit-Mangion, and N. Cressie. 2019. *Spatio-temporal Statistics with R*. Chapman and Hall/CRC Press, Boca Raton, FL.

Williams, P. and M. Hooten. 2016. Combining statistical inference and decisions in ecology. *Ecological Applications*, 26:1930–1942.

Williams, P., M. Hooten, J. Womble, G. Esslinger, and M. Bower. 2018. Monitoring dynamic spatio-temporal ecological processes optimally. *Ecology*, 99:524–535.

Williams, P., M. Hooten, J. Womble, G. Esslinger, M. Bower, and T. Hefley. 2017. An integrated data model to estimate spatio-temporal occupancy, abundance, and colonization dynamics. *Ecology*, 98:328–336.

Wood, S. 2004. Stable and efficient multiple smoothing parameter estimation for generalized additive models. *Journal of the American Statistical Association*, 99: 673–686.

Wright, W., K. Irvine, and T. Rodhouse. 2016. A goodness-of-fit test for occupancy models with correlated within-season revisits. *Ecology and Evolution*, 6: 5404–5415.

Xing, J. and X. Qian. 2017. Bayesian expectile regression with asymmetric normal distribution. *Communications in Statistics—Theory and Methods*, 46:4545–4555.

Yang, Y. and S. Tokdar. 2017. Joint estimation of quantile planes over arbitrary predictor spaces. *Journal of the American Statistical Association*, 112:1107–1120.

Yu, K. and R. Moyeed. 2001. Bayesian quantile regression. *Statistics and Probability Letters*, 54:437–447.

Yu, K. and J. Zhang. 2005. A three-parameter asymmetric Laplace distribution and its extension. *Communication in Statistics—Theory and Methods*, 34:1867–1879.

Zamo, M. and P. Naveau. 2018. Estimation of the continuous ranked probability score with limited information and applications to ensemble weather forecasts. *Mathematical Geosciences*, 50:209–234.

Zellner, A. 1986. On assessing prior distributions and Bayesian regression analysis with g-prior distributions. In *Bayesian Inference and Decision Techniques*. Elsevier, New York.

Zhang, Y., Z. Ghahramani, A. Storkey, and C. Sutton. 2012. Continuous relaxations for discrete Hamiltonian Monte Carlo. In *Advances in Neural Information Processing Systems*, pp. 3194–3202.

Zipkin, E., J. Royle, D. Dawson, and S. Bates. 2010. Multi-species occurrence models to evaluate the effects of conservation and management actions. *Biological Conservation*, 143:479–484.

Probability Distributions

DISCRETE PROBABILITY DISTRIBUTIONS

Bernoulli: $\theta \sim \text{Bern}(p)$ for $\theta \in \{0,1\}$ and $p \in (0,1)$

 PMF: $[\theta|p] = p^\theta (1-p)^{1-\theta}$
 Mean: $E(\theta|p) = p$
 Variance: $\text{Var}(\theta|p) = p(1-p)$

Binomial: $\theta \sim \text{Binom}(N,p)$ for $\theta \in \{0,1,\ldots,N\}$, $N \in \{1,2,\ldots\}$, and $p \in (0,1)$

 PMF: $[\theta|N,p] = \binom{N}{\theta} p^\theta (1-p)^{1-\theta}$
 Mean: $E(\theta|N,p) = Np$
 Variance: $\text{Var}(\theta|N,p) = Np(1-p)$

Poisson: $\theta \sim \text{Pois}(\lambda)$ for $\theta \in \{0,1,\ldots\}$ and $\lambda \in \Re^+$

 PMF: $[\theta|\lambda] = \frac{\lambda^\theta e^{-\lambda}}{\theta!}$
 Mean: $E(\theta|\lambda) = \lambda$
 Variance: $\text{Var}(\theta|\lambda) = \lambda$

Negative Binomial: $\theta \sim \text{NB}(\mu,N)$ for $\theta \in \{0,1,\ldots\}$ and $\mu,N \in \Re^+$

 PMF: $[\theta|\mu,N] = \frac{\Gamma(\theta+N)}{\Gamma(N)\theta!} \left(\frac{N}{N+\mu}\right)^N \left(1-\frac{N}{N+\mu}\right)^\theta$
 Mean: $E(\theta|\mu,N) = \mu$
 Variance: $\text{Var}(\theta|\mu,N) = \mu + \frac{\mu^2}{N}$

CONTINUOUS PROBABILITY DISTRIBUTIONS

Uniform: $\theta \sim \text{Unif}(a,b)$ for $\theta \in (a,b)$ and $a,b \in \Re$

 PDF: $[\theta|a,b] = \frac{1}{b-a}$
 Mean: $E(\theta|a,b) = \frac{a+b}{2}$
 Variance: $\text{Var}(\theta|a,b) = \frac{(b-a)^2}{12}$

Beta: $\theta \sim \text{Beta}(\alpha,\beta)$ for $\theta \in (0,1)$ and $\alpha,\beta \in \Re^+$

 PDF: $[\theta|\alpha,\beta] = \frac{\Gamma(\alpha+\beta)}{\Gamma(\alpha)\Gamma(\beta)} \theta^{\alpha-1}(1-\theta)^{\beta-1}$
 Mean: $E(\theta|\alpha,\beta) = \frac{\alpha}{\alpha+\beta}$
 Variance: $\text{Var}(\theta|\alpha,\beta) = \frac{\alpha\beta}{(\alpha+\beta)^2(\alpha+\beta+1)}$

Exponential: $\theta \sim \text{Exp}(\lambda)$ for $\theta, \lambda \in \mathfrak{R}^+$

> **PDF:** $[\theta|\lambda] = \lambda e^{-\lambda\theta}$
> **Mean:** $E(\theta|\lambda) = \frac{1}{\lambda}$
> **Variance:** $\text{Var}(\theta|\lambda) = \frac{1}{\lambda^2}$

Gamma: $\theta \sim \text{Gamma}(\alpha, \beta)$ for $\theta, \alpha, \beta \in \mathfrak{R}^+$

> **PDF:** $[\theta|\alpha,\beta] = \frac{\beta^\alpha}{\Gamma(\alpha)}\theta^{\alpha-1}e^{-\beta\theta}$
> **Mean:** $E(\theta|\alpha,\beta) = \frac{\alpha}{\beta}$
> **Variance:** $\text{Var}(\theta|\alpha,\beta) = \frac{\alpha}{\beta^2}$

Inverse Gamma: $\theta \sim \text{IG}(q, r)$ for $\theta, q, r \in \mathfrak{R}^+$

> **PDF:** $[\theta|q,r] = \frac{1}{r^q\Gamma(q)}\theta^{-(q+1)}e^{-\frac{1}{r\theta}}$
> **Mean:** $E(\theta|q,r) = \frac{1}{r(q-1)}$
> **Variance:** $\text{Var}(\theta|q,r) = \frac{1}{r^2(q-1)^2(q-2)}$

Normal: $\theta \sim \text{N}(\mu, \sigma^2)$ for $\theta, \mu \in \mathfrak{R}$ and $\sigma^2 \in \mathfrak{R}^+$

> **PDF:** $[\theta|\mu,\sigma^2] = \frac{1}{\sqrt{2\pi\sigma^2}}e^{-\frac{1}{2\sigma^2}(\theta-\mu)^2}$
> **Mean:** $E(\theta|\mu,\sigma^2) = \mu$
> **Variance:** $\text{Var}(\theta|\mu,\sigma^2) = \sigma^2$

Multivariate Normal: $\boldsymbol{\theta} \sim \text{N}(\boldsymbol{\mu}, \boldsymbol{\Sigma})$ for $\boldsymbol{\theta}_{p\times1}, \boldsymbol{\mu}_{p\times1} \in \mathfrak{R}^p$ and $\boldsymbol{\Sigma}_{p\times p}$ symmetric and positive definite

> **PDF:** $[\boldsymbol{\theta}|\boldsymbol{\mu}, \boldsymbol{\Sigma}] = |2\pi\boldsymbol{\Sigma}|^{-\frac{1}{2}}e^{-\frac{1}{2}(\boldsymbol{\theta}-\boldsymbol{\mu})'\boldsymbol{\Sigma}^{-1}(\boldsymbol{\theta}-\boldsymbol{\mu})}$
> **Mean:** $E(\boldsymbol{\theta}|\boldsymbol{\mu}, \boldsymbol{\Sigma}) = \boldsymbol{\mu}$
> **Covariance:** $\text{Cov}(\boldsymbol{\theta}|\boldsymbol{\mu}, \boldsymbol{\Sigma}) = \boldsymbol{\Sigma}$

Index

Note: Page numbers in *italic* and **bold** refer to figures and tables, respectively.

Printed and bound by CPI Group (UK) Ltd, Croydon, CR0 4YY

23/10/2024

01778223-0018